Law, Governance and Technology Series

Issues in Privacy and Data Protection

Volume 24

Series editors

Pompeu Casanovas, UAB Institute of Law and Technology,
Bellaterra, Barcelona, Spain
Giovanni Sartor, European University Institute, Florence, Italy

Sub Series editor

Serge Gutwirth, Law, Science, Technology and Society (LSTS),
Vrije Universiteit Brussels (VUB), Brussel, Belgium

The *Law, Governance and Technology Series* is intended to attract manuscripts arising from an interdisciplinary approach in law, artificial intelligence and information technologies. The idea is to bridge the gap between research in IT law and IT-applications for lawyers developing a unifying techno-legal perspective. The series will welcome proposals that have a fairly specific focus on problems or projects that will lead to innovative research charting the course for new interdisciplinary developments in law, legal theory, and law and society research as well as in computer technologies, artificial intelligence and cognitive sciences. In broad strokes, manuscripts for this series may be mainly located in the fields of the Internet law (data protection, intellectual property, Internet rights, etc.), Computational models of the legal contents and legal reasoning, Legal Information Retrieval, Electronic Data Discovery, Collaborative Tools (e.g. Online Dispute Resolution platforms), Metadata and XML Technologies (for Semantic Web Services), Technologies in Courtrooms and Judicial Offices (E-Court), Technologies for Governments and Administrations (E-Government), Legal Multimedia, and Legal Electronic Institutions (Multi-Agent Systems and Artificial Societies).

More information about this series at http://www.springer.com/series/13087

Serge Gutwirth · Ronald Leenes · Paul De Hert
Editors

Data Protection on the Move

Current Developments in ICT
and Privacy/Data Protection

 Springer

Editors
Serge Gutwirth
Law, Science, Technology and Society (LSTS)
Vrije Universiteit Brussel
Brussels
Belgium

Paul De Hert
Law, Science, Technology and Society (LSTS)
Vrije Universiteit Brussel
Brussels
Belgium

Ronald Leenes
Tilburg Institute for Law, Technology, and Society
Tilburg University
Tilburg
The Netherlands

ISSN 2352-1902 ISSN 2352-1910 (electronic)
Law, Governance and Technology Series
ISSN 2352-1929 ISSN 2352-1937 (electronic)
Issues in Privacy and Data Protection
ISBN 978-94-017-7375-1 ISBN 978-94-017-7376-8 (eBook)
DOI 10.1007/978-94-017-7376-8

Library of Congress Control Number: 2015954596

Springer Dordrecht Heidelberg New York London
© Springer Science+Business Media Dordrecht 2016
This work is subject to copyright. All rights are reserved by the Publisher, whether the whole or part of the material is concerned, specifically the rights of translation, reprinting, reuse of illustrations, recitation, broadcasting, reproduction on microfilms or in any other physical way, and transmission or information storage and retrieval, electronic adaptation, computer software, or by similar or dissimilar methodology now known or hereafter developed.
The use of general descriptive names, registered names, trademarks, service marks, etc. in this publication does not imply, even in the absence of a specific statement, that such names are exempt from the relevant protective laws and regulations and therefore free for general use.
The publisher, the authors and the editors are safe to assume that the advice and information in this book are believed to be true and accurate at the date of publication. Neither the publisher nor the authors or the editors give a warranty, express or implied, with respect to the material contained herein or for any errors or omissions that may have been made.

Printed on acid-free paper

Springer Science+Business Media B.V. Dordrecht is part of Springer Science+Business Media (www.springer.com)

Preface

At the moment of writing this preface—June 2015—the reform process of European data protection law, which was officially launched at the end of January 2012, was still ongoing. While in generic terms the objective is still the same as in the eighties of the past century, namely warranting a European-wide high level of protection of personal data in order to ensure the free circulation of personal data and stimulating business in the digital single market, the points of discussion and disagreement are still manifold.

On the basis of a document of no less than 200 pages (Doc. 9565/15 of 11/06/2015), on 15 June 2015 The EU Council of Ministers agreed upon a "general approach" with regard to the new data protection regulation that mandates the presidency of the Council to engage in the next steps of the procedure and more particularly to enter into negotiations with representatives of the European Parliament. In a press release Latvia's minister for justice Dzintars Rasnačs said: "Today we have moved a great step closer to modernized and harmonized data protection framework for the European Union. I am very content that after more than 3 years of negotiations we have finally found a compromise on the text. The new data protection regulation, adapted to the needs of the digital age, will strengthen individual rights of our citizens and ensure a high standard of protection".

On 24 June 2015 the first "trilogue" negotiations will start, bringing together representatives of the European Commission, the European Council and the European Parliament in order to move forward to the joint adoption of the new piece of legislation by the Council and the Parliament, as is foreseen by the procedure, still known as "co-decision" procedure. The European Parliament already adopted a compromise text on 13 March 2014 which was inspired by EP member Jan Philip Albrecht, who will lead the delegation of the European Parliament in the trilateral negotiations. It is highly probable that the negotiations will be no sinecure, since there are quite some divergences and discrepancies between the viewpoints of the involved actors, and the interests at stake are very high (as it already was when the Directive of 1995 was elaborated). Even though the incoming Luxembourg Presidency announced that it will speed up the process in order to round it off before the end of the year, we believe that our next and ninth International Conference on Computers, Privacy and Data Protection (CPDP 2016), like the three former editions, will be held in the light or in the shadow of an ongoing reform process, with still quite some uncertainties as to its outcomes.

The book you have opened is one of the products of the 8th edition of the annual Brussels-based International Conference on Computers, Privacy and Data Protection (CPDP 2015), which took place, six months ago, on 21, 22 and 23 January 2015, again in the famous Les Halles, in Schaerbeek, at the heart of Brussels. The 3-day conference provided 70 panels, workshops and special sessions, with 415 speakers from academia, the public and private sectors, and civil society from 43 countries, more than 1000 participants. Under these circumstances, and indeed given the fact that the reform process of European data protection is still very actual, being reconsidered, debated and indeed, pushed and pulled in many directions by the many interested parties and forces they mobilize, CPDP2015 turned into an extremely timely, colourful and challenging happening. That is why, in hindsight, its title, which traditionally is adopted as the book's title, was right on the spot: *Data Protection on the Move*.

The conference addressed many privacy and data protection issues in its 70 panels. Far too many topics to be listed here. We suffice here with highlighting some of them (in random order).

- The Right to be Forgotten in view of the Google Spain ruling. One of the important decisions in 2014 regarding data protection was the so-called Google Spain ruling (C-131/12 (ECLI:EU:C:2014:317)) delivered by the CJEU in May 2014. It was heavily discussed in the media, academia and at CPDP. In this volume, the chapter by Yod-Samuel Martin and Jose M. Del Alamo (Chapter "Forget About Being Forgotten: From the Right to Oblivion to the Right of Reply") deals with a technical measure to cope with situations that may call for invoking the Right to be Forgotten.
- Jurisdictional issues. The Internet spans the entire globe and knows no borders, yet legal systems are based on territoriality. Given the significant differences in data protection regimes both in general and in the domain of law enforcement, this leads to fundamental and practical issues. Chapter "Global Views on Internet Jurisdiction and Trans-Border Access" written by Cristos Velasco, Julia Hörnle and Anna-Maria Osula addresses the jurisdiction of law enforcement authorities under international law.
- Privacy by design, engineering privacy into the Internet. Privacy by design was prominently put on the agenda by former Ontario Privacy Commissioner Ann Cavoukian. The Data Protection Directive called for technical measures already, but the upcoming General Data Protection Regulation is likely to take PbD to the next level. Tilman Frosh, Sven Schäge, Martin Goll, and Thorsten Holz provide a practical (technical) implementation of privacy by design in the context of payment for recharging electrical vehicles (Chapter "On Locational Privacy in the Absence of Anonymous Payments"). Bibi van den Berg (Chapter "Mind the Air Gap: Preventing Privacy Issues in Robotics") calls for taking privacy

seriously even before starting the design of domestic robots which, in her view, should in many cases result in an air gap between robot and the Internet (stand-alone robots).
- Privacy self-management. The data protection legal framework is there to protect individuals against privacy infringements, but individuals themselves also have a responsibility (and capabilities) in protecting their privacy. Privacy self-management is one of the forms this can take. Tobias Matzner and Philip Masur (Chapter "Do-It-Yourself Data Protection—Empowerment or Burden?") discuss whether individuals can and should play a more active role in managing their own data.
- Health care. Health care is one of the areas where large amounts of (sensitive) personal data are being processed. Obviously, this needs to be done in a responsible and legally compliant manner. Different CPDP panels addressed the use of health data and the design of privacy preserving healthcare systems. Chapter "Development Towards a Learning Healthcare System—Experiences with the Privacy Protection Framework of the TRANSFoRm Project" (Wolfgang Kuchinke, Christian Ohmann, Robert Verheij, Evert-Ben van Veen and Brendan Delaney) describe a graphic privacy model to improve the privacy characteristics of a learning health system.
- Data retention. Data retention of traffic data in view of law enforcement as mandated by the Data Retention Directive has always been a topic of heated debate. The CJEU ruling in the Digital Ireland case of 8 April 2014 has not settled the debate. The consequences of the ruling and the future of data retention were discussed in a panel and the topic surfaced in numerous other panels.
- Anonymity and pseudonymity. Both of these are supposed to limit potential privacy issues. In theory, they do, but developments such as Big and Open Data may limit the protection they offer because, by combining data from different sources, it increasingly turns out to be possible to identify "anonymous" individuals. Chapter "A Precautionary Approach to Big Data Privacy" in this volume (Arvind Narayanan, Joanna Huey and Edward F. Felten on the risks re-identification offers in the context of Big Data) pays attention to these issues.
- Surveillance. Surveillance is one of the big recurring themes at CPDP. This year, special attention has been placed at specific topics, such as monitoring the net for violent extremist material, tracking people's browsing habits on for instance porn sites and the surveillance of LGBT+ people.
- Privacy attitudes. Much attention is always being paid to data subjects at CPDP, also with regard to their attitudes and experiences. A panel at the CPDP 2015 conference paid attention to these in the context of ubiquitous surveillance. Chapter "The Context-Dependence of Citizens' Attitudes and Preferences Regarding Privacy and Security", written by Michael Friedewald, Marc van Lieshout, Sven Rung and Merel Ooms, presents factors that determine citizens' perceptions of concrete security technologies and surveillance practices based on extensive survey.
- Big Data. Big Data and Data Science will have a profound impact on privacy and data protection given that invaluable insights can be derived from large

data sets containing information about (identifiable) individuals. Panels were devoted to topics, such as (government) policy, opportunities and risks in health care and disease control. Chapters "Visions of Technology: Big Data Lessons Understood by EU Policy Makers in Their Review of the Legal Frameworks on Intellectual Property Rights, Access to and Re-use of PSI and the Protection of Personal Data" (Hans Lammerant and Paul De Hert, addressing EU Big Data policy), "A Precautionary Approach to Big Data Privacy" (Arvind Narayanan, Joanna Huey and Edward F. Felten on the risks re-identification offers in the context of Big Data) and "Is the Human Rights Framework Still Fit for the Big Data Era? A Discussion of the ECtHR's Case Law on Privacy Violations Arising from Surveillance Activities" (Bart van der Sloot on the adequacy of the human rights framework in the Big Data era) deal with some of the issues.

- Privacy and data protection as a business opportunity. Data protection compliance and privacy are often seen as cost factors in offering products and services in the information age. There are signs that the tide is changing in this respect. Partly as a result of the Snowden revelations, people seem to be looking for more privacy friendly solutions and companies are increasingly using privacy as a product/service differentiator. Chapter "Privacy and Innovation: from Disruption to Opportunities" (Marc van Lieshout) discusses the challenges and opportunities of privacy in innovation.
- Robotics and drones. Robots are entering the market place in serious numbers. Drones and domestic robots are the forerunners of these. Equipped with a multitude of sensors (audio, visual) and connectivity, they pose privacy and data protection issues that should be addressed urgently. A well-attended evening panel at De Markten discussed the privacy and data protection issues of robots. Chapters "Mind the Air Gap: Preventing Privacy Issues in Robotics" (Bibi van den Berg) and "The Impact of Domestic Robots on Privacy Data Protection, and the Troubles with Legal Regulation by Design" (Ugo Pagallo) draw attention to these issues and provide (directions of) solutions.
- Accountability. One of the key elements in the upcoming EU data protection framework is accountability. Data controllers and processors will have to be more open about their personal data processing. Processes and policies will have to be documented and on request be provided to the regulator and other stakeholders (including the data subjects).
- Data protection and privacy and other rights (IPR, freedom of speech). Sometimes, it seems as if data protection occupies its own space in debate and policy. They are both human rights, for instance enshrined in the EU charter (articles 7 and 8) and as such lessons may be learned from comparison with other rights, such as freedom of expression/speech and intellectual property. A look at other rights is also important because privacy/data protection sometimes conflicts with other rights. Several panels addressed what can be learned from other domains and the conflicts between rights.
- The open call panels. Every year, CPDP puts out an open call for papers, primarily aimed at junior (Ph.D. students) and senior academics. After a rigorous peer-reviewed process, this year 18 papers were selected for presentation (out

of 43 submissions!). This volume contains 8 out of the 18 presented papers: Chapter "Europe V. Facebook: An Imbroglio of EU Data Protection Issues", by Liana Colonna on Europe V. Facebook; Chapter "On Locational Privacy in the Absence of Anonymous Payments", by Tilman Frosh, Sven Schäge, Martin Goll, and Thorsten Holz on locational privacy in electronic vehicle charging; Chapter "Could the CE Marking be Relevant to Enforce Privacy by Design in the Internet of Things?", by Eric Lachaud on CE marking as a potential mechanism to enforce privacy by design in the Internet of Things; Chapter "Behavioural Advertising and the New 'EU Cookie Law' as a Victim of Business Resistance and a Lack of Official Determination" by Christiana Markou on behavioural advertising in the view of the "EU cookie law"; Chapter "Forget About Being Forgotten: From the Right to Oblivion to the Right of Reply", by Yod-Samuel Martin and José M. Del Alamo on the "Right of Reply" as a partial alternative to the Right to be Forgotten; Chapter "Privacy Failures as Systems Failures: A Privacy-Specific Formal System Model—A Systemic and Multi-perspective Approach" by Antony Morton on a Privacy-Specific Formal System Model to analyse privacy failures in systems; Chapter "A Precautionary Approach to Big Data Privacy", by Arvind Narayanan, Joanna Huey and Edward F. Felten on a precautionary approach to Big Data privacy; and Chapter "Metadata, Traffic Data, Communications Data, Service Use Information… What Is the Difference? Does the Difference Matter? An Interdisciplinary View from the UK" by Sophie Stalla-Bourdillon, Evangelia Papadaki, and Tim Chown discussing the differences between the terms metadata, traffic data, communications data, service use information, which are sometimes used as almost synonymous, but which are not.

As part of the open call sessions, a Best Student Paper Award is granted to the best paper written (primarily) by a young researcher, Ph.D. student or even master's student. This year, we have decided to award two papers the Best Student Paper Award: Tilman Frosch, Sven Schäge, Martin Goll and Thorsten Holz for their paper entitled "On Locational Privacy in the Absence of Anonymous Payments" and Yod-Samuel Martin and Jose M. Del Alamo for their paper entitled "Forget about Being Forgotten: From the Right to Oblivion to the Right of Reply".

Eight chapters ("Europe V. Facebook: An Imbroglio of EU Data Protection Issues", "On Locational Privacy in the Absence of Anonymous Payments", "Could the CE Marking be Relevant to Enforce Privacy by Design in the Internet of Things?", "Behavioural Advertising and the New 'EU Cookie Law' as a Victim of Business Resistance and a Lack of Official Determination", "Forget About Being Forgotten: From the Right to Oblivion to the Right of Reply", "Privacy Failures as Systems Failures: A Privacy-Specific Formal System Model—A Systemic and Multi-perspective Approach", "A Precautionary Approach to Big Data Privacy" and "Metadata, Traffic Data, Communications Data, Service Use Information… What Is the Difference? Does the Difference Matter? An Interdisciplinary View from the UK") originate from responses to the conference's call for papers and have thus already in their full form been presented during the conference. The remaining chapters ("Mind the Air Gap: Preventing Privacy Issues in Robotics", "The Context-Dependence of Citizens' Attitudes and Preferences Regarding

Privacy and Security", "Development Towards a Learning Healthcare System—Experiences with the Privacy Protection Framework of the TRANSFoRm Project", "Big Data in Governmental ICT Policies: A Comparison Between the EU and the US", "Privacy and Innovation: From Disruption to Opportunities", "Do-It-Yourself Data Protection—Empowerment or Burden?", "The Impact of Domestic Robots on Privacy Data Protection, and the Troubles with Legal Regulation by Design", "Is the Human Rights Framework Still Fit for the Big Data Era? A Discussion of the ECtHR's Case Law on Privacy Violations Arising from Surveillance Activities" and "Global Views on Internet Jurisdiction and Transborder Access") were submitted by some of the conference's invited speakers in the months following the conference. All papers have been subjected to a peer-reviewed process, which has further improved the quality of the papers.

In previous CPDP volumes, the various chapters were organized thematically. This year, it is difficult to do so due to the wide variety of topics covered. Therefore, we have decided to present them in alphabetical order of the first author.

The CPDP conference has grown over the years to become one of the biggest venues for privacy scholars, policy makers, regulators, practitioners, industry and civil society. In 2015, we were able to welcome over a thousand participants. The 3-day conference offered participants 70 panels, and workshops and special sessions with 415 speakers from academia, public and private sectors and civil society. The current volume can only offer a very small part of what the conference has to offer. Nevertheless, the editors feel the current volume represents a very valuable set of papers describing and discussing contemporary privacy and data protection issues.

All the chapters of this book have been peer-reviewed and commented on by at least two referees with expertise and interest in the subject matters. Since their work is crucial for maintaining the scientific quality of the book, we would explicitly take the opportunity to thank the following for their commitment and efforts:

Norberto Andrade, Josep Balash, Solon Barocas, Lejla Batina, Colin Bennett, Bibi van den Berg, Michael Birnhack, Franziska Boehm, Maurizio Borghi, Rosamunde Van Brakel, Marlou Brokx, Lee Bygrave, Colette Cuijpers, Els Debusser, Sari Depreeuw, Niels van Dijk, Claudia Diaz, Simone Fischer-Hübner, Antonella Galetta, Raphaël Gellert, Gloria González Fuster, Nathalie Grandjean, Kristrun Gunnarsdottir, Seda Gürses, Dara Hallinan, Rob Heyman, Mireille Hildebrandt, Dennis Hirsch, Joris van Hoboken, Jaap Henk Hoepman, Chris Hoofnagle, Kristina Irion, Paulan Korenhof, Eleni Kosta, Giovanni Livraga, Valerio Lubello, Marin Luisa, Aleecia M. McDonald, Lucas Melgaço, Maartje Niezen, Monica Palmirani, Martin Pekárek, Jo Pierson, Bart Preneel, Nadezhda Purtova, Charles Raab, Rowena Rodrigues, Kjetil Rommetveit, Ira Rubinstein, Pierangela Samarati, Joseph Savirimuthu, Sarah Spiekermann, Dimitra Stefanatou, Kees Stuurman, Ivan Szekely, Peggy Valcke, Katja de Vries, Diane Whitehouse, David Wright and Tal Zarsky.

13 July 2015

Serge Gutwirth
Ronald Leenes
Paul De Hert

Contents

Mind the Air Gap 1
Bibi van den Berg

Europe Versus Facebook: An Imbroglio of EU Data Protection Issues 25
Liane Colonna

**The Context-Dependence of Citizens' Attitudes
and Preferences Regarding Privacy and Security** 51
Michael Friedewald, Marc van Lieshout, Sven Rung and Merel Ooms

On Locational Privacy in the Absence of Anonymous Payments 75
Tilman Frosch, Sven Schäge, Martin Goll and Thorsten Holz

**Development Towards a Learning Health System—Experiences
with the Privacy Protection Model of the TRANSFoRm Project** 101
Wolfgang Kuchinke, Christian Ohmann, Robert A. Verheij,
Evert-Ben van Veen and Brendan C. Delaney

**Could the CE Marking Be Relevant to Enforce Privacy
by Design in the Internet of Things?** 135
Eric Lachaud

Visions of Technology 163
Hans Lammerant and Paul De Hert

Privacy and Innovation: From Disruption to Opportunities 195
Marc van Lieshout

**Behavioural Advertising and the New 'EU Cookie Law'
as a Victim of Business Resistance and a Lack
of Official Determination** 213
Christina Markou

Forget About Being Forgotten 249
Yod-Samuel Martin and Jose M. del Alamo

Do-It-Yourself Data Protection—Empowerment or Burden? 277
Tobias Matzner, Philipp K. Masur, Carsten Ochs and Thilo von Pape

Privacy Failures as Systems Failures: A Privacy-Specific Formal System Model... 307
Anthony Morton

A Precautionary Approach to Big Data Privacy 357
Arvind Narayanan, Joanna Huey and Edward W. Felten

The Impact of Domestic Robots on Privacy and Data Protection, and the Troubles with Legal Regulation by Design.................. 387
Ugo Pagallo

Is the Human Rights Framework Still Fit for the Big Data Era? A Discussion of the ECtHR's Case Law on Privacy Violations Arising from Surveillance Activities 411
Bart van der Sloot

Metadata, Traffic Data, Communications Data, Service Use Information... What Is the Difference? Does the Difference Matter? An Interdisciplinary View from the UK 437
Eric Lachaud

Global Views on Internet Jurisdiction and Trans-border Access........ 465
Cristos Velasco, Julia Hörnle and Anna-Maria Osula

Contributors

Tim Chown Ph.D. is a lecturer in electronics and computer science at the University of Southampton and an active member of its Web Science Doctoral Training Centre. His background is computer networking, including network security, which he has been teaching for over 15 years. He has recently taken leadership at Southampton for the creation and management of the new Cyber Security M.Sc., certified by GCHQ. His most recent publications include those in legal journals covering topics such as metadata handling and deep packet inspection, and legal approaches to combating malware. He has strong industrial connections with ISPs and vendors through standardization activity, being a co-chair and directorate member in the IETF, on the PC for the UKNOF (UK ISP group) and co-chair of the UK IPv6 Council. Tim has been a PI or participant in multiple EU projects over the past 15 years and was one of the original design team for the eduroam wireless roaming federation. Email: tjc@ecs.soton.ac.uk

Liane Colonna is a doctoral student at Stockholm University's Institute for Law and Informatics (IRI) writing about the legal regulation of data mining from an EU-US perspective. Email: liane.colonna@juridicum.su.se

Paul de Hert is professor of law at the Faculty of Law and Criminology of Vrije Universiteit Brussel. He is the director of the research group on Fundamental Rights and Constitutionalism (FRC) and senior member of the research group on Law, Science, Technology and Society (LSTS). Paul de Hert is also an associated professor of law and technology at the Tilburg Institute for Law, Technology, and Society (TILT). Email: paul.de.hert@vub.ac.be

Jose M. del Alamo Ph.D. is an associate professor in the Department of ICT Systems Engineering at the Universidad Politécnica de Madrid (DIT-UPM) where he teaches object-oriented programming, algorithms and data structures, software engineering, system architecture and distributed computing topics at undergraduate, postgraduate and doctorate levels. He has been involved in research activities from 2005, when he joined DIT-UPM. His research work focuses on issues related to personal data management including personal data disclosure, the Right to be Forgotten, identity and privacy management, and considering these aspects to advance the software and systems engineering methodologies applying approaches such as Privacy by Design and Privacy by Default. He is the UPM representative in different working groups at several standardization bodies such as W3C Social Networking Incubator and W3C Privacy Interest Group, OASIS Privacy Management Reference Model and OASIS Privacy by Design for Software Engineers working groups. Email: jmdela@dit.upm.es

Brendan Delaney is a professor of medical informatics and decision making at Imperial College London. He has previously held chairs in primary Care Research at Birmingham University Brendan's research interests span a wide range of quantitative methods in clinical informatics, health services research, including meta-analysis, modelling, randomized controlled trials and the use of diagnostic information. Over recent years, his work has focused mainly on clinical research informatics and the integration of clinical and research data and systems. He was awarded the John Fry Award for Research by the Royal College of General Practitioners in 2004. He is a scientific director for the TRANSFoRm Project. Email: Brendan.delaney@kcl.ac.uk

Edward W. Felten is deputy US chief technology officer at the White House. He is on leave from Princeton University, where he is the Robert E. Kahn Professor of Computer Science and Public Affairs, and the founding director of Princeton's Center for Information Technology Policy. In 2011–2012, he served as the first chief technologist at the US Federal Trade Commission. His research interests include computer security and privacy, and technology law and policy. He has published more than 100 papers in the research literature, and two books. His research on topics such as Internet security, privacy, copyright and copy protection, and electronic voting has been covered extensively in the popular press. He is a member of the National Academy of Engineering and the American Academy of Arts and Sciences and is a fellow of the ACM. He has testified before the House and Senate committee hearings on privacy, electronic voting and digital television. In 2004, Scientific American magazine named him to its list of fifty worldwide science and technology leaders. Email: felten@cs.princeton.edu

Michael Friedewald D. Eng.is a senior researcher and head of the ICT research group at the Fraunhofer Institute for Systems and Innovation Research ISI in Karlsruhe, Germany. He has a background in electrical engineering, economics and science and technology studies and received a doctorate in electrical engineering from RWTH Aachen University. He has been responsible for numerous studies on privacy and trust related to emerging sciences and technologies. He is currently co-ordinating the FP7 PRISMS projects dealing with citizens' perceptions of privacy and security. Email: michael.friedewald@isi.fraunhofer.de

Tilman Frosch is a postdoctoral researcher at Ruhr-Universität Bochum. He is a member of Horst Goertz Institute for IT Security, where his recent research interests centre around locational privacy and usable secure communications. He received his Ph.D. from Ruhr-Universität Bochum in 2015. Email: tilman.frosch@rub.de

Contributors

Martin Goll is a graduate engineer in IT security. He obtained his diploma from the Ruhr-University Bochum in 2014 and works as developer for VMRay GmbH. The focus of his work is on automated malware analysis with virtualization techniques. However, he retains an interest in novel, efficient implementations of cryptographic schemes. Email: martin.goll@rub.de

Thorsten Holz is a professor in the Faculty of Electrical Engineering and Information Technology at Ruhr-University Bochum, Germany. His research interests include systems-oriented aspects of secure systems, with a specific focus on applied computer security. Currently, his work concentrates on bots/botnets, automated analysis of malicious software, and studying latest attack vectors. He received the Dipl.-Inform. degree in computer science from RWTH Aachen, Germany (2005), and the Ph.D. degree from University of Mannheim (2009). Prior to joining Ruhr-University Bochum in April 2010, he was a postdoctoral researcher in the Automation Systems Group at the Technical University of Vienna, Austria. In 2011, Thorsten received the Heinz Maier-Leibnitz Prize from the German Research Foundation (DFG). Email: thorsten.holz@rub.de

Joanna Huey is the associate director of Princeton's Center for Information Technology Policy, which takes an interdisciplinary approach to addressing the interaction of digital technologies and society. Prior to joining CITP, she clerked for the Honorable Michael Boudin, worked as a business associate at Goodwin Procter, and co-founded Casetext, a Y Combinator-backed start-up. She holds an A.B. in physics and math from Harvard College, an M.P.P. in science and technology policy from the Harvard Kennedy School, and a J.D. from Harvard Law School, where she was president of the Harvard Law Review. Email: joanna.huey@gmail.com

Wolfgang Kuchinke is a trained biologist with a Ph.D. in molecular biology with expertise in clinical pharmacology. At the Coordination Centre for Clinical Trials (KKS) of the Heinrich Heine University Düsseldorf, Germany, he managed clinical trials and has focused since on information technology using the Internet for clinical research, conducting the evaluation and implementation of electronic data capture (EDC) systems, using CDISC standards to facilitate interoperability between different clinical trial systems. Recently, he is involved in facilitating regenerative biotherapies trials, developing legal and ethical frameworks for EU projects and participating in creating the clinical research infrastructure for ECRIN. Email: kuchinke@med.uni-duesseldorf.de

Eric Lachaud is currently a Ph.D. candidate at TILT, the Tilburg Institute for Law, Technology, and Society, Tilburg University, in the Netherlands. His research focuses on the possible contribution of certification schemes to data protection regulation. Email: eric.lachaud@outlook.com

Hans Lammerant has studied philosophy (VUB 1996) and law (VUB/UIA 2004). He also has a candidate degree in industrial engineering (1992) and is currently finalizing a master in statistics (UGent). Previously, he has worked in civil society organizations on peace and human rights issues. In his research, he focuses on the effect of new developments in Data Science and statistics on surveillance and privacy. More generally, he is interested in how technological developments and globalization influence the development of new forms of exercising power and how this impacts law. Email: hans.lammerant@vub.ac.be

Marc van Lieshout M.Sc.is a senior scientist at TNO. Since January 2015, he is acting as business director of the Privacy and Identity Lab, a Dutch knowledge centre in which TNO, Radboud University, Tilburg University and SIDN join forces to study privacy and Identity, offering concrete solutions. He has been acting as programme manager of TNO's programme on Societal impact of ICT in the period 2005–2012. He is vice-chair of IFIP Working Group 9.2. He has been visiting scientist at JRC-IPTS during July 2008–July 2009. He has been engaged in several international projects related to the assessment and evaluation of national and international activities related to privacy and data protection, for the European Commission, the European Parliament and national departments. His research interests are on innovation policy and privacy, including regulatory, business and societal processes. Email: Marc.vanLieshout@tno.nl

Christiana Markou was born in Nicosia, Cyprus, and holds an LLB (Hons.) from the University of Sheffield (UK) and an LLM in international, commercial and European law from the same university. She was called to the Cyprus Bar Association in 2001 and has been practicing law since then at Markou-Christodoulou and Polycarpou LLC, of which she is the founding director. She also holds a Ph.D. degree from the University of Lancaster (UK). Her thesis was on EU cyber consumer law, specifically on online software agents and EU consumer and privacy law. Since 2012, she holds the post of lecturer at the European University Cyprus teaching consumer protection, private international law and internal market law. She writes mainly on consumer protection, e-commerce and privacy issues. Email: c.markou@euc.ac.cy

Yod Samuel Martín is a researcher at the Department of ICT Systems Engineering at the Universidad Politécnica de Madrid (DIT-UPM), where he is currently pursuing his Ph.D. on telematics systems engineering. His research work focuses on different categories of non-functional software and service requirements, especially on the categories of accessibility and privacy, understood from different points of view, and considering the role of "prosumers" and user-generated context in relation to these requirements. He is concerned with not only the technical perspectives (requirements engineering, design methodologies, quality attributes and user experience), but also their relation with and impact on non-technical realms such as digital rights and corporate social responsibility. Regarding his research line on privacy, his work focuses on issues related to personal data management including personal data disclosure, the Right to be Forgotten, identity and privacy management, and engineering methodologies such as Privacy by Design and Privacy by Default. Email: samuelm@dit.upm.es

Philipp K. Masur is a research assistant and Ph.D. candidate at the Department of Media Psychology at the University of Hohenheim, Germany. He studied communication science, philosophy and economics at the University of Mainz,

Germany, and at the Macquarie University in Sydney, Australia. His research focuses on computer-mediated communication and online privacy. In his dissertation, he analyses how social media users perceive situational factors of privacy and in how far this perception influences their privacy behaviour and self-disclosure. Email: philipp.masur@uni-hohenheim.de

Tobias Matzner Ph.D. is postdoctoral research associate at the International Centre for Ethics in the Sciences and Humanities at the University of Tübingen in Germany. His research focuses on questions of subjectivity and identity at the intersection of social and technological developments and the political implications of algorithms and Big Data. He holds a Ph.D. in philosophy and a graduate degree (Diploma) in computer science, both from the Karlsruhe Institute of Technology. Email: tobias.matzner@uni-tuebingen.de

Anthony Morton is a research student in the Information Security Research Group at the Department of Computer Science, University College London (UCL), UK. He commenced his Ph.D. in 2010, having gained an M.Sc. in Information Security at UCL. He is a chartered engineer through the British Computer Society and also holds an M.B.A. in technology management, an M.Sc. in computing for commerce and industry and an M.A. in classical studies, all from The Open University. His Ph.D. research focuses on the influence of the privacy behaviour of technology services—consisting of a technology platform and providing organization—on the construction of people's privacy concerns. Prior to commencing his studies at UCL, he was employed in the IT industry for 25 years in software development, technical management, solution architecture and consultancy roles. Email: anthony.morton.09@ucl.ac.uk

Arvind Narayanan is an assistant professor of computer science at Princeton. He studies information privacy and security and has a side interest in technology policy. His research has shown that data anonymization is broken in fundamental ways, for which he jointly received the 2008 Privacy Enhancing Technologies Award. Narayanan leads the Princeton Web Transparency and Accountability Project, which aims to uncover how companies are collecting and using our personal information. He also studies the security and stability of Bitcoin and cryptocurrencies. Narayanan is an affiliated faculty member at the Center for Information Technology Policy at Princeton and an affiliate scholar at Stanford Law School's Center for Internet and Society. Email: arvindn@CS.Princeton.EDU

Carsten Ochs is since January 2014 postdoctoral researcher at the Institute for Sociology (Sociological Theory Department), University of Kassel, working for the BMBF research project "Privacy Forum". Prior to that he held postdoctoral positions at Technical University Darmstadt (European Center for Security and Privacy by Design and Sociology Department/Research Project "Internet Privacy: A Culture of Privacy and Trust" funded by the National Academy of Science and Engineering). In June 2011, he completed his Ph.D. studies at the Graduate Centre for the Study of Culture (GCSC), Justus-Liebig-University, Giessen/sociology department, before Carsten Ochs had attended the master programme "Interactive Media: Critical Theory and Practice" at

Goldsmiths College/Centre for Cultural Studies, London (2004–2005); and as undergraduate, he was trained in cultural anthropology, sociology and philosophy at Goethe University Frankfurt. Email: carsten.ochs@uni-kassel.de

Christian Ohmann has a Ph.D. in mathematics and an interim examination in medicine. He was the head of the Coordination Centre for Clinical Trials (KKS) at the Medical Faculty of the Heinrich Heine University Düsseldorf, Germany. He is currently the German representative and chair of the Network Committee of ECRIN-ERIC. He has major competence and experience in the field of clinical research/clinical trials and clinical research informatics. Email: Christian.Ohmann@med.uni-duesseldorf.de

Merel Ooms M.Sc.is research scientist at TNO. She has a background in sociology and policy research. As a researcher, she has been involved in several European and Dutch national projects on topics related to the relationship between technology and human behaviour. Her research interests focus mainly on privacy, sustainable energy and social innovation. On these topics, she collaborated in the European FP7 projects PRISMS and SI-DRIVE. For PRISMS, she has contributed to a number of research papers and chapters in which she is responsible for the quantitative analysis. Email: merel.ooms@tno.nl

Ugo Pagallo is a former lawyer and current professor of jurisprudence at the Department of Law, University of Turin (Italy). He is author of ten monographs, numerous essays in scholarly journals and book chapters, chief editor of the *Digitalica* series published by Giappichelli in Turin and co-editor of the AICOL series by Springer. Member of the European RPAS Steering Group (2011–2012), of the Group of Experts for the Onlife Initiative set up by the European Commission (2012–2013), of the ethical committee of the CAPER project, supported through the EU Seventh Framework Programme for Research and Technological Development (2013–2014), and expert for the evaluation of proposals in the Horizon 2020 robotics programme (2015). He is a faculty fellow at the Center for Transnational Legal Studies in London, U.K. (2013, 2008); and current NEXA Fellow at the Center for Internet and Society at the Polytechnic University of Turin. Email: ugo.pagallo@unito.it

Evangelia Papadaki is a Ph.D. student at the University of Southampton, UK. She is a member of the Web Science Institute and the Institute for Law and the Web. She holds a degree in law (University of Athens, Greece), an LLM in cyberlaw (University of Leeds, UK) and a M.Sc. in Web Science (University of Southampton, UK). She is primarily interested in Internet governance and the regulation of cybercrime. Her current research focuses on the role of Internet intermediaries in improving cybersecurity and, in particular, on the EU legal framework surrounding their security obligations, the technologies that need to be implemented to effectively safeguard users' personal data, and the incentive structure defined by their business model.

Sven Rung M.Sc.is a junior researcher of the ICT research group at the Fraunhofer Institute for Systems and Innovation Research ISI in Karlsruhe, Germany. He has a background in statistics and economics, and therefore, he is involved in projects with focus on socio-economic and statistical analysis (e.g. surveys). He is also a Ph.D. student at Hohenheim University, Germany. For PRISMS, he has contributed to a number of research papers and chapters in which he is responsible for the quantitative analysis. Email: sven.rung@isi.fraunhofer.de

Sven Schäge is since 2014 a postdoctoral researcher at Ruhr-UniversitätBochum. His research interests include all topics around provable and practice-oriented cryptography, in particular digital signatures and key exchange. While not doing research, he works as a interdisciplinary coordinator where he supports the Ph.D. students of the DFG Research Training Group UbiCrypt to get involved into interdisciplinary research cooperations in the field of IT security. From 2011 to 2014, he was a research associate (postdoctoral researcher) at University College London working on cryptography & security. He received his Ph.D. (on "Efficient and Provably Secure Signature Schemes in the Standard Model") from Ruhr-Universität Bochum in 2011. Email: sven.schaege@rub.de

Sophie Stalla-Bourdillon Ph.D.is an associate professor in IT law and director of the Institute for Law and the Web. She specializes in IT-related issues and in particular the impact of traditional bodies of law and fundamental rights and liberties upon online communications. She has been researching and writing on the roles and responsibilities of Internet intermediaries in the law enforcement process such as Internet service providers, Web 2.0 platforms, search engines, on the legal implications of deep packet inspection practices implemented by Internet service providers, and on the role of hosting providers in relation to malicious web pages. Sophie has recently co-authored a book on Privacy versus Security published by Springer. She is currently leading the legal team in FP7 ICT OPTET (on operational trustworthiness enabling technologies). She is part of the task force working on data privacy and cloud computing for the E-infrastructure Leadership Council. Email: s.stallabourdillon@soton.ac.uk

Thilo van Pape Ph.D.is associate researcher at the Institut für Kommunikationswissenschaft, Universität Hohenheim, Germany. His research interests lie in the diffusion and appropriation of new communication technologies as well as uses and consequences of mobile and online communication. Email: thilo.vonpape@uni-hohenheim.de

Evert-Bert van Veen has a Ph.D. and works as a lawyer in the healthcare sector. He is a senior consultant at MedLawconsult and publishes regularly. His core competencies cover medical research, innovative therapies, European law and administrative law, privacy regulations, healthcare claims, etc. Email: eb.vanveen@medlaw.nl

Dr. Bibi van den Berg is an associate professor at eLaw, the Center for Law and Digital Technologies at University Leiden's Law School. Van den Berg has an M.A. and Ph.D. in philosophy, both from Erasmus University in Rotterdam. After completing her Ph.D., she held a postdoctoral position at the Tilburg Institute for Law, Technology and Society (TILT), for three years. In 2012, she transferred to Leiden for her current position. Van den Berg's research and teaching focus on several themes: (1) cybersecurity, (2) regulating human behaviour through the use of technologies (techno-regulation and nudging), (3) regulation and governance of/on the Internet, (4) privacy and identity and (5) robotics and artificial intelligence. Van den Berg is also a member of the Dutch Cyber Security Council. Email: b.van.den.berg@law.leidenuniv.nl

Bart van der Sloot studied law and philosophy in the Netherlands. He currently works part-time for the Netherlands Scientific Council for Government Policy (WRR) and at the Institute for Information Law of the University of Amsterdam (IvIR) where he currently aims to develop a novel understanding of privacy, in which privacy is not defined as the right of the individual, but as the virtue of the state. Bart is also the coordinator of the Amsterdam Platform for Privacy Research (APPR), which incorporates about 70 researches from the University of Amsterdam, who in their daily research and teaching focus on privacy-related issues. They do so from different perspectives, such as law, philosophy, economics, informatics, medicine, communication studies, and political science. For more info: www.appr.uva.nl/en. For the Amsterdam Privacy Conference 2015, please check: www.apc2015.net. Email: b.vandersloot@uva.nl

Robert Verheij has received a Ph.D. in social sciences. He is the project leader of NIN-GP at NIVEL, the Netherlands Institute for Health Services Research, and has been involved in the eHID project. He was responsible for developing a tool for assessing the quality of data of electronic health records and for handling the legal issues of the use of these data. Email: r.verheij@nivel.nl

Mind the Air Gap

Preventing Privacy Issues in Robotics

Bibi van den Berg

Abstract The market for domestic and service robots, which can help consumers in their homes, is growing rapidly. Privacy scholars have warned that the deployment of robots in the home can lead to serious privacy risks, since these robots come equipped with sensors that register the environment in which they operate, including the human beings present therein. Moreover, many modern robots are, or will soon become, connected to the internet. This means that they can pass on any data they record, and that they can also be hacked by outsiders. The privacy and security risks are significant with this novel type of technology. In this chapter I argue that attaching robots to the internet does indeed lead to serious privacy and security risks. But I will also argue that there is a straightforward solution that can contribute to eliminating many of these risks, which is left unaddressed in the current debate. Borrowing a term from the field of cybersecurity I will argue that consumer robots ought to be 'air gapped', that is they ought *not* to be connected to the internet or the cloud—expect in a few exceptional cases where network connections are a critical requirement for robots to be able to function properly or effectively. I will explain what air gaps are, how they are used, and which strengths and weaknesses they have. Next, I will critically assess their use as a strategy to prevent privacy and security issues in domestic and service robots. I will argue that it is important to have a debate on the networked character of robots today, since we can now still *prevent* privacy and security problems in this novel technology, rather than having to remedy them once they are mass-marketed. This is why I propose to have a debate on privacy *before* design.

Keywords Robots · Privacy · Data protection

B. van den Berg (✉)
Fac. der Rechtgeleerdheid, eLaw, Universiteit Leiden, Steenschuur 25,
2311 ES Leiden, The Netherlands
e-mail: b.van.den.berg@law.leidenuniv.nl

1 Introduction

In recent years the sale of consumer robots has been on a sharp increase. The International Federation of Robotics (IFR) has calculated that in 2013 alone 4 million of such robots were sold worldwide, an increase of 28 % over the year before.[1] It predicts that this growth will continue between 2014 and 2017, and estimates that in that three-year period "*almost 23.9 million units*" will be sold worldwide "*with an estimated value of US$ 6.5 billion*".[2] Especially the market for robots that can assist the elderly or disabled in their homes, also called domestic or service robots, is expected to grow exponentially, not just in the next few years but also in the longer term. This prediction is made in light of demographic changes, most importantly falling birth rates and increasing life expectancy in many countries,[3] which result in significant societal changes such as ageing societies. If more people can live independently in their own homes thanks to the services provided by domestic or service robots this reduces healthcare costs and increases individuals' welfare, or so the argument goes.

However, scholars point out that bringing robots into the home has significant consequences for individuals' privacy (cf. Calo 2012; Coeckelbergh 2009; Denning et al. 2009; Sharkey and Sharkey 2010; Sharkey 2009). This is so for several reasons. Most importantly, robots are progressively connected to the internet and/or make use of cloud services. In this chapter I will argue that hooking up robots to the internet does indeed lead to serious privacy risks. Privacy advocates, therefore, are quite right in pleading for the need for privacy-awareness in designers and for the use of privacy by design principles (cf. Cavoukian 2009, 2010; Rubinstein 2011). They are also correct in proposing legal and regulatory solutions to prevent and remedy potential privacy breaches.

Having said that, as I will show there is a simple and effective solution to preventing, or at the very least minimizing, privacy problems in relation to personal and domestic robots, which both privacy advocates and technology designers have failed to see. Borrowing a term from the field of cybersecurity I will argue that consumer robots ought to be 'air gapped', that is they ought *not* to be connected to the internet or the cloud—expect in a few exceptional cases where network connections are a critical requirement for robots to be able to function properly or effectively. The number of cases in which is this so, I will argue below, is in fact very, very limited. All other robots, in my view, ought to be standalone, autonomous machines, not connected to outside networks. If robots cannot communicate the data they store as they navigate through and operate within our homes, and if outsiders cannot get direct access to this data, no real danger to our privacy will exist.

I will argue that the time for deciding to air gap robots is now, and that urgent action is required. While domestic and personal robots are a rapidly growing

---

[1] See http://www.ifr.org/service-robots/statistics/ (last accessed on 22 May 2015).
[2] Ibid.
[3] See e.g. http://www.who.int/ageing/about/facts/en/ (last accessed on 22 May 2015).

market worldwide, the vast majority of robotic systems, tools and applications that are on the horizon are still in laboratories or in early design phases. Making fundamental choices for the future development of this field, for example with respect to balancing privacy protection and security on the one hand and efficiency and usability on the other, therefore, is still feasible and worthwhile. However, the window of opportunity for making such choices is closing and therefore we must make firm decisions now. Rather than remedying privacy and security problems through regulation or the application of privacy by design principles *after* consumer robots have truly hit the mass market this chapter can be considered a plea for privacy before design. As the chapter will show simply reconsidering some of the fundamental reasons for choices in technology design, and implementing elementary solutions like the use of air gaps can sometimes make grave (privacy and other) concerns—and significant problems down the road—go away even before they have had a chance to materialise.

2 Privacy Concerns for Domestic Robots

Domestic and service robots are robots that are used in the home to conduct a variety of household chores. At CES2015,[4] one of the largest technology trade shows in the world, it was clear that such robots are among the fastest growing sectors in consumer electronics (cf. Green 2015; Kelion 2015). Domestic and service robots come in many shapes, sizes, guises and varieties, and also with different degrees of functionality. Some are very specialist and can only conduct one specific type of activity. Think of robotic lawnmowers, pool cleaners, vacuum cleaners and barbecue cleaners, but also of robots that carry objects around the house or keep people company. Others are more versatile and can conduct a set of different tasks. For example, one robot presented at CES2015 can *"take control of internet-controlled smart devices*—[one can give it verbal instructions] *to turn lights, music and heating on or off—use it as a teaching aid for* [one's] *children, or take advantage of its health check software to help care for elderly relations"* (Green 2015). Having said that, all experts agree that all-round, fully functional robot butlers or housekeepers who can conduct the same variety of tasks that their human counterparts can undertake are still far beyond the horizon.

There are several reasons why domestic and service robots are becoming increasingly popular. Over the past decades the price of such robots has dropped steadily, making them a viable option for a growing group of consumers worldwide. Moreover, as we have seen societal changes such as the emergence of ageing societies and increasing pressure on healthcare systems in many countries have

---

[4]CES is an international, annual event on consumer electronics that is held in Las Vegas (USA). CES2015 took place from 6 until 9 January 2015. See http://www.cesweb.org for more information on this event. (last accessed on 28 May 2015).

greatly contributed to their popularity. Having one or more robotic helpers in the home to take over menial tasks is convenient, efficient and cost-effective.

However, inviting robots into our private spaces also raises important issues, most importantly with respect to privacy and security. Both scholars from the field of law and from the field of robotics design have expressed concern with regard to robots' entrance into the home in recent years (cf. Calo 2012; Coeckelbergh 2009; Denning et al. 2009; Sharkey and Sharkey 2010; Sharkey 2009; Wallach 2008). Their main concern lies with the fact that robots are equipped with all sorts of sensors, cameras, and microphones. These devices enable the robot to register its environment, to record images and video, to register movement and sounds and to find its way through its surroundings. Without these pieces of equipment robots would not be able to operate independently in complex environments. But at the same time such sensors and recording devices may cause privacy risks, especially in the home. After all, robots not only record, process and store information about their environment, but also about the human beings that are present therein, at any given moment in time. These human beings may not always be aware that this is the case.

Moreover, domestic robots find themselves in an environment that is normally considered personal and private, where less stringent social norms and rules of etiquette apply and where human beings can display so-called 'backstage behavior' (Meyrowitz 1985).[5] Allowing robots to enter our homes and register their environment, including the human beings in that environment entails that backstage behaviours may be captured and stored by these machines.

What's more, what makes the privacy risks of letting robots into the home even graver is the fact that they may access domestic spaces that are traditionally considered to be very private, such as bedrooms and bathrooms, and that would traditionally only be accessible to a very limited set of human beings other than the occupants. While there, they may register data about individuals in exceptionally private or compromising situations, especially when such robots have care tasks for example for the elderly, for children or for sick people.[6] Since domestic robots can be operational around the clock, they could turn into veritable 'surveillance machines', registering highly sensitive data about human beings in highly private places.

The final, and definitive step in reasoning about privacy problems generated through the deployment of robots in the home environment is that robots are increasingly connected to the internet, and hence may pass on, or provide access to, all of the potentially personal and sensitive data they record, process and store. Moreover, a growing number of robots, including domestic and service robots, aim to use the cloud for the storage of both a robot's programs and for all of the data it records. One common argument legitimating the development of so-called

---

[5]More broadly, also see Goffman (1959, 1968, 1982, 1997).

[6]One line of research into robotics suggests that in the future these machines may even become sexual partners (cf. Levy 2009). Obviously, this takes potential privacy risks to new heights.

'cloud robotics' (Guizzo 2011) is the idea that it may greatly increase the amount and variety of tasks a single robot can complete. This works as follows. For each (set of) tasks, a robot requires separate programming, specific software programmes that define what the task is and how to execute it. Due to the fact that robots operate in very complex environments these programmes tend to be highly complicated, involving large numbers of lines of code, and they are difficult to develop. Developing robots that can fulfil a wide variety of functions, therefore, is costly and very time-consuming. Moreover, because robots have limited processing and storage capacity, they can only store a restricted set of programmes in their memory to begin with. These two issues combined lead to the fact that standalone, autonomous robots are limited in the amount of tasks they can do simply because their developers can only implement a limited number of different programmes in their memory. Learning new tasks, moreover, is not an option for such robots. Cloud robots seek to overcome these limitations, for example by storing software for robots in the cloud. The *RoboEarth* project,[7] an EU-funded project that ran from 2010 until 2014, sought to provide a cloud platform from which robots themselves could download software updates or even entirely new software packages in order to increase their functionality. The underlying idea was that once one robot has been taught to do something, then other robots could benefit by expanding their capacities, thus speeding up the development of 'all purpose' robots and making individual robots infinitely more versatile. Robots could thus use the internet and/or the cloud to learn, and to overcome their physical and 'mental' limitations in the real world. In the words of Weng and Zhao: *"networked robots can reduce the uncertainty of their physical world with resources from the virtual world, or cyberspace."* (2012: 62)

Cloud robotics obviously necessitates that robots are connected to the internet. So what are the privacy and security risks that may arise due to the emergence of cloud robotics and of plugging robots into the internet? In fact, two different types of such risks arise. First, robots may be hacked or otherwise deliberately compromised by outsiders, for example because these outsiders seek access to the (personal) data of individuals stored in the robots' memory or aim to record information live when under their command (cf. Denning et al. 2009).[8] Many of

---

[7]See http://roboearth.org (Last accessed on 16 June 2015).

[8]If a robot is attached to a properly secured home network (using e.g. proper firewalls and encryption for data transfer), many of the security threats that exist for such robots (e.g. man-in-the-middle attacks, password cracks and brute force attacks) disappear (Denning et al., 2009). However, it is safe to assume that many consumers do not secure their home networks properly, which leads to a variety of vulnerabilities that can easily be exploited, for example not changing the factory settings or default password for networked appliances (see for a broader discussion on this issue Aitel 2013; Farwell and Rohozinski 2011; Schneier 2004). As Denning et al. point out: when using robots in the home *"the supposition that the robot is secure is based upon the assumption that users will correctly configure and administer secure encryption on their networks."* (2009: 5). But there is no reason to assume that security slip-ups will not occur and hence robots are susceptible to security breaches.

the security risks we know in other devices, be they phones, tablets, webcams, computers etcetera, also apply to robots. They are vulnerable to the same types of breaches, exploits and attacks. It is likely that security breaches will occur in relation to domestic and service robots in the future, because of the types of data stored on such machines—as we have seen they may contain images or videos of individuals in potentially sensitive or compromising situations. Since robots can record (personal) data about individuals over longer periods of time, they can deliver rich, detailed data sets on their living patterns and expose their daily habits, but also gather data regarding individuals' *"house layout and where* [their] *valuables are"* (Hardy 2014). Therefore, it may be (financially or otherwise) advantageous to hack into such robots.

Second, aside from outsiders who gain access to robots in the home, the latter may also actively pass on data themselves. In some cases, this may be perfectly legitimate on the surface of things. For example, domestic robots may be brought into the home to monitor specific (sets of) health parameters of patients, and to communicate measurements to healthcare practitioners who can thus keep an eye on patients' health status. In this case, it is obvious that such data must be passed on via a network and hence the robot must have access to the internet or another network technology for it to be able to function properly. As long as proper security and privacy protections are in place[9] in such robots the risk of privacy violations could be minimal. This entails a combination of implementing technical measures, such as the use of encryption and privacy by design principles, as well as framing robots within the existing legal landscape, most importantly for data protection.[10] As for the latter, scholars have argued that robots fall under the same legal data protection frameworks as many other networked computational devices (Pagallo 2015), and as long as these frameworks are respected proper privacy protection with respect to domestic robots is in place. However, in light of the privacy and security breaches that we have encountered in relation to myriad other new networked technologies in the past decades it seems likely that security and privacy protections, either technical or legal, will not always suffice to prevent incidents and violations. All domestic and service robots that are connected to the

---

[9]Think of, among other things, ensuring that individuals give their consent for transmission, that they are able to access and rectify the data, ensuring that proper protections with respect to the transmission and storage of data are in place, that there is a clear differentiation and definition of roles with respect to access to the data (which people, in which roles, gets access to which (types of) data), and that data is only shared and (re)used for legimimate and limited purposes.

[10]Note that although privacy and data protection are often used interchangeably there are important differences between these two concepts (cf. De Hert and Schreuders 2001). Data protection seeks to contribute to a proper protection of privacy, but it also has other goals, such as strengthening the internal market in the EU and ensuring free flow of information. To these ends, data protection rules describe the conditions under which data may be shared and processed, and how this ought to be done in a way that promotes transparency and accountability. Privacy, by contrast, has a strong experiential component and focuses not only on the protection of data, but also on individuals' private space and physical integrity (cf. Nissenbaum 2004). In this chapter the focus is on privacy rather than data protection.

internet may lead to privacy and security risks, because they may pass on too much or the wrong information, pass it on to unwanted audiences, without consent or without clearly defined purposes and so on and so forth.

If we follow the steps in the arguments presented above it is understandable that scholars warn for privacy and security risks, especially in relation to domestic and service robots. Privacy scholars' concerns are valid and ought to be taken seriously. And both technology designers and regulators ought to look into means to prevent or minimize privacy and security risks as well as consider potential remedies should breaches occur nevertheless. Again, if we follow the arguments presented here, then privacy and security concerns are indeed a grave matter. But *should* we follow these arguments? Are they correct in their analysis of the problem? Or are there alternative possibilities that these arguments disregard?

3 Why Privacy Need Not Be a Problem: Unravelling the Arguments

As we have seen above, privacy and security problems arise in relation to domestic and service robots because these robots enter the home and may register anything and everything in that home using their sensing devices, or so the argument begins. This includes human beings in all of their private and vulnerable states. When we look more closely at this argument it is immediately evident that *in and of itself*, in fact, the idea that robots register us, and store data about us in our homes, is not a reason for concern, let alone for privacy or security concerns. After all, when other human beings come into our homes they, too, 'register' and 'store' information about us using their 'sensing devices'. Of course one could argue that most humans do not stay in our homes for very long periods of time, and could thus not be considered 'surveillance machines'. This is a valid argument. However, there is one significant exception: healthcare professionals may come into our homes in times of sickness and spend considerable amounts of time with us in precisely the same private spaces, and engage with us in exactly the same very sensitive and potentially compromising situations as domestic and service robots would. We have no concerns, generally, over the fact that such healthcare professionals would violate our privacy or security by viewing us in such a fragile state, or in such private parts of the home. We are also not concerned about healthcare professionals sharing the information they have 'stored' about us in such situations with others outside the home. We simply rely on their sense of propriety and their professional ethics to keep what is private private. So what is behind our concerns over robots gathering (personal) data? Let's unpack these concerns to see what they are all about.

First of all, there is the fact that we have concerns over the data collection and data processing capacities of robots. These worries, I argue, may stem from two sources. First of all, we may find the idea that robots record and register data using their sensors eerie, uncanny, because robots can store detailed and extensive sets of data about the environment surrounding them, using a wide variety of sensors, some of which do not resemble the 'sensing devices' we ourselves are equipped with, such as ultrasonic sensors or GPS. While on the surface robots resemble humans in the fact that they come equipped with a set of 'sensory organs' to help them navigate and respond to the world that surrounds them, at the same time the 'organs' that robots have deliver them quite different data sets, and humans may feel uncanny about the types and the multitude of data that is collected and processed about the environment, and their personae in it, through all of these channels. In this case, one could argue, humans may feel uncomfortable about robots collecting data because robots are quite a bit like humans, but differ from them in some essential respect (see also Bryant 2004; Mori 1970; Van den Berg 2011).

Interestingly, at the same time humans may feel uncomfortable about robots collecting massive amounts of data due to the fact that robots are 'mere machines', devices, and hence may disappear into the background of their everyday surroundings just like other computing devices do, which may lead them to underestimate the amount of data they record and store about them. It is easy to see how this may happen: if a device is continuously present in one's home environment one may gradually start to forget, ignore or downplay the ingenuity of the device and the extent of data collection that it may engage in. A sense of creepiness can arise in moments when human beings suddenly become aware of the fact that a robot is present recording some situation, just as it would when they realise someone is looking in through a window or when they become aware that someone is present in a room that they did not realise was there before. In this case, then, the sense of eeriness is not due to an awareness of the amount of information that is continuously collected by robots, but rather the reverse: it emerges when the occupants of a home suddenly become aware of this fact while having taken for granted the robot for a longer period of time. In both cases, however, eeriness, a sense of awkwardness and creepiness may be the result.

While it is important to realise that individuals may feel eerie or awkward about robots and the data they may collect in their homes, in and of itself eeriness is unrelated to privacy or security concerns. In fact, neither of these reasons has anything to do with privacy or security (also see Thierer 2013). After all, they refer to the function and capabilities of the robot (reason 1), or humans' tendency to take devices for granted (reason 2). So while these are valid concerns to look into for technology developers, they ought to be ignored in discussions on privacy and security, for they simply have no bearing on these issues.

A second, and much more important reason why privacy and security concerns arise in relation to domestic and service robots, as we have seen, is because these machines are connected to the internet and, in some cases, the cloud. This means that they have a connection to the outside world, and hence may (inadvertently) share, or provide access to, information that they have collected inside the home. As said, this is a real and valid concern. As a matter of fact, this is the real issue on which any discussion on privacy and security in relation to domestic robots ought to focus. However, one simple yet elementary question is not raised in discussions on privacy and robotics: *why* is this the case? Why are robots connected to the internet in the first place? And what would happen if we *did not* connect them to the internet? Would this entail that many, if not all, privacy and security problems simply disappear?

The answer to the latter question, I argue, is a resounding yes. I view the development of networked and cloud robotics as part of a larger movement that we can label the 'internet of things' (cf. ITU 2005).[11] The key idea behind this movement is that any sensor, appliance, device or system that we create is hooked into the internet. There are myriad valid reasons for doing so. For one, end users and consumers may greatly benefit attaching all of the devices in their homes to outside networks. This makes devices and systems remotely accessible, so that they can be operated at a distance. Think of using one's mobile phone to raise the temperature at home via the smart meter while in transit from work. Or of being able to access files and folders on devices at a distance. Or of being able to monitor (home) security via one's mobile phone. Remote access to systems and devices raises efficiency, is fun and 'comes in handy' for end users.

Technology companies, for their part, have their own reasons why they are keen to connect devices to the internet. The data they can collect from these systems could potentially be turned into gold, especially when data streams from many different devices are combined into profiles that reveal end users living patterns.[12] What's more, internet access to devices in consumers' homes enables businesses to update software remotely. This means they can put products into the market with much shorter development cycles. After all, if the software is not entirely fault-free before the product enters the market this is not a serious problem, because it can be updated after it has been sold without significant hassle for the consumer (e.g. having to take the product back to the store). In highly competitive markets, where business cycles are extremely short, an opportunity such as remote access provides businesses with just the kind of competitive edge they may need to stay ahead of their competitors.

Obviously, these are all arguments that have some merit. Having said that, at the same time the reason why I am labelling the internet of things as a *movement* is because the trend towards networking devices, sensors and systems almost appears to have taken on a life of its own, whereby we now uncritically, automatically outfit *every* technological piece of equipment we create with the potential to connect to networks, without explicitly, consciously asking ourselves whether this is necessary for its functioning, whether it could also operate adequately without an internet connection, and whether or not such a connection is even desirable. As a matter of fact, hardly any consumer technology now enters the market *without* an internet connection, or at least the possibility thereof. It seems we have come to

---

[11]The same, or a very similar, movement is known under the banner of Ambient Intelligence (cf. Aarts and Encarnação 2006; ISTAG 2003; Ruyter and Aarts 2004; Van den Berg 2010a, b), ubiquitous computing (cf. Edwards and Grinter 2001; Greenfield 2006; Weiser 1991) or pervasive computing (cf. McCullough 2004).

[12]More on this below.

think of hooking up devices to the internet as a default way of using them, regardless of their role or functionality in our everyday lives. Consumer electronics and machines as diverse as coffee machines,[13] baby phones,[14] televisions,[15] ovens,[16] vacuum cleaning robots,[17] light switches,[18] cars[19] and bicycles[20] now all come equipped with wifi or some other means to connect to the internet. It is far from clear whether this is always necessary or even useful. Surely being able to remotely access one's coffee machine so that it can deliver a cup of coffee just as one enters one's home does not deliver that much of a consumer benefit over pressing a button once one gets there? The same applies to remotely activating the oven or switching on the lights: what is the real benefit, the real added value of being able to do so via the internet using one's smart phone?[21] These products are marketed as convenient and fun, and while consumers may indeed find it 'fun' to switch on the coffee machine or the oven just before they arrive at home, the most urgent question, of course, is: at what price can they be offered this (minor bit of) 'fun' and convenience? Is the limited amount of efficiency and enjoyment that is gained by hooking these devices and machines into the internet really worth the security and privacy risks this inadvertently entails? Do consumers realise, or have enough information and enough of an understanding to realise, what the potential consequences of such a connection are, and can they make proper, weighed decisions in this respect? While switching on the thermostat or the lighting may seem

---

[13]See for example http://www.amazon.com/Mr-Coffee-Wifi-Enabled-Coffeemaker-BVMC-PSTX91WE/dp/B00LUFSSWG (Last accessed on 22 June 2015).

[14]See for example http://www.amazon.com/Otium-Microphone-Surveillance-Monitoring-Smartphone/dp/B00WE055HW (Last accessed on 22 June 2015).

[15]See for example http://www.amazon.com/Samsung-UN40J5500-40-Inch-1080p-Smart/dp/B00U5ZT8OO (Last accessed on 22 June 2015).

[16]See for example http://www.lg.com/us/ranges-ovens/lg-LRE3027ST-electric-range (Last accessed on 22 June 2015).

[17]See for example http://www.amazon.com/Ameribot-Vacuum-Cleaning-Robot-Allergies/dp/B00ZKXP6ME (Last accessed on 22 June 2015).

[18]See for example http://www.amazon.com/Belkin-WeMo-Switch-Enabled-Compatible/dp/B00DGEGJ02 (Last accessed on 22 June 2015).

[19]See for example the 2015 BMW 3 series, which comes equipped with 4G: http://www.autoexpress.co.uk/bmw/3-series/91367/bmw-3-series-2015-engine-tech-and-styling-tweaks (Last accessed on 22 June 2015).

[20]See for example http://connectedcycle.com (Last accessed on 22 June 2015).

[21]When reading a draft version of this chapter one of the reviewers correctly pointed out that being able to switch on or off light when one is not at home also leads to *more* security in some respects: it can help decrease chances of break-ins. While this is true, of course, an internet connection or remote access is not required to accomplish this. For the past decades many people have come to use physical timers, to be plugged into wall sockets, to improve their home security. While such a system may be less flexible than an app on one's smart phone the simplicity and security of such devices has great benefits over light switches that can be accessed remotely. As with many examples in this chapter, it is all about finding the right balance between convenience, efficiency and fun on the one hand, and security and privacy on the other.

like an innocent, handy gimmick to end users, how many of them are fully aware of the vulnerabilities this entails in terms of security and privacy, and would they still feel the same if they could make a more informed choice for or against such functionality?

These questions are all the more urgent when one considers that the home of the near future, in which the internet of things has materialised, is filled to the brim with a multitude of different devices and sensors that each record their own data stream. When combined, a highly textured and detailed image of end users lives' and life styles may emerge, leading to an amplified version of the privacy and security concerns discussed above (cf. Hildebrandt and Van Dijk 2012; Hildebrandt 2013; Mayer-Schönberger and Cukier 2013; Morozov 2013; Zarsky 2003). It is not primarily single, standalone networked devices as such that lead to privacy and security concerns—although they may do as well—but especially the totality of such devices combined. Each device, then, that is added to the network in a consumer's home increases the chances of privacy and security risks, both in and of itself, and in relation to all the other devices that are active in that environment. Hence, each and every device that human beings bring into the home and connect to the internet ought to be critically assessed: does it really need to be connected to the internet in order for it to function properly? Can it do without an internet connection? And in whose interest is this internet connection?

As we have seen, companies often provide electronic devices such as coffee makers and vacuum cleaning robots with an internet connection so that software updates can be installed remotely and these products can be marketed more quickly. And yes, it also means that the owner can switch on his coffee machine remotely. One could argue, however, that this latter benefit is so small that it ought to be outweighed by the security and privacy concerns such an internet connection always entails. And it is obvious that the argument of providing any and all devices with an internet connection primarily so that businesses can sell their (sloppily programmed) products more quickly really does not stand up against scrutiny. Businesses ought to market their products once they reach sufficient quality standards, and not a day sooner, especially when the price paid for improperly functioning technology lies squarely on the shoulders of consumers: it is their homes that these devices enter, and it is their security and privacy that are at stake. This is all the more so since proper software design involves designing security measures, and rapid market times are detrimental to the implementation of especially this kind of measures.

Ensuring that software on a device works properly in all respects before the product goes to market seems like an obvious and straightforward requirement, which as a by-product would eliminate the need for internet connections for a vast number of different devices, systems and appliances that now come equipped, or will soon be equipped with such a connection. As for the customers and their wishes: should a device *not* come with an internet connection and become operable remotely I imagine most consumers would never even miss such an option, especially if the same device (coffee makers, light switches, cars, ovens etcetera) in previous generations was not connected to a network either. Consumers will not miss what they did not have in the past.

And even if consumers, in the end, decided they *did* want to buy a product with an internet connection, they should at least be able to do so on the basis of informed choice: after having received proper information on the benefits and risks this entails.[22] Moreover, consumers ought to be able to choose between products with and without an internet connection, so that those who value their privacy and security over more functionality can buy an 'offline' version of a product, while those seeking convenience can choose the 'online' version—again, after being properly informed about the benefits and risks of each. Now consumers are neither informed, nor have a choice, and, what is worse, products on the market today cannot utilise their full functionality without an internet connection, as we have seen above, for instance, in the discussion on cloud robotics. Thus, technology push ensures that consumer products uncritically develop in one single direction: towards ever more connectivity.

This is the reason why it is essential that we have this debate about internet connectivity and the internet of things today rather than tomorrow, when gradually ever more new generations of consumer electronics will be marketed with wifi or other network technologies on board. To jumpstart this discussion I will discuss a key concept from the field of cyber security: that of the 'air gap'.

4 Mind the Air Gap: Prevention Rather Than Cure

Security specialists have struggled with networked systems' vulnerabilities, and the opportunities for exploitation these offered, since the 1980s.[23] One potential method to help protect computer systems that contain vital or sensitive information is the use of so-called 'air gaps' (cf. Byres 2013; Kello 2013; Kim 2014; Lachow 2015; Rid 2012; Singer and Friedman 2013). As Thomas Rid explains, an air gap entails *"that* [a] *secure network is physically, electrically, and electromagnetically separated from insecure networks"* (2012: 21). Or in the words of Sven Herpig: using air gaps *"reflects the disconnections of critical and vital networks from broader networks which are ultimately connected to the Internet"* (Herpig 2013: 168). By not attaching vital systems to the internet, and by completely separating them from networks that are connected to the internet, these systems cannot be attacked from the outside, since there is no connection to the external world. The notion of air gaps is relevant for the protection of critical infrastructures, such as (nuclear) power plants, transportation networks, (tele)communications networks etcetera. It may also be used to protect the 'crown jewels' of both businesses and

---

[22]Of course this is a requirement in data protection legislation as well.

[23]See for example Clifford Stoll's book on one of the first large-scale international hunts for a hacker that invaded numerous computer systems in universities, government organisations and even NASA in the USA before getting caught: (Stoll 1989).

government organisations, for example networks that contain military, strategically or commercially sensitive information.[24]

The operative idea behind 'air gapping', then, is that one can prevent cyber security incidents, or diminish the risk thereof, through the physical separation of critical networks from other networks and systems that are attached to the internet. Because of this separation attacks on these systems or networks, or infections thereof, become very difficult (cf. Kello 2013). As Herpig notes, using air gaps *"is a rather simple and straightforward strategy. Critical devices which do not need to be connected to larger networks, especially the Internet, should not be connected. Also, in times of emergency, networks might be air-gapped temporarily to mitigate damage"* (2013: 169). Using air gaps is part of the larger idea of so-called 'network segmentation': the idea that one divides one's information and networks into segments that are (physically and virtually) isolated from one another so that *"a successful cyber attack cannot spread laterally across the entire network"* (Aitel 2013: 58).

Obviously, using air gaps does not lead to 100 % security—no system or network is ever 100 % secure, and any security measure, in the end, can be breached. Air gaps, too, can be bypassed, and examples of such breaches do exist. The Stuxnet worm and breaches into the classified (and air gapped) mail system of the US Department of Defense are examples in case (cf. Aitel 2013; Appazov 2014; Clarke and Knake 2010; Farwell and Rohozinski 2011; Kim 2014; Lachow 2015; O'Harrow 2013). So what are the weaknesses of using air gaps? First, using air gaps in organisations does not stop insiders, most importantly employees from breaching security, either intentionally or accidentally. If disgruntled employees want to attack a system, they will find ways to do so even if there is an air gap. Relatedly, air gaps do not protect against so-called 'sneakernet threats' or 'sneakernet effects' (cf. Byres 2013; Rid 2012). This means that employees may infect secure, air-gapped systems by plugging USB drives or CD-roms into them. If they have previously used these devices on systems that were connected to the internet they may (accidentally or intentionally) spread malware from one network to the next. When this happens, these employees form the physical link between secure and non-secure networks and, as such, close the loop between these networks, effectively eliminating the air gap. In fact, this is precisely how both the Stuxnet worm and the Department of Defense hacker got access to air-gapped systems (cf. Clarke and Knake 2010; Farwell and Rohozinski 2011; Lachow 2015; Zetter 2014).

A second argument that is often raised against the use of air gaps is the fact that they may lead to less efficiency or effectiveness. The line of reasoning here is that in recent decades elements of the critical infrastructure, such as power companies, water and sewage management systems, and telecommunications networks have all been attached to the internet because this greatly increases the efficiency of

---

[24]Note that separating critical systems from systems that are connected to the internet (and hence the outside world), does not entail that the former cannot be networked at all. Critical systems can use a wide variety of different types of networks to communicate or share information amongst themselves. There is, however, no connection to the outside world at any point in such a network.

operating such systems. Operators can access elements of the infrastructure over distance—switch on or off a pump, open or close a valve, connect or disconnect communication lines at a remote location—which reduces costs and raises effectiveness.[25] Disconnecting such critical elements from the internet, or air gapping them, would thus lead to a significant decrease in efficiency for such critical infrastructures. Moreover, it would also entail that power plants and the like would lose novel possibilities that have emerged in the wake of networking their systems. As Singer and Friedman explain: "*Power companies that don't link up* [...] *may be less vulnerable, but they can't run 'smart' power grids that save both money and the environment*" (2013: 159, also see Byres 2013). So while it would be beneficial for critical infrastructures to air gap the vital elements of the networks they control in order to improve security, at the same time doing so would have a negative impact on the degree to which they could exploit advances in data collection and analytics as well as on realising truly smart power grids, water management systems, traffic systems and so on and so forth. Finding the right 'balance' is not really an option here, because by definition air gaps will only work when they are applied and maintained religiously—or else they will not work at all.

So we have seen that air gaps, in the past, have been proposed as a remedy against cybersecurity threats, and that several arguments have been raised against their effectiveness. Let's look a little more critically at the latter. The first argument against the use of air gaps is that it does not prevent security breaches because human beings (insiders) may accidentally or intentionally bridge them. The underlying message here is: air gaps do not provide a guarantee that no security breaches will occur. And because they do not provide 100 % security protection the idea of using air gaps ought to be abandoned.

The latter, of course, is a *non sequitur*. It is like saying: even with a proper lock on your door thieves might still break into your house. Hence, using locks does not work and ought to be abandoned as a protective measure. This is obviously flawed reasoning. If we would not use locks on our doors the chances of thieves entering our homes would increase dramatically. The lock does not act as a 100 % guarantee that no thief will ever enter the building—if the treasures stored inside are attractive enough (as was the case in the Stuxnet and in de DoD affairs!) then some persistent person at some point may indeed find a way to break or circumvent that lock. But using a lock surely makes it a lot more difficult to enter the building, and acts as a deterrence to the vast majority of not so persistent thieves who will decide to seek treasures elsewhere. Moreover, it seems silly to require of any single security measure that it should guarantee protection against any and all potential security breaches. No single measure could ever provide that kind of protection. As I have argued above, there is no such thing as 100 % failsafe protection. The trick in improving security is to create combinations of protective

---

[25]Note, of course, that if operators can operate such systems over a distance, so can attackers. The security of such systems, therefore, is of critical importance, and using air gaps has been considered a valid means in improving this security, although in recent years this idea has come under attack.

measures that lead to the best possible defence against attacks, while realising that none—not even in combination—will protect against any and all possible threats. Demanding that air gaps provide guaranteed safety, and casting aside their deployment because no such guarantee can be offered—ever—is not only unrealistic, but it is also unfortunate: it may entail that we toss out the child with the bathwater.

The second argument against using air gaps, as we have seen, is that their implementation undermines the efficiency of the critical infrastructures in which they are often used, and that it turns back time to a point where novel advantages of massive data collection and analytics cannot be exploited by these businesses. While this is a valid point, many experts on critical infrastructures point out that industrial control (ICS) systems, also known as SCADA systems, ought never to have been attached to the internet in the first place. The reason why this is so is because these industrial control systems *"are, by nature, insecure"* (Aitel 2013: 56). These systems were mostly developed in the 1960s and 1970s, a time in which issues of cybersecurity were non-existent or very low on the agenda (Aitel 2013). As Appazov points out: *"These technological dinosaurs were never designed to interface with massive corporate intranets that put SCADA systems within reach of the Internet and all its cyber pathogens..."* (2014: 18). Retrospectively implementing cybersecurity measures on these systems is difficult due to their complexity (Rid 2012). And replacing them with novel, cyber-proof versions is costly, since these systems are often highly specialist and tailored exactly to the processes and requirements of individual industries. Raising the question, therefore, whether it is wise to keep critical systems connected to the internet when viewed from the perspective of risk management is defensible. Deciding for or against such a connection then becomes a discussion regarding the trade-off between increasing efficiency at the (potentially grave) expense of security—imagine a hacker opening a sluice and flooding an urbanised piece of land—or increasing security at the (admittedly significant) expense of losing the ability to gather and process data and work more cost-effectively. How this trade-off is weighed is for politicians, regulators, policy makers and the industries to decide. The question is, however, whether this argument is even relevant in relation to the discussion in this chapter: using 'air gaps' in consumer technologies, such as for example domestic and service robots. I will return to this below. But let us first see what would happen if we were to apply the idea of using air gaps to household technologies that are part of the internet of things, including household robots.

5 Air Gaps and Domestic and Service Robots: A Look at the Issues

Imagine an 'air gapped' domestic or service robot, i.e. a robot that is *not* connected to the internet, that can operate autonomously, independently and does not need the cloud or a connection to the outside world in order to function. Such a robot would operate as a standalone machine that is unable to communicate the data it stores to the outside world, and incapable of being compromised for data theft by third parties. What does an 'unconnected' robot look like? Can it be functional?

The obvious answer is: well yes, wasn't that the plan in the first place?[26] Ironically, while it may seem like a big step to *not* attach robots to the internet, in fact when we look at the definition of a robot, and the design parameters that have fuelled robots' development over the last decades, they were always intended to be *autonomous, independently operating* systems. As a matter of fact, the term 'robot', which was coined by the Czech playwright Karel Capek, means 'self-laborer' (cf. Benford and Malartre 2007: 101), i.e. a machine that can operate independently, free from (human or other) intervention, and—by extension—also free from networks or outside attachments. As Peter W. Singer explains:

> "Robots are machines that are built upon what researchers call the 'sense-think-act' paradigm. That is, they are man-made devices with three key components: 'sensors' that monitor the environment and detect changes in it, 'processors' or 'artificial intelligence' that decides how to respond, and 'effectors' that act upon the environment in a manner that reflects the decisions, creating some sort of change in the world around a robot. When these three parts act together, a robot gains the functionality of an artificial organism." (Singer 2009: 67; also see Calo 2012: 1)

Together being able to register the environment, to make decisions on the basis of perceptions of that environment, and to bring about changes in the environment on the basis of these decisions are necessary and sufficient conditions to demarcate a robot from other technical devices (compare Denning et al. 2009). What the definition of robots thus reveals is that these were traditionally perceived as machines that ought to be capable of sensing, processing and acting autonomously, without intervention of outsiders. Read also: without connections to outsiders or the outside world.

Now, with the rise of the idea of cloud robotics we have apparently let go of two of these key characteristics. The 'think' part of the paradigm that Singer describes is now delegated to the cloud. A robot's processing capacity is no longer (fully) stored locally within the machine, but rather in the cloud, which the robot can access via the internet. What's more, the 'act' part of the paradigm, that is the number and type of actions a robot can fulfil, is no longer static because it can download new functions, new capabilities, from the cloud in the form of extra software. And as we have seen above in the discussion on the internet of things, one added bonus in attaching robots to the internet is that software developers can upgrade them remotely, and hence improve the original programs that run on the machine remotely after they have been sold.

Let's look at all three of these changes in turn. First, there is the issue of externalising the processing capacity of robots. While this obviously leads to an

---

[26]As explained on page 4 of this chapter, there are domestic and service robots that require an internet connection in order for them to exploit their full functionality. Most importantly this involves domestic robots that will monitor specific (sets of) health parameters of patients. These robots must be connected to the internet so that they can share the data they have collected with healthcare practitioners for the purpose of monitoring, analysis and—if necessary—intervention. Obviously, unhooking this type of robot from the internet would be detrimental to its proper functioning. However, for the vast majority of domestic and service robots this does not apply. It is this majority that will be the focus of this rest of this chapter.

increase in efficiency and is more cost-effective, there is no reason to believe that robots could not function effectively if all of their processing power would be on board. Ample processing capability is available in the market today to equip standalone, autonomous, 'offline' domestic and service robots with enough power to function properly in almost all of the tasks they are required to fulfil. This is especially so since these household robots are often single- or limited-purpose machines. They are not required to complete a multitude of different chores and tasks, but only need to be exceptionally good at doing one specific task (vacuum cleaning, lawn-mowing etcetera) or at most at a few different types of tasks (serving food and drinks, and making small-talk).

The argument behind cloud robotics and projects such as *RoboEarth*[27] of course is that in the (near) future robots could dramatically increase their set of functionality, i.e. learn a lot of new things, by downloading novel software from the cloud. This is where the expansion of the 'act' part of the paradigm comes in. What is interesting about this idea is that it builds on the silent, widespread assumption that eventually (all?) robots will become all-purpose machines, or at least many-purpose machines. But is this really the case? Will we no longer use a number of single-purpose robots (vacuum cleaners, pool cleaners, garden watering robots) in the future, but opt for one single robot butler who can do all of these things and more instead? First, we have already seen that researchers and robotics companies consistently point out that full-fledged robot butlers are still far beyond the horizon. But if this is the case, then why are we already attaching our current generation of robots into the internet? This is like equipping today's generation of cars with wings, because flying cars may become a reality at some point in the future. Why not wait until robots gradually become more multi-functional, and only start implementing internet connectivity when it becomes truly relevant or even a real requirement for the robot's proper functioning?

Building or buying a real robot butler is a deep-seated dream for many. In fact, it is one of the central visions for the future of domestic and service robots. Such a vision cannot be ignored lightly. At the same time, if the last 60 years of the history of Artificial Intelligence have taught us anything it is that creating 'real intelligence' (whatever that may be) is far more difficult than anyone could have imagined, that even creating limited (i.e. task-specific) intelligence is challenging despite decades of research and development, and that the only areas of applied Artificial Intelligence that have, in fact, spawned great commercial success were those in which robots conduct a very limited, relatively simple set of chores—most notably domestic robots such as lawnmowers and pool cleaners. On the basis of these learnings I imagine that single- or limited-functionality machines will be with us for some time to come. And this need not even be a bad thing: it might be just as simple to have a set of cheap, limited-functionality robots to hoover one's home, water the garden and wake the kids than to have one single, very expensive full-purpose one. As an added bonus this also creates less dependency on that single machine.

---

[27]See p. 5 of this chapter.

Second, one of the points that cloud robotics enthusiasts consistently gloss over is the fact that expanding the 'act' part of the paradigm requires several things: a robot's set of software programs needs to be expanded—this is where the cloud and its library of programs comes in. But in order for a robot to expand its functionality, it may also need sensing devices (input) and actuators (output) that it does not necessarily have on board. Multiplying functionality through different software programs by downloading them via the internet sounds easy, sensible and efficient, but it can only lead to a real expansion in a robot's functionality if it has the hardware to effectuate such functions. In simple terms: a Roomba vacuum cleaner can download all the software it wants, but it will never be able to serve drinks because it lacks arms and other actuators. So why equip a Roomba with an internet connection? Downloading new software would be useless for such a machine if it could not actually use that software. Expanding the 'act' part of the paradigm is much less straightforward than it is often made out to be, then.

All in all, we may tentatively conclude that the rise of cloud robotics should be viewed primarily against the background of cost saving and the need for increased efficiency for the companies that make robots. Moreover, I argue that the unquestioned assumptions that underpin this paradigm may be in need of some critical reflection. On the basis of the analysis above it is curious, to say the least, that robotics designers have come to believe that cloud robotics is the *only* viable solution for the future development or robots, that robots could not function properly, effectively or efficiently without being hooked into the internet. It seems, in fact, that robotics designers may have uncritically stumbled into the same implicit reasoning that drives much of the development of the internet of things: that future technologies must be connected to the internet, by default, regardless of whether or not this adds to, or is even relevant to their functionality.

It is this implicit reasoning that this chapter seeks to challenge. From the above we may conclude that it is not easy to find legitimate grounds, technical or otherwise, to justify the rise of cloud and/or internet-connected robotics. By contrast, from the perspective of privacy protection and security there are, in fact, valid reasons why it may be wise to decide that robots *need* not and *should* not be attached to the internet. Networking robots will lead to more problems than it solves.

Where there is no networked connection, the vast majority of all privacy and security problems simply go away. After all, as we have seen the origin of privacy and security concerns in relation to robotics is not primarily that robots gather and store data about people and their environment as they go through the motions of the tasks they are assigned. Rather, it is the *passing on* of that data, or the potential *access* that others, notably humans, may have to such data, that forms the real focus of concern, and rightly so. By removing this possibility, then, privacy and security risks, by and large, will be eliminated as well.[28]

---

[28] This claim was empirically tested, and is confirmed, by Tamara Denning et al. in their article entitled 'A spotlight on security and privacy risks with future household robots: Attacks and lessons' (2009). The researchers tested a variety of different types of attacks, ranging from remote detection by an attacker to leaking login credentials and from eavesdropping on the audio-visual stream a robot registers to moving the robots etcetera. None of these attacks is possible when a robot is not connected to the internet.

Having said that, in the previous section we discussed a number of shortcomings of, or weaknesses in the use of air gaps. We have seen that even systems that are not connected to the internet are vulnerable in terms of security. Would the shortcomings of 'air gapping' be applicable to domestic and service robots as well? If so, is not attaching robots to the internet really a solid solution for increasing security and improving privacy protection?

6 The Weaknesses of Air Gaps Revisited

In Sect. 4 we discussed the merits and the weaknesses of the use of air gaps in relation to improving cybersecurity for critical infrastructures, such as electrical and nuclear plants, transport networks, water and sewage facilities, and so on and so forth. Let's return to the weaknesses discussed there to see whether they would also apply to 'air gapped' domestic and service robots, and if so, whether or not this would have an impact on the usefulness of this notion to the latter domain.

The first argument against the use of air gaps, as we have seen, was the fact that physically separating vital or critical systems from non-critical systems, and only attaching the latter to the internet (and hence the outside world), does not eliminate the issue of insider threats or the so-called sneakernet effect. If insiders want to do harm to critical systems, they will find ways to do so, even if there is a physical separation between these systems and those connected to the internet. The term sneakernet effects is used to describe one way in which insiders may go about infecting or damaging critical systems, for example through the use of USB sticks or other digital information carriers. This term is used for both intentional and accidental exploits.

When we look at the issue of insider threats in relation to domestic and service robots it is immediately apparent that this argument is not relevant to this type of device. End users will not deliberately attack their own systems. On the other hand, accidentally causing privacy or security issues through the sneakernet effect is, in fact, a realistic threat for domestic or service robots. For example, it is conceivable that robots without an internet connection will still be equipped with one or more different types of ports for digital information carriers,[29] so that their software can be upgraded, either by end users themselves or by vendors or repair shops. The sneakernet effect may arise whenever an individual plugs an information carrier into the robot. If the information carrier—let's say a USB stick—contains malware, this may (accidentally) infect the robot. Alternatively, malware may be used to harvest data from the robot, which can be sent back to its creator once the same carrier is plugged into a networked computer afterwards. As with critical infrastructures, here, too, individuals may unintentionally expose their own robots to security threats by closing the loop and crossing the physical divide between the robot and

---

[29]Think of, e.g., a USB port or a DVD/CD-rom drive.

the internet. Having said that, while the sneakernet effect is a realistic threat for the security of domestic and service robots, at the same time it is safe to assume that the probability of the materialisation of this risk is not very high. If a robot comes equipped with properly functioning software and has a limited set of functions and abilities—as is the case with domestic and service robots today and will be the case for the foreseeable future—then chances are that end users will not be inclined to even consider upgrading their software. This is all the more likely since such robots traditionally do not come with screens and buttons that allow for much interaction or 'tinkering' with the machine. We may assume, therefore, that if a Roomba or a pool-cleaning robot breaks down or displays software errors, the owner will not attempt to attempt to fix it or to replace the software him/herself, but will take it back to the vendor or a specialist store instead. There, too, the sneaker effect could arise, of course. But the the chance of that occurring could be minimised, as Bruce Schneier points out, by introducing liability for vendors, repair and upgrade stations (2004: 2). This way these parties will be very motivated to keep their own computer systems and networks as safe and secure as possible, thus diminishing the risk of accidentally infecting domestic and service robots.

The second weakness of using air gaps that we encountered above is that it decreases efficiency and may hinder the use of the full potential of massive data collection and data analytics. The smart grid, smart cities, smart transportation networks, and smart logistics all cannot exist if critical and non-critical systems are not both connected to the internet. Hence using air gaps hinders innovation and is economically unattractive. But does the same apply to domestic and service robots? Would not connecting them to the internet also lead to less efficiency, or would it harm their functionality? We have already seen the answer to this question above: it need not. In principle, robots are intended to be stand-alone, self-sufficient, autonomous and intelligent machines that can function fully without intervention from humans. Admittedly, we may lose some efficiency by not connecting robots to the internet, in the sense that such systems cannot be upgraded easily or automatically. It may seem clear by now, however, that in my view this does not outweigh the privacy and security risks that arise when robots are connected to the internet. As for the second part of this argument, regarding the exploitation of the innovative potential of cloud computers: yes, robots will undeniably mature into full-fledged butlers at a lower speed when they cannot 'learn' new behaviours by downloading available modules from the cloud. But as I have pointed out above, the age of robotic butlers is decades away, at best, and therefore we need not attach robots to the internet today. This seems like a small price to pay.

All in all, we may conclude that the benefits of not hooking robots into the internet far outweigh the disadvantages. To be clear: I am not claiming that all privacy and security risks will disappear, in full and forever, by implementing this step. Privacy and security risks can have different causes and hence may still occur even if robots are used as stand-alone, autonomous, unconnected machines only. Moreover, as has been pointed out above already no single measure can ever be 100 % effective in terms of protection. Having said that, any such measure or solution need not be perfect for it to be exceptionally good. What this chapter has aimed to do is describe precisely one such solution: 'air gapping' robots. In comparison to other solutions proposed, it may in fact be the most effective way of ensuring proper security and privacy protection.

7 Conclusion: A Plea for Privacy *Before* Design

We have seen in the introduction to this chapter (page 4) that legal scholars argue that privacy problems in domestic and service robots can be mitigated, by and large, when robots are considered to fall under existing privacy and data protection legislation. In recent years the notion of *privacy by design* has also become widespread as an approach to protecting privacy. Privacy by design *"prescribes that we build privacy directly into the design and operation, not only of technology, but also of operational systems, work processes, management structures, physical spaces and **networked** infrastructure"* (Cavoukian 2010: 248 [emphasis added]). It is an approach that asks of technology designers, developers and companies that they seek to give privacy its proper place in every networked system, architecture and specification they develop (also see Rubinstein 2011). Whenever companies design, develop or create new technologies they ought to consider privacy protection one of the essential design parameters from the very start. It must be *"incorporated into **networked** data systems and technologies, by default"* (Cavoukian 2009: 1 [emphasis added]). That is the central argument of the privacy by design movement. Cavoukian has pointed out that this entails, among other things, that privacy measures must be implemented proactively, not reactively, i.e. after privacy issues have arisen. Moreover, it means that *"personal data are automatically protected in any given IT system or business practice"* (2009: 2). More specifically, this involves respecting data protection requirements such as purpose specification and collection limitation.

The ideas behind the privacy by design approach make sense and point towards vital ways in which technology developers and designers can contribute to the prevention of future privacy risks. However, when we view this approach against the background of the discussion in this chapter we may wonder whether privacy by design (or privacy by default) takes protection far enough. If we look at the way in which privacy by design is defined above it involves 'embedding' privacy into architectures, and notably *networked* (!) infrastructures or *networked* (!) data systems. It focuses on proactively implementing such measures, i.e. by default. This is a valid and important perspective for all technologies that operate in networked ways—after all, for systems that gather and process data and are connected to the internet proper privacy protection is key.

However, the privacy by design perspective still builds on the assumption that systems *will be networked*, and hence that implementing privacy protective measures is necessary. As we have seen in this chapter, there are solutions available, in some cases, and with respect to some technologies, that diminish the risk of privacy violations and security breaches to such a degree that implementing (further) protective measures—by default or retrospectively—becomes unnecessary. Using air gaps, not attaching systems to the internet if this is not absolutely necessary for their functioning, is one vital example thereof. If we choose not to network these systems, then privacy by design becomes irrelevant.

This is why I propose to take the debate on privacy protection one step further than that of privacy by design. I argue for privacy *before* design rather than privacy *by* design. Privacy before design entails that technology designers and developers are asked to consider carefully, in the laboratory stage of developing novel technologies, whether the benefits of making them networked will outweigh the risks that this entails, both with respect to privacy and security. If it is absolutely necessary for such technologies to be attached to the internet, in order to maximise functionality, then of course they should be. Privacy by design standards must then be applied. But if an internet connection is not critical, then I argue that technology designers and developers ought to reconsider implementing one. The way forwards, toward a high-tech future need *not* necessarily, inevitably be one of internet connectivity for every sensor, every device, every system, and every network. Nor every coffee maker, light switch or… robot. We will not hamper 'innovation' by making different choices with respect to the internet of things—this builds on a misunderstanding of what innovation really is, and views it as a single line running from here towards one specific instantiation of the future. Questioning that vision is not a rear-guard action, nor is it driven by fear of change. We can, and should have critical debates amongst technology designers, business leaders, regulators and law makers, politicians and the general public about the form and functionality of our future technologies, and of the requirements this entails in terms of, for example, the balance between connectivity, security and privacy protection. Using 'air gaps', choosing not to connect some technologies to the internet, is one proposal that can be used as input for such a debate. As said, it is not about hampering innovation. It is simply a matter of common sense.

Bibliography

Aarts, E., and J.L. Encarnação. 2006. Into ambient intelligence. In true visions: The emergence of ambient intelligence, 1–17. Berlin: Springer.
Aitel, D. 2013. Cybersecurity essentials for electric operators. *Electricity Journal* 26(1): 52–58. doi:10.1016/j.tej.2012.11.014.
Appazov, A. 2014. Legal aspects of cybersecurity. Copenhagen.
Benford, G., and E. Malartre. 2007. *Beyond human: Living with robots and cyborgs*, 1st ed. New York: Forge.
Bryant, D. 2004. The uncanny valley: Why are monster-movie zombies so horrifying and talking animals so fascinating?
Byres, E. 2013. The air gap: SCADA's enduring security myth. *Communications of the ACM* 56(8): 29. doi:10.1145/2492007.2492018.
Calo, R. 2012. Robots and privacy. In *Robot ethics: The ethical and social implications of robotics*, ed. P. Lin, K. Abney, and G.A. Bekey, 386 pages. Cambridge: MIT Press.
Cavoukian, A. 2009. Privacy by design: The 7 foundational principles.
Cavoukian, A. 2010. Privacy by design: The definitive workshop. A foreword. *Identity in Information Society (IDIS)* 3: 247–251. doi:10.1007/s12394-010-0062-y.
Clarke, R.A, and R.K. Knake. 2010. Cyber war: The next threat to national security and what to do about it. HarperCollins e-Books. doi:10.1080/09546553.2011.533082.

Coeckelbergh, M. 2009. Health care, capabilities, and AI assistive technologies. *Ethical Theory and Moral Practice* 13(2): 181–190.
Denning, T., C. Matuszek, K. Koscher, J.R. Smith, and T. Kohno. 2009. *A spotlight on security and privacy risks with future household robots: Attacks and lessons.* In *Proceedings of the 11th International Conference on Ubiquitous Computing (UbiComp '09)*, 1–10. ACM.
Edwards, W.K., and R.E. Grinter. 2001. At home with ubiquitous computing: Seven challenges. In *Ubicomp 2001*, ed. G.D. Abowd, B. Brumitt, and S.A.N. Shafer, 256–272. Berlin: Springer.
Farwell, J.P., and R. Rohozinski. 2011. Stuxnet and the future of cyber war. *Survival: Global Politics and Strategy* 53(1): 23–40. doi:10.1080/00396338.2011.555586.
Goffman, E. 1959. *The presentation of self in everyday life.* Garden City: Doubleday.
Goffman, E. 1963. *Behavior in public places: Notes on the social organization of gatherings.* New York: Free Press of Glencoe.
Goffman, E. 1968. Information control and personal identity. *Stigma: Notes on the management of spoiled identity*, 57–129. Harmondsworth: Penguin.
Goffman, E. 1982. Interaction ritual: Essays on face-to-face behavior (1st Panthe.). New York: Pantheon Books.
Goffman, E. 1997. The self and social roles. In *The Goffman reader*, ed. C.C. Lemert, and A. Branaman, 35–43. Cambridge: Blackwell Publishers.
Green, T. 2015. CES 2015: What kinds of robots will people actually buy? *Robotics Business Review*. http://www.roboticsbusinessreview.com/article/ces_2015_what_kinds_of_robots_will_people_actually_buy. Retrieved 26 May 2015.
Greenfield, A. 2006. *Everyware: The dawning age of ubiquitous computing.* Berkeley: New Riders.
Guizzo, E. 2011. Robots with their heads in the clouds. *IEEE Spectrum* 48(3): 17–18. doi:10.1109/MSPEC.2011.5719709.
Hardy, Q. 2014. The robot in the cloud: A conversation with Ken Goldberg. NYTimes.com, October 25.
Hert, P. De, and Schreuders, E. 2001. The Relevance of Convention 108.European Conference on Data Protection on Council of Europe Convention 108 for the protection of individuals with regard to automatic processing of personal data: present and future. ed. / The Council Of Europe; The Bureau Of The Inspector General Of Poland For Personal Data Protection. Council of Europe, 2001. p. 63–76 (DP Conf (2001) Reports).
Herpig, S. 2013. Anti-war era: The need for proactive cyber security. In *Cyber security and privacy*, ed. M. Felici 182, 165–176. Berlin: Springer. doi:10.1007/978-3-642-41205-9.
Hildebrandt, M. 2013. The rule of law in cyberspace?
Hildebrandt, M., and N. Van Dijk. 2012. Customer profiles: The invisible hand of the Internet. In *Databases: The promises of ICT, the hunger for information, and digital autonomy*, ed. G.M. Munnichs, M. Schuijff, and M. Besters, 62–72. The Hague: Rathenau Instituut.
ISTAG. 2003. *Ambient Intelligence: From vision to reality.* Brussels: Information Society Technologies Advisory Group.
ITU. 2005. The internet of things. ITU internet reports—executive summary. International Telecommunication Union.
Kelion, L. 2015. CES 2015: The robots moving into your house. *BBC News*. http://www.bbc.com/news/technology-30708953. Retrieved 26 May 2015.
Kello, L. 2013. The meaning of the cyber revolution: Perils to theory and statecraft. *International Security* 38(2): 7–40. doi:10.1162/ISEC_a_00138.
Kim, D.-Y. 2014. Cyber security issues imposed on nuclear power plants. *Annals of Nuclear Energy* 65: 141–143. doi:10.1016/j.anucene.2013.10.039.
Lachow, I. 2015. The Stuxnet enigma: Implications for the future of cybersecurity. *Journal of International Affairs* 1: 118–127.
Levy, D. 2009. *Love & sex with robots: The evolution of human-robot relations*, 1st ed. London: Duckworth Overlook.
Mayer-Schönberger, V., and K. Cukier. 2013. *Big data: A revolution that will transform how we live, work and think.* London: John Murray (Publishers).
McCullough, M. 2004. *Digital ground: Architecture, pervasive computing, and environmental knowing.* Cambridge: MIT Press.

Meyrowitz, J. 1985. *No sense of place: The impact of electronic media on social behavior*. New York: Oxford University Press.
Mori, M. 1970. The uncanny valley (trans: Karl, F., MacDorman and Takashi Minato). *Energy* 7(4): 33–35.
Morozov, E. 2013. *To save everything click here: Technology, solutionism and the urge to fix problems that don't exist*. London (UK): Penguin Books.
Nissenbaum, Helen. 2004. Privacy as Contextual Integrity. *Washington Law Review* 79(119):119–59.
O'Harrow Jr., R. 2013. Zero day: The threat in cyberspace (Kindle edi.). New York: Diversion Books.
Pagallo, U. 2015. Teaching 'consumer robots' respect for informational privacy: A legal stance on HRI. In *Human-robot interactions: Principles, technologies and challenges*, ed. D. Coleman, 35–56. New York: Nova Science Publishers.
Rid, T. 2012. Cyber war will not take place. *Journal of Strategic Studies* 35(1): 5–32. doi:10.1080/01402390.2011.608939.
Robotics exhibits grow by 25 percent at the 2015 International CES. 2015. CES Press Release. http://www.cesweb.org/News/Press-Releases/CES-Press-Release.aspx?NodeID=62713b0b-6742-469b-935e-926629a11413. Retrieved 26 May 2015.
Rubinstein, I.S. 2011. Regulating privacy by design. *Berkeley Technology Law Journal* 26(3): 1409–1456.
Ruyter, B. De, and E. Aarts. 2004. Ambient intelligence: Visualizing the future. In *Proceedings of the Working Conference on …*, 203–208.
Schneier, B. 2004. *Secrets & lies: Digital security in a networked world*, 2nd ed. Indianapolis (IN): Wiley Publishers.
Sharkey, N. 2009. The robot arm of the law grows longer. *Computer* 112–115.
Sharkey, N., and A. Sharkey. 2010. The crying shame of robot nannies. *Interaction Studies* 11(2): 161–190.
Singer, P.W. 2009. *Wired for war: The robotics revolution and conflict in the twenty-first century*. New York: Penguin Press.
Singer, P.W., and A. Friedman. 2013. Cybersecurity and cyberwar: What everyone needs to know.
Stoll, C. 1989. The cuckoo's egg: Tracking a spy through the maze of computer espionage (Kindle edi.). New York: Doubleday.
Thierer, A. 2013. The pursuit of privacy in a world where information control is failing. *Harvard Journal of Law and Public Policy* 36(2): 410–455.
Van den Berg, B. 2010a. Ambient intelligence: Wat, wie, en… waarom? *Computerrecht* 6.
Van den Berg, B. 2010b. *The situated self: Identity in a world of ambient intelligence*, vol. 6. Nijmegen: Wolf Legal Publishers.
Van den Berg, B. 2011. The uncanny valley everywhere? On privacy perception and expectation management (Chapter 15). In *Privacy and identity management for life*. 6th IFIP WG 9.2, 9.6/11.7, 11.4, 11.6/PrimeLife International Summer School, Helsingborg, Sweden, August 2–6, …, 352, 178–191.
Wallach, W. 2008. Implementing moral decision making faculties into computers and robots. *Artificial Intelligence & Society* 22(4): 463–475.
Weiser, M. 1991. The computer for the 21st century. *Scientific American* 265(3): 66–76.
Weng, Y.H., and S.T.H. Zhao. 2012. The legal challenges of networked robotics: From the safety intelligence perspective. In *AI approaches to the complexity of legal systems: Models and ethical challenges for legal systems, legal language and legal ontologies, argumentation and software agents*, ed. M. Palmirani, U. Pagallo, P. Casanovas, and G. Sartor, 7639 LNAI, 61–72. Heidelberg: Springer. doi:10.1007/978-3-642-35731-2_4.
Zarsky, T.Z. 2003. 'Mine your own business!': Making the case for the implications of the data mining of personal information in the forum of public opinion. *Yale Journal of Law Technology* 5(1): 1–57.
Zetter, K. 2014. Countdown to zero day: Stuxnet and the launch of the world's first digital weapon (Kindle edi.). New York: Crown Publishers.

Europe Versus Facebook: An Imbroglio of EU Data Protection Issues

Liane Colonna

Abstract In this paper, the case *Europe versus Facebook* is presented as a microcosm of the modern data protection challenges that arise from globalization, technological progress and seamless cross-border flows of personal data. It aims to shed light on a number of sensitive issues closely related to the case, which namely surround how to delimit the power of a European Data Protection Authority to prevent a specific data flow to the US from the authority of the European Commission to find the entire EU-US Safe Harbor Agreement invalid. This comment will also consider whether the entire matter might have been more clear-cut if Europe-versus-Facebook had asserted its claims against Facebook US directly pursuant to Article 4 of the EU Data Protection Directive, rather than through Facebook Ireland indirectly under the Safe Harbor Agreement.

Keywords Facebook · Data protection · Extraterritoriality · European data protection authority (DPA)

1 Introduction

Today, threats to privacy and data protection no longer stem solely from within the boundaries of the single state and from the State's use of personal data in public administration and law enforcement.[1] Rather, threats emerge in a highly connected and technologically complex world where a myriad of public and private actors participate, perhaps even unwittingly, in creating them (see Footnote 1).

---

[1]Colin J. Bennett, and Charles D. Raab, *The Governance of Privacy: Policy Instruments in Global Perspective* (The MIT Press, 2006).

L. Colonna (✉)
Skeppargatan 55, Stockholm 11459, Sweden
e-mail: liane.colonna@juridicum.su.se

Furthermore, personal data is no longer held on mainframe computers within easily identifiable organizations that reside and operate within the borders of modern territorial states. Instead, data flows around the globe with indifference to national boundaries, and it is stored on servers located in random jurisdictions.

Against this background, the case *Europe versus Facebook* is illuminating and is presented as a microcosm of the modern data protection challenges that arise from globalization, technological progress and seamless cross-border flows of personal data. In this case comment, the ruling by the Irish High Court in *Europe versus Facebook* is summarized, critically analyzed and commented on from the perspective of European law. It aims to shed light on a number of sensitive issues closely related to the case which namely surround how to delimit the power of a European Data Protection Authority ("DPA") to prevent a specific data flow to the US from the authority of the European Commission to find the entire EU-US Safe Harbor Agreement invalid. This comment will also consider whether the entire matter might have been more clear-cut if Europe-versus-Facebook had asserted its claims against Facebook US directly pursuant to Article 4 of the EU Data Protection Directive, rather than through Facebook Ireland indirectly under the Safe Harbor Agreement.

The outline of this comment is as set forth. First, an overview of the Safe Harbor Program will be provided in order to make evident the legal framework from which *Europe versus Facebook* case arises. Second, the facts and the procedural history of the case will be reviewed. Then, a commentary will ensue about the authority of a DPA to suspend data flows to an organization participating in the Safe Harbor Program. This discussion is organized around the following legal texts: the Safe Harbor Program, the EU Data Protection Directive and the EU Charter of Fundamental Rights. Finally, an alternate perspective of the case will be presented that will question whether Europe-versus-Facebook would have a stronger claim if it had based its case around Article 4 of the EU Data Protection Directive rather than the Safe Harbor Program.

2 The Safe Harbor Program

The EU Data Protection Directive prohibits data transfer to third countries, such as the US, unless there an "adequate" level of protection for the transferred data can be assured.[2] The purpose of the adequacy requirement is simple in that "if controllers in a Member State transferred data to a third country that failed to protect personal data, then the Members State's protection of personal data would be

---

[2]Article 25(1) of Directive 95/46/EC of the European Parliament and of the Council of 24 October 1995 on the protection of individuals with regard to the processing of personal data and on the free movement of such data Official Journal L 281, 23/11/1995 P. 0031–0050 (hereinafter referred to as "Data Protection Directive").

effectively lost once the Member State transferred the data to the third country."[3] In essence, the adequacy requirement is a mechanism to ensure that there are no loopholes found in the high level of protection of personal data provided by the Directive.[4]

On the basis of Article 25(6), the European Commission may conclude that a third country ensures an adequate level of protection by reason of its domestic law or of the international commitments it has entered.[5] The adoption of a (comitology) Commission decision based on Article 25(6) of the Directive involves a multi-step process where, for example, a proposal is set forth by the Commission and an opinion by Member States' DPAs and the European Data Protection Supervisor (EDPS) is issued within the framework of the Article 29 Working Party.[6] The effect of such a decision is that personal data can flow from the EU to that third country without any further safeguard being necessary.[7] Importantly, Member States are obligated to respect an adequacy finding by the Commission under Article 25(6) as well as Article 4(3) of the Lisbon Treaty, which contains the "duty of loyal cooperation."[8]

In 2000, the Commission found that the Safe Harbor Agreement, a bilateral agreement made between the EU and the US, constitutes a sufficient mechanism to safeguard the rights of European data subjects in accordance with the adequacy requirement.[9] Under the Safe Harbor Agreement, US businesses are afforded the opportunity to self-certify their compliance with seven principles laid down in the Agreement (e.g. notice, choice, access, and enforcement) in order to permit the systemic and free flow of data to them from the EU without any conflicts arising under the Data Protection Directive. After certification, a participating organization is under no requirement to obtain prior approval of data transfers as the

---

[3]Patrick J. Murray, "The Adequacy Standard Under Directive 95/46/EC: Does U.S. Data Protection Meet This Standard?," 21 *Fordham International Law Journal* 932, 964–965 (1998).

[4]Article 29 Data Protection Working Party, Opinion 8/2010 on applicable law (Dec. 16, 2010), http://ec.europa.eu/justice/policies/privacy /docs/wpdocs/2010/wp179_en.pdf.

[5]Article 25(6).

[6]For more information about this procedure see, European Commission, Commission decisions on the adequacy of the protection of personal data in third countries retrieved at http://ec.europa.eu/justice/data-protection/document/international-transfers/adequacy/index_en.htm.

[7]European Commission, Commission decisions on the adequacy of the protection of personal data in third countries retrieved at http://ec.europa.eu/justice/data-protection/document/international-transfers/adequacy/index_en.htm.

[8]Article 25(6) ("The Commission may find…that a third country ensures an adequate level of protection … by reason of its domestic law or of the international commitments it has entered into … for the protection of the private lives and basic freedoms and rights of individuals. Member States shall take the measures necessary to comply with the Commission's decision." (emphasis added); Article 4(3) Treaty of Lisbon.

[9]Commission decision of July 26, 2000 pursuant to Directive 95/46/EC of the European Parliament and of the Council on the adequacy of the protection provided by the safe harbour privacy principles and related frequently asked questions issued by the US Department of Commerce, 2000/520/EC.

approval will either be waived or automatically granted. The significance of the Commission's decision cannot be understated as it provides a critical mechanism to facilitate the seamless, predictable, and consistent transfer of personal data from the EU to US.[10]

With respect to "onward transfers," the Safe Harbor Program provides that "to disclose information to a third party, organizations must apply the Notice and Choice Principles."[11] According to the Article 29 working party, this means that if a US company enrolled in the Safe Harbor Program wants to "communicate" data to a third party acting as a controller it "shall inform the data subject about the onward transfer to the third party, offering the opportunity to the data subject to consent (opt-out) to such onward transfer where data is to be used for 'a purpose incompatible with the purpose(s) for which it was originally collected."[12]

The Safe Harbor Program includes several derogations. Its principles may be limited:

> to the extent necessary to meet national security, public interest, or law enforcement requirements...provided that, in exercising any such authorization, an organization can demonstrate that its non-compliance with the Principles is limited to the extent necessary to meet the overriding legitimate interests furthered by such authorization...[13]

If there is an alleged breach of the principles set forth in the Safe Harbor Agreement, then there are four ways that a complainant may obtain relief. First, because violations of the Safe Harbor principles are considered to be acts of unfair or deceptive trade practices,[14] the US Federal Trade Commission (the "FTC") can bring an action against the entity that has allegedly failed to comply with the agreement pursuant to its Sect. 5 authority granted to it under the FTC Act of 1914.[15]

---

[10] There are other ways to permit the transfer of personal data from the EU to the US such as reliance on Binding Corporate Rules but the Safe Harbor Agreement is the only mechanism to permit to free flow and systematic transfer of personal data between the EU and the US; for more see, Liane Colonna, "Article 4 of the EU Data Protection Directive and the irrelevance of the EU-US Safe Harbor Program?" 4(3) *International Data Privacy Law* (Oxford 2014).

[11] U.S.-EU Safe Harbor Overview http://www.export.gov/safeharbor/eu/eg_main_018476.asp.

[12] Working Document on surveillance of electronic communications for intelligence and national security purposes, WP 228 adopted on 5 Dec 2014.

[13] Safe Harbor Principles http://www.export.gov/safeharbor/eu/eg_main_018475.asp.

[14] Safe Harbor Enforcement Overview retrieved at http://export.gov/safeharbor/eu/eg_main_018481.asp; *see also*, Joanna, Kulesza, "Walled Gardens of Privacy or 'Binding Corporate Rules?': A Critical Look at International Protection of Online Privacy," 34 *University Arkansas Little Rock Law Review* 747 (2012); Robert R Schriver, "You Cheated, You Lied: The Safe Harbor Agreement and Its Enforcement by the Federal Trade Commission," 70 *Fordham Law Review* 2777 (May 2002).

[15] Damon Greer, "Safe harbor - A Framework That Works," 1(3) *International Data Privacy Law* 143 (2011).

Second, an affected individual can, at least theoretically, bring a direct action in the US courts under federal or state law prohibiting unfair and deceptive acts.[16] An affected individual may also seek recourse through the EU Data Protection Panel, a body competent for investigating and resolving complaints lodged by individuals for alleged infringement of the Safe Harbor Privacy Principles. This option, however, is only available if the US entity has explicitly made the choice to rely on the EU Data Protection Panel rather than on independent recourse mechanisms.[17]

Third, EU national data protection authorities (DPAs), who believe that the principles of the Safe Harbor Agreement are in breach, may suspend data flows to Safe Harbor certified companies in specific cases pursuant to the authority granted to them in Article 3(1)(b) of the Commission Decision on the Safe Harbor principles. Specifically, this provision gives the DPAs the possibility to suspend data flows in either cases where the government body in the US has determined that the company is violating the Safe Harbor principles or where there is a substantial likelihood that the Safe Harbor principles are being violated and where the continuing transfer would create an imminent risk of grave harm to data subjects.[18] Two

---

[16]Arguably, an individual can bring a claim under federal or state law prohibiting unfair and deceptive acts. However, at the moment, the only two clear enforcement bodies with jurisdiction to hear claims concerning the Safe Harbor Agreement are the FTC (which while covering commerce in general excludes financial services, transport, telecommunications, among others, from its jurisdiction) and the US Department of Transportation. See Safe Harbor Enforcement Overview retrieved at http://export.gov/safeharbor/eu/eg_main_018481.asp (stating, "Where an organization relies in whole or in part on self-regulation in complying with the Safe Harbor Privacy Principles, its failure to comply with such self-regulation must be actionable under federal or state law prohibiting unfair and deceptive acts or it is not eligible to join the safe harbor); see also, Caspar Bowden and Judith Rauhofer, "Protecting Their Own: Fundamental Rights Implications for EU Data Sovereignty in the Cloud," *Edinburgh School of Law Research Paper No. 2013/28* (2013)(stating, "Failure to comply with the safe harbor principles can result in enforcement proceedings by the US Federal Trade Commission and direct action by affected individuals in the US courts.).

[17]Data Protection Panel (related to FAQs 5 and 9 issued by the US Department of Commerce, and annexed to Commission Decision 2000/520/EC on the adequacy of the 'safe harbor' privacy principles) 25 July 2005 retrieved at http://ec.europa.eu/justice/policies/privacy/docs/adequacy/information_safe_harbour_en.pdf (stating, "Is the data protection panel competent to investigate all the complaints that derive from an alleged infringement of the Safe Harbour principles? No, the data protection panel does not have competence to investigate all the complaints that derive from an alleged infringement of the Safe Harbour principles. In certain cases, individuals will have other recourse mechanisms…"); see additionally footnote 65 in Ioanna Tourkochoriti, "The Transatlantic Flow of Data and the National Security Exception in the European Data Privacy Regulation: In Search for Legal Protection Against Surveillance", 36 *University of Pennsylvania Journal of International Law* 459 (Winter 2014).

[18]Communication from the Commission to the European Parliament and the Council on the Functioning of the Safe Harbour from the Perspective of EU Citizens and Companies Established in the EU COM(2013) 847 (27 November 2013), available at http://ec.europa.eu/justice/data-protection/files/com_2013_847_en.pdf (hereinafter referred to as "Commission's Safe Harbor Communication"); see also, PRISM and Data Protection for EU Citizens, *The Society for Computers and Law* (June 6, 2013), http://www.scl.org/site.aspx?i=ne32989.

examples of where DPAs retain powers to intervene are provided by the Commission:

> EU authorities retain powers to intervene... if a private sector dispute resolution body found that a company had made serious violations of the principles, but the company contested the finding and the case was referred to the FTC, the EU authorities could suspend data transfers to that company until the matter was resolved. Also for example, if evidence of non-compliance accumulates and the relevant US enforcement body is not doing its job properly and if letting transfers continue risks causing grave harm to data subjects, EU authorities can once again suspend transfers. The Commission could subsequently change the 'safe harbor' decision to exclude an ineffective US enforcement body.[19]

Fourth, the Commission can, acting in accordance with the examination procedure set out in Regulation 182/2011, suspend, amend or revoke the entire Agreement at any time.[20] This is particularly true if there is a systemic failure on the US side to ensure compliance with the principles.[21] The Commission explains: "If the US authorities failed to take the action necessary to (remedy compliance with the Safe Harbor principles and to ensure effective redress mechanisms), the Commission could reverse its decision to grant the 'safe harbor' arrangement 'adequate protection' status."[22] In the event that either the Commission or a Member State blocks a data transfer to the US, there is a requirement under Article 26(d) to inform one another of this decision.[23]

On November 27, 2013, the European Commission published an analysis of the EU-US Safe Harbor Framework.[24] The purpose of the study was to reassess the Safe Harbor Framework in light of four different factors: the exponential increase in worldwide data flows; the critical importance of data flows for the transatlantic economy; the rapid growth of the number of companies in the US adhering to the Safe Harbor scheme; and, the revelations the US national intelligence services had developed a sweeping surveillance system targeted at non-American persons located outside of the United States (see Footnote 21). The study analyzed the

---

[19] The European Commission. How will the "safe harbor" arrangement for personal data transfers to the US work? retrieved at http://ec.europa.eu/justice/policies/privacy/thridcountries/adequacy-faq1_en.htm.

[20] Commission's Safe Harbor Communication of 27 November 2013 (stating, "As recalled in the current Safe Harbour Decision, it is the competence of the Commission—acting in accordance with the examination procedure set out in Regulation 182/2011—to adapt the Decision, to suspend it or limit its scope at any time, in the light of experience with its implementation.").

[21] Commission's Safe Harbor Communication of 27 November 2013.

[22] The European Commission. How will the "safe harbor" arrangement for personal data transfers to the US work? http://ec.europa.eu/justice/policies/privacy/thridcountries/adequacy-faq1_en.htm.

[23] Article 26(d) (stating, "The Member States and the Commission shall inform each other of cases where they consider that a third country does not ensure and adequate level of protection .... Member States shall take the measures necessary to prevent any transfer of data to [this country].").

[24] European Commission Press Release, European Commission calls on the U.S. to restore trust in EU-U.S. data flows (Brussels, 27 November 2013) retrieved at http://europa.eu/rapid/press-release_IP-13-1166_en.htm.

extent to which the fundamental principles of the Safe Harbor Framework—transparency of adhering companies' privacy policies, incorporation of the Safe Harbor principles in companies' privacy policies, and transparency, evidence of compliance and limited enforcement—are being adhered to in actuality (see Footnote 21). While the Commission validated the continued viability of the Safe Harbor Framework in this study, it simultaneously made a number of recommendations where the Framework could be improved. These recommendations generally involve adding greater transparency to the program and strengthening enforcement mechanisms (see Footnote 21).

In its review, particular attention was paid by the Commission to its concerns that data transferred to the US by Safe Harbor certified companies may undermine the data protection rights of Europeans when their data subsequently transferred to the US government for national intelligence purposes (see Footnote 21). It stated:

> The large scale nature of (US Surveillance programs) may result in data transferred under Safe Harbor being accessed and further processed by US authorities beyond what is strictly necessary and proportionate to the protection of national security as foreseen under the exception provided in the Safe Harbor Decision.[25]

Pursuant to this concern, it made a specific recommendation that the national security exception foreseen by the Safe Harbor Decision is used only to an extent that is strictly necessary or proportionate in order to help restore trust in the Program (see Footnote 21).

3 Factual and Legal Background

Europe-versus-Facebook is an Austrian non-profit organization represented by Maximilian Schrems. It alleges that the Irish Data Protection Commissioner, pursuant to its obligation to protect the rights of individuals originating from Article 8 of the EU Charter of Fundamental Rights and the EU Data Protection Directive, should suspend data transfers from Facebook Ireland Limited to its parent company in the US, Facebook Inc.[26] While these transfers are *prima facie* authorized pursuant to the EU-US Safe Harbor Agreement, Europe-versus-Facebook contends that, based on the revelations from Edward Snowden that Facebook provided the US government direct access to all personal data of its users to use in a massive electronic surveillance program entitled PRISM, there is no possibility that Facebook can demonstrate its actual compliance with the principles set forth in the program (see Footnote 26). In other words, Europe-versus-Facebook contends that

---

[25]COM(2013) 847 Communication from the Commission to the European Parliament and the Council on the functioning of the safe Harbor from the perspective of EU citizens and companies established in the EU, 27 November 2013.

[26]Complaint against Facebook Ireland Ltd to the Data Protection Commissioner of Ireland (25 June 2013) retrieved at http://www.europe-v-facebook.org/prism/facebook.pdf.

the operation of the Safe Harbor agreement in this particular context is defunct, that the transfer of data to the US government is not in accordance with any exceptions under the agreement and, as such, the Irish Data Protection Commissioner must act pursuant to the authority given to it under Article 3 (1) of the Safe Harbor agreement to suspend the data flow from Facebook's subsidiary to its parent organization. Importantly, it must be noted that Europe-versus-Facebook does not challenge the validity the Commission's Safe Harbor decision but rather it challenges a specific transfer of data that occurs within the framework.

The Irish Data Protection Commissioner determined that Europe-versus-Facebook's claims were "frivolous and vexatious" because the data transfers between Facebook Ireland and its US parent properly fell within the scope of the Safe Harbor agreement.[27] The Commissioner found that there was no basis to investigate the claim because Facebook Inc. had self-certified with the Safe Harbor Program and the EU Commission had issued a formal decision that all data transfers made via the Safe Harbor Program met the adequacy requirement set forth in the EU Data Protection Directive (see Footnote 27). In short, because the Commissioner found that it was "statutorily bound" by the 2000 Commission decision "even where such data is accessed by national security authorities in the United States", it could not conduct its own investigation of the matter (see Footnote 27).

After the claim was rejected by Irish Data Protection Commissioner, the case was brought to the Irish High Court for judicial review.[28] Upon review, Europe-versus-Facebook, represented by Schrems, asserted that the decision by the Irish DPA was unlawful.[29] The Irish DPA, however, maintained that it was bound by the terms of the finding by the European Commission that the Safe Harbor framework provides an adequate level of protection for personal data.[30]

In its decision, the High Court acknowledged that monitoring global communications may be essential for the US to discharge its global security responsibilities.[31] At the same time, however, it noted that the system of oversight of law enforcement data access in the US is problematic, at least from an EU perspective, as "oversight

---

[27]Pleading Documents in relation to High Court Judicial Review between Maximilian Schrems (applicant) and Data Protection Commissioner (respondent) (Record No. 2013/765 JR), Affidavit of B Hawkes sworn 16 December 2013 retrieved at http://europe-v-facebook.org/JR_First_Response_DPC.pdf.

[28]*Schrems v. Data Protection Commissioner* (No.2), [2014] IEHC 351 (2014).

[29]Pleading Documents in relation to High Court Judicial Review between Maximilian Schrems (applicant) and Data Protection Commissioner (respondent) (Record No. 2013/765 JR), Grounding Affidavit of Applicant sworn 21October2013_REDACTED retrieved at https://dataprotection.ie/docimages/documents/Grounding%20Affiadvit_Applicant%20%5Bredacted%5D.PDF.

[30]Pleading Documents in relation to High Court Judicial Review between Maximilian Schrems (applicant) and Data Protection Commissioner (respondent) (Record No. 2013/765 JR), Affidavit of B Hawkes sworn 16 December 2013 retrieved at http://europe-v-facebook.org/JR_First_Response_DPC.pdf.

[31]*Schrems v. Data Protection Commissioner* (No.2), [2014] IEHC 351 (2014).

is not carried out on European soil" and in circumstances where the data subject may make submissions and where the review makes references to EU law.[32]

The Court continued its decision by emphasizing that Europe-versus-Facebook's claim concerned the manner in which the DPA applied the Safe Harbor regime—and not to the validity of the Safe Harbor regime in its entirety.[33] In other words, the Court suggested that while the Safe Harbor program may be problematic because it lacks an independent oversight body that acts in accordance with the principles of EU law, the validity of the Safe Harbor Agreement as such was outside the scope of its decision. The Court was explicit that the scope of its review was strictly limited to a question of whether the Irish Commissioner was bound to the Commission's Safe Harbor Decision.[34]

Acknowledging that the Irish Commissioner may not be able to look beyond the Commission's Safe Harbor Decision, the Court stated:

> ...if the Commissioner cannot look beyond the European Commission's Safe Harbour Decision of July 2000, then it is clear that the present application for judicial review must fail ... (because)... the Commission has decided that the US provides an adequate level of data protection and, ... (section) 11(2)(a) of the 1998 Act (which in turn follows the provisions of Article 25(6) of the 1995 Directive) ties the Commissioner to the Commission's finding. In those circumstances, any complaint to the Commissioner concerning the transfer of personal data by Facebook Ireland (or, indeed, Facebook) to the US on the ground that US data protection was inadequate would be doomed to fail.[35]

In short, the Court concluded that if the Irish DPA is not allowed to disregard the Commission's 2000 decision then Europe-v-Facebook's complaint must fail.

On 18 June 2014 Ireland's High Court asked the CJEU whether a DPA is absolutely bound by a European Commission decision from 2000. Specifically, it asked the CJEU to clarify:

> Whether in the course of determining a complaint which has been made to an independent office holder who has been vested by statute with the functions of administering and enforcing data protection legislation that personal data is being transferred to another third country (in this case, the United States of America) the laws and practices of which, it is claimed, do not contain adequate protections for the data subject, that office holder is absolutely bound by the Community finding to the contrary contained in Commission Decision of 26 July 2000 (2000/520/EC) having regard to Article 7 and Article 8 of the Charter of Fundamental Rights of the European Union (2000/C 364/01), the provisions of Article 25(6) of Directive 95/46/EC notwithstanding? Or, alternatively, may the office holder conduct his or her own investigation of the matter in the light of factual developments in the meantime since that Commission Decision was first published?[36]

---

[32]*Schrems v. Data Protection Commissioner* (No.2), [2014] IEHC 351 (2014) para. 62.

[33]*Schrems v. Data Protection Commissioner* (No.2), [2014] IEHC 351 (2014) para. 62.

[34]*Schrems v. Data Protection Commissioner* (No.2), [2014] IEHC 351 (2014) para. 62 (stating, "It must be stressed, however, that neither the validity of the 1995 Directive nor the Commission Decision providing for the Safe Harbour Regime are, as such, under challenge in these judicial review proceedings.").

[35]*Schrems v. Data Protection Commissioner* (No.2), [2014] IEHC 351 (2014) para. 66.

[36]*Schrems v. Data Protection Commissioner* (No.2), [2014] IEHC 351 (2014) para. 71.

4 Comment and Analysis

4.1 Article 3 of the Safe Harbor Agreement

As noted above, Article 3 of the Commission Decision on the Safe Harbor principles affords national DPAs the opportunity to suspend data flows to an organization participating in the Safe Harbor Program "in order to protect individuals with regard to the processing of their personal data."[37] First, a specific suspension is authorized in cases where "the government body in the (US)(such as the FTC) … has determined that the organisation (sic) is violating the Principles…"[38] Second, a specific suspension is authorized where:

> … there is a substantial likelihood that the Principles are being violated; there is a reasonable basis for believing that the enforcement mechanism concerned is not taking or will not take adequate and timely steps to settle the case at issue; the continuing transfer would create an imminent risk of grave harm to data subjects; and the competent authorities in the Member State have made reasonable efforts under the circumstances to provide the organisation (sic) with notice and an opportunity to respond (see Footnote 38).

In the event that a DPA decides to suspend a specific data flow then it must "inform the Commission without delay."[39] The Commission may, on the basis of this knowledge, take actions to reverse or suspend the Safe Harbor Agreement (see Footnote 39). Here, it is evident that a DPA may suspend a specific data flow from the EU to a participating organization in the US but that any decision to nullify or modify the entirety of the Safe Harbor Agreement lies with the Commission.[40]

One question that arises is how does Article 3 of the Commission Decision on the Safe Harbor principles interact with Annex 1 of the Safe Harbor Decision which provides that adherence to the Principles may be limited, if justified by national security, public interest, or law enforcement requirements or by statute,

---

[37] 2000/520/EC: Commission Decision of 26 July 2000 pursuant to Directive 95/46/EC of the European Parliament and of the Council on the adequacy of the protection provided by the safe harbour privacy principles and related frequently asked questions issued by the US Department of Commerce retrieved at http://eur-lex.europa.eu/LexUriServ/LexUriServ.do?uri=CELEX:32000D0520:EN:HTML.

[38] Article 3, Commission decision of July 26, 2000 pursuant to Directive 95/46/EC of the European Parliament and of the Council on the adequacy of the protection provided by the safe harbour privacy principles and related frequently asked questions issued by the US Department of Commerce, 2000/520/EC.
Safe Harbor Enforcement Overview retrieved at http://export.gov/safeharbor/eu/eg_main_018481.asp.

[39] Article 3(4) 2000/520/EC: Commission Decision of 26 July 2000 pursuant to Directive 95/46/EC of the European Parliament and of the Council on the adequacy of the protection provided by the safe harbour privacy principles and related frequently asked questions issued by the US Department of Commerce retrieved at
http://eur-lex.europa.eu/LexUriServ/LexUriServ.do?uri=CELEX:32000D0520:EN:HTML.

[40] See the procedure referred to in Article 31 of the Data Protection Directive for how the Commission may go about suspending the entirety of the Safe Harbor Agreement.

government regulation or case-law. Specifically, Annex 1 states: "Adherence to these Principles may be limited: (a) to the extent necessary to meet national security, public interest, or law enforcement requirements..."[41]

There seems to be an ambiguity as to which organization determines whether the national security requirements demanded of a technology company like Facebook falls within the ambit of this exception: is it the role of the national DPA or the Commission? In other words, may a national DPA determine that a specific data flow must be prevented and that any assertion on behalf of the US company that it is acting in accordance with the national security exemption is not justified or is this an issue that rests solely within the domain of the Commission? After all, if, for example, the Irish DPA is able to determine that Facebook is not justified under the national security exemption to provide personal data to the US government and suspends the specific data flow based on Article 3 then, it would logically follow that a large number of other technology firms would be blocked by this precedent from asserting the same argument. The effect would practically be nullify the Safe Harbor Agreement or to substantially limit its application to only those companies not cooperating with the US government in the PRISM program. At the very least, a legal precedent would be set which would require some thoughtful political deliberation.

4.2 The EU Data Protection Directive

The Data Protection Directive ("Directive"), adopted in 1995, is broad in scope, applying to all "processing" of "personal data," except personal data processing which relates to purely personal or household activities, law enforcement activities, and activities concerning public security, defense and State security.[42] The Directive affords rights to the data subject such as the right to access his/her personal data and to rectify his/her data.[43] It also imposes responsibilities on individuals and organizations that process personal data such as to ensure that personal data is processed fairly and lawfully, collected for specified, explicit and legitimate purposes, and protected against accidental or unlawful destruction.[44]

The Directive does not apply to "processing operations concerning public security, defense, State security (including the economic well-being of the State when

---

[41] Annex 1, 2000/520/EC: Commission Decision of 26 July 2000 pursuant to Directive 95/46/EC of the European Parliament and of the Council on the adequacy of the protection provided by the safe harbour privacy principles and related frequently asked questions issued by the US Department of Commerce retrieved at http://eur-lex.europa.eu/LexUriServ/LexUriServ.do?uri=CELEX:32000D0520:EN:HTML.

[42] *See*, Article 4 Data Protection Directive.

[43] *See*, Article 12 Data Protection Directive.

[44] *See*, Article 6 and Article 17 Data Protection Directive.

the processing operation relates to State security matters) and the activities of the State in areas of criminal law."⁴⁵ This provision echoes Article 4(2) TEU and reflects the division of competences between the EU and the Member States.⁴⁶ While the Directive "does not regulate data processing by the law enforcement authorities and the intelligence services" it does "govern the transmission of personal data from data controllers and processors when they are ordered to submit information to intelligence services and law enforcement authorities."⁴⁷

It is further important to mention that where a processing activity falls within the scope of the Directive, it is still possible, pursuant to Article 13 of the Directive, for a Member State to adopt legislative measures to restrict the scope of the obligations and rights provided for in Directive when such a restriction constitutes a necessary measures to safeguard, among others, national security, defense and public security.⁴⁸ According to the CJEU in *Osterreichischer Rundfunk*, each of the exceptions included in Article 13 of that Directive must adhere to the proportionality principle and "cannot be interpreted as conferring legitimacy on an interference with the right to respect for private life contrary to Article 8 of the (European Convention of Human Rights)."⁴⁹ In short, there are general national security exemptions set forth in Article 3(2) TEU and Article 3 of the Directive as well as a specific provision in the Directive, Article 13, which allows for certain safeguards to be excluded for reasons of national security.

With respect to "external transfers", the Directive sets forth an entire chapter on the subject. As noted at the outset, the general rule is that external transfers are only permitted where an "adequate" level of protection for the transferred data can be assured.⁵⁰ Enrollment in the Safe Harbor Program is one way to meet the adequacy requirement. The external transfer rules in the Directive are silent with respect whether the adequacy requirement can be derogated from on national security grounds.

The Directive requires that each Member State both promulgate a national law that complies with its terms and establish one or more DPAs to enforce the rights set forth in the national law.⁵¹ Pursuant to Article 28, these DPAs should "act with complete independence in exercising the functions entrusted to them."⁵² They

---

⁴⁵Article 3(2) Data Protection Directive.

⁴⁶Article 4(2) Data Protection Directive states "national security remains the sole responsibility of each Member State"; *see further*, Working Document on surveillance of electronic communications for intelligence and national security purposes, WP 228 adopted on 5 December 2014.

⁴⁷Working Document on surveillance of electronic communications for intelligence and national security purposes, WP 228 adopted on 5 December 2014.

⁴⁸Article 13 Data Protection Directive.

⁴⁹C-465/00, Rechnungshof v Österreichischer Rundfunk, judgment of 20 May 2003 (para. 91).

⁵⁰Article 25(1) of Data Protection Directive.

⁵¹Peter Swire, "Of Elephants, Mice, and Privacy: International Choice of Law and the Internet," 32 *International Law* 991 (1998).

⁵²Article 28(1) Data Protection Directive.

should also be endowed with "investigative powers", "effective powers of intervention" and "the power to engage in legal proceedings" where the national law implementing the Directive has been violated.[53] In addition to administrative remedies before the national DPAs, Article 22 of the Directive sets forth the general obligation for Member States to provide judicial remedies available before the ordinary courts or tribunals.[54]

Article 3 of Commission Decision on the Safe Harbor principles expressly provides that it is "(w)ithout prejudice to (the DPAs') powers to take action to ensure compliance with national provisions adopted pursuant to provisions other than Article 25 of (the EU Data Protection Directive)."[55] Accordingly, this provision suggests that, independent from the powers they have under the Safe Harbor decision, DPAs have the competence to suspend data flows to the US if the general principles set forth in the EU Data Protection Directive are not being followed by a company enrolled in the Safe Harbor program.[56] This argument is further supported by reference to Article 28 of the Directive, which emphasizes that DPAs should act independently and possess meaningful enforcement powers.

Again, while the DPA may only be able to suspend a specific data flow under the authority given to it by the Directive—and not invalidate the Safe Harbor Agreement as such—the practical effect of suspending the data flow from Facebook Ireland to Facebook Inc. would be to substantially limit the Safe Harbor Agreement, at least as it appertains to the transnational companies that cooperate with the US government in the PRISM program. This is because all transfers that occur within the framework of the PRISM program could be prevented on grounds that the principles of the EU Data Protection Directive are not being respected.

---

[53] Article 28(3) Data Protection Directive.

[54] Article 22 Data Protection Directive (stating, "Without prejudice to any administrative remedy for which provision may be made, inter alia before the supervisory authority referred to in Article 28, prior to referral to the judicial authority, Member States shall provide for the right of every person to a judicial remedy for any breach of the rights guaranteed him by the national law applicable to the processing in question.").

[55] 2000/520/EC: Commission Decision of 26 July 2000 pursuant to Directive 95/46/EC of the European Parliament and of the Council on the adequacy of the protection provided by the safe harbour privacy principles and related frequently asked questions issued by the US Department of Commerce retrieved at http://eur-lex.europa.eu/LexUriServ/LexUriServ.do?uri=CELEX:32000D0520:EN:HTML; *see further*, Commission's Safe Harbor Communication of 27 November 2013 (stating, "Independently of the powers they enjoy under the Safe Harbour (sic) Decision, EU national data protection authorities are competent to intervene, including in the case of international transfers, in order to ensure compliance with the general principles of data protection set forth in the 1995 Data Protection Directive.").

[56] Commission's Safe Harbor Communication of 27 November 2013 (making clear that "Independently of the powers they enjoy under the Safe Harbour (sic) Decision, EU national data protection authorities are competent to intervene, including in the case of international transfers, in order to ensure compliance with the general principles of data protection set forth in the 1995 Data Protection Directive.").

4.3 The EU Charter

In 2000, the EU proclaimed the Charter of Fundamental Rights of the European Union ("Charter"). The Charter became legally binding as EU primary law, pursuant to Article 6(1) of the TEU, when the Lisbon Treaty came into force on 1 December 2009.[57] Article 7 sets forth the right to privacy.[58] Additionally, the EU Charter recognizes the right to data protection as an independent right in Article 8, which reads:

> Everyone has the right to the protection of their personal data. Such data must be processed fairly for specified purposes and on the basis of the consent of the person concerned or some other legitimate basis laid down by law. Everyone has the right of access to their data, and the right to have it rectified. Compliance with these rules shall be subject to control by an independent authority.[59]

It is important to emphasize that Article 8 not only explicitly mentions a right to data protection, but also refers to key data protection principles. It is further worth highlighting that the Charter requires that an independent authority will ensure compliance with the principles set forth in Article 8.

Like Article 7 of the ECHR, the right to protection of personal data provided under Article 8 of the Charter is not an absolute right. Rather, it is one that is subject to restriction and "must be considered in relation to its function in society."[60] More specifically, any limitation on the exercise of the right must meet the requirements of Article 52(1), namely that the restriction is "provided for by law", respects the essence of the right, and, "subject to the principle of proportionality", is "necessary and genuinely meet(s) objectives of general interest recognised (sic) by the Union or the need to protect the rights and freedoms of others."[61]

As a result of the national security exemption set forth in Article 3(2) TEU the scope of application of the Charter is limited. Nevertheless, Articles 7 and 8 still apply to EU institutions and bodies and all the activities of Member States when they implement Union law.[62] It is further important to mention that the CJEU has recently confirmed in *Pfleger* the application of the Charter to state measures derogating from the fundamental rights.[63]

---

[57] *See* consolidated versions of the European Communities (2012), Treaty on European Union, OJ 2012 C 326; and of European Communities (2012), TFEU, OJ 2012 C 326.

[58] European Charter of Fundamental Rights (2000).

[59] Article 8, European Charter of Fundamental Rights (2000).

[60] Joined cases C-92/09 and C-93/09, *Volker and Markus SChecke GbR and Hartmut Eifert v. Land Hessen*, judgment of 9 November 2010, para. 48.

[61] Article 52(1) European Charter of Fundamental Rights.

[62] Working Document on surveillance of electronic communications for intelligence and national security purposes, WP 228 adopted on 5 dec 2014.

[63] C-390/12, *Pfleger and Others*, judgment of 30 April 2014 (confirming that the use by a Member State of a derogation provided for by EU law in order to justify a limitation of a fundamental right guaranteed by the Treaty must be regarded as "implementing Union law" within the meaning of Article 51 of the Charter).

Like all public authorities, the DPAs are bound to respect the EU Charter when they apply EU law. Accordingly, Article 7 and 8 of the EU Charter are a source of authority by which a DPA may suspend the flow of personal data from an EU company to a US company enrolled in the Safe Harbor program. On this basis, the Irish DPA could be required to suspend the data flow from Facebook Ireland to Facebook Inc. because of its impact on the fundamental rights to privacy and data protection in the EU.

On April 8 the Court of Justice of the European Union (CJEU) announced its judgment in *Digital Rights Ireland,* which may shed light on the compatibility between the Safe Harbor Agreement and the EU Charter.[64] The Court, raising concerns that the aggregation of meta-data may "allow very precise conclusions to be drawn concerning the private lives" of individuals[65] and that the retention of personal data "is likely to generate in minds of the persons concerned the feeling that their private lives are the subject of constant surveillance",[66] held that the EU Data Retention Directive was invalid. It reasoned that the EU Data Retention Directive did not contain appropriate safeguards in accordance with EU fundamental rights law. The Court concluded:

> … (the EU Data Retention Directive) does not lay down clear and precise rules governing the extent of the interference with the fundamental rights enshrined in Articles 7 and 8 of the Charter. It must therefore be held that (the EU Data Retention Directive) entails a wide-ranging and particularly serious interference with those fundamental rights in the legal order of the EU, without such an interference being precisely circumscribed by provisions to ensure that it is actually limited to what is strictly necessary.[67]

One particular safeguard that the Court found was lacking was the existence of an administrative authority or court to oversee the access and use of the data retained by national authorities pursuant to the Data Retention Directive.[68] Other safeguards that were deficient concerned the absence of an objective time period for which the data could be retained[69] as well as limits on the access and use to any data retained (see Footnote 68).

In light of the principles elaborated by the court in *Digital Rights Ireland,* it is apparent that the Safe Harbor Program may not satisfy the requirements of

---

[64]Joined cases C-293/12 and C-594/12, *Digital Rights Ireland and Seitlinger and Others*, judgment of 8 April 2014.

[65]Joined cases C-293/12 and C-594/12, *Digital Rights Ireland and Seitlinger and Others*, judgment of 8 April 2014, para. 27.

[66]Joined cases C-293/12 and C-594/12, *Digital Rights Ireland and Seitlinger and Others*, judgment of 8 April 2014, para. 37.

[67]Joined cases C-293/12 and C-594/12, *Digital Rights Ireland and Seitlinger and Others*, judgment of 8 April 2014, para 65.

[68]Joined cases C-293/12 and C-594/12, *Digital Rights Ireland and Seitlinger and Others*, judgment of 8 April 2014, paras. 60–62.

[69]Joined cases C-293/12 and C-594/12, *Digital Rights Ireland and Seitlinger and Others*, judgment of 8 April 2014, paras. 64–65.

Articles 7 and 8 of the Charter. This is alluded to in the High Court decision of June 18, 2014 where the Court the stated that:

> While the FISA Court doubtless does good work, the FISA system can at best be described as a form of oversight by judicial personages in respect of applications for surveillance by the US security authorities. Yet the very fact that this oversight is not carried out on European soil and in circumstances where the data subject has no effective possibility of being heard or making submissions and, further, where any such review is not carried out by reference to EU law are all considerations which would seem to pose considerable legal difficulties.[70]

In other words, the Irish High Court seemed to suggest that the lack of an independent authority—on EU territory—with oversight over the data being used may require that the Safe Harbor Agreement be nullified pursuant to EU fundamental rights law.[71]

A question, however, that arises in connection with the DPA's ability to invoke the Charter as a tool to invalidate data transfers occurring under the Safe Harbor Program is how this invocation of fundamental rights law should interact with Article 25(6) of the EU Data Protection Directive. Article 25(6) provides:

> the Commission may find, in accordance with the procedure referred to in Article 31 (2), that a third country ensures an adequate level of protection..., by reason of its domestic law or of the international commitments it has entered into ... for the protection of the private lives and basic freedoms and rights of individuals. *Member States shall take the measures necessary to comply with the Commission's decision.* (emphasis added).

Here, the question becomes whether the DPA must comply with the Commission's adequacy decision even where such compliance may result in an infringement of the fundamental rights of data subjects afforded to them by Article 8 of the EU Charter. On one hand, this obligation seems flow naturally from the Charter, particularly where Article 8 expressly refers to the obligation of an independent authority to ensure compliance with the principles set forth therein. On the other hand, as noted by Kuner, "If each DPA is allowed to override Commission adequacy decisions based on its individual view of what the Charter of Fundamental Rights requires, then there would be no point to such decisions in the first place."[72] Removing the ability of the Commission to make adequacy determinations would not only mean that Member States would have to conduct their own assessments but it may also undermine a common approach to the adequacy procedure.

---

[70]*Schrems v. Data Protection Commissioner* (No.2), [2014] IEHC 351 (2014).

[71]*For more see*, Christopher Kuner, "The data retention judgment, the Irish Facebook case, and the future of EU data transfer regulation," *Concurring Opinions Blog* (19 June 2014) retrieved at http://www.concurringopinions.com/archives/2014/06/the-data-retention-judgment-the-irish-facebook-case-and-the-future-of-eu-data-transfer-regulation.html (stating, "...the logical consequence of the Court's statement in Digital Rights Ireland would seem to be that fundamental rights law requires oversight of data processing by the DPAs also with regard to the data of EU individuals that are transferred to other regions.").

[72]Christopher Kuner, "The data retention judgment, the Irish Facebook case, and the future of EU data transfer regulation," *Concurring Opinions Blog* (19 June 2014).

5 Additional Issues

5.1 What if There Is no Transfer?

Schrems' assumption that there is an international transfer of data from Facebook Ireland to Facebook Inc. must be questioned. While it is very possible that Facebook Ireland sends personal data about its users to Facebook Inc., it is equally likely that the personal data arrives in the US by alternative means. For example, EU users may upload their data directly to US servers without any intermediary in Europe. If a data transfer were found to exist in this instance, however, then the restrictions of Article 25 would apply any time information is loaded onto and made accessible via the Internet in direct contravention of the *Lindqvist* decision.[73] Pragmatically, the CJEU reasoned in *Lindqvist* that finding the existence of a transfer in such situations would effectively require Article 25 to apply any time personal data was loaded onto an Internet page making it a provision of general application and, therefore, such a conclusion should not be reached.[74]

Because the Safe Harbor program presupposes that there must be two separate actors in its conception of transfer (a data controller in the EU and a data controller in the US) and the existence of an EU data controller may be lacking (if Facebook Ireland is not considered the data controller nor is the individual user considered the data controller) then the entire Safe Harbor Framework may be inapplicable.[75] This would create a gap in fundamental rights protection, at least from the perspective of the EU citizen, who may be unable to seek effective redress from a European DPA in the event his/her rights have been trespassed upon by Facebook Inc. due to lack of jurisdiction over Facebook Inc. and is further unlikely to be able to assert a claim in the US due to limitations in US constitutional and statutory law.

5.2 What if Facebook Inc. Must Comply with the Directive Pursuant to Article 4?

As noted at the outset of this paper, Schrems decision to bring an action against Facebook Ireland under the Safe Harbor Agreement rather than against Facebook

---

[73]*See*, Case C-101/1, *Criminal Proceedings Against Bodil Lindqvist*, judgment 6 November 2003.

[74]Case C-101/1, *Criminal Proceedings Against Bodil Lindqvist*, judgment 6 November 2003, para. 69.

[75]For more see, Liane Colonna, "Article 4 of the EU Data Protection Directive and the irrelevance of the EU-US Safe Harbor Program?" 4(3) *International Data Privacy Law* (Oxford 2014).

Inc. under Article 4 can be questioned. Article 4 sets forth the territorial scope of the directive and, if provided a broad interpretation, could remedy any perceived gaps in the protection afforded by the Safe Harbor Agreement. Under Article 4, Member States must apply the national provisions implementing the directive to the processing of personal data, where "the processing is carried out in the context of the activities of an establishment of the controller on the territory of the Member State", and where "the controller is not established on Community territory and, for purposes of processing personal data makes use of equipment, automated or otherwise, situated on the territory of the said Member State."[76]

The draft Regulation, similar to the Directive, will apply "to the processing of personal data in the context of the activities of an establishment of a controller or a processor in the Union."[77] The "use of equipment" test, however, has been completely abandoned in favor of two new standards: the "offering of goods or services" standard and the "targeting" standard. These two new standards will broaden the extraterritorial scope of EU data protection law considerably.

Recently, the CJEU announced its decision in *Google Spain*, which provides clarification on how Article 4(1) should be interpreted, as well as Article 3(1) of the draft Regulation which, retains similar language.[78] The case concerned whether Google Inc., in its capacity as an online search engine, is required to block unwanted personal data appearing in search results.[79] With respect to applicable law, the central question was whether Spanish data protection law (implementing the EU Data Protection Directive) applies to Google Inc. where: (1) its search engine is operated exclusively by Californian-based Google Inc. and (2) where it has a subsidiary in Spain, Google Spain SL, for the sole purpose of acting as a commercial representative for its advertising functions (see Footnote 79).

Google Spain and Google Inc. asserted that the Directive should be inapplicable to Google Inc. because the processing of personal data at issue—namely that associated with Google's search engine operations—was carried out exclusively

---

[76]Article 4(1)(a)–(c) Data Protection Directive.

[77]Article 3, Proposal for a Regulation of the European Parliament and of the Council on the Protection of Individuals with Regard to the Processing of Personal Data and on the Free Movement of Such Data (General Data Protection Regulation), COM (2012) 11 final (25 Jan. 2012).

[78]Case C-131/12 *Google Spain SL Google Inc. v Agencia Española de Protección de Datos (AEPD) Mario Costeja González,* judgment of 13 May 2014 (concerning the application of EU data protection law to data processing outside the EU); for information about the intra-EU application of national data protection laws under the Directive please see Case C-230/14 *Weltimmo s.r.o. v Nemzeti Adatvédelmi és Információszabadság hatóság,* request for a preliminary ruling from the Kúria (Hungary) lodged on 12 May 2014; see also, Christopher Kuner, The Court of Justice of the EU Judgment on Data Protection and Internet Search Engines, *LSE Law, Society and Economy Working Papers 3/2015,* London School of Economics and Political Science Law Department (making the relevant distinction between the application of EU data protection law to data processing outside the EU versus inside the EU).

[79]Case C-131/12 *Google Spain SL Google Inc. v Agencia Española de Protección de Datos (AEPD) Mario Costeja González,* judgment of 13 May 2014.

by Google Inc. in the US. In other words, the fact that Google Spain did not carry out any activities directly related to the indexing or storage of information meant that the processing in question could not possibly be considered to take place "in the context of the activities" of the establishment. They emphasized that the only processing carried out at Google Spain concerned the promotion and sales of advertisements.

First, the Opinion of the CJEU determined that Google Spain constitutes an establishment of Google Inc. Pointing to recital 19 of the Directive, the Court stated that an "establishment on the territory of a Member State implies the effective and real exercise of activity through stable arrangements" and that "the legal form of such an establishment, whether simply (a) branch or a subsidiary with a legal personality, is not the determining factor."[80]. Because it was undisputed that Google Spain both engaged in the effective and real exercise of activity through stable arrangements in Spain and constituted a subsidiary of Google Inc. on Spanish territory, the Court quickly concluded that Google Spain was an "establishment" within the meaning of Article 4(1)(a) (see Footnote 80).

Second, even though Google Spain merely promotes and sells online advertising space for Google Inc., the CJEU found that the personal data processed at Google Spain was in the "context of the activities" of that establishment with the consequence that the EU Data Protection Directive was applicable to Google Inc. The CJEU reasoned that the advertising activities of Google Spain were "inextricably linked" to Google Inc.'s search engine.[81] This is because, without the sales and promotion of the online advertising space, the search engine would not be economically profitable (see Footnote 79). As further evidence of the interdependence between the advertising activities of Google Spain and the search engine operations of Google Inc., the CJEU pointed to the fact that the advertisements and search adverts are displayed on the same page (see Footnote 79).

In its decision the CJEU emphasized that the meaning of the words "in the context of the activities" of the establishment "cannot be interpreted restrictively."[82] The Court reasoned that a broad interpretation should be afforded to these words in light of the objective of the directive: to "(ensure) effective and complete protection of the fundamental rights and freedoms of natural persons."[83]

---

[80]Case C-131/12 *Google Spain SL Google Inc. v Agencia Española de Protección de Datos (AEPD) Mario Costeja González,* judgment of 13 May 2014, para. 48–49.

[81]Case C-131/12 *Google Spain SL Google Inc. v Agencia Española de Protección de Datos (AEPD) Mario Costeja González,* judgment of 13 May 2014, para. 56.

[82]Case C-131/12 *Google Spain SL Google Inc. v Agencia Española de Protección de Datos (AEPD) Mario Costeja González,* judgment of 13 May 2014, para. 53 (stating, "Furthermore, in the light of the objective of Directive 95/46 of ensuring effective and complete protection of the fundamental rights and freedoms of natural persons, and in particular their right to privacy, with respect to the processing of personal data, those words cannot be interpreted restrictively.").

[83]Case C-131/12 *Google Spain SL Google Inc. v Agencia Española de Protección de Datos (AEPD) Mario Costeja González,* judgment of 13 May 2014, para. 53.

As a result *Google Spain*, it is logical to conclude that Facebook US could be required to adhere to the Directive pursuant to Article 4. The decision suggests that the processing activities that take place "in the context" of an EU establishment, even if the activities just have a seemingly peripheral connection to a US company, are sufficient for the Directive to apply to the US company. It effectively turns the "in the context of activities" test into a "direct link to economic activities" test.

The fact that Facebook Ireland, an ostensible establishment of Facebook US, provides memberships to Facebook could trigger the Directive, even if these memberships are provided for free. After all, the memberships offered via Facebook Ireland have a direct link to the economic profitability to the parent company in the US: they could even be thought of as the raison d'etre of the business. That said, it is possible that the CJEU could limit the holding of *Google Spain* to the specific facts therein, which namely concerned an issue of search engine operations.

If, however, Facebook US is required to follow the Directive by virtue of Article 4, then the entire *Europe versus Facebook* case could be flipped: instead of asking whether Facebook Ireland is compliant with the Directive (and then examining whether the criteria of the Safe Harbor Program are met when it transfers data from the EU to its US parent) the question could be whether Facebook US is compliant with the Directive when it transfers data from its servers to the US government. If the technology firm is obligated to comply with the Directive under Article 4, then the firm can only transfer the data to the US government if it can ensure an adequate level of protection pursuant to Article 25 (i.e. it must frame the transfer through one of the transfer tools provided for in the Directive). Because it is doubtful that the US Government offers an adequate level of data protection in the context of PRISM, DPAs could act, pursuant to their authority to ensure that the principles of the EU Data Protection Directive are complied, to suspend the data flow.

In response to a DPA's claims that it is in breach of data transfer rules when they transfer data to the US government, however, Facebook could assert that the entire matter is outside of the scope of the Directive as it concerns national security.[84] Additionally, even if EU law and the Directive are found to be applicable, it could assert that it is permitted to transfer data pursuant to the "important ground of public interest exception."[85] Specifically, Article 26(1)(d) states:

> By way of derogation from Article 25 and save where otherwise provided by domestic law governing particular cases, Member States shall provide that a transfer or a set of transfers of personal data to a third country which does not ensure an adequate level of protection within the meaning of Article 25 (2) may take place on condition that ... the transfer is necessary or legally required on important public interest grounds... (see Footnote 85)

---

[84] Article 3 Data Protection Directive (stating, "This Directive shall not apply to the processing of personal data: in the course of an activity which falls outside the scope of Community law, such as ... processing operations concerning public security, defence (sic), State security...").

[85] Article 26 Data Protection Directive.

Furthermore, Facebook may also be able to rely on the national security exception under Article 13(1) of the Directive, which reads:

> Member States may adopt legislative measures to restrict the scope of the obligations and rights provided for in Articles 6 (1), 10, 11 (1), 12 and 21 when such a restriction constitutes a necessary measures to safeguard: (a) national security...

There are a few points to make with regard to these assertions. First, the scope of the national security exemptions provided for in Article 3(2) of the Directive and Article 4(2) TEU are unclear.[86] This particularly true in light of the growing involvement of the private sector in national security matters, the result of which has been a blurring of the legal rules. In two widely criticized opinions, the CJEU has found that the transfer of PNR data from airlines to US law enforcement authorities was outside the scope of EU data protection law while the retention of telecommunications data by private companies for the purpose of fighting against terrorism and/or serious crime is subject to EU data protection law.[87] In the wake of these decisions, it is unsurprising that the Working Party has called for the scope of the national security exemption to "be clarified in order to give legal certainty regarding the scope of application of EU law" because "(t)o date, no clear definition of the concept of national security has been adopted by the European legislator, nor is the case law of the European courts conclusive."[88]

Likewise, the national security derogation provided for in Article 13 is equally vague, affording Member States broad latitude to enumerate the cases where restrictions on the obligations and rights provided to the data subject under the Directive may be imposed. The consequence of this provision is that some of the most serious privacy threats, such as excessive surveillance activities, remain outside of the reach of the harmonized EU legal framework. In other words, the safeguards and the guarantees provided for in the Directive such as the prevention of further processing for incompatible purposes, clear conditions for onward transfers, and judicial redress mechanisms may not exist in cases where the national security derogation is invoked.[89]

The second point to be made is that Article 13(1) of the Directive does not expressly refer to the data transfer rules set forth in Article 25. This raises a

---

[86]*See also*, Article 4(2) of the Treaty of the European Union (TEU)(imposing a national security exemption).

[87]*Compare* Joined Cases C-317/04 and C-318/04, *EP v Council and Commission (PNR)*, judgment of 30 May 2006 *with* Case C-301/06, *Ireland v. Parliament and Council,* judgment of 10 February 2009.

[88]Article 29 Working Party, Opinion 04/2014 on surveillance of electronic communications for intelligence and national security purposes, 819/14/EN WP 215 (Adopted on 10 April 2014) retrieved at http://ec.europa.eu/justice/data-protection/article-29/documentation/opinion-recommendation/files/2014/wp215_en.pdf.

[89]*For more see*, Opinion of the European Data Protection Supervisor on the future development of the area of freedom, security and justice (4 June 2014) https://secure.edps.europa.eu/EDPSWEB/webdav/site/mySite/shared/Documents/Consultation/Opinions/2014/14-06-04_Future_AFSJ_EN.pdf.

question about whether the adequacy requirement set forth in Article 25 may be subject to restriction in the national security context, especially the national security interests of a third country. While Article 26(1)(d) expressly provides for an exception based on "important public interests grounds" this, at least arguably, must be considered something different from the "national security" exception set forth in Article 13(1).[90] Further clarification on how the important ground of public interest exception set forth in Article 26(1)(d) should be understood, particularly in comparison to the "national security" exception set forth in Article 13, is lacking.

The third point concerns whether an entity subject to the Directive may invoke the Article 26(1)(d) exception where it is US public interests at stake and not necessarily EU public interests that require a transfer be made. In other words, there is an ambiguity in the law as to whether Facebook would be able to assert that the public interests of the US requires that the EU's adequacy requirement be inapplicable to it, particularly when there are no joint EU interests at stake. This seems unlikely based on the recent comments from the Article 29 Working party. Not only has the Article 29 Working party commented that "…the national security (exemption)…only applies to the national security of an EU Member State, and not to the national security of a third country" (see Footnote 88) but it has also stated that "(s)ince the adequacy instruments are primarily intended to offer protection to personal data originating in the EU, they should never be implemented to the detriment of the level of protection guaranteed by EU rules and instruments governing transfers."[91]

Finally, even if Facebook US could properly invoke an exception to the adequacy requirement, it is unlikely that the systemic access provided by it to data to the US government in the PRISM program would constitute a narrowly tailored exception under EU law. Indeed, the CJEU has emphasized that, "the protection of the fundamental right to privacy requires that derogations and limitations in relation to the protection of personal data must apply only in so far as is strictly necessary."[92]

---

[90] *See generally*, Working Document on surveillance of electronic communications for intelligence and national security purposes, WP 228 adopted on 5 dec 2014 stating ("…national security needs to be distinguished from the security of the European Union, but also from State security, public security and defence (sic)".

[91] Article 29 Working Party, Opinion 04/2014 on surveillance of electronic communications for intelligence and national security purposes, 819/14/EN WP 215 (Adopted on 10 April 2014) retrieved at http://ec.europa.eu/justice/data-protection/article-29/documentation/opinion-recommendation/files/2014/wp215_en.pdf; see further Working Document on surveillance of electronic communications for intelligence and national security purposes, WP 228 adopted on 5 dec 2014 (stating, "the Working Party points out that the national security exemption has to be interpreted to reflect the competence of the EU vis-à-vis the Member States and not as a general exemption from EU data protection requirements of all activities requested by third countries in the name of national security.).

[92] Case C-473/12, *Institut professionnel des agents immobiliers (IPI) v Geoffrey Englebert and Others, Judgment of the Court,* judgment 7 November 2013, para. 39.

Because the adequacy requirement seeks to ensure that core EU data protection values are not rendered meaningless after data leaves the US,[93] and the ongoing and systematic access to data provided by Facebook to the US seems to lack adherence to many of these principles such as purpose limitation and data minimization, it is hard to imagine that a limitation on the adequacy requirement in this context could be viewed as necessary and proportionate in a democratic society.

5.3 Can the DPAs Enforce Their Decisions?

The issue of enforcement jurisdiction may help to explain why Schrems ultimately chose to bring an action against Facebook Ireland under the Safe Harbor Agreement rather than against Facebook US under Article 4. Indeed, a fundamental issue with the expansion of the Directive beyond EU territory is that, pursuant to Article 28(6) of the Directive, the jurisdiction of a DPA is only valid within the border of the DPA's respective Member State.[94] That is, the Directive is difficult to enforce against foreign data controllers because DPA's are restricted in their actions to the territory in which they operate. Arguably, the existence of enforcement jurisdiction should be a prerequisite for the application of EU data protection laws in international cases: without effective enforcement jurisdiction the law is merely hypothetical and only has a "bark" effect.[95]

The Article 29 Working Party interprets Article 28(6) to mean:

> ...national data protection authorities are competent to supervise the implementation of the data protection legislation on the territory of the Member State where they are established. But if the law of another Member State were applicable on its territory, the enforcement powers of the DPA would not be limited: the applicable law criteria of the Directive foresee the possibility that a DPA is empowered to verify and intervene on a processing operation that is taking place on its territory even if the law applicable is the law of another Member State.[96]

---

[93]Els De Busser and Gert Vermeulen, "Towards a coherent EU policy on outgoing data transfers for use in criminal matters? The adequacy requirement and the framework decision on data protection in criminal matters. A transatlantic exercise in adequacy," *EU and International Crime Control* (GOFS Research Paper Series, 2010).

[94]Article 28(6) of the EU Directive provides: "Each supervisory authority is competent, whatever the national law applicable to the processing in question, to exercise, on the territory of its own Member State, the powers conferred on it in accordance with (this Directive)."

[95]Lee Bygrave, Determining Applicable Law Pursuant to European Data Protection Legislation, 16 Computer L. & Sec. Rev. 252, 255 (2000).

[96]Article 29 Working Party, Working document on determining the international application of EU data protection law to personal data processing on the Internet by non-EU based websites, WP 56, 30 May 2002, http://ec.europa.eu/justice/policies/privacy/docs/wpdocs/2002/wp56_en.pdf.

In other words, the Article 29 Working Party suggests that a DPA is able to exercise its powers when the data protection law of another place applies to the processing of personal data so long as the "processing operation" occurs where the DPA is established.[97]

Article 28(6) is murky because, in a world of ubiquitous data processing, it can be very hard to determine that a certain "data processing operation" is occurring in a single and distinct territory. This is where the close nexus between Article 4 (applicable law) and Article 28(6) (DPA jurisdiction) becomes evident. As noted by Kuner, "(i) n practice, national data protection authorities often equate jurisdiction and applicable law."[98]

One reason why DPAs are quick to assert jurisdiction over a matter, particularly where the data controller is a non-EU entity, is that it affords the EU complainant with optimal protection of their human rights.[99] DPAs want to make sure that EU residents are not deprived of the protection of their national law once they are transferred outside their territory—even if the effect is only symbolic. As noted by the Working Party, "The external scope of EU law is an expression of its capacity to lay down rules in order to protect fundamental interests within its jurisdiction." (see Footnote 96) Another reason why DPAs may assert jurisdiction is because of a perceived superiority of their approach.[100]

Accordingly, the EU can attempt to regulate data processors with only remote connections to the EU. It may, however, discover that there are limitations to the regulation. This is because the jurisdiction of a DPA is confined to the respective Member State where that DPA is physically situated and the enforcement of a judgment issued by a DPA against a controller with no establishment in the EU is difficult to effectuate where the controller has no assets or establishment within the EU. These reasons help to explain why Schrems may have made the strategic decision to bring an action against Facebook Ireland under the Safe Harbor Agreement rather than against Facebook US under Article 4 of the Directive.

---

[97]Article 29 Working Party, Working document on determining the international application of EU data protection law to personal data processing on the Internet by non-EU based websites, WP 56, 30 May 2002, http://ec.europa.eu/justice/policies/privacy/docs/wpdocs/2002/wp56_en.pdf. (stating, "the applicable law criteria of the Directive foresee the possibility that a DPA is empowered to verify and intervene on a processing operation that is taking place on its territory even if the law applicable is the law of another Member State.").

[98]Christopher Kuner, Internet Jurisdiction and Data Protection Law: An International Legal Analysis http://www.justice.gov.il/NR/rdonlyres/0C7218F7-8B6D-4A62-9E45-7BFB80621E32/26405/ConflictKuner_article.pdf.

[99]Bernhard Maier, How Has the Law Attempted to Tackle the Borderless Nature of the Internet?, *International Journal of Law and Information Technology* vol. 18, no. 2, Oxford: Oxford University Press 2010.

[100]Edward R. McNicholas, *Privacy and Surveillance Legal Issues. Leading Lawyers on Navigating Changes in Security Program Requirements and Helping Clients Prevent Breaches* (Aspatore, 2014).

6 Conclusion

At the heart of the *Europe versus Facebook* case is one of the most central questions in European data protection reform: how to safeguard European data from the perceived normative inadequacies raised by large-scale, US national intelligence, data-mining programs. A central ambition of this paper has been to point out two different policy choices that can be made to uphold EU data protection principles, at least until greater harmonization of the substantive legal rules occurs between the EU and the US.

The first approach is referred to as the adequacy and the "little cheese" approach. Here, the EU may safeguard its principles by imposing restrictions on local controllers of personal data a duty to only transfer such data to those countries where the data can be adequately protected.[101] This route, however, is problematic insofar as it could discourage foreign companies from establishing local subsidiaries/offices in the EU and it may not fully respect the concept of separate corporate identities, especially where a subsidiary is blamed for the bad actions of its parent (see Footnote 101). Additionally, the adequacy framework can be criticized as creating an unduly complex system, being under inclusive especially with respect to the processing of data for law enforcement purposes and establishing a political fiction that some countries provide "adequate" data protection while others do not.[102]

The second approach is referred to as the Article 4 and the "big-enchilada" approach. Here, the EU may safeguard its principles by imposing restrictions directly on the foreign wrongdoer who, for example, transfers personal data about EU residents to foreign intelligence agencies (see Footnote 101). While this approach may be more efficient to the extent that it is the central bad actor that is the one that is targeted, it may, nonetheless, present several challenges. These challenges include, first, the fact that many foreign data controllers are unlikely to know they have EU data protection obligations and, second, several important international law norms of jurisdiction may be violated where the EU seeks to take action against foreign companies in relation to activities that take place outside of its territorial borders.[103] Additionally, there is no guarantee that EU law can be enforced where it is asserted over an entity that operates outside of its territory, especially when one considers the limitation to the jurisdiction of DPAs set forth in Article 28 of the Directive.

---

[101]Uta Kohl, *Jurisdiction and the Internet Regulatory Competence over Online Activity* (Cambridge University Press, 2010).

[102]Liane Colonna, The New Proposal to Regulate Data Protection in the Law Enforcement Sector: Raises the Bar but Not High Enough. *IRI-memo*, Nr. 2/2012 available at http://www.juridicum.su.se/iri/docs/IRI-PM/2012-02.pdf.

[103]*See*, Christopher Kuner, "Internet Jurisdiction and Data Protection Law: An International Legal Analysis (Part 1)," *International Journal of Law and Information Technology*, Vol. 18, p. 176 (2010).

Transfers of personal data are a critical element of the transatlantic relationship. However, US national intelligence data mining programs, such as PRISM, threaten to undermine the free flow of data from the EU to the US because the EU, including its citizens and member states, are deeply concerned that the fundamental rights of Europeans are not being safeguarded sufficiently in this context.[104] Against this background, the EU can select one of aforementioned policy options but the reality is that no single jurisdiction can protect the privacy of its citizens without reliance on public and private organizations that operate outside of its borders.[105] What is truly needed is global governance of data protection in the form of a binding international data protection agreement which can prevent the gaps in data protection legislation created by a lack of harmonization in the law and, second, facilitate global data flows through the availability of universally applicable compliance rules.[106]

---

[104]7 Report on the Findings by the EU Co-chairs of the Ad Hoc EU-US Working Group on Data Protection accompanying the Communication from the Commission to the European Parliament and the Council on "Rebuilding Trust in EU-US Data Flows" (COM(2013) 846 final)—http://ec.europa.eu/justice/dataprotection/files/report-findings-of-the-adhoc-eu-us-working-group-on-data-protection.pdf.

[105]Colin J. Bennett, and Charles D. Raab, *The Governance of Privacy: Policy Instruments in Global Perspective* (The MIT Press, 2006).

[106]Christopher Kuner, An International Legal Framework for Data Protection: Issues and Prospects. *Computer Law & Security Review*, Vol. 25, pp. 307–317 (2009); *see more generally*, The Madrid Privacy Declaration (3 November 2009) retrieved at http://thepublicvoice.org/TheMadridPrivacyDeclaration.pdf (calling for "…the establishment of a new international framework for privacy protection, with the full participation of civil society, that is based on the rule of law, respect for fundamental human rights, and support for democratic institutions.").

The Context-Dependence of Citizens' Attitudes and Preferences Regarding Privacy and Security

Michael Friedewald, Marc van Lieshout, Sven Rung and Merel Ooms

Abstract This paper considers the relationship between privacy and security and, in particular, the traditional "trade-off" paradigm that argues that citizens might be willing to sacrifice some privacy for more security. Academics have long argued against the trade-off paradigm, but these arguments have often fallen on deaf ears. Based on data gathered in a pan-European survey we discuss which factors determine citizens' perceptions of concrete security technologies and surveillance practices.

Keywords Privacy · Public opinion · Security · Trade-off

1 Introduction

The relationship between privacy and security has often been understood as a zero-sum game, whereby any increase in security would inevitably mean a reduction in the privacy enjoyed by citizens. A typical incarnation of this thinking is the all-too-common argument: "If you have got nothing to hide you have got nothing to fear". This trade-off model has, however, been criticised because it approaches privacy

---

M. Friedewald (✉) · S. Rung
Fraunhofer Institute for Systems and Innovation Research ISI, Breslauer Straße 48, 76139 Karlsruhe, Germany
e-mail: Michael.friedewald@isi.fraunhofer.de

S. Rung
e-mail: sven.rung@isi.fraunhofer.de

M. van Lieshout · M. Ooms
Strategy and Policy Department, The Netherlands Organisation for Applied Science (TNO), Van Mourik Broekmanweg 6, 2628 XE Delft, The Netherlands
e-mail: Marc.vanLieshout@tno.nl

M. Ooms
e-mail: merel.ooms@tno.nl

and security in abstract terms and because it reduces public opinion to one specific attitude, which considers surveillance technologies to be useful in terms of security but potentially harmful in terms of privacy.[1] Whilst some people consider privacy and security as intrinsically intertwined conditions where the increase of one inevitably means the decrease of the other. There are also other views: There are those who are very sceptical about surveillance technologies and question whether their implementation can be considered beneficial in any way. Then there are people who do not consider monitoring technologies problematic at all and do not see their privacy threatened in any way by their proliferation. Finally there are those who doubt that surveillance technologies are effective enough in the prevention and detection of crime and terrorisms to justify the infringement of privacy they cause.[2]

Insight in the public understanding of security measures is important for decision makers in industry and politics who are often surprised about the negative public reactions showing that citizens are not willing to sacrifice their privacy for a bit more potential security. On the back of this the PRISMS project aimed to answer *inter alia* the question: When there is no simple trade-off between privacy and security perceptions, what then are the main factors that affect public assessment of the security and privacy implications of specific security technologies, of specific security contexts and of specific security-related surveillance practices?

The PRISMS project has approached this question by conducting a large-scale survey of European citizens. This is, however, not simply a matter of gathering data from a public opinion survey, as such questions have intricate conceptual, methodological and empirical dimensions. Citizens are influenced by a multitude of factors. For example, privacy and security may be experienced differently in different political and socio-cultural contexts. In this paper, however, our focus will be on the survey results, not their interpretation from different disciplinary perspectives.

2 Measuring People's Perceptions of Security Technologies

Researchers investigating the relationship between privacy and security have to deal with the so-called privacy paradox: It is well known that while European citizens are concerned about how the government and private sector collect data about citizens and consumers, these same citizens seem happy to freely give up personal

---

[1]Vincenzo Pavone and Sara Degli Esposti, "Public Assessment of New Surveillance-Oriented Security Technologies: Beyond the Trade-Off between Privacy and Security," *Public Understanding of Science* 21, no. 5 (2012). Daniel J. Solove, *Understanding Privacy* (Cambridge, MA.: Harvard University Press, 2008).

[2]Reinhard Kreissl et al., "Surveillance: Preventing and Detecting Crime and Terrorism," in *Surveillance in Europe*, ed. David Wright and Reinhard Kreissl (London. New York: Routledge, 2015).

and private information when they use the Internet. This "paradox" is not really paradoxical but represents a typical value-action gap, which has been observed in other fields as well.[3]

Measuring privacy and security perceptions thus has to deal with problems similar to ecopsychology at the beginning of the environmental movement in the 1970s: What is the relationship between general values and concrete (environmental) concerns and how do they translate into individual behaviour? In PRISMS we have been inspired by the "theory of planned behaviour" (TPB) that suggests that if people evaluate the suggested behaviour as positive (attitude), and if they think their significant others want them to perform the behaviour (subjective norm), this results in a higher intention and they are more likely to behave in a certain way.[4] TBP is a positivist approach as it assumes that there are rules structuring the way people think and these "social facts", as Durkheim is calling them, can be verified by scientific observation and experimentation.[5] We are aware of the fact that this assumption has been critised by other epistemological perspectives such as critical school, cultural studies and STS, which are highlighting that attitudes and values may be situationally determined rather than stable dispositions and that a number of context factors may limit individual choice.[6] On the other hand a high correlation of attitudes and subjective norms to behavioural intention, and subsequently to behaviour, has been confirmed in many studies.[7]

Another issue to be considered is that people can (and do) understand concepts such as privacy and security in very different ways, and that they often only have a vague idea how security technologies work and what kind and how much information they collect. Nonetheless people often are voicing (even strong) opinions.

---

[3]E.g. in the context of environmentalism consumers often state a high importance of environmental protection that is not reflect in their actual behaviour. See Anja Kollmuss and Julian Agyeman, "Mind the Gap: Why Do People Act Environmentally and What Are the Barriers to Pro-Environmental Behavior?" *Environmental Education Research* 8, no. 3 (2002).

[4]One of the most successful (and most criticized) application of TPB is the so-called "Technology Acceptance Model" and its extension, the "Unified Theory of Acceptance and Use of Technology", which simplifies the TPB approach by eliminating the direct consideration of attitudes because they are difficult or impossible to measure. They are very popular methods in computer science assess the acceptance of human computer interface designs. Cf. Viswanath Venkatesh et al., "User Acceptance of Information Technology: Toward a Unified View," *MIS Quarterly* 27, no. 3 (2003); Fred D. Davis, "Perceived Usefulness, Perceived Ease of Use, and User Acceptance of Information Technology," ibid.13 (1989).

[5]Emile Durkheim, *The Rules of Sociological Method [1895]*, trans. Steven Lakes (New York et al.: The Free Press, 1982).

[6]Cf. for instance Andrew J. Cook, Kevin Moore, and Gary D. Steel, "Taking a Position: A Reinterpretation of the Theory of Planned Behaviour," *Journal for the Theory of Social Behaviour* 35, no. 2 (2005).

[7]Icek Ajzen and Martin Fishbein, "The Influence of Attitudes on Behavior," in *The Handbook of Attitudes*, ed. Dolores Albarracin, Blair T. Johnson, and Mark P. Zanna (Mahwah, NJ: Erlbaum, 2005).

2.1 Operationalization of Privacy

Privacy is a concept that is not only hard to measure but also difficult to define. It is, however, a key lens through which many new technologies, and most especially new surveillance or security technologies, are critiqued. Although a widely accepted definition of privacy remains elusive, there has been more consensus on a recognition that privacy comprises multiple dimensions, and some privacy theorists have attempted to create taxonomies of privacy intrusions or problems—for instance Debbie Kasper or Daniel Solove.[8] However, the outlining of privacy problems or intrusions does little to provide an overarching framework that would ensure that individuals' rights are proactively protected.

On the other hand operationalising privacy as a positive right focuses on preventing harms rather than providing redress. Roger Clarke outlined specific elements of individual privacy that ought to be protected. His very popular taxonomy distinguished four types of privacy, but is no longer adequate to capture the range of potential privacy issues, which must be addressed.[9] In PRISMS we are using an extensions of Clarke's typology developed by Finn et al. who suggest seven different types of privacy that ought to be protected and that receive different attention and valuation in practice. Such a detailed taxonomy helps to overcome the problem that privacy is too abstract as a concept and therefore helps to deal with the fact that people can (and do) understand the term in very different ways. The seven types of privacy comprise[10]:

1. *Privacy of the person* encompasses the right to keep body functions and body characteristics (such as genetic codes and biometrics) private. This aspect of privacy also includes non-physical intrusions into the body such as occur with airport body scanners.
2. *Privacy of behaviour* and action includes sensitive issues such as sexual preferences and habits, political activities and religious practices. However, the notion of privacy of personal behaviour concerns activities that happen in public space *and* private space.
3. *Privacy of communication* relates to avoiding the interception of communications, including mail interception, the use of bugs, directional microphones, telephone or wireless communication interception or recording and access to e-mail messages.

---

[8]Debbie V. S. Kasper, "The Evolution (or Devolution) of Privacy," *Sociological Forum* 20, no. 1 (2005); Solove, *Understanding Privacy*.

[9]Roger Clarke, "Introduction to Dataveillance and Information Privacy, and Definitions of Terms," Xamax Consultancy.

[10]Rachel L. Finn, David Wright, and Michael Friedewald, "Seven Types of Privacy," in *European Data Protection: Coming of Age*, ed. Serge Gutwirth, et al. (Dordrecht: Springer, 2013)., p. 7–9.

4. *Privacy of data and image* includes protecting an individual's data from being automatically available or accessible to other individuals and organisations and that people can "exercise a substantial degree of control over that data and its use".[11]
5. *Privacy of thoughts and feelings* is the right not to share ones thoughts or feelings or to have those thoughts or feelings revealed. Privacy of thought and feelings can be distinguished from privacy of the person, in the same way that the mind can be distinguished from the body.
6. *Privacy of location and space* means that individuals have the right to move about in public or semi-public space without being identified, tracked or monitored. This conception of privacy also includes a right to solitude and a right to privacy in spaces such as the home, the car or the office.
7. *Privacy of association* (including group privacy) is concerned with people's right to associate with whomever they wish, without being monitored.

For the PRISMS survey we have developed a battery of eight questions that are covering all these seven aspects. A factor analysis has shown that all the answers to all these questions are highly correlated and can therefore be grouped into one construct labelled "privacy attitude".[12]

2.2 Operationalization of Security

The concept of security is at least as difficult to approach as privacy. Researchers have stated that the "multidimensional nature of security results in both a society and industry that has no clear understanding of a definition for the concept of security. Moreover the current concepts of security are so broad as to be impracticable".[13]

According to Fischer and Green's reference work "security implies a stable, relatively predictable environment in which an individual or group may pursue its ends without disruption or harm and without fear of such disturbance or injury".[14] Such a definition is also picked up in the context of European policy makers. The European Committee on Standardisation's working group 161 defines that "security is the condition (perceived or confirmed) of an individual, a community, and organisation, a societal institution, a state, and their assets (such as goods,

---

[11]Clarke, "Introduction to Dataveillance and Information Privacy, and Definitions of Terms".

[12]Michael Friedewald et al., "Privacy and Security Perceptions of European Citizens: A Test of the Trade-Off Model," in *Privacy and Identity 2014, IFIP AICT, Vol. 457*, ed. Jan Camenisch, Simone Fischer-Hübner, and Marit Hansen (Heidelberg, Berlin: Springer, 2015).

[13]David J. Brooks, "What Is Security: Definition through Knowledge Categorization," *Security Journal* 23, no. 3 (2009).

[14]Robert J. Fischer and Gion Green, *Introduction to Security*, 7th ed. (Amsterdam, Boston: Butterworth-Heinemann, 2004).

infrastructure), to be protected against danger or threats such as criminal activity, terrorism or other deliberate or hostile acts, disasters (natural and man-made)".[15] Security is thus negatively defined as the absence of insecurity. Perfect objective security thus implies the absence of any threat. Even if this was achieved today it remains open to societal negotiations of new threats in the future.

Over the years this view has partly been replaced by that of risk management and loss prevention. The latter does no longer focus on dangers and hazards and replaces them by risks, which have an (un)certainty to occur. They are based on the assumption that risks cannot totally be prevented and damages and losses will occur anyway. The focus of this approach is no longer directed towards the source of the damage but towards the management of the effects with the goal to minimise the adverse effects for those affected.

It is also difficult to delineate the content of "security". The discourse in the media and among (EU) policy makers is often narrowed down to issues of terrorism, crime and, increasingly, border security. For the general public, organisations, companies and states, however, security is usually much more, including socio-economic conditions, health or cultural security. Therefore we are using a broad definition, in order not to exclude any of these perspectives.

We have identified seven general types of security contexts and the accompanying measures to safeguard and protect these contexts[16]:

1. *Physical security* deals with physical measures designed to safeguard the physical characteristics and properties of systems, spaces, objects and human beings.
2. *Socio-economic security* deals with economic measures designed to safeguard the economic system, its development and its impact on individuals.
3. *Radical uncertainty* security deals with measures designed to provide safety from exceptional and rare violence/threats, which are not deliberately inflicted by an external or internal agent, but can still threaten drastically to degrade the quality of life.
4. *Information security* deals with measures designed to protect information and information systems from unauthorized access, use, disclosure, disruption, modification, perusal, inspection, recording or destruction.
5. *Political security* deals with the protection of acquired rights, established institutions/structures and recognised policy choices.
6. *Cultural security* deals with measures designed to safeguard the permanence of traditional schemas of language, culture, associations, identity and religious practices while allowing for changes that are judged to be acceptable.

---

[15]European Committee on Standardisation (CEN), BT/WG 161, cited in Carlos Martí Sempere, "The European Security Industry: A Research Agenda," (Berlin: German Institute for Economic Research, 2010).

[16]Monica Lagazio, "The Evolution of the Concept of Security," *The Thinker*, September 2012.

7. *Environmental security* deals with measures designed to provide safety from environmental dangers caused by natural or human processes due to ignorance, accident, mismanagement or intentional design, and originating within or across national borders.

It has also to be taken into account that security can be an individual or a collective issue. As in the case of privacy we have designed two batteries of questions to address the wide spectrum of meanings. Again factor analysis has shown—though not as unambiguously as for privacy—that all the items within the two batteries correlated and can therefore be grouped into two constructs labelled "general security" and "personal security" attitudes. Finally we have shown elsewhere that there is no statistically significant correlation between the security and the privacy attitudes.[17]

2.3 Vignettes as a Tool for Contextualisation

To address this ambiguity and context dependence of the central concepts the PRISMS survey is working with so called anchoring vignettes that are used when survey respondents may understand survey questions in different ways, due to the abstractness of the presented concepts (privacy, security), their complexity (security technologies and practices) and because they come from different cultures. Vignettes translate theoretical definitions of complicated concepts in presenting hypothetical situations and asking respondents questions to reveal their perceptions and values.[18]

In PRISMS we have developed eight different vignettes that are covering all seven types of privacy. Since our aim is to scrutinize how citizens assess the implications of specific security technologies our focus is limited to those types of security that are technologically supported, in particular by surveillance-oriented security technologies. This implies that vignettes are mainly covering applications such as the fight against public disorder, criminality and terrorism and additionally some commercial applications. We have also made sure that the vignettes cover virtual as well as physical applications, which are operated by public as well as private sector organisations (see Fig. 1).

---

[17]Friedewald et al., "Privacy and Security Perceptions of European Citizens: A Test of the Trade-Off Model.".

[18]Andrey Pavlov, "Application of the Vignette Approach to Analyzing Cross-Cultural Incompatibilities in Attitudes to Privacy of Personal Data and Security Checks at Airports," in *Surveillance, Privacy, and the Globalization of Personal Information: International Comparisons*, ed. Elia Zureik, et al. (Montreal, Kingston: McGill-Queen's University Press, 2010). Gary King and Jonathan Wand, "Comparing Incomparable Survey Responses: Evaluating and Selecting Anchoring Vignettes" *Political Analysis* 15 (2007).

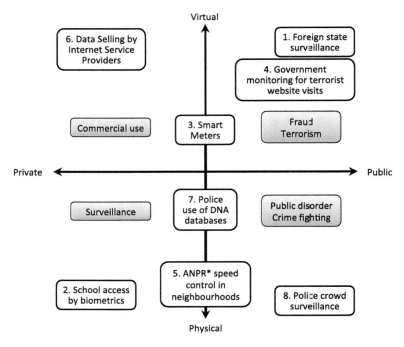

Fig. 1 Classification of the vignettes (*ANPR* Automatic number plate recognition)

The PRISMS vignettes are very short narratives of 50–100 words. They have been validated and refined though sixteen focus groups in eight representative EU countries.[19] In this way it was ensured that the vignettes are understood uniformly in different languages and that they do not cause extreme reactions that would conceal the perceptions to be measured. Two of the vignettes also had (slightly) different wording to test if it makes a difference if citizens assess security practices that are directly affecting them or if they are asked to assess a practice in general (a company selling your data vs. a company selling customers' data). In the vignette about police monitoring crowds the situation was slightly varied, in one version surveillance takes place at a football match while in the other version participants of a political demonstration are monitored. For each of the vignettes citizens were asked two identical questions (the complete set of vignettes and questions can be found in the annex).

- *Question 1:* "To what extent, if at all, do you think that [actors] should or should not [do this]" with answer options on a 5-point Likert-scale ranging from "definitely should" to "definitely should not", and

---

[19]Carolina Haita and Daniel Cameron, "Privacy or Security: A False Choice? European Citizens' Perceptions of Privacy, Personal Data, Surveillance and Security," *Understanding Society* (2014).

- Question 2: "Do you think the [actor] doing this...
  - ... helps to protect people's rights and freedoms
  - ... threatens people's rights and freedoms
  - ... has no impact on people's rights and freedoms"
  - (don't know)

2.4 Data Collection

Fieldwork took place between February and June 2014. The survey company Ipsos MORI conducted around 1000 telephone interviews in each EU member states except Croatia[20] (27,195 in total) amongst a representative sample (based on age, gender, work status and region) within each country (see Table 1). For economic reasons each interviewee was presented only four randomly selected vignettes, resulting in approx. 13,600 responses for each vignette (500 per country).[21]

Table 1 Survey composition

		Responses	Percent (%)
Total		27.195	100
Gender	Male	12.566	46
	Female	14.629	54
Age	16–24	2.793	10
	25–34	4.006	15
	35–44	4.704	17
	45–54	4.960	18
	55–59	2.435	9
	60–64	2.305	8
	65–74	3.643	13
	75+	2.294	8
Work status	Working	13.775	51
	Unemployed or in education	5.788	21
	Retired	7.209	27
Geographic area[a]	Big city	6.535	24
	Suburban area or small city	12.833	47
	Rural area	7.748	28

[a]Self-assessment by interviewees (answer categories: 1a big city; 2a suburbs or outskirts of a big city, 2b town or small city, 3a country village, 3b farm or home in the countryside)

---

[20]Croatia had not acceded to the EU at the time of the project planning.

[21]For those vignettes with alternative wording the sample was halved again to 6800 responses in total or 250 responses per country.

3 Descriptive Results

In an introductory question we asked citizens to what extent they think that an institutional actor should or should not implement the given security practice. The answers were measured on a 5-point Likert-scale ranging from "definitely should" to "definitely should not" (see Fig. 2).

About half of the vignettes produced a rather clear positive or negative assessment. For instance more than two thirds of the respondents agreed that "Police surveilling football match", "ANPR speed control in neighbourhoods" and "Monitoring terrorist website visits" should be used to protect security. On the other side of the spectrum more than 80 % of the respondents thought that "ISPs selling customer data" should not take place.

The rest of the vignettes, however, did not produce equally clear results. While still a majority of respondents were in favour of "Police surveillance at demonstrations" and against "Foreign state surveillance" the remaining three vignettes had about as many supporters as opponents. Especially the usage of smart meter data did not only have almost as many positive as negative votes, it also had the highest number of undecided respondents.

In the second question citizens were asked if the practice described in a vignette is having an impact of people's rights and freedoms. This question was intentionally formulated in general terms since security practices do not only have an impact on the right to data protection but can also include other rights and freedoms such as freedom of decision, freedom of association or freedom of movement that citizens would not automatically regard as part of an extended privacy concept that is used in the PRISMS survey (cf. Sect. 2.1).

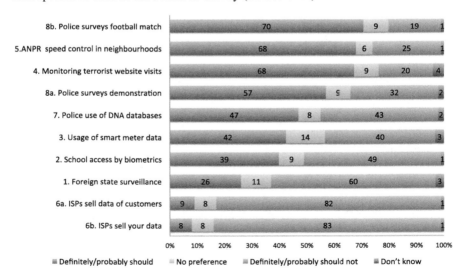

Fig. 2 Question 1. To what extent, if at all, do you think that [an institution] should or should not…?

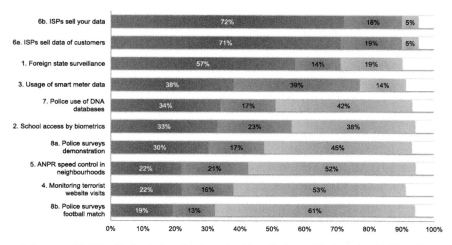

Fig. 3 Question 2. Do you think the _____ doing this (impacts people's rights and freedoms)?

Again the survey shows a wide spectrum of differences between the vignettes though the results are more evident than in the previous question (see Fig. 3). The first observation is that in all but one cases only a minority of the respondents (between 12 and 24 %) think that the described practice is helping to protect people's rights and freedoms.

On the other side of the spectrum citizens think that the practices describe in the vignettes about "ISPs selling customer data" are having the biggest negative impact on their personal freedoms. Many other vignettes, however, are not regarded as very momentous. For these vignettes only a minority (between 19 and 34 %) see a threat for people's rights and freedoms. A particularly interesting case is again the vignette on smart meters with almost 40 % of positive and 40 % of negative answers. It shows that there is no societal consensus about this technology yet and that citizens often do not trust that private companies operate responsibly towards their customers' interests (from those interviewees that think that energy companies should not collect smart meter data almost 50 % also stated that they do not trust businesses). Finally the assessment of "foreign state surveillance" shows that a majority of citizens feels this practice is harming, and only 14 % think it is helping protect people's freedom. With 10 % this vignette is having the highest number of "don't know" answers.

All in all, European citizens are rather critical of surveillance oriented security measures in the sense that their assessment differs widely depending on the context, purpose and implementation of a specific measure.

Already on the basis of the descriptive statistics it becomes clear that there is a distinction between security technologies and practices operated by public and private sector institutions. Even in spite of the obscure role that European authorities (mainly intelligences services) have played in the NSA spying scandal citizens still have more trust that public authorities do respect their rights to privacy and data protection rather than profit-oriented companies (which are often branches of multinational corporations).

The figures also show that citizens are especially critical with regard to purely virtual forms of surveillance. There is opposition against covert surveillance practices and secondary use or disclosure of data, especially for commercial purposes.

4 Determinant of Citizen's Acceptance of Specific Surveillance Oriented Security Technologies

Elsewhere we have already demonstrated that there is no strong correlation between the privacy and security constructs and thus no simple trade-off between security and privacy attitudes of European citizens.[22] In the following section we are presenting the analysis of a selection of factors that determine citizens' assessment of the systems/practices outlined in the vignettes. It makes clear that there is no simple impact of specific factors in the assessment of concrete cases of security technologies and surveillance practices.

4.1 Methodology

To answer the research questions and to empirically test our theoretical assumptions we conducted a series of ordered logistic regressions. The introductory question for each vignette was defined as the dependent variable and regressed by two different sets of independent variables. The list of independent categorical variables included 'age', 'gender', 'education', 'political orientation', 'level of privacy activism'[23] and 'experience with privacy infringements'.[24] For the virtual surveillance cases we used 'intensity of Internet use' as an additional variable. For the physical surveillance cases 'work status' and 'living environment' were added.[25]

---

[22]Friedewald et al., "Privacy and Security Perceptions of European Citizens: A Test of the Trade-Off Model."

[23]To measure "privacy activism" we asked citizens if they had actively taken steps to protect their personal information. Answer categories included: refuse to give information, ask company to delete information, ask company not to disclose information, deliberatively give incorrect information etc. Citizens who answered they had taken at least two of the given possibilities were considered as "privacy active".

[24]Some of these items are, however, themselves determined by basic control variables (e.g. age or education level).

[25]Consequently the respective cells in the table are empty (n/a).

Table 2 is a simple overview of our results. The first column shows the independent variables. The cells indicate the results of the regression analysis for all 8 (10) vignettes. We have not used numerical results as these would not be directly comparable across vignettes. Instead, we have opted for a visual indication of the direction and strength of the correlation:

- A single plus mark (+) indicates that respondents to which the marked factor applies are up to twice as likely to accept the respective surveillance practice as the reference group.
- A double plus mark (++) indicates that respondents to which the marked factor applies are more than twice as likely to accept the respective surveillance practice as the reference group.
- A single minus mark (−) indicates that that respondents to which the marked factor applies are between half as likely and as likely, to accept the respective surveillance practice as the reference group.
- The double minus mark (−−) indicates that respondents to which the marked factor applies are less than half as likely to accept the respective surveillance practice as the reference group.
- Finally, a grey cell shading indicates of a correlation of statistical significance ($p < 0.05$). The variables relating to the cells shaded grey can be assumed to have a "significant influence" on the vignette.[26]

For instance: The likelihood that young adults (16–34) assess the practice of data selling by ISPs positively ("Actors should be doing this") has a [++] and is thus more than twice as high as for the reference group of the older adults (60+). For middle age (34–59) this likelihood [+] is still higher than for reference group of older adults but lower than for the young adults. In summary this means that the younger people are the more positive they are about this practice.

4.2 Results

The analysis shows that there are only a few factors, which play an important role in all cases. Not surprisingly these include the constructs describing citizens' *privacy and security attitudes*. Firstly in most cases there is a strong positive correlation between worries about personal security and support for a security practice. The support is stronger for the cases of physical surveillance than for virtual surveillance practices, which means that people tend to accept security practices when they come close to personal concerns, are understandable and do not affect them personally. Secondly there is an even stronger correlation between privacy worries and the non-acceptance of a security practice.

---

[26]More details on methodology can be found in Michael Friedewald et al., "Report on the Analysis of Survey Results," (2015).

Table 2 Factors influencing the personal perception of different security practices

	Public	Virtual surveillance ------->		Private		Public	Physical surveillance ------->			Private
Vignette Factor	1. Foreign government surveillance	4. Monitoring terrorist website visits	3. Usage of smart meter data	6b. ISPs sell your data	6a. ISPs sell data of customers	8b. Police surveys football match	8a. Police surveys demonstration	5. ANPR speed control	7. Police use of DNA databases	2. School access by biometrics
Avg. acceptance level	3,6	2,28	2,92	4,27	4,22	2,28	2,72	2,17	2,97	3,26
More worried about personal security	+	++	+	++	+	++	++	++	++	++
More worried about privacy	--	-	--	-	--	--	--	--	--	--
High trust in institutions	+	+	++	++	++	+	++	++	+	+
Young adults (16-34)[a]	+	-	++	++	++	-	-	+	+	+
Middle age (35-59)[a]	+	-	+	+	+	-	-	+	+	+
Male[b]	-	-	+	+	+	+	+	-	+	-
Secondary education 1†[c]	+	+	-	+	+	+	+	+	+	+
Secondary education 2 and post secondary education†[c]	+	+	-	+	+	+	+	+	+	+
Unemployed and in education[d]	n/a	n/a	n/a	n/a	n/a	-	+	-	-	+
No Internet use[e]	+	+	+	n/a	++	n/a	n/a	n/a	n/a	-
Occasional Internet use‡[e]	-	-	-	+	+	n/a	n/a	n/a	n/a	-
Living in big cities[f]	n/a	n/a	n/a	n/a	n/a	+	+	n/a	-	+
Living in suburbs/small cities[f]	n/a	n/a	n/a	n/a	n/a	+	-	-	-	-
Political left[g]	-	-	-	-	-	-	--	-	-	-
Political center[g]	-	-	-	-	-	-	-	-	-	-
No privacy activism[h]	+	+	+	+	+	-	+	+	+	+
Low privacy activism[h]	+	-	+	+	-	+	+	-	-	+
Privacy never invaded	-	-	-	+	-	+	-	+	-	+

▲ **Avg. acceptance level**: Scale from 1 (definitely should) to 5 (definitely should not)
Correlations and significance: ++ strong positive correlation/+ positive correlation/– negative correlation/– – strong negative correlation. *Grey-shaded cells indicate a significant correlations ($p < 0.05$)*
Reference groups: [a]Old age (60+), [b]female, [c]higher education (ISCED level > 5), [d]retired, [e]regular internet use (once per week or more), [f]rural area (country villages, farms or countryside), [g]political right, [h]high privacy activism (people who have actively protected their privacy more than one time)
†Secondary education stage 1: ISCED-11 levels 1–2; secondary education stage 2 and post secondary education: ISCED-11 levels 3–5
‡Occasional Internet use: Less than once per week

The third factor that has a significant positive correlation with citizens' support for a security practice is their *trust in institutions*.[27] It is clearly visible that the perceived trustworthiness of an authority, organisation or company operating a security system has a positive effect on citizens' acceptance. This supports recent discussions about the importance of trust for the assessment of risks and benefits and the acceptability of technologies.[28] According to these discussions trust reduces the complexity people need to face. Instead of making rational judgements based on knowledge, trust is employed to select actors who are trustworthy and whose opinions can be considered accurate and reliable. People having trust in the authorities and management responsible for the technology perceive less risk than people who lack that sense of trust in those members, although some studies seem to suggest that this is not always the case.[29]

Other factors do not show an equally clear picture and are more difficult to interpret, either because the correlations with the assessment of the vignettes are not always statistically significant or even have effects in different directions.

Gender for instance has a significantly positive correlation in three and a significantly negative correlation in four of the cases. Men tend to reject surveillance practices by public authorities more than those of private sector. This is in line with the fact that, according to our survey, men have less trust in public authorities than in public sector and less trust in institutions in general than women.

Age is an interesting factor inasmuch as it has been recently shown that the younger generation is not generally valuing privacy differently from older citizens.[30] The assumption that this also leads to a more critical assessment of surveillance practices by youngsters is not supported by the survey results. Rather, the likelihood that young adults (16–34) found a surveillance practice acceptable is higher than that of middle-aged people and much higher than that of older citizens. This correlation, however, is not significant for all the vignettes. Young adults only found the monitoring of websites in search of terrorists a less acceptable practice.

---

[27]The importance of trust (or distrust) has been a familiar factor in explaining privacy attitudes since the earliest surveys by Alan Westin. See for instance: Susannah Fox et al., "Trust and Privacy Online: Why Americans Want to Rewrite the Rules," (Washington, DC: Pew Internet & American Life Project, 2001); Stephen T. Margulis, Jennifer A. Pope, and Aaron Lowen, "The Harris-Westin Index of General Concern About Privacy: An Exploratory Conceptual Replication," in *Surveillance, Privacy, and the Globalization of Personal Information: International Comparisons*, ed. Elia Zureik, et al. (Montreal, Kingston: McGill-Queen's University Press, 2010); Wainer Lusoli et al., *Pan-European Survey of Practices, Attitudes and Policy Preferences as Regards Personal Identity Data Management*, JRC Scientific and Policy Report EUR 25295 (Luxembourg: Publication Office of the European Union, 2012).

[28]Timothy C. Earle and George Cvetkovich, *Social Trust: Toward a Cosmopolitan Society* (Westport, Conn.: Praeger, 1995).

[29]Richard J. Bord and Robert E. O'Connor, "Determinants of Risk Perceptions of a Hazardous Waste Site," *Risk Analysis* 12 (1992).

[30]Mary Madden et al., "Teens, Social Media, and Privacy," (Washington, DC: Pew Research Center, 2013); Wainer Lusoli et al., "Young People and Emerging Digital Services: An Exploratory Survey on Motivations, Perceptions and Acceptance of Risks," (Luxembourg: Office for Official Publications of the European Communities, 2009).

Based on qualitative research Pavone et al. and others suggest that a possible explanation might be that older citizens who made experience with European authoritarian regimes, are more distrustful, whereas younger people, who had not lived in surveillance states are less concerned.[31]

In general the survey has shown that the *educational level* is positively correlated with the valuation of privacy and negatively correlated with the valuation of security. In concrete cases, however, education only seems to have a weak influence on the acceptance of a surveillance measure. For most of our vignettes one can state that the higher the education the less likely it is that one is willing to accept a surveillance practice. This indicates that the more knowledge and understandings of the context people have the more critical they are. These observations, however, are only significant in some of the cases. This is an interesting complement to the findings about privacy since people with a higher education have a significantly higher appreciation for their privacy than those with an intermediate or low level of education.

It has sometimes been suggested that people living in big cities are more worried about their security and thus more supportive to physical security measures than citizens' living in small cities, suburbs or even in rural areas. Our survey results do not fully confirm this hypothesis. Residents of big cities are only significantly more supportive to vignette on "school access by biometrics". On the other side, their support for the police use of DNA databases is significantly lower. For all other cases we could not show a significant correlation. The situation is similarly mixed for smaller cities and suburbs. It is in line with the observation that the people least in danger are most afraid.[32] More important than the fear of crime seems to be the perceived usefulness and effectiveness of concrete measures.[33]

Political orientation has a weak effect on the assessment. Citizens with a left wing or liberal orientation are less likely to accept surveillance than those who consider themselves conservatives or right wing.

It could not be shown that work status, intensity of Internet use and experience with privacy infringements is influencing citizens' assessment of surveillance based security technologies in a significant way.

In summary one can say that people who are not worried at all about being monitored (do not mind being under surveillance), have lower education, are relatively young, and prefer conservative over liberal thinking.

---

[31]Vincenzo Pavone, Sara Degli Esposti, and Elvira Santiago, "Key Factors Affecting Public Acceptance and Acceptability of SOSTs," (The SurPRISE consortium, 2015), p. 139. Iván Székely, "Changing Attitudes in a Changing Society? Information Privacy in Hungary, 1989–2006," in *Surveillance, Privacy, and the Globalization of Personal Information: International Comparisons*, ed. Elia Zureik, et al. (Montreal, Kingston: McGill-Queen's University Press, 2010).

[32]Kristof Verfaillie et al., "Public Assessments of the Security/Privacy Trade-Off: A Criminological Conceptualization," (PRISMS Project, 2013).

[33]Kreissl et al., "Surveillance: Preventing and Detecting Crime and Terrorism."

5 Discussion of Results and Conclusions

Our analysis of the questions that aimed to measure European citizens' attitudes towards specific examples of surveillance technologies and practices had the following main results:

- Trust in the operating institution is the essential factor for the acceptability of a security practice. The important role of trust, in people, institutions as well as the whole societal environment, is regularly confirmed in surveys.[34] The SurPRISE project, for instance, confirmed clearly that "the more people trust scientific and political institutions ... the more acceptable a technology would be." In their explanatory model institutional trust is the strongest positive influence factor for acceptability of surveillance oriented security technologies.[35] The PACT project on the other side stresses the strong impact that distrust has on the likelihood that citizens' reject a given security measure.[36]
- Openness has a positive effect on the willingness of citizens to accept security practices. This can be understood on different levels:
  - Citizens tend to accept security practices when they are convinced that a security measure is necessary, proportional and effective.
  - This is easier when a security practice is embedded in a context that citizens are familiar with and where they understand who is surveying whom and how.
  - As a result the surveillance activity should not be covert but perceivable for the citizen and communicated in a responsible way by the operator.
  - Understanding and acceptance is also a question of proper knowledge and education—though not only in one way. While education contributes to understanding technicalities and complexity of a security practices it also drives critical reflections.[37]

---

[34]Baldo Blinkert, "Unsicherheitsbefindlichkeit als ‚sozialer Tatbestand'. Kriminalitätsfurcht und die Wahrnehmung von Sicherheit und Unsicherheit in Europa," *Monatsschrift für Kriminologie und Strafrechtsreform* 93, no. 2 (2010); Fox et al., "Trust and Privacy Online: Why Americans Want to Rewrite the Rules."; Dina Hummelsheim, "Subjektive Unsicherheit und Lebenszufriedenheit in Deutschland: Empirische Ergebnisse einer repräsentativen Bevölkerungsbefragung," in *Sichere Zeiten? Gesellschaftliche Dimensionen der Sicherheitsforschung*, ed. Peter Zoche, Stefan Kaufmann, and Harald Arnold (Münster: Lit Verlag, 2015).

[35]Pavone, Esposti, and Santiago, "Key Factors Affecting Public Acceptance and Acceptability of SOSTs," p. 135–136.

[36]Sunil Patil et al., "Public Perception of Security and Privacy: Results of the Comprehensive Analysis of PACT's Pan-European Survey," (Cambridge, UK: RAND Corporation, 2014), p. v.

[37]SurPRISE also confirmed most these observations. Cf. Pavone, Esposti, and Santiago, "Key Factors Affecting Public Acceptance and Acceptability of SOSTs," p. 154–155.

- All these factors also involve an inherent risk for manipulation, since a security practice can be designed to create false trust among citizens to be accepted.
- On the downside it can also be stated that many citizens do not care about surveillance that does not negatively affect them personally but only others.[38]

For the design and introduction of security measures it is useful to consider some of the main determinants, since poorly designed measures can consume significant resources without achieving either security or privacy while others can increase security at the expense of privacy. However, since there is no natural trade-off between privacy and security, carefully designed solutions can benefit both privacy and security.

Law enforcement and government officials often heavily weight security.[39] On the other hand we have shown in our analysis of the vignettes that citizens' opinions on security measures vary, and are influenced by some crucial factors. Apart from trust in the operating agency or company we could observe mainly four different types of reactions (cf. Fig. 4):

- Citizens may consider a measure as useless to enhance security, and at the same time invasive for their privacy (Quadrant 1). Such a situation has to be absolutely avoided.
- Citizens may consider a measure useless to enhance security but with no risk for their privacy (Quadrant 3)
- Citizens may consider a measure as useful in terms of security, but privacy invasive (Quadrant 2).
- Finally, citizens may consider a measure both useful to increase security and with no risk for their privacy (Quadrant 4).

As Fig. 4 classifies citizens' reactions it does not (always) have to reflect the real effectiveness of a security measure and its real impact on privacy. Considering the importance of trust for the acceptability and acceptance the responsible parties should aim to reconcile the perceived and real impacts. Potential for conflicts can be mainly found at the border to quadrants 2 and 3 where citizens fear an invasion of their privacy or perceive a technology as ineffective.[40] The diagram represents a logical ordering of responses of citizens towards security and privacy measures rather than being the result of empirical study. These responses are generally based

---

[38]SurPRISE concludes "the more participants perceive SOSTs to be targeted at others rather than themselves, the more likely they are to find a SOST more acceptable". Ibid., p. 138.

[39]It is quite telling that in the most recent Eurobarometer study on Europeans' attitudes towards security the focus is strongly on terrorism, cybercrime, organized crime and insecurity of the EU's external borders trustworthiness of security agencies and their measures are not even mentioned. Cf. TNS Opinion & Social, "Europeans' Attitudes Towards Security," (Brussels, 2015).

[40]Pavone, Esposti, and Santiago, "Key Factors Affecting Public Acceptance and Acceptability of SOSTs."; Gregory Conti, Lisa Shay, and Woodrow Hartzog, "Deconstructing the Relationship between Privacy and Security," *IEEE Technology and Society Magazine* 33, no. 2 (2014).

Fig. 4 Mapping of the perceived risk-benefit for privacy and security (based on Conti et al. 2014)

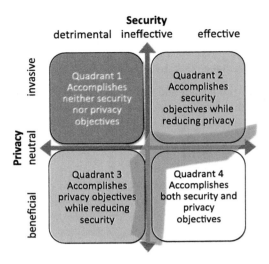

upon perceptions rather than rational assessments. They are influenced by a multitude of factors as we found them in the survey. Trust in institutions is one, the perceived self-interest is another, the measure being overt or covert a potential third. In that sense the diagram quite neatly presents a kind of heuristics for elaborating decision-making processes that try to overcome the barriers we sketch (the boundaries between the quadrants).

Especially for these cases PRISMS has developed a participatory and discursive technique[41] that can help decision-makers in industry, public authorities and politics to implement security measures, which raise fewer concerns in the population and are thus more acceptable along the lines stated in many policy documents.[42]

Acknowledgments This work was carried out in the project "PRISMS: Privacy and Security Mirrors" co-funded from the European Union's Seventh Framework Programme for research, technological development and demonstration under grant agreement 285399. For more information see: http://prismsproject.eu

---

[41]As the empirical basis PRISMS has defined a structural model that describes the relationship of the main constructs in greater detail. This will be a translation of the theory of planned behaviour into a survey based empirical model. Cf. Friedewald et al., "Report on the Analysis of Survey Results."; Marc van Lieshout, Anne Fleur van Veenstra, and David Barnard-Wills, "The PRISMS Decision Support System," (2015).

[42]The most notable is maybe the European Union's "Stockholm programme" that states "[t]he challenge will be to ensure respect for fundamental freedoms and integrity while guaranteeing security in Europe" European Council, "The Stockholm Programme—an Open and Secure Europe Serving and Protecting the Citizens," *Official Journal of the European Union*, 4.5.2010 2010, p. 4.

Appendix: The Vignettes

1. *Foreign government surveillance*

   An international disaster relief charity has been sending a monthly newsletter by email to its supporters. The people who run the charity find out through the media that a foreign government has been regularly capturing large amounts of data on citizens of other countries by monitoring their emails. The foreign government says it needs to monitor some communications to help keep its citizens safe and that the main purpose is to focus on terrorism. The charity's officials are unsure whether this means their supporters' personal information is no longer confidential.

2. *School access by biometrics*

   At a local primary school a new system for getting into the school has been installed. All pupils, teachers, parents, other family members and other visitors have to provide their fingerprints on an electronic pad to identify themselves in order to enter or leave the school.

3. *Usage of smart meter data*

   A power company has decided to offer smart meters to all its consumers. Smart meters enable consumers to use energy more efficiently by allowing them to see how much they are using through a display unit. The data recorded by smart meters allows power companies to improve energy efficiency and charge lower costs. They also enable power companies to build up a more detailed picture of how their customers use energy. It also enables the companies to find out other things, like whether people are living at the address, or how many people are in the household.

4. *Monitoring terrorist website visits*

   A student is doing some research on extremism and as part of his work he visits websites and online forums that contain terrorist propaganda. When his parents find out they immediately ask him to stop this type of online research because they are afraid security agencies such as the police or anti-terrorism bodies will find out what he has been doing and start to watch him.

5. *Speed control in neighbourhoods by ANPR*

   Michael lives in a suburban neighbourhood, where his children like to play outside with their friends. However, his street is a short cut for commuters who drive faster than the speed limit. In response to complaints from residents, the local authority decides to install automatic number plate recognition (ANPR) systems, which identify and track all vehicles and calculate their average speed. This allows those who drive too fast to be prosecuted.

6. *ISP Data*

Companies offering services on the Internet want to sell information about [(a) your (b) their customers] Internet use to advertisers and other service providers so the information can be used to create more personal offers and deals. This would include the searches you conduct and the websites you visit. Your provider says the information they sell will be anonymous.

7. Use of DNA databases by police

James voluntarily provided a sample of his DNA to a company that carries out medical research. DNA contains the genetic pattern that is uniquely characteristic to each person. He then learns that the research company has been asked to disclose all their DNA samples to police for use in criminal investigations. Samples of DNA can be used to understand potential health problems but also to identify people and to make inferences about who they are related to.

8. *Crowd surveillance by police*

Version a "Demonstration": Claire is an active member of an environmental group, and is taking part in a demonstration against the building of a new nuclear plant. The police monitor the crowd in various ways to track and identify individuals who cause trouble: they use uniformed and plain-clothes police, CCTV, helicopters and drones, phone tapping, and try to find people on social media.

Version b "Football": David is a football fan who regularly attends home matches. The police monitor the crowd in various ways to track and identify individuals who cause trouble: through uniformed police and plain-clothes police, CCTV, by using helicopters and drones, tapping phones, and by trying to find people on social media.

Bibliography

Ajzen, Icek, and Martin Fishbein. 2005. The influence of attitudes on behavior. In *The handbook of attitudes*, ed. Dolores Albarracin, Blair T. Johnson, and Mark P. Zanna, 173–221. Mahwah, NJ: Erlbaum.

Blinkert, Baldo. 2010. Unsicherheitsbefindlichkeit als, sozialer Tatbestand'. Kriminalitätsfurcht und die Wahrnehmung von Sicherheit und Unsicherheit in Europa. *Monatsschrift für Kriminologie und Strafrechtsreform* 93(2): 106–125.

Bord, Richard J., and Robert E. O'Connor. 1992. Determinants of risk perceptions of a hazardous waste site. *Risk Analysis* 12: 411–416.

Brooks, David J. 2009. What is security: Definition through knowledge categorization. *Security Journal* 23(3): 225–239.

Clarke, Roger. 2006. Introduction to dataveillance and information privacy, and definitions of terms. Chapman, Australia: Xamax Consultancy, 7 Aug 2006. http://www.rogerclarke.com/DV/Intro.html.

Conti, Gregory, Lisa Shay, and Woodrow Hartzog. 2014. Deconstructing the relationship between privacy and security. *IEEE Technology and Society Magazine* 33(2): 28–30 (Summer).

Cook, Andrew J., Kevin Moore, and Gary D. Steel. 2005. Taking a position: A reinterpretation of the theory of planned behaviour. *Journal for the Theory of Social Behaviour* 35(2): 143–154.

Davis, Fred D. 1989. Perceived usefulness, perceived ease of use, and user acceptance of information technology. *MIS Quarterly* 13(3): 319–339.

Durkheim, Emile. 1982. *The rules of sociological method [1895]* (Steven Lakes, Trans.). New York: The Free Press.

Earle, Timothy C., and George Cvetkovich. 1995. *Social trust: Toward a cosmopolitan society.* Westport, CT.: Praeger.

European Council. 2010. The Stockholm Programme—An open and secure Europe serving and protecting the citizens. *Official Journal of the European Union* C115: 1–38 (4 May 2010).

Finn, Rachel L., David Wright, and Michael Friedewald. 2013. Seven types of privacy. In *European data protection: coming of age*, ed. Serge Gutwirth, Ronald Leenes, Paul De Hert, and Yves Poullet, 3–32. Dordrecht: Springer.

Fischer, Robert J., and Gion Green. 2004. *Introduction to security*, 7th ed. Amsterdam, Boston: Butterworth-Heinemann.

Fox, Susannah, Lee Rainie, John Horrigan, Amanda Lenhart, Tom Spooner, and Cornelia Carter. 2001. Trust and privacy online: Why Americans want to rewrite the rules. Washington, DC: Pew Internet & American Life Project. http://www.pewinternet.org/files/old-media//Files/Reports/2000/PIP_Trust_Privacy_Report.pdf.pdf.

Friedewald, Michael, Marc van Lieshout, Sven Rung, Merel Ooms, and Jelmer Ypma. 2015. Privacy and security perceptions of European citizens: A test of the trade-off model. In *Privacy and identity 2014, IFIP AICT* (Vol. 457), ed. Jan Camenisch, Simone Fischer-Hübner and Marit Hansen (pp. 39–53). Heidelberg, Berlin: Springer.

Friedewald, Michael, Marc van Lieshout, Sven Rung, Merel Ooms, and Jelmer Ypma. 2015. Report on the analysis of the PRISMS survey. PRISMS Project, Deliverable 10.1. http://prismsproject.eu.

Haita, Carolina, and Daniel Cameron. 2014. Privacy or security: A false choice? European citizens' perceptions of privacy, personal data, surveillance and security. *Understanding Society* (July 2014): 12–16.

Hummelsheim, Dina. 2015. Subjektive Unsicherheit und Lebenszufriedenheit in Deutschland: Empirische Ergebnisse einer repräsentativen Bevölkerungsbefragung. In *Sichere Zeiten? Gesellschaftliche Dimensionen der Sicherheitsforschung*, ed. Peter Zoche, Stefan Kaufmann, and Harald Arnold, 67–89. Münster: Lit Verlag.

Kasper, Debbie V. S. 2005. The evolution (or devolution) of privacy. *Sociological Forum* 20(1): 69–92.

King, Gary, and Jonathan Wand. 2007. Comparing incomparable survey responses: Evaluating and selecting anchoring vignettes. *Political Analysis* 15: 46–66.

Kollmuss, Anja, and Julian Agyeman. 2002. Mind the gap: Why do people act environmentally and what are the barriers to pro-environmental behavior? *Environmental Education Research* 8(3): 239–260.

Kreissl, Reinhard, Clive Norris, Marija Krlic, Leroy Groves, and Anthony Amicelle. 2015. Surveillance: Preventing and detecting crime and terrorism. In *Surveillance in Europe*, ed. David Wright, and Reinhard Kreissl, 150–210. London, New York: Routledge.

Lagazio, Monica. 2012. The evolution of the concept of security. *The Thinker* 43(9): 36–43.

Lusoli, Wainer, Margherita Bacigalupo, Francisco Lupiañez, Norberto Andrade, Shara Monteleone, and Ioannis Maghiros. 2012. *Pan-European survey of practices, attitudes and policy preferences as regards personal identity data management*. JRC Scientific and Policy Report EUR 25295, Luxembourg: Publication Office of the European Union.

Lusoli, Wainer, Caroline Miltgen, Ramón Compañó, and Ioannis Maghiros. 2009. *Young people and emerging digital services: An exploratory survey on motivations, perceptions and acceptance of risks*. IPTS Technical Report EUR 23765 EN, Luxembourg: Office for Official Publications of the European Communities.

Madden, Mary, Amanda Lenhart, Sandra Cortesi, Urs Gasser, Maeve Duggan, Aaron Smith, and Meredith Beaton. 2013. Teens, social media, and privacy. Washington, DC: Pew Research Center. http://pewinternet.org/Reports/2013/Teens-Social-Media-And-Privacy.aspx.

Margulis, Stephen T., Jennifer A. Pope, and Aaron Lowen. 2010. The Harris-Westin index of general concern about privacy: An exploratory conceptual replication. In *Surveillance, privacy, and the globalization of personal information: International comparisons*, ed. L. Elia Zureik, Lynda Harling Stalker, Emily Smith, David Lyon, and Yolande E. Chan, 91–109. Montreal, Kingston: McGill-Queen's University Press.

Martí Sempere, Carlos. 2010. *The European security industry: A research agenda*. Economics of Security Working Paper 29. Berlin: German Institute for Economic Research, February 2010.

Patil, Sunil, Bhanu Patruni, Hui Lu, Fay Dunkerley, James Fox, Dimitris Potoglou, and Neil Robinson. 2014. Public perception of security and privacy: Results of the comprehensive analysis of PACT's Pan-European Survey. PACT Project, Deliverable 4.2, Cambridge, UK: RAND Corporation, 2014. http://www.rand.org/content/dam/rand/pubs/research_reports/RR700/RR704/RAND_RR704.pdf.

Pavlov, Andrey. 2010. Application of the vignette approach to analyzing cross-cultural incompatibilities in attitudes to privacy of personal data and security checks at airports. In *Surveillance, privacy, and the globalization of personal information: International comparisons*, ed. L. Elia Zureik, Lynda Harling Stalker, Emily Smith, David Lyon, and Yolande E. Chan, 31–45. Montreal, Kingston: McGill-Queen's University Press.

Pavone, Vincenzo, and Sara Degli Esposti. 2012. Public assessment of new surveillance-oriented security technologies: beyond the trade-off between privacy and security. *Public Understanding of Science* 21(5): 556–572.

Pavone, Vincenzo, Sara Degli Esposti, and Elvira Santiago. 2015. Key factors affecting public acceptance and acceptability of SOSTs. SurPRISE Project, Deliverable 2.4. http://surprise-project.eu/wp-content/uploads/2015/02/SurPRISE-D24-Key-Factors-affecting-public-acceptance-and-acceptability-of-SOSTs-c.pdf.

Solove, Daniel J. 2008. *Understanding Privacy*. Cambridge, Mass.: Harvard University Press.

Székely, Iván. 2010. Changing attitudes in a changing society? Information privacy in Hungary, 1989–2006. In *Surveillance, privacy, and the globalization of personal information: International comparisons*, ed. L. Elia Zureik, Lynda Harling Stalker, Emily Smith, David Lyon, and Yolande E. Chan, 150–170. Montreal, Kingston: McGill-Queen's University Press.

TNS Opinion & Social. 2015. *Europeans' attitudes towards security*. Special Eurobarometer 432, Brussels.

van Lieshout, Marc, Anne Fleur van Veenstra, and David Barnard-Wills. 2015. The PRISMS decision support system. PRISMS Project, Deliverable 10.2. http://prismsproject.eu.

Venkatesh, Viswanath, Michael G. Morris, Gordon B. Davis, and Fred D. Davis. 2003. User acceptance of information technology: Toward a unified view. *MIS Quarterly* 27(3): 425–478.

Verfaillie, Kristof, Evelien van den Herrewegen, Jenneke Christiaens, and Serge Gutwirth. 2013. Public assessments of the security/privacy trade-off: A criminological conceptualization. PRISMS Project, Deliverable 4.1. http://prismsproject.eu.

On Locational Privacy in the Absence of Anonymous Payments

Tilman Frosch, Sven Schäge, Martin Goll and Thorsten Holz

Abstract In this paper we deal with the situation that in certain contexts vendors have no incentive to implement anonymous payments or that existing regulation prevents complete customer anonymity. While the paper discusses the problem also in a general fashion, we use the recharging of electric vehicles using public charging infrastructure as a working example. Here, customers leave rather detailed movement trails, as they authenticate to charge and the whole process is post-paid, i.e., are billed after consumption. In an attempt to enforce transparency and give customers the information necessary to dispute a bill they deem inaccurate, Germany and other European countries require to retain the ID of the energy meter used in each charging process. Similar information is also retained in other applications, where Point of Sales terminals are used. While this happens in the customers' best interest, this information is a location bound token, which compromises customers' locational privacy and thus allows for the creation of rather detailed movement profiles. We adapt a carefully chosen group signature scheme to match these legal requirements and show how modern cryptographic methods can reunite the, in this case, conflicting requirements of transparency on the one hand and locational privacy on the other. In our solution, the user's

---

T. Frosch (✉) · S. Schäge · M. Goll · T. Holz
Horst Görtz Institute for IT-Security, Ruhr-University Bochum, Bochum, Germany
e-mail: tilman.frosch@rub.de

S. Schäge
e-mail: sven.schaege@rub.de

M. Goll
e-mail: martin.goll@rub.de

T. Holz
e-mail: thorsten.holz@rub.de

© Springer Science+Busines Media Dordrecht 2016
S. Gutwirth et al. (eds.), *Data Protection on the Move*,
Law, Governance and Technology Series 24, DOI 10.1007/978-94-017-7376-8_4

identity is explicitly known during a transaction, yet the user's *location* is concealed, effectively hindering the creation of a movement profile based on financial transactions.

Keywords Locational privacy · Big data and profiling · Privacy enhancing/friendly technologies · Security and privacy by design · Electric vehicle charging infrastructure · Point of sales infrastructure

1 Introduction

Blumberg and Eckersley define locational privacy as "the ability of an individual to move in public space with the expectation that under normal circumstances their location will not be systematically and secretly recorded for later use".[1] In a world of Big Data, where any fact about an individual's life, once revealed, will potentially be stored indefinitely, it is important to limit the data that is created or revealed in the first place. While completely anonymous systems would be desirable in many cases from a customer's side, legitimate business interests on the side of the vendor may prevent the adoption of a technical solution that relies on complete anonymity. In the context of financial transactions, the prevalent academic approach to protecting a user's locational privacy is to protect the user's *identity* and thus indirectly conceal their location. Various anonymous electronic cash (e-cash) schemes have been published[2] since Chaum published his seminal paper *Blind Signatures for Untraceable Payments*[3] in 1982. However, none has been (widely) adopted. Besides posing technical hurdles, e-cash often makes it hard for the vendor to walk the established path of resolving a dispute with a customer on front of a court of law, as the customer is not known—although many schemes reveal the customer's identity in case of double spending, but only then. Anonymous payment schemes also forfeit the option of post-paid good and services, where the customer needs to be billed and thus is typically known. Finally, there may be applications where regulations and legal restrictions prohibit the

---

[1] Andrew J. Blumberg and Peter Eckersley, *On Locational Privacy, and How to Avoid Losing it Forever*, technical report (Electronic Frontier Foundation, 2009), accessed February 4, 2013, https://www.eff.org/wp/locational-privacy.

[2] E.g. David Chaum, "Security without identification: transaction systems to make big brother obsolete," *Commun. ACM* 28, no. 10 (October 1985): 1030–1044, ISSN: 0001-0782, doi:10.1145/4372.4373, http://doi.acm.org/10.1145/4372.4373; David Chaum, Amos Fiat, and Moni Naor, "Untraceable Electronic Cash" in *Advances in Cryptology—CRYPTO* (1988); Stefan Brands, "Electronic cash systems based on the representation problem in groups of prime order" in *CRYPTO* (1993); Jan L. Camenisch, Jean-Marc Piveteau, and Markus A. Stadler, "An efficient electronic payment system protecting privacy," in *ESORICS* (1994).

[3] David Chaum, "Blind Signatures for Untraceable Payments," in A*dvances in Cryptology: Proceedings of CRYPTO '82* (1982).

customer from being anonymous. Vendors in this market will be unable to provide anonymous payment services to their customers.

Under the premise that the customer must be identifiable, we thus must conceptionally deviate from the widespread paradigm of anonymizing customers in privacy-enhancing payment and billing systems. Instead of obscuring or removing identity information, in our solution, the user's identity is explicitly known during a transaction, yet the user's *location* is concealed. Our approach effectively hinders the creation of a movement profile based on financial transactions. We use the increasingly relevant example of re-charging electric vehicles and paying for energy on the go to showcase our approach. We do not exclude the possibility that our approach can be adapted to other settings that require the customer to be known during such a transaction.

Why Electric Vehicles? The electric vehicle (EV) scenario offers several interesting constraints. First of all, the proliferation of vehicles and infrastructure is limited, but rapidly increasing. Market research predicts up to 3.4 million annual world-wide sales of plug-in hybrid (PHEV) and battery electric vehicles (BEV) in 2020.[4] While we are aware that most people leave a cornucopia of movement traces due to their use of existing technology (e.g., cell phones), we think that technical solutions for emerging fields, like electric mobility, should be designed with privacy in mind.

Second, at least for the time being, the capacity of most electric vehicle batteries is rather limited, thus most EVs require relatively frequent charging using the growing network of charging stations (CSs)—the European Commission aims at 795,000 public charging stations throughout the EU by the year 2020.[5] The increasing availability of public charging stations is positive and necessary for the success of EVs. However, in combination with the need to charge frequently, it renders vehicle movement profiles more detailed than those derived from fossil fuel not paid with cash.

Third, cash is not an option for almost all utilities. In most parts of the developed world, utilities deliver energy either based on a subscription (post-paid) or pre-paid model where the customer's name is known. In contrast to the current network of fuel stations, EV charging infrastructure is much more distributed, which makes cash logistics prohibitively expensive.

Fourth, the sales of electric energy are tightly regulated in many countries. Many of these regulations aim at making the market more transparent to the customer. However, when applied to the relatively new EV scenario some of these requirements can compromise the locational privacy of the customer.

---

[4]Pike Research, Electric Vehicle Market Forecasts, http://www.pikeresearch.com/research/electric-vehicle-market-forecasts, 2013, accessed January 29, 2013.

[5]cars21.com, EU proposes minimum of 8 million EV charging points by 2020, http://beta.cars21.com/news/view/5171, 2013, accessed January 29, 2013.

Contributions. In this paper, we propose a system to authenticate non-anonymous transactions, while preserving users' locational privacy. We use the example of electric vehicle charging, as it offers several interesting constraints. More precisely, we make the following contributions:

- We adapt a carefully chosen group signature scheme without compromising its strong security properties to allow for full compliance with regulations and legal requirements. These requirements were identified with the help of experts in the field of commercial law and energy law. The privacy mechanisms protecting the user's location data are very strong: not only is it impossible to decide whether a user has charged her vehicle at a specific CS, it is even impossible to decide whether a user has *ever* been charging at one or several CSs more than once.
- Our solution is complete, in that it covers the charging process from after authentication to providing all information necessary for the clearing process. It closely fits existing clearing and billing structures and can be implemented efficiently on a large-scale.
- To the best of our knowledge, we are the first to also offer an implementation of a practical charging and billing system for electric vehicles that provides strong protection of the customer's locational privacy. Our implementation performs well even on the limited hardware of a CS, while we are able to process more than one million charging processes per hour using off-the-shelf hardware at the backend (BE), thus providing a cost-effective way to process billing information from a large network of CSs.

2 Overview

The system we propose consists of three main phases: (1) authenticating the customer, (2) authenticating the tuple of customer identity and energy consumption data, and (3) transmitting this data to a clearing house, all without compromising the customer's locational privacy. In the following, we first lay out the problem space before presenting our scheme.

2.1 Problem Space

We define the problem space as follows: Electric utility companies that are honest but curious and want to learn about past, present, and future locations of vehicles, or any entity obtaining (billing) records from utilities, can infer a movement profile for every customer, based on these records. Under the assumption that

(a) the creation of movement profiles without explicit consent of the subject is undesirable and the existence of unnecessary data is to be avoided,
(b) anonymous payments are not an option,

(c) the solution should integrate well with existing billing infrastructure and processes

we explore how the creation of movement profiles can be prevented, while integrity, authenticity, and, in parts, the confidentiality of the data transmitted between a point of sale (i.e., a CS) and a backend is provided. Conceptually, we thus must deviate from the widespread paradigm of anonymizing customers in privacy-enhancing payment and billing systems. Instead, our approach to this problem is to *anonymize locations*, i.e., to cryptographically ensure that charging station locations cannot be linked to customers' identities and timestamps. In this context we identify three core issues:

One way to cryptographically bind a customer identity to metering data are digital signature algorithms, as they achieve non-repudiation. However, the location where a charging process took place can be directly inferred, classical digital signatures not only guarantee the authenticity of the signed data, but also authenticate entities, i.e., the respective charging station (Issue 1). Location-bound tokens, like the identifier of the energy meter used for the measurement, naturally compromise the customer's locations, but utilities are legally required to retain this information in many European countries (Issue 2). The location of a transaction can also be inferred from network-based identifiers (Issue 3), primarily the charging station's IP address, e.g., by correlating BE server access logs with billing data timestamps.

Furthermore, an entity may have access to the network that connects the backend or a CS to the Internet. Such an attacker might try to infer the origin of a message by using a timing side channel, but must be unable to attribute the connection to a specific user.

We assume that all attackers are computationally bound and accordingly unable to break computationally hard cryptographic primitives. Attacks against the point of sales itself are out of scope of this paper.

2.2 Approach

We address *Issue 1* by employing a group signature scheme with strong security properties that provides very efficient verification procedures for large numbers of signatures as a central building block of our system. The scheme allows for the conditional identification of a signer, while in the default case allowing him to remain anonymous. For every entity that is not in possession of the so-called opening key, the actual signer of a message is indistinguishable from every other potential signer within the same group. Thus, while a customer's transaction is always linked to his customer account, our system guarantees unlinkability with respect to location and time of a transaction.

We address *Issue 2* by modifying the signature scheme such that information that is required by law or regulations, but would compromise the customers' locational privacy, is also only conditionally available. In normal operation this information is as strongly protected, as the signer's identity itself. In case of a legal dispute, where this information must be produced by the utility in front of a court of law, such that an independent entity can assess the proper calibration of the energy meter, the identifier of charging station and energy meter can be revealed by a trusted third party. Legally required information for Germany was identified with the help of our colleagues from the faculty of law, who specialize in commercial law and energy law. However, we present a generalized approach, i.e., the exact datum required in the respective jurisdiction is secondary: if the information is location-bound, it can be afforded the aforementioned strong cryptographic protection. Thus, our approach is adaptable and usable in arbitrary national and international contexts.

We are aware that in being compliant with legal regulation, our system also depends on legal protection: A high legal hurdle must be placed before the identification of a signer (i.e., the respective CS) and the disclosure of location-bound tokens. This could mean, for example, that a court order or the customer's consent is required, not only in case of a dispute between customer and vendor, but especially in the context of criminal law.

We address *Issue 3* by anonymizing the sender of billing-relevant information on the network level. As the communication between charging station and backend is not highly time critical, we could in principle use high-latency mix networks, such as Mixminion[6] or Mixmaster.[7] However, as network availability is an issue, we chose to use the, at the time being, most popular anonymity network, which increasingly offers good redundancy due to its high number of nodes: the Tor network.[8] As Tor does not provide protection against exploiting timing side channels, especially in the presence of low traffic volume, we also discuss how these kinds of attacks can be mitigated in our application context. Please note however that Tor is only one tool in this context and could be replaced by another anonymity network.

In summary, the authentication and charging process we propose is as follows (cf. Fig. 1):

1. The CS authenticates towards the customer and vice versa. The CS retains the authenticated customer identity.
2. Upon successful authentication, the CS's power outlet is unlocked and/or put on-line. Charging begins as soon as the EV is connected.
3. When the power-line connection between the CS and the EV is interrupted, the CS generates a tuple containing all information required for the billing process (i.e., the authenticated customer identity stored from Step 1, the amount of

---

[6]George Danezis, Roger Dingledine, and Nick Mathewson, "Mixminion: Design of a type III anonymous remailer protocol," in *IEEE Symposium on Security and Privacy*, (2003).

[7]Ulf Möller et al., *Mixmaster Protocol | Version 2*, http://www.abditum.com/mixmaster-spec.txt, 2003.

[8]Roger Dingledine, Nick Mathewson, and Paul Syverson, "Tor: the second-generation onion router," in *13th USENIX Security Symposium* (2004).

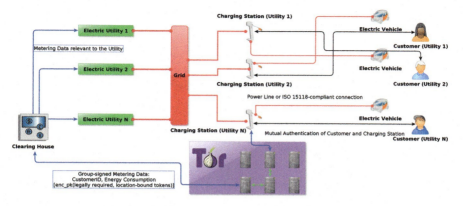

Fig. 1 Charging and transmission of metering data (including roaming use case)

energy provided to the user, a timestamp indicating the beginning of the charging process and a timestamp indicating its end). Each location-bound token that is legally required is encrypted to the single entity in possession of the opening key, a trusted third party, denoted as the opener. The tuple is signed using the group secret xi of the respective CS and the data is transmitted to the BE via the Tor network. To ensure confidentiality of the transmitted data, we establish a TLS tunnel between CS and BE prior to transmission. Our approach also addresses the relevant use case of customers roaming between energy providers, which we detail in Sect. 2.3.

2.3 Roaming

While a significant part of customers can still only charge at CSs owned by the utility they have a contract with, roaming is desired by most market participants. In Fig. 1 the concept is represented by the introduction of a clearing house. Following the example of the banking and telecommunications sector, at least two parallel efforts[9] are already under way in the energy sector to establish a clearing house to back a roaming-enabled charging infrastructure for electric vehicles. As the clearing house aggregates and verifies metering data from all the CSs, it is capable to provide either only data clearing or also financial clearing to the associated electric utility companies, which in turn allows each utility's customers to roam freely between all other utilities cooperating with the clearing house.

---

[9]https://www.e-clearing.net/; http://www.hubject.com.

3 System Design

In this section we describe all processes that constitute our system. Group signature schemes are an essential part of our approach and thus we explain below how we utilize and adapt this concept and why we chose the *eXtremely Short Group Signature* (XSGS) scheme.

3.1 Group Signatures and XSGS

The idea of group signature schemes has first been introduced by Chaum and van Heyst in 1991.[10] A group signature scheme is a digital signature scheme that (additionally) provides a (strong) form of sender-anonymity. Unlike in classical signature schemes where each signature is produced by a single signer, in a group signature scheme each signature is produced on behalf of a group. For the verifier it is easy to check whether the signature has been produced by one of the current *group members*. However, finding out who exactly produced the group signature is impossible. Intuitively, the larger the group is, the better are the anonymity guarantees provided for each group member—an ideal property for our scenario.

Anonymity: Pseudonyms vs. Group Signatures. Group signatures provide a very strong form of anonymity that is usually referred to as unlinkability: it is not only impossible to map a signature to its creator—this could be achieved by pseudonyms alone. Unlinkability also implies that no one, except for a dedicated trusted party called opener, is able to decide whether two group signatures have been produced by the same signer. We believe that for our application this property is crucial.[11] When using pseudonyms for CSs alone to protect the user's locational privacy, the verifier could easily build up customer profiles for every CS which, with more and more user-dependent billing data, could possibly be narrowed down to a single CS. In this way one could easily reveal the true CSs behind the pseudonyms. As a consequence, the verifier could easily follow where and when each user charged its vehicle. Group signatures on the other hand do not even reveal whether two signatures belong to the same CS. So users who constantly charge their vehicle at the same CS are indistinguishable from those who travel a lot and often use CSs that they have never visited before.

Design Features of the XSGS Scheme. Group signatures vary in the extent of functionality they offer and in the security guarantees they provide for group members and verifiers. In our work, we utilize the *XSGS scheme* by Delerable and Pointcheval.[12] The XSGS scheme is an extended variant of the well-known group

---

[10] David Chaum and Eugène van Heyst, "Group Signatures" in *EUROCRYPT* (1991), 257–265.

[11] We recall once again that user identities have to be known to the verifier for a proper billing process. Thus it is not possible to anonymize user identities in the bills.

[12] Cécile Delerable and David Pointcheval, "Dynamic Fully Anonymous Short Group Signatures" in *VIETCRYPT* (2006), 193–210.

signature scheme by Boneh, Boyen and Shacham (BBS) which achieves very high efficiency with respect to both signature size and speed.[13] It modifies the BBS scheme in two ways. First, it adds improved protection of group members against collusions of (corrupted) members who try to frame a user. In XSGS, even if the issuer itself is corrupted and takes part in that collusion, its honest group members cannot be framed. Second, XSGS guarantees unlinkability of signatures to even hold against an adversary that can convince the opener to open all other signatures. BBS does in general not cover such attacks (not even when the adversary may convince the opener only once). As a theoretical benefit of these extensions, the XSGS scheme can be proven secure in the very strong security model of Bellare, Shi, and Zhang.[14] We believe that these extended properties of XSGS are necessary in our application. In particular, they allow to implement the issuer at the same place as the (only) verifier (i.e., the clearing house), without risking the CS's anonymity. In the selected context this property implies that the clearing house may act as group manager *and* initial verifier, removing administrative and computational work load from the participating energy providers, without compromising the systems' security guarantees.

Support for Batch Verification. An important design restriction of our solution is that we consider a single verifier that has to verify a huge amount of signatures. The group members, on the other hand, do only have to generate a moderate amount of signatures each day. Thus our group signature scheme should ideally feature very fast verification procedures. Kim et al. showed that XSGS supports batch verification.[15] For security reasons, the combination process is setup in such a way that adversaries cannot produce a combination of invalid signatures which pass the batch verification test.[16]

Dynamic Groups. XSGS allows for dynamic groups, i.e., group members can be added and removed without re-initializing the whole scheme. Also, member joins do not require updates of the group public key *GPK*. We stress that if member joins do not require modifications of *GPK*, it is necessary to modify the group public key when revoking users. In the setting at hand, where the system is likely to expand, not needing to recalculate the *GPK* for every new CS improves overall system performance. Instead, the system-wide update of *GPK* is only required when a CS is removed. However, even then the approach for updating the *GPK* underlying XSGS is very efficient. It is based on dynamic accumulators.[17]

---

[13]Dan Boneh, Xavier Boyen, and Hovav Shacham, "Short Group Signatures" in *CRYPTO* (2004), 41–55.

[14]Mihir Bellare, Haixia Shi, and Chong Zhang, "Foundations of Group Signatures: The Case of Dynamic Groups" in *CT-RSA* (2005), 136–153.

[15]Kitae Kim et al., "Batch Verification and Finding Invalid Signatures in a Group Signature Scheme," I. J. *Network Security* 13, no. 2 (2011): 61–70.

[16]The batch verifier of Kim et al. uses the so-called small exponent test. Mihir Bellare, Juan A. Garay, and Tal Rabin, "Fast Batch Verification for Modular Exponentiation and Digital Signatures" in *EUROCRYPT* (1998), 236–250.

[17]Jan Camenisch and Anna Lysyanskaya, "Dynamic Accumulators and Application to Efficient Revocation of Anonymous Credentials" in *CRYPTO* (2002), 61-76; Lan Nguyen, "Accumulators from Bilinear Pairings and Applications," in *CT-RSA* (2005), 275–292.

3.2 Bootstrapping the System

Before we can start authenticating users, charging vehicles, and securely transmitting energy consumption data, we have to set up the infrastructure. The clearing house acts as the group manager within the XSGS scheme. It can add a new CS to the group by issuing a user certificate (credential) *UCert* to CS. A CS with a valid *UCert* is also referred to as a group member. The clearing house can also revoke the ability of group members to sign on behalf of the group. An entity sufficiently independent of the clearing house serves as the opener. In our scenario N electric utilities choose to cooperate by utilizing a certain clearing house. Each utility i provides m_i charging stations to the public.

In order to bootstrap the XSGS scheme, the group manager first needs to generate the group (curve) parameters of a bilinear group (including group descriptions, generators, and pairing specification). Technically, the bilinear group consists of two elliptic curve groups $\mathbf{G_1}$ and $\mathbf{G_2}$ of prime order p with random generators $G_1 \in \mathbf{G_1}$ and $H, G_2 \in \mathbf{G_2}$ and the description of a non-degenerated bilinear pairing $e : G_1 \times G_2 \to G_t$ such that $e(G_1^a; G_2^b) = e(G_1; G_2)^{ab}$ for every $a, b \in \mathbf{Z}_p$. For more details we refer to Boneh, Boyen and Shaham.[18] Next it generates a secret Diffie-Hellman key $IK \in \mathbf{Z}_p$ (called issuer key) with its corresponding public key $= G_2^{IK}$.

The issuer key IK is used to generate certificates for new group members. Given these values, the opener generates a private key of a chosen-ciphertext secure encryption system, the opening key OK. The corresponding public encryption key is denoted as OPK. The public key OPK is used in the signing process of the group signature scheme to encrypt the signer's certificate *UCert*. This enables the opener to reveal which CS has actually created a given group signature. On a technical level OK consist of two independent secret keys of an ElGamal encryption system. OPK contains the corresponding public keys. It is well known that ElGamal is only chosen-plaintext secure. However, the system applies the well-known Naor-Yung transformation[19] which encrypts a given message under both ElGamal keys resulting in ciphertext Z_1 and Z_2. Additionally, it generates a NIZK proof P of equality of plaintexts in Z_1 and Z_2. The ciphertext Z consist of $Z = (Z_1; Z_2; P)$. The group public key GPK consist of the parameters of the bilinear group, W, and OPK. Besides these values we also require that a public RSA modulus n is available to all parties. This value is generated by a trusted third party. The corresponding secret key is deleted. The setup procedure is depicted in Fig. 2.

---

[18]Boneh, Boyen, and Shacham, "Short Group Signatures." in CRYPTO (2004).

[19]Moni Naor and Moti Yung, "Universal One-Way Hash Functions and their Cryptographic Applications," in *STOC* (1989).

Fig. 2 Setup phase

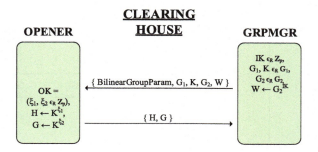

3.3 Setting up New Charging Stations

Each new CS must join the group before it can sign metering data. Now that group manager and opener are set up, the group manager can add new charging stations to the group. Note that all charging stations, independent of the utility that operates them, will be members of the same group.

The group manager starts the join process by transmitting the *GPK* to the CS. The CS draws its private signing key $UK \in Z_p$ and computes a commitment $C = H^{UK}$ of *UK*. Then it sends C together with a NIZK proof of knowledge of *UK* to the group manager. On successful verification of this proof, the group manager selects a random signing key $x \in Z_p$ for the CS and calculates the group member identifier

$$A = (G_1 \cdot C)^{\frac{1}{IK+x}} \Leftrightarrow e(A, W \cdot G_2^x) = e(G_1 \cdot C, G_2) \qquad (1)$$

The values A and x constitute the certificate *UCert* of the CS. Intuitively, *UCert* is a digital signature over x that can only be computed with the help of *IK*. The group manager first sends A to the CS and proves that it knows a corresponding x that fulfills the above equation.

Knowing that its communication partner can indeed issue certificates, the CS produces a classical signature S using its *USK* over A as $S = Sign_{USK}(A)$ and sends $(S, cert_{CS})$ to the issuer. This pair is important when resolving disputes as it binds the anonymous certificate *UCert* to a concrete CS that can be identified via the classical PKI. If the signature is valid, the group manager sends x to the CS and registers the entry $(UCert, C, cert_{CS}, S)$ in a database. Now since $C = H^{UK}$ and *UK* is known to the CS we get that

$$A = \left(G_1 \cdot H^{UK}\right)^{\frac{1}{IK+x}} \Leftrightarrow e(A, W \cdot G_2^x) = e\left(G_1 \cdot H^{UK}, G_2\right) \qquad (2)$$

The join process is depicted in Fig. 3.

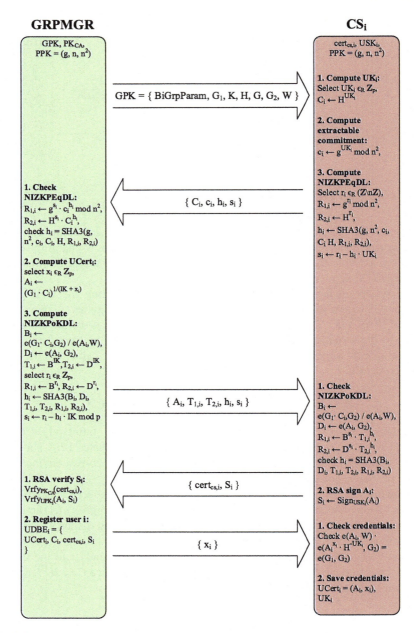

Fig. 3 Join procedure

3.4 Decommission of Charging Stations

Occasionally it may be necessary to remove a CS from the group, be it because it is replaced by a CS of a newer generation or to deal with a compromise. We consider the revocation of a group member's credentials to be a less frequent event than the joining of a new member. Thus, while *UCert* and *UK* remain unchanged upon the joining of a new member, removing a member from the group requires that all remaining group members receive information on how to re-calculate their group identifiers *A*.

Assume the group manager wants to revoke a CS with

$$UCert' = (A', x').$$

First, it publishes an updated version of the *GPK*. For example G_1, G_2, and H are substituted by $G_1^* = G_1^{\frac{1}{IK+x'}}$, $G_2^* = G_2^{\frac{1}{IK+x'}}$, and $H^* = H^{\frac{1}{IK+x'}}$. each group member with $UCert = (A,x)$ and secret key *UK* except for the one to be revoked has to update its group identifier

$$A^* = A^{\frac{1}{IK+x'}}$$

To this end it is sufficient that the group manager simply publishes x'.

$$A^* = A^{\frac{1}{IK+x'}} = \left(G_1^* \cdot H^{*UK} \cdot A^{-1} \right)^{\frac{1}{(x-x')}}. \tag{3}$$

Next, each charging station computes a new signature $S = Sign_{USK}(A^*)$ over the new group member identifier A^* and sends it to the group manager. The group manager verifies S^* from each CS and, on success, updates the existing database entries with the new values for A^*, C^* and S^*. Note that the CSs do not have to save an incremental revocation list of all revoked members to decide on the validity of newly signed metering data. However, it might be necessary for the group manager to retain a limited set of old group credentials for the time span that the respective jurisdiction sets for the resolution of disputes concerning past charging processes. The revocation process is depicted in Fig. 4.

3.5 Ensuring Authenticity of Metering Data

When the charging process is terminated (i.e., the cable connection between EV and CS is severed), the CS creates a message *M* consisting of the authenticated customer identity, the amount of energy consumed by the customer, two timestamps marking the beginning and the end of the charging process, and a string

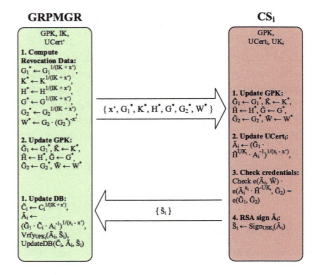

Fig. 4 Revocation procedure

that identifies the utility owning the CS. As discussed above, legal regulations often requires transmission and storage of the identifier (*meterID*) of the calibrated energy meter or of other certified components of a point of sale.

These identifiers would reveal the physical location of the transaction. To avoid this, we have to adapt the group signature scheme slightly. Instead of being sent in the clear, the *meterID* is encrypted using the opener's encryption key *OPK* before being added to M. In the same way other location-critical information can be incorporated into the group signature. Only the opener can decrypt these values using its secret decryption key *OSK*. We stress that while the *meterID* is always encrypted with the opener's public key and never transmitted in the clear, it is not necessary to prove that the correct *meterID* has been incorporated into the ciphertext. The opener can uniquely identify the CS and any incorrect information of a CS on its *meterID* can thus easily be revealed. As sketched above, CS's group signature s on M consists of an encryption Z of *UCert* and a message-dependent NIZK proof showing that CS knows a valid *UCert* with corresponding *UK* which fulfill Eq. 2 and that *UCert* has been encrypted correctly under public key *OPK* in the ciphertext Z (which is part of s).

Intuitively, these types of message-dependent proofs work like signatures. Generating them on a messages M requires the creator to know A, x, and *UK*. They are often referred to as signatures of knowledge.[20] The entire signing process is depicted in Fig. 5. For more details on the computations of the group signature, we refer to the literature.[21]

---

[20]Melissa Chase and Anna Lysyanskaya, "On Signatures of Knowledge," in *CRYPTO* (2006), 78–96.

[21]Kim et al., "Batch Verification and Finding Invalid Signatures in a Group Signature Scheme"; Delerable and Pointcheval, "Dynamic Fully Anonymous Short Group Signatures."

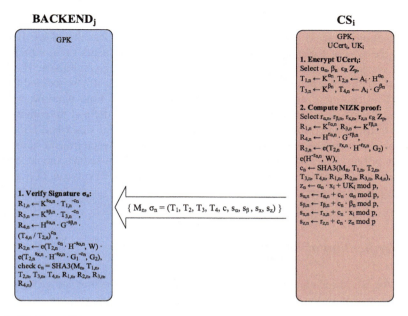

Fig. 5 Sign procedure

3.6 Transmission of Metering Data

To prevent the disclosure of the CS's network location, the CS first connects to the Tor network and establishes a routing circuit. It then starts a TLS session with the backend (BE) and in the process verifies the certificate presented by BE. We use a ciphersuite based on Ephemeral Diffie-Hellman (DHE) with CBC-MAC, as it offers perfect forward secrecy and because of its cryptographic security properties: it has recently be shown to be provably secure in a strong security model.[22] We rely on TLS to guarantee that each transmission from a CS reaches the backend. Although Tor provides sender anonymity, a possible timing side channel exists: if there is only sporadic network traffic within the system of CSs and BE, an attacker observing both the network at a CS and the BE could correlate these events with charging timestamps (somehow obtained) from the clearing house. As the transmission of billing relevant data is not time-critical in the example of EV charging, we can prevent correlation as follows:

---

[22]Tibor Jager et al., "On the Security of TLS-DHE in the Standard Model" in *Advances in Cryptology—CRYPTO* (2012).

Each CS is scheduled to send a transmission of a given size once per 15 min. Each charging process results in one message of typically less than 1000 Byte. If we fix the size of the transmission, for instance, at 5 kB, it fits several messages (M, s). We pad each transmission with random data to the maximum size. If a message (M, s) does not fit in the current transmission anymore, it is scheduled for the next. If no charging process has been finished within the time window, we just transmit the string empty and pad the transmission to the defined maximum size. As all transmissions are of equal size and are encrypted as described above, an attacker observing the network is unable to distinguish between transmissions that contain billing data and those that do not. The BE acknowledges the successful submission by sending the string ACK and a timestamp. We rely on TLS for the authenticity of the reply.

3.7 Verification of Metering Data

When the BE at the clearing house has received (M, s) it verifies the group signature s by checking the NIZK proof with respect to the GPK and thus determines whether the consumption data that is bound to the identity of a customer is valid. For details on the computations, we refer to.[23] If the signature does not verify it simply discards the message as it cannot stem from a CS within the group. On success, the signed tuple M is passed on to the clearing service for processing. As there is one central verifier in the system that verifies all metering data, batch verification of group signatures offers a significant efficiency gain.

3.8 Dispute Resolution

In the case of a dispute, the opener can craft a non-repudiable publicly verifiable proof of the actual creator of a given group signature. The opener will act so only upon the request of a judge or with the consent of the customer. Note that even after a message M_i has been subject to the opening process, it is impossible to decide, whether a CS who signed M_i also signed a different message M_j, i.e., the location of other, potentially unrelated charging events remains hidden. To open the signature s, the opener uses its secret opening key OK to decrypt the ciphertext Z and obtain the certificate *UCert* of the signer. Next it uses its access to the registration database to obtain *UPK* and S which correspond to *UCert*. From this

---

[23]Kim et al., "Batch Verification and Finding Invalid Signatures in a Group Signature Scheme"; Delerable and Pointcheval, "Dynamic Fully Anonymous Short Group Signatures."

information she computes a publicly verifiable NIZK proof that *UCert* is actually encrypted in *Z*. Together with the database entry *A*, *certCS*, *S* this convincingly reveals the identity of the signer in a non-reputable way.

4 Evaluation

In this section, we describe how we evaluated our prototype implementation. We also present an overview of the performance results obtained both for the various operations of the XSGS scheme and the transmission of data from a CS to the BE. For implementation details, please refer to Appendix 1; for the choice of cryptographic parameters, please refer to Appendix 2.

4.1 Evaluation Environment

We aimed at evaluating our approach in a realistic environment. Thus, we implemented XSGS and tested the creation of signed messages, the setup process for adding new charging stations, and the procedure to decommission charging stations on a prototype of a CS for EVs built at our department. The CS contains an inexpensive industrial-grade Intel Atom platform (CS_1, cf. Table 1) as control unit that interacts with the energy flow control subsystems within the CS and acts as a front-end to the user. Additionally, we evaluated our implementation on a Freescale i.MX53, which is an implementation of an ARM A8 core. Comparable platforms to both variants can be found in CSs in the market or under development today.

As BE we chose an Intel server platform (cf. Table 1). We used this platform to evaluate all XSGS operations typically performed by the group manager, opener, judge, or any entity that wishes to verify a signature. We also created signatures and performed join operations as a comparison to the measurements on the actual CS. While the Tor network is widely used and considered usable for non-time critical applications, we also used this platform to evaluate if latency and throughput are acceptable in our application scenario.

Table 1 Evaluation environment

	Hardware platform	OS
CS_1	Intel Atom D2550, 1 GB RAM	Ubuntu 12.04
CS_2	Free scale i.MX53, 1 GB RAM	Ubuntu 10.04
BE	Intel Xeon X5650, 2 GB RAM	Ubuntu 12.04

4.2 Evaluation Results

We performed the setup procedure required for adding a new CS 100 times. The computations necessary on the CS are performed on average in 757.4 ms on CS_1 and 1077.3 ms on CS_2, while the computations on the *BE* took 55 ms on average. Accordingly, we performed 100 decommission procedures: on average, the computations performed on CS_1 take 49.0 ms (resp. 77.8 ms on CS_2), the computations on the *BE* take 20 ms. We also performed 100 dispute resolution procedures on the *BE*: on average opening a message takes 8.2 ms, while judging takes 6.9 ms.

We evaluate the time required to prepare a message to transmit the metering data to the *BE*. Preparing a message 1000 B (taken from/dev/urandom) takes 28.5 ms on average on CS_1; on CS_2 the process takes 41.5 ms. Preparing a message that allows for batch verification on the BE takes slightly longer: 28.8 ms on CS_1 or 43.1 ms on CS_2. For a message size up to 100,000 B message creation takes less than 33 ms on CS_1 and 54.2 ms on CS_2. Figure 6 shows that the size of the message only has a limited impact on the time required to create a valid signature, as we only sign a hash of the message. Creating a signed message of one million bytes takes 66.7 ms on average on CS_1 and 161.1 ms on CS_2. These results show that ensuring the authenticity of messages by means of group signatures is feasible on the limited hardware found in a CS. Even more so, as we only need to generate one signature for each charging process.

Being able to batch verify messages offers a significant performance increase. While a CS will typically only create one message every few minutes or every few hours, each message has to be verified by the *BE*. The verification of a normal message takes 30 ms, a single batch-enabled one is verified in about the same time. Figure 7 shows that verification time increases linearly with the amount of messages. Standard verification allows for processing 41 messages per second on the BE, while batch verification allows for processing of 93 messages in the same time. When comparing the time required for verifying one thousand messages, batch verification is about 2.3 times faster. In a worst case scenario, where a batch contains so many invalid signatures that it is faster to verify each individual message, we can still process 147,600 messages per hour using a single CPU core. As the process can be parallelized at will, a comparable server with eight CPUs cores instead of one is sufficient for processing more than one million messages per hour.

As transmission times vary due to network latency, we evaluate the network performance separately: We used iperf[24] to measure whether the Tor network offers enough bandwidth for transmitting metering data from the CS to the BE. We controlled that the bandwidth between the host running the iperf server and the one running the client is not the limiting factor and repeated our measurements at various times of the day, building a new Tor circuit for each iteration. We were able to transfer a minimum of 373 kbit per second and a maximum of 1.07 Mbits per second through the Tor network. While the actual throughput may vary depending on the time of day and the chosen circuit, our evaluation shows that it is reasonable to

---

[24]http://iperf.sourceforge.net/.

assume that we can transfer metering data through the Tor network, especially as the communication between CS and BE is not subject to real-time requirements. Also note that the BE is not affected by Tor's limited bandwidth, as there is no need to obscure the BE's location and only CSs communicate via Tor. We expect that anyone willing to operate a large-scale commercial system, that relies on a anonymity network like Tor, will need to contribute to the infrastructure of the respective anonymity network to increase dependability and throughput. Thus, as a positive side-effect, a large-scale application of the respective anonymity network would strengthen the overall availability and resources of this anonymity network.

In summary, we found that our approach performed well on all tested platforms and, most importantly, is fast enough for our application.

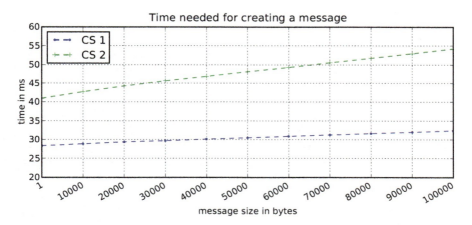

Fig. 6 Time required for message creation by size

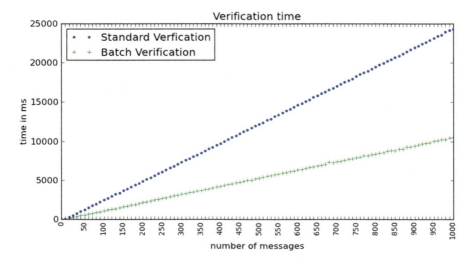

Fig. 7 Time required for verification by #messages received

5 Discussion

We now discuss possible attacks against both the authenticity of billing-relevant data and against the user's locational privacy.

5.1 Malicious Customer

While our system is well equipped to counter attackers with capabilities as described in Sect. 2.1, there exists the theoretical possibility that an attacker, who is a valid customer in the system, could force a CS_1 offline before a revocation of a different CS_2 takes place. Thus, CS_1 does not realize that the group credentials have changed and must be recomputed. The attacker then authenticates herself and charges her EV at CS_1, which is possible as user authentication works offline. The CS signs the metering data with its current credentials. At some point in the future, when the CS is online again, it transmits the data to the BE. It will then also receive new group credentials and will be able to create valid messages, as during the revocation process. Still, the BE will discard the delayed metering data from the CS as it has been generated with the old credentials. Hence, the attacker was able to charge for free in the meantime. There are at least two counters to this attack. First, if the CS is up and running again, it may simply re-sign all the unsent metering data with the updated credential. Second, if the CS is for some reason not able to continue signature generation (for example if a trusted key storage is broken), we can still retain old credentials for verification and use the old group signature to bill the customer correctly.

5.2 Tracking and Localization Attacks

Ma et al. show that if a set of traces of time and corresponding location of mobiles nodes exist, where '[t]he traces are anonymous in that the true identity of a participant has been replaced by a random and unique identifier',[25] a small amount of side information is sufficient for an attacker to infer the true identity of a user. The work of de Montjoye et al.[26] supports these claims and shows that even datasets

---

[25]Chris Y.T. Ma et al., "Privacy vulnerability of published anonymous mobility traces," in *MobiCom '10* (2010).

[26]Yves-Alexandre de Montjoye et al., "Unique in the Crowd: The privacy bounds of human mobility", *Scientific Reports*, 2013, http://www.nature.com/srep/2013/130325/srep01376/full/srep01376.html

with coarse traces provide little anonymity, in such that four spatio-temporal points are enough to uniquely identify 95 % of the individuals.

However, neither approach is applicable to our system. Due to the nature our approach, no spatio-temporal data points, let alone location tracks, are available to any entity except the opener. Thus Krumm's inference attacks,[27] which aim at de-anonymizing entities from anonymous or pseudonymous location tracks, are not applicable in the setting at hand. We do not conceal the identity of the user, but cryptographically protect their location. To thwart attacks by third parties (i.e., non-legitimate receivers of transmitted data), all information is transmitted encrypted with a provably secure TLS variant. Thus the attacker needs to be a legitimate receiver of the data, i.e., the clearing house or a utility. Both receive the following information: customer A of utility B consumed N kWh of energy, starting from timestamp X, ending at timestamp Y. Every location-bound token, like the CS's public key and the *meterID*, is encrypted only to the opener and thus never leaked to any other party. Thus, the only entity able to access location data at will is the opener, who is explicitly trusted. Given the exposed position of the opener as a trusted third party, it is mandatory for this party to be independent from all other parties (i.e., from vendors, customers, intermediaries, law enforcement etc.) in a commercial deployment of the system. However, the concrete instantiation of this trusted third party is both an organizational and a political question, which is beyond the scope of this technical paper.

As we necessarily need to exclude any trusted third party from the group of potential attackers, the data available to an adversary thus does not contain the location of the user, nor can the attacker use the amount of energy consumed to infer the distance the user has driven between two charging events, due to external factors that influence power consumption, like driving style, speed, etc. Shokri et al.[28] propose a metric to quantify the performance of a location privacy protection mechanism (LPPM). Our systems applies location hiding as an online LPPM in a distributed architecture, i.e., we only look at the current event at the time of its creation and hide all location-bound information by encrypting it to the opener. As argued above, while records of user interaction exists for billing purposes, they do not contain any spatio-temporal locations or references to such data. An adversary, who knows the location of every CS, may determine the location where the EV could have been charged with a high accuracy (as it was necessarily at the location of a CS), but he is unable to achieve a high correctness as to where the EV was actually charged.

---

[27]John Krumm, "Inference Attacks on Location Tracks", in *Pervasive Computing* (Pervasive 2007).

[28]Reza Shokri et al., "Quantifying Location Privacy," in *2011 IEEE Symposium on Security and Privacy (SP)* (May 2011), doi:10.1109/SP.2011.18

6 Related Work

Locational privacy has been recognized as being desirable as early as 1996.[29] Its importance has been recognized for example in the field of pervasive computing[30] and also in the context of location-based mobile applications.[31] The importance of location privacy in the context of transportation is underlined by numerous publications that aim at preserving location privacy in various applications like vehicular communication systems,[32] ticketing for public transport systems,[33] and electronic road toll collection. In the latter context, Balasch et al.[34] use commitments that do not reveal information on the user's location, while relying on a disjoint audit system based on spot checking cameras. In the audit system, user locations are sporadically but routinely linked to identity information. Meiklejohn et al.[35] follow closely the PrETB construction by Balasch et al., but also include malicious colluding users into their threat model. Our approach, in contrast, does not require the routinuous linking of users' locations and identities. We reserve this extreme measure to singular occurrences, where a vendor can argue an initial suspicion of misuse. Chen et al.[36] propose the use of a group signature scheme to enhance the users' privacy, by hiding a user's identity within a group while

---

[29] Ian Jackson, "Anonymous addresses and confidentiality of location", in *Information Hiding* (1996).

[30] Alastair R. Beresford and Frank Stajano, "Location privacy in pervasive computing", *IEEE Pervasive Computing 2*, no. 1 (March 2003): 46–55, issn: 1536-1268, doi: 10.1109/MPRV.2003.1186725.

[31] Raluca Ada Popa et al., "Privacy and accountability for location-based aggregate statistics", in *ACM CCS* (2011).

[32] Jean-Pierre Hubaux, Srdjan Capkun, and Jun Luo, "The security and privacy of smart vehicles," *Security & Privacy, IEEE 2*, no. 3 (2004): 49–55; Florian Dötzer, "Privacy Issues in Vehicular Ad Hoc Networks," in *Privacy Enhancing Technologies* (2005); Julien Freudiger et al., "Mix-zones for location privacy in vehicular networks," in *Win-ITS* (2007); K. Sampigethaya et al., "AMOEBA: Robust Location Privacy Scheme for VANET," *IEEE Journal on Selected Areas in Communications 25*, no. 8 (October 2007): 1569–1589, issn: 0733-8716, doi: 10.1109/JSAC.2007.071007; Zhendong Ma, *Location Privacy in Vehicular Communication Systems: a Measurement Approach* (PhD thesis, University of Ulm, 2011).

[33] Thomas S. Heydt-Benjamin et al., "Privacy for Public Transportation", in *Privacy Enhancing Technologies* (2006); Erik-Oliver Blass et al., "PSP: private and secure payment with RFID," in *WPES* (2009); Foteini Baldimtsi et al., "Pay as you go," in *HotPETs* (2012).

[34] Josep Balasch et al., "PrETP: Privacy-Preserving Electronic Toll Pricing," in 19th USENIX Security Symposium (2010).

[35] Sarah Meiklejohn et al., "The Phantom Tollbooth: Privacy-Preserving Electronic Toll Collection in the Presence of Driver Collusion," in *20th USENIX Security Symposium* (2011).

[36] Xihui Chen et al., "A Group Signature Based Electronic Toll Pricing System," in *ARES* (2012).

ensuring data integrity and authenticity. Popa et al. [37] anonymize vehicles on the move by using random identifiers (tags) to prevent a server from linking user locations, effectively hiding their identity. However, the authors did not implement the proposed solution and fail to evaluate the feasibility of their approach in the given scenario. A limited amount of publications have considered locational privacy in the context of e-mobility so far: Chao Li[38] implement a merchant entity of the Compact e-Cash scheme[39] aimed at a charging station. Liu et al.[40] propose an anonymous electronic payment scheme that supports two-way anonymous payments. Stegelmann's and Kesdogan's approach[41] aims at providing locational privacy in the presence of a smart grid that actively manages EVs as energy buffers. We are not aware of an implementation that allows to evaluate the practicality. While their design incorporates optional anonymity revocation, it relies on an anonymous electronic cash scheme for billing. None of these approaches can be used when anonymous electronic payments are not tolerated by legislation or even just undesired by the vendor.

7 Conclusion

In this paper, we introduced a system that enables locational privacy for financial transactions in the absence of anonymous payments. We focused on the example of re-charging electric vehicles and are able to protect the customer's locational privacy during the whole charging process. Our system also fully supports all requirements needed to bill the customer after the charging process and enables users to roam between different CSs provided by different electric utilities. As such, it covers all relevant aspects required for the charging of EVs. The basic idea of our approach is to adapt a group key signature scheme to the tightly regulated setting of selling electric energy as means of propulsion. We described all protocol steps and outlined how the system can be deployed in practice. In an empirical evaluation, we also demonstrated that the solution has a low overhead and can scale to millions of charging processes per hour (even on off-the-shelf hardware).

---

[37]Raluca Ada Popa, Hari Balakrishnan, and Andrew Blumberg, "VPriv: protecting privacy in location-based vehicular services," in USENIX Security Symposium (2009).

[38]Chao Li, *Anonymous Payment Mechanisms for Electric Car Infrastructure*, (master's thesis, LU Leuven, 2011).

[39]Jan Camenisch, Susan Hohenberger, and Anna Lysyanskaya, "Compact E-Cash," in *Advances in Cryptology—EUROCRYPT* (2005).

[40]Joseph Liu et al., "Enhancing Location Privacy for Electric Vehicles (at the right time)," in *ESORICS* (2012).

[41]Mark Stegelmann and Dogan Kesdogan, "Design and Evaluation of a Privacy-Preserving Architecture for Vehicle-to-Grid Interaction," in *EuroPKI* (2012).

Appendix 1: Implementation Details

The current source code is a makefile project, written in C. We chose the language C, as the external routines and the libraries we rely on are also written in C, hence the whole project and its dependencies are written in one language. We implemented XSGS as a library. This XSGS library uses the GNU Multiple Precision Arithmetic Library1[42] for the basic arithmetic operations, the Pairing-Based Cryptography Library2[43] (PBC) for the curve and pairing-based arithmetic operations, the optimized reference implementation of the authors for the SHA3 hash algorithm (Keccak3) and the OpenSSL Library4 for RSA signature and certificate support.

At compile time one can choose between the TCMalloc Library[44] for a fast and multithreaded malloc() or the GNU C Library memory allocation, which will be linked to the XSGS library.

Appendix 2: Cryptographic Parameters

The PBC library defines a variety of pairing types, of which our XSGS implementation uses either type D, F, or G, respectively. The type can be chosen at compile time. The group order is ~300 bits, the curve parameters are as follows: $r \geq 160$, $q \geq 1024/k$, $k = 6$ (type D) 12 (type F) 10 (type G).

Where Paillier's operations are used, the modulus is of 1024 bit; RSA can by chose at compile time to use key lengths of either 1024, 2048, or 4096. The cryptographic hash function used throughout the XSGS implementation is the SHA3 contest winner Keccak with 256 bit hash length.

Bibliography

Balasch, Josep, Alfredo Rial, Carmela Troncoso, Christophe Geuens, Bart Preneel, and Ingrid Verbauwhede. 2010. PrETP: Privacy-preserving electronic toll pricing. In *19th USENIX Security Symposium*.

Baldimtsi, Foteini, Gesine Hinterwalder, Andy Rupp, Anna Lysyanskaya, Christof Paar, and Wayne Burleson. 2012. Pay as you go. In *HotPETs*.

Bellare, Mihir, Juan A. Garay, and Tal Rabin. 1998. Fast batch verification for modular exponentiation and digital signatures. In *EUROCRYPT*, 236–250.

Bellare, Mihir, Haixia Shi, and Chong Zhang. 2005. Foundations of group signatures: The case of dynamic groups. In *CT-RSA*, 136–153.

---

[42]https://gmplib.org/.
[43]https://crypto.stanford.edu/pbc/.
[44]https://code.google.com/p/gperftools/.

Beresford, Alastair, R., and Frank Stajano. 2003. Location privacy in pervasive computing. *IEEE pervasive computing 2*, 1 (Mar 2003): 46–55. ISSN: 1536-1268. doi:10.1109/M PRV.2003.1186725.

Blass, Erik-Oliver, Anil Kurmus, Refik Molva, and Thorsten Strufe. 2009. PSP: Private and secure payment with RFID. In *WPES*.

Blumberg, Andrew, J., and Peter Eckersley. 2009. On locational privacy, and how to avoid losing it forever. Technical report. *Electronic frontier foundation*. https://www.eff.org/wp/locational-privacy. Accessed 4 Feb 2013.

Boneh, Dan, Xavier Boyen, and Hovav Shacham. 2004. Short group signatures. In *CRYPTO*, 41–55.

Brands, Stefan. 1993. Electronic cash systems based on the representation problem in groups of prime order. In *CRYPTO*.

Camenisch, Jan L., Jean-Marc Piveteau, and Markus A. Stadler. 1994. An efficient electronic payment system protecting privacy. In *ESORICS*.

Camenisch, Jan, Susan Hohenberger, and Anna Lysyanskaya. 2005. Compact e-Cash. In *Advances in cryptology—EUROCRYPT*.

Camenisch, Jan, and Anna Lysyanskaya. 2002. Dynamic accumulators and application to efficient revocation of anonymous credentials. In *CRYPTO*, 61–76.

cars21.com. 2013. EU proposes minimum of 8 million EV charging points by 2020. http://beta.cars21.com/news/view/5171. Accessed 29 Jan 2013.

Chao Li. 2011. Anonymous payment mechanisms for electric car infrastructure. *Master's thesis*, LU Leuven.

Chase, Melissa, and Anna Lysyanskaya. 2006. On signatures of knowledge. In *CRYPTO*, 78–96.

Chaum, David. 2013. Blind signatures for untraceable payments. In *Advances in cryptology: Proceedings of CRYPTO '82. 1982. Security without identification: Transaction systems to make big brother obsolete*. Communication ACM 28, 10 (Oct 1985): 1030–1044. ISSN: 0001-0782. doi:10.1145/4372.4373. http://doi.acm.org/10.1145/4372.4373. Accessed 23 Jan 2013.

Chaum, David, Amos Fiat, and Moni Naor. 1988. Untraceable electronic cash. *In Advances in cryptology—CRYPTO*.

Chaum, David, and Eugne van Heyst. 1991. Group signatures. In *EUROCRYPT*, 257–265.

Chen, Xihui, Gabriele Lenzini, Sjouke Mauw, and Jun Pang.2012. A group signature based electronic toll pricing system. In *ARES*.

Danezis, George, Roger Dingledine, and Nick Mathewson. 2003. Mixminion: Design of a type III anonymous remailer protocol. In *IEEE Symposium on Security and Privacy*.

Delerable, Ccile, and David Pointcheval. 2006. Dynamic fully anonymous short group signatures. In *VIETCRYPT*, 193–210.

Dingledine, Roger, Nick Mathewson, and Paul Syverson. 2004. Tor: The second-generation onion router. In *13th USENIX Security Symposium*.

Dtzer, Florian. 2006. Privacy issues in vehicular Ad Hoc networks. In *Privacy enhancing technologies*.

Freudiger, Julien, Maxim Raya, Mrk Flegyhzi, Panos Papadimitratos, et al. 2007. Mix-zones for location privacy in vehicular networks. In *Win-ITS*.

Heydt-Benjamin, Thomas S., Hee-Jin Chae, Benessa Defend, and Kevin Fu. Privacy for public transportation. In *Privacy enhancing technologies*.

Hubaux, Jean-Pierre, Srdjan Capkun, and Jun Luo. 2004. The security and privacy of smart vehicles. *Security and Privacy, IEEE 2*, 3: 49–55.

Jackson, Ian. 1996. Anonymous addresses and confidentiality of location. In *Information hiding*.

Jager, Tibor, Florian Kohlar, Sven Schge, and Jrg Schwenk. 2012. On the security of TLS-DHE in the standard model. In *Advances in cryptology—CRYPTO*.

Kim, Kitae, Ikkwon Yie, Seongan Lim, and Daehun Nyang. 2011. Batch verification and finding invalid signatures in a group signature scheme. *I. J. Network Security 13* 2: 61–70.

John Krumm. 2007. Inference attacks on location tracks. In Pervasive computing (Pervasive 2007).

Liu, Joseph, Man Au, Willy Susilo, and Jianying Zhou. 2012. Enhancing location privacy for electric vehicles (at the right time). In *ESORICS*.

Ma, Chris Y.T., David K.Y. Yau, Nung Kwan Yip, and Nageswara S.V. Rao. 2010. Privacy vulnerability of published anonymous mobility traces. In *MobiCom '10*.

Ma, Zhendong. 2011. Location privacy in vehicular communication systems: A measurement approach. *Ph.D. dissertation*, University of Ulm, Ulm.

Meiklejohn, Sarah, Keaton Mowery, Stephen Checkoway, and Hovav Shacham. 2011. The phantom tollbooth: Privacy-preserving electronic toll collection in the presence of driver collusion. In *20th USENIX Security Symposium*.

Möller, Ulf, Lance Cottrell, Peter Palfrader, and Len Sassaman. 2003. Mixmaster protocol I Version 2. http://www.abditum.com/mixmaster-spec.txt.

Montjoye, Yves-Alexandre de, Csar A. Hidalgo, Michel Verleysen, and Vincent D. Blondel. 2013. Unique in the crowd: The privacy bounds of human mobility. *Scientific Reports*. http://www.nature.com/srep/2013/130325/srep01376/full/srep01376.html.

Naor, Moni, and Moti Yung. 1989. Universal one-way hash functions and their cryptographic applications. In *STOC*, 33–43.

Nguyen, Lan. 2005. Accumulators from bilinear pairings and applications. In *CT-RSA*, 275–292.

Popa, Raluca Ada, Hari Balakrishnan, and Andrew Blumberg. 2009. VPriv: Protecting privacy in location-based vehicular services. In *USENIX Security Symposium*.

Popa, Raluca Ada, Andrew J Blumberg, Hari Balakrishnan, and Frank H Li. 2011. Privacy and accountability for location-based aggregate statistics. In *ACM CCS*.

Research, Pike. Electric Vehicle Market Forecasts. 2013. http://www.pikeresearch.com/research/electric-vehicle-market-forecasts. Accessed 29 Jan 2013.

Sampigethaya, K., Mingyan Li, Leping Huang, and R. Poovendran. 2007. AMOEBA: Robust location privacy scheme for VANET. *IEEE Journal on Selected Areas in Communications* 25, 8 (Oct 2007): 1569–1589. ISSN: 0733-8716. doi:10.1109/JSAC.2007.071007.

Shokri, R., G. Theodorakopoulos, J. Le Boudec, and J. Hubaux. 2011. Quantifying location privacy. In *2011 IEEE Symposium on Security and Privacy (SP), May 2011*. doi:10.1109/SP.2011.18.

Stegelmann, Mark, and Dogan Kesdogan. 2012. Design and evaluation of a privacy-preserving architecture for vehicle-to-grid interaction. In *EuroPKI*.

Development Towards a Learning Health System—Experiences with the Privacy Protection Model of the TRANSFoRm Project

Wolfgang Kuchinke, Christian Ohmann, Robert A. Verheij, Evert-Ben van Veen and Brendan C. Delaney

Abstract The connection of clinical care with clinical research is the main purpose of the Learning Health System (LHS) integrating scientific information, informatics, and patient care. The LHS generates new medical knowledge as a by-product of the care process. For this purpose, the aggregation of data from Electronic Health Records (EHR), Case Report Forms (CRF), web questionnaires with other data sources like primary care databases and genetic data repositories is necessary for research purposes. This joining of healthcare and research processes results in challenges for the privacy protection framework of the LHS. Based on an exploration of EU legal requirements for data protection and privacy, different data access policies of data provider organizations as well as existing privacy

---

W. Kuchinke (✉)
Coordination Centre for Clinical Trials, Heinrich-Heine-University Düsseldorf, Universitätsklinikum, Geb.14.75, Moorenstr. 5, 40225 Düsseldorf, Germany
e-mail: kuchinke@med.uni-duesseldorf

C. Ohmann
European Clinical Research Infrastructures Network (ECRIN), Prinz-Georg-Str. 51, 40477 Düsseldorf, Germany
e-mail: Christian.Ohmann@med.uni-duesseldorf.de

R.A. Verheij
NIVEL, Netherlands Institute for Health Services Research, Otterstraat 118-124, 3513 CR Utrecht, The Netherlands
e-mail: R.Verheij@nivel.nl

E.-B. van Veen
MedLawConsult, Javastraat 76, 2584 AS Den Haag, The Netherlands
e-mail: eb.vanveen@medlaw.nl

B.C. Delaney
Medical Informatics and Decision Making, Imperial College London, South Kensington Campus, London SW7 2AZ, UK
e-mail: brendan.delaney@imperial.ac.uk

frameworks of research projects, basic privacy principles and privacy requirements were extracted. Based on privacy principles and legal requirements a graphical model to display privacy protection requirements was created. This graphical model is based on concepts of requirements engineering and can be used like a model kit to create new privacy frameworks and to ease knowledge exchange with stakeholders of the LHS. Our model is built upon the concept of three privacy zones (Care Zone, Non-Care Zone and Research Zone) representing areas where similar legal requirements and rules apply. These zones contain databases, data transformation operators, such as data linkers and privacy filters and graphs to indicate the data flow necessary for research processes. The aim of the model is to help arrange its components in a way that creates a risk gradient for the data flow from a zone of high risk for patient identification to a zone of low risk. The model is applied to the analysis of several general clinical research usage scenarios and two research use cases from the TRANSFoRm project (finding patients for clinical research and linkage of databases). Both use cases represent different data collection aspects of the LHS. The model was used during discussions with data managers from the NIVEL Primary Care Database in the Netherlands and validated by representing an approved research case of using primary care data employing NIVEL services. Experiences with the graphic privacy model used to improve the privacy framework of TRANSFoRm and with the presentation of the model to LHS stakeholders and the research community are discussed.

Keywords Learning health system · Data sharing · Privacy · Legal requirements · Legislation · Zones · Data protection directive · Code of conduct · Model building · Graphic representation · Data linking

1 Introduction

1.1 Learning Health System

The connection of clinical care with clinical research is the main objective of the Learning Health System (LHS), an international initiative that aims to establish an advanced healthcare system able to integrate scientific information, informatics, and patient care. The LHS generates new medical knowledge as a by-product of the care process with the aim to improve patient's health and safety.[1,2] For this purpose, the routinely and secure aggregation of data from diverse data sources

---

[1]Leigh Anne Olsen., Dana Aisner, J. Michael McGinnis, The Learning Healthcare System: Workshop Summary (IOM Roundtable on Evidence-Based Medicine), Institute of Medicine, Washington D.C.: The national Academies Press (US), 2007.
[2]Charles P. Friedman, Adam K. Wong, David Blumenthal, "Achieving a Nationwide Learning Health System." Sci. Transl. Med. 2, (2010): 57cm29.

plays a major part of the LHS, in addition to the conversion of data into knowledge that can be employed in decision support systems.[3] Each participant in the process of the LHS, clinician, patient, or researcher, is both, a consumer and a producer of data. This interaction makes the LHS a highly collaborative structure[4] and a challenge for applying legal requirements. Recent LHS-related efforts have focused on integrating clinical research and patient care workflows through the joint employment of Electronic Health Records (EHR), Personal Medical Records (PMR) and Case Report Forms (CRF) resulting in the need for robust protection of personal data and integration of privacy protection and trust into the data flow within the LHS.[5] This joining of healthcare and research is realised in the LHS in the double role of the doctor as carer and as researcher, as well as in data sharing between EHR (care) and CRF (research). In the LHS different forms of trust are joint, the special trust relationship between patient and doctor in the care context and trust in the proper research process, where the focus is on following the research protocol. The consequence of connecting healthcare with research context is that in the LHS all regulations, rules, guidelines and standards relevant for both areas apply.

1.2 The TRANSFoRm Project: On the Way to an European Learning Health System

TRANSFoRm (Translational Medicine and Patient Safety in Europe)[6] is a project funded partially by the European Commission[7] that develops the digital infrastructure for a "Learning Health System" (LHS) in Europe.[8] Originating in the US by

---

[3]For the fully established LHS, not only regulations for health care, clinical research, epidemiological research, but also for medical devices apply.

[4]Diane J. Skiba, "Informatics and the Learning Healthcare System." Nursing Education Perspectives. 32, (5), (2011): 334–336.

[5]Claudia Grossmann, Brian Powers, J. Michael McGinnis, editors. Digital Infrastructure for the Learning Health System: The Foundation for Continuous Improvement in Health and Health Care: Workshop Series Summary. Institute of Medicine (US), Washington D.C.: National Academies Press (US), 2011.

[6]"TRANSFoRm project," www.transformproject.eu, accessed March 1, 2015.

[7]The TRANSFoRm project has received funding from the European Union's Seventh Framework Programme for research, technological development and demonstration under grant agreement no 247787 [TRANSFoRm].

[8]Sarah N. Lim Choi Keung, Lei Zhao, Theodoros N. Arvanitis, Vasa Curcin, Brendan Delaney, Jean-François Ethier, Anita Burgun, Mark McGilchrist, Piotr Brodka, Wlodzimierz Tuliglowicz, Anna Andreasson. "TRANSFoRm: Implementing a Learning Healthcare System in Europe through Embedding Clinical Research into Clinical Practice." Paper presented at the 48th Hawaii International Conference on System Sciences, Hawaii, US, January 6, 2015.

the Institute of Medicine (IOM),[9] the concept of the LHS was taken up by TRANSFoRm and adapted to European needs including European legal requirements. TRANSFoRm integrates data collection in primary care practices, data mining in primary care databases and decision support with the aim to improve both patient safety and the conduct and volume of clinical research. The project builds on the realization that IT systems in primary care settings (e.g., general practices) represent a large and valuable source of electronic clinical data at patient level; it doesn't consider secondary care data or hospital data, even though, this kind of data is important for the LHS, too.

To support the development of a framework for privacy compliant data flow for the TRANSFoRm project, a generic graphic model was needed to enable the representation of core privacy and confidentiality concepts for the use of health data for research purposes in an easily understandable way. The privacy protection landscape, and here especially the one for medical research within the LHS, has become so heterogeneous and interdependent that we felt the need for a form of graphical support to enable and improve discussions with the different stakeholders involved in data sharing in the TRANSFoRm clinical use cases. Such a graphical model has to be able to distinguish the various phases of the research data flow from primary data sources until data reaches the researcher for analysis. It should be applicable to all commonly found research scenarios to further the understanding of privacy protection requirements covering the different privacy needs for research with primary care data, EHRs, clinical research databases (Case Report Form based data), and data stored in genetic and cancer registries. Here we report our experiences with this graphical model used in discussions with stakeholders of the LHS and used to create the privacy framework of TRANSFoRm. The goal is to demonstrate that a graphical way to depict data privacy requirements applied to concrete data flow and data linkage needs in TRANSFoRm research use cases is useful to find solutions for privacy protection of the LHS.

1.3 Diverse Data Sources and Diverse Legal Requirements for Privacy Protection in the Care and Research Domain

The TRANSFoRm project focuses on primary care in physician practises and the employment of the EHR as central device, but it allows also for data sharing with patient data registries, study data bases, death registries, and cancer registries. It was one of the first results of the TRANSFoRm project to realise that, although being subject to the EU Data Protection Directive (DPD), each of the data providers may apply differences in the privacy and data security rules and data access

---

[9]See: Web site about The Learning Health Care System in America: http://www.iom.edu/activities/quality/learninghealthcare.aspx.

and protection policies.[10] There may for example be differences in the rights of a physician as researcher to re-use care data, in the involvement of an Ethics Committee or Data Protection Committee, differences in best practices for medical conduct, and in the definition of data processing and anonymisation. In the context of research as part of the LHS, clinical trials are subject to drug laws and the GCP Directive; in both data protection plays an important part.

The standardisation efforts by the implementation of the DPD has not prevented that national legal requirements have to be considered for medical research. In the UK, for example, the current national regulatory framework consists of the Common law recognising an obligation of confidence that arises within the particular relationship between doctor and patient and the Data Protection Act 1998 (DPA) that was introduced in response to the EU DPD 1995. In addition, the Human Rights Act 1998, Health and the Social Care Act 2001 (England and Wales), demanding that all research using identifiable patient data requires the consent of the individuals involved, the Research Governance Framework for Health and Social Care for all research involving identifiable NHS patient data, requiring that research undertaken in the NHS can only take place following approval by the NHS REC. In case personal patient data is involved, an approval by a "Caldicott Guardian" is required. Additional guidance on confidentiality and the use of medical data in research has been issued by organisations like the British Medical Association, the General Medical Council and the Medical Research Council.[11] The confidentiality of patient records in the practice forms part of the ancient Hippocratic Oath, and was always central to the ethical tradition of medicine. This form of confidentiality is in line with the legal requirements of the Data Protection Acts in different EU member states, under which personal data must be obtained for a specified purpose, and must not be disclosed to any third party.

For the European LHS the DPD sets the stage for developing a privacy framework. The right of data protection is a fundamental right stated in Article 8 of the EU Charter of fundamental rights. According to this provision, everyone has the right to the protection of their own personal data. Such data must be processed fairly and for a specified purpose and either on the basis of consent of the person concerned or some other legitimate basis being laid down by national laws. In addition, all data contained in medical documentation and in the EHR should be considered to be "sensitive personal data" being subject to the special data protection rules on the processing of sensitive information contained in Art. 8 of the DPD. According to Art. 8 (2) (c) of the DPD, the processing of sensitive personal data can be justified if it is necessary to protect the vital interests of the data subject. But it may be the case that only those healthcare professionals who are

---

[10]Christian Ohmann, Wolfgang Kuchinke, Helen Corley. Deliverable 3.2: A review of the European clinical trial and data confidentiality legal frameworks. TRANSFoRm, March 31, 2011.

[11]Data Protection and Medical Research. By the Parliamentary Office of Science and Technology (London). Postnote 235, 2005. http://www.parliament.uk/documents/post/postpn235.pdf.

involved in the patient's treatment are allowed to access medical records and may use this data. In addition, Art. 8 (2) (a) of the DPD allows the use of data when the data subject has given his/her explicit consent to the processing of those data.

In the context of clinical trials, anonymisation of patient records and freely given informed consent are the foundations for ethically conducted medical research. Clinical trials by definition deal with health data, which are sensitive data. All activities in a clinical trial are highly regulated, which often increases the complexity of international trials that collect personal data of thousands of patients.[12] The Good Clinical Practice guideline ICH GCP[13] considers privacy and requires that the confidentiality of records that could identify subjects should be protected, respecting the privacy and confidentiality rules in accordance with the applicable regulatory requirement(s). In clinical trials, investigators are considered data controllers and have to ensure that their respective national data protection law applies.

The legal requirements as demanded by the DPD and the different national data protection laws are quite clear, but interpreting and applying data protection laws is not straightforward when the treating physician acts as a researcher, when care data is re-used for research purposes unrelated to the initial disease, when genetic profiling is involved and when results from research must be returned to the patient. Here legal requirements are complemented by ethical demands.

Building trust among all stakeholders of the LHS infrastructure, but in particular addressing the needs of patients and study participants, is vital for its proper functioning. Therefore, for the TRANSFoRm project the creation of a sustainable privacy framework was essential. But the development of such a privacy framework for the LHS confronted us with the need of an in-depth discussion of the meaning of privacy and its reach in the LHS. Privacy is still a controversial concept, with different interpretations. In TRANSFoRm, we had defined two research use cases, an epidemiological cohort study on Diabetes treatment and genetic risk factors, and a randomised clinical trial about different GORD treatment regimes. Both use cases required solutions for data access and linkage of different data sources, like primary care data repositories, EHR data bases, clinical research databases, genetic and cancer registries. We approached the legal issues not top-down from the lawyer's point of view, but rather bottom-up from the one of the researcher, analysing privacy requirements associated with actual data flows in both TRANSFoRm use cases. For us, the LHS is an information generating system, requiring data collection and sharing not only for individual patients, but also

---

[12]Kristof Van Quathem. Controlling personal data—The case of clinical trials. Data Protection Compliance Advisor, Covington and Burling, 2005. https://www.cov.com/~/media/files/corporate/publications/2005/10/oid64167.ashx.

[13]ICH E6: Good Clinical Practice: Consolidated guideline, CPMP/ICH/135/95, 1996.

for population health and research studies. The overall situation is one of complexity, diversity, and constant change.[14]

2 Methods

In Europe, the US, and elsewhere in the world data privacy protection frameworks exist that define in the form of general rules and sometimes specific requirements, how to protect personal data (e.g., EU Data Protection Directive,[15] HIPAA (Health Insurance Portability and Accountability Act),[16] OECD (Organization for Economic Co-operation and Development) Privacy Framework,[17] APEC (Asia-Pacific Economic Cooperation),[18] Madrid resolution 2009,[19] and US/EU Safe Harbour Agreement.[20] Some of the international frameworks, like OECD and APEC, are rather general, whereas the EU data protection Directive and HIPAA address specifically health data and are often concrete in their legal requirements. Not only in Europe, but also in the US, a discussion of legal interoperability exists. For example, in the US the interpretation of HIPAA rules has not been uniform and reconciliation with other federal regulations is often difficult.[21] These data privacy protection frameworks were analysed[22] to extract guiding principles handling

---

[14]Ian Foster, Building a secure Learning Health System. In: Digital infrastructure for the Learning Health System. Workshop series summary. Eds.: Claudia Grossmann, Brian Powers, and J. Michael McGinnis, Institute of Medicine, The National Academies Press, Washington, D.C., USA, 161–165.

[15]EU Directive 95/46/EC of the European Parliament and of the Council of 24 October 1995 on the protection of individuals with regard to the processing of personal data and on the free movement of such data. Official Journal of the European Communities, 1999; No. L281/31-281/39.

[16]Health Insurance Portability and Accountability Act of 1996, Public Law 104-191, Report 104-726, 104th Congress (1996).

[17]"The OECD Privacy Framework", OECD Paris, France (2013), accessed March 3, 2015. http://www.oecd.org/sti/ieconomy/oecd_privacy_framework.pdf.

[18]APEC Secretariat, APEC privacy framework. Singapore, 2005. ISBN 981-05-4471-5.

[19]International Standards on the Protection of Personal Data and Privacy. The Madrid Resolution. Spanish Data Protection Agency, Madrid, 2009.

[20]US-EU Safe Harbor Framework Documents. US Federal Register, July 24, 2000, accessed March 3, 2015. http://export.gov/safeharbor/eu/eg_main_018493.asp.

[21]Sharyl J. Nass, Laura A. Levit, Lawrence O. Gostin, editors, Beyond the HIPAA privacy rule: enhancing privacy, improving health through research. Institute of Medicine. Washington D.C. (US): The National Academies Press, 2009.

[22]For example, there are conditions explained in the Safe Habour framework, where the processing of sensitive data is allowed by providing explicit (opt-in) consent, or for example, is in the vital interests of the data subject, or is being carried out in the course of legitimate activities by a foundation, association, etc. See: http://export.gov/safeharbor/eu/eg_main_018375.asp, accessed March 3, 2015.

confidentiality and data privacy issues. The extracted principles were rated for their importance for research purposes. Because the mentioned regulatory frameworks are different in nature and content, drawing a common set of principles is challenging. The problem is that legal statements give way to interpretation. In this way, legal requirements have been applied differently in the various data access and processing policies of data providers. First, we analysed the legal landscape, extracting applicable rules. We build mainly on the statements of the Data Protection Directive (DPD).[23] Second, general principles were extracted based on the interpretation of legal rules and application of these rules in policies. Third, we created a graphic model to be able to express these principles and rules with graphs and symbols.

With the LHS we have a system that is highly complex and the definition of privacy is far from clear and may change with the data flow. To address this problem, we used concepts from requirements engineering, like structural requirements and activity diagrams and adapted them for the LHS. To make interdependencies of legal requirements more easily understandable, we developed a graphic method to represent the data flow in both use cases, then gradually include more and more legal requirements and assess their influence on the data flow. During discussions with various stakeholders, we became aware that a formal graphic representation would be helpful to depict the influence of legal requirements on the activities in the LHS. In software engineering, such graphic ways to represent and analyse data flows, requirements, and components as diagrams are a common instrument.[24] In this context, we searched for some form of graphic representation to enable a joint understanding between different data providers and data consumers and their different privacy requirements.

A formal description in the form of Unified Modeling Language (UML)[25] activity diagrams[26] of the data flow of two clinical use cases was developed allowing a structured representation of data confidentiality and data privacy issues. To support this approach a graphic model to display privacy and data security requirements and specifics of database access and linkage was created. As a starting point the graphic symbols were used to depict the data flow of two TRANSFoRm use cases; first, prediction of individual type 2 Diabetes complication risks using single nucleotide polymorphisms (SNPs) and EHR data; second, patient-reported outcomes and long term risk on-demand versus continuous use of proton pump

---

[23]EUR-Lex: Directive 95/46/EC, http://eur-lex.europa.eu/legal-content/EN/ALL/?uri=CELEX:31995L0046.

[24]Doug Rosenberg and Matt Stephens. Use Case Driven Object Modeling with UML. Theory and Practice. APress, Berkely, USA, 2007.

[25]Unified Modelling Language™ (UML) Resource Page, Object Management Group; accessed March 3, 2015. http://www.uml.org.

[26]Unified Modeling Language (UML) is a general-purpose modeling language in the field of software engineering, which is designed to provide a standard way to visualize the design of a system (Wikipedia).

inhibitors in gastroesophageal reflux disease (GORD). The model was enriched by additional symbols to adapt UML to our data flow/data linkage needs. We used new symbols as little as possible, but introduced specific ones for privacy functions (filters), users, databases, and data linkage using one-way and two-way coding. The graphic model was created by defining the system in question (healthcare research domain), identifying relevant elements and features (based on the workflow and data flow of TRANSFoRm clinical use cases), defining risks to privacy and finally by combining this information and elements to build a conceptual model. Our model avoids mapping every single pseudonymisation or obfuscation method or access restriction technique for privacy filters in its diagrams. We have included only basic components and high-level concepts (e.g. privacy filter, zones), but no detailed anonymisation methods. Nonetheless, any insertion of a specific privacy tool (e.g. anonymisation algorithm) in the privacy filter by the user of our model should be done according to a risk analysis.

3 Results

3.1 Common Privacy Principles and Privacy Requirements of the LHS

The analysis of existing data privacy protection frameworks and guidelines (Table 1) resulted in the extraction and formulation of privacy principles that were condensed around a number of key terms relevant for our project (Table 2): policies, responsibility of the treating physician, data chain, data flow requirements, patient questionnaires, explicit consent, trust, indirectly identifiable data, third parties, pseudonymisation, data controller and contractual agreements. These key terms are a component of most privacy frameworks and seem to present essential high level concepts under which a multitude of different and always improving technical realizations exist.

But privacy frameworks have been criticised that they are becoming less able to protect privacy and privacy frameworks establishes for EU projects (e.g. ACGT, p-medicine) often apply very stringent approaches to control their research data flow, requiring combinations of explicit consent, restrictive definitions of anonymisation, data encryption and data usage contracts.[27, 28, 29, 30]. For some

---

[27]Martin Enserink, Gilbert Chin. The end of privacy. Science 347(6221), 2015: 490–491.

[28]Tene, Omer. "Privacy - The next generations".

[29]Privacy trends 2014. Privacy protection in the age of technology. 2014 Ernst & Young Global Limited.

[30]Chris Pounder. Why the APEC Privacy Framework is unlikely to protect privacy. http://www.bakercyberlawcentre.org/ipp/apec_privacy_framework/0710_pounder.pdf.

Table 1 Overview of regulations and guidelines considered for the analysis of legal requirements (guidelines are only examples)

Regulations and guidelines	Key points
Directive 95/46/EC (DPD)	Protection of personal data. The Directive says nothing on medical research explicitly; its implications for the processing of personal data for medical research must be inferred from the general processing of personal data
Directive 97/66/EC	Privacy in the telecommunications sector
Directive 2002/58/EC	Protection of privacy in the electronic communications sector
Regulation (EC) 45/2001	Processing of personal data by Community institutions and bodies
Treaty on the European Union	Fundamental rights, as guaranteed by the European Convention
European Convention	Article 8: Respect for privacy and family life
EU Charter	Article 8: Protection of personal data. Such data must be processed fairly, for specified purposes and on the basis of the consent of the person concerned or some other legitimate basis laid down by law
Commission decision of 5 February 2010	Transfer of personal data to data processors in third countries
Case law C-518/07	Differences between data protection by public bodies and by non-public bodies
Status of implementation of Directive 95/46	Descriptions of situation in EU Member States
Promoting data protection by Privacy Enhancing Technologies (PETs)	PETs should be developed and more widely used
Declaration of Helsinki, 2008	It is the duty of physicians who participate in medical research to protect the life, health, dignity, integrity, right to self-determination, privacy, and confidentiality of personal information of research subjects
ECRIN SOP: Hernández R. and Sanz WS.: Personal data protection. ECRIN GE-SOP ØØ2-VØ.1.Draft VØ.1	Fairly and lawfully processing, transparency, limited purpose; collection should be adequate, relevant and not excessive for the purpose, etc.
Report of the Care Record Development Board Working Group on the Secondary Uses of Patient Information, 2007	Uses of patient-identifiable data for purposes other than direct patient care, involving an honest broker, a trusted custodian of the data who has the responsibility to implement systems of access
Department of Health: Confidentiality NHS Code of Practice (2003)	Reasonable efforts should be made to ensure that patients understand how their information is to be used. The use of anonymised data is preferable for research purposes. The use of identifiable patient information to support research normally requires explicit consent
NHS Connecting for Health: summary responses to the consultation on the additional uses of patient data	Wherever possible, the best available technologies should be used to improve security and enhance confidentiality. Where researchers are to have access to identifiable patient information, there must be mechanisms of accreditation and accountability

(continued)

Table 1 (continued)

Regulations and guidelines	Key points
Code of Professional Conduct, The Code of Medical Ethics (examples)	Good medical practice (General Medical Council, UK): doctors work in partnership with patients and respect their rights to privacy and dignity Berufsordnung für die in Deutschland tätigen Ärztinnen und Ärzte—MBO-Ä 1997 (Germany)
Code of Conduct for health research (examples)	FMWV Code of Conduct for Health Research (The Netherlands) Code of Practice for Research (King's College London, UK)

research questions, such a restrictive approach can result in difficulties aligning legal requirements with research needs. Therefore, a more flexible approach is needed; and it was our aim to develop our graphic method to aid the creation of more flexible and suitable structures able to guarantee both, privacy of patient data and research with as little restrictions as possible.

It must be considered that the term "privacy framework" can have several meanings. General guidelines exist, like the OECD and APEC privacy frameworks; the ISO/IEC 29100:2011[31] standard is a privacy framework, which specifies a common privacy terminology; and the European privacy framework DPD[32] is a legal framework. Additional project specific frameworks (e.g. PRIPARE,[33] TRANSFoRm,[34] SurPRISE,[35] ACGT,[36] p-medicine,[37] SMART,[38] @neurist project,[39] MediGRID,[40] GenoMatch[41]) exist. In general, these frameworks contain general privacy principles and differ in the degree in which they provide specifications. All of these project specific frameworks were analysed to find concepts and

---

[31] www.iso.org/iso/iso_catalogue/catalogue_tc/catalogue_detail.htm?csnumber=45123.

[32] The European privacy legislation consists of two parts: first, the Data Protection Directive (95/46/EC) sets the common ground across Europe; second, all member states have an own "adequate" level of data protection ensured by the Directive.

[33] http://pripareproject.eu/.

[34] Brendan Delaney, Paul van Royen, Adel Taweel, et al.: TRANSFoRm: Requirements analysis for the learning healthcare system. Abstract. In AMIA 2011 Summit on Clinical Research Informatics, San Francisco, CA: USA, p. 7, 2011.

[35] http://surprise-project.eu/.

[36] http://acgt.ercim.eu/.

[37] http://p-medicine.eu/project/in-brief/.

[38] SMART 2007/0059: Study on the legal framework for interoperable eHealth in Europe, INFSO eHealth Legal, Version: 1.5, Issued on: 15/09.

[39] IT infrastructure for the management, integration and processing of data associated with the diagnosis and treatment of cerebral aneurysm and subarachnoid hemorrhage.

[40] Feasibility and usefulness of grid services in medicine and life sciences, www.medigrid.de/.

[41] http://www.tembit.de/bereiche/healthcare-software/genomatch-university/.

Table 2 Privacy principles of privacy protection frameworks

Privacy principles and their descriptions	
Policies and interest of the patient	Policies should apply to patients participating in research projects or whose data will be used for research purposes The interests of patients who will benefit from the research should be considered The population as a whole has also a right that medical research will be performed from which it might profit
Responsibility of the treating physician	Use of patient data for the best possible treatment of patients Obligation to raise these care standards and to use patient data acquired in the medical care context for medical research
Data chain, data flow	The data flow in the LHS has to be analysed continuously, checking for enrichment, merging or linking of data from diverse data sources Data are combined to acquire sufficient statistical power
Data flow requirements	PETs should be used in the data chain as much as possible The data chain should be transparent to the patient; it should be as short as possible Whenever possible, anonymised data should be used To retain their research value, data should be used coded-anonymous It should always be clear who is the data controller Use of Data Transfer Agreements (DTA)
Explicit consent	Should be obtained whenever direct or indirect identifiable data are used by a third party Consideration that consent to treatment, participation in a trial, or re-use of data for statistical research has a different meaning Exceptions on the consent principle for statistics and health research may be possible
Trust	Personal data is used wisely and diligently in the context of medical research to improve the conditions of all patients. Researchers should not try to retrieve the identity of patients
Indirectly identifiable data	The increased use of genetic data for research results in the need for increased privacy protection and trust
Pseudonymisation	For third parties pseudonymised data may be classified as anonymous, if adequate conditions hold (secure coding, no direct or indirect re-identification)
Data controller	Each data controller of a database ensures compliance of the processes with applicable national/regional data protection legislations
Contractual agreements	These should regulate the transfer of data in the data chain; databases should have data access and usage policies
Data sharing	Under conditions of data protection, data should be shared as widely as possible within the scientific community

mechanisms that might by applicable to the LHS. The TRANSFoRm privacy framework for the two use cases is a collection of principles, rules and requirements that was developed and improved with the help of the graphic model and was mapped to the security infrastructure of the project. It is still in development.

Applying the extracted privacy requirements to the LHS and integrating research requirements into the care domain resulted in several insights. One is that the physician as part of the LHS has the additional obligation to raise care standards by using patient data acquired in the care context for medical research. Nonetheless, this is not a legal requirement, but part of the ethical obligations, laid down in Code of Medical Ethics.[42]

To retain their value for research, data should be coded-anonymous to avoid falling under the scope of the DPD. In this context, each data controller of a database or repository has the responsibility to ensure legal compliance of the data processes. Data collection as part of the LHS, takes place at the medical practice by using the EHR for direct data input or using the already collected data, the Case Report Forms (CRF) for clinical trials data and web questionnaires for patient reported outcome (PRO) data. In TRANSFoRm use cases, this data is enriched by data coming from primary care databases and genetic data repositories. Normally, research and care have been seen as conceptually and practically distinct areas and have had distinct oversight regimes. Recently, a new moral framework for both care and research of the LHS has been developed that merges both area.[43] We are going another way, because we think that the distinction between care and research is well founded, we keep both areas as specific zones. In this context, the EU Data Protection Directive knows different requirements for processing of data in the context of a treatment relationship[44] and in the context of scientific reasons.[45] Because genetic data can be indirectly identifiable, special measures to ensure privacy should be employed, the DPD speaks of additional safeguards.[46] Technically this may be the deletion or obfuscation of data not necessary for research, as well as the strict control of data exchange between partners. Explicit consent from the research subject or patient is a necessary legal requirement for

---

[42]For example, AMA's Code of Medical Ethics, see: http://www.medicalassistantcertification.org/medical-ethics/.

[43]The Hastings Center Report. 45 (1): The Hastings Center, Garrison, NY, USA, 2015.

[44]Art. 8 (3): … for the purposes of preventive medicine, medical diagnosis, the provision of care or treatment or the management of health-care services, and where those data are processed by a health professional subject under national law or rules ….

[45]Art. 6 (1): Further processing of data for historical, statistical or scientific purposes shall not be considered as incompatible provided that Member States provide appropriate safeguards.

[46]Art 6 (1) (e): kept in a form which permits identification of data subjects for no longer than is necessary for the purposes for which the data were collected or for which they are further processed.

the processing of personal data[47]; it should only be obtained when direct or indirect identifiable data are used by a third party. In contrast, consent for participating in a clinical trial requires 'informed consent', but consent for the re-use of data requires 'explicit consent'. The DPD also contains the provision of the vital interest of the patient[48] and it can be discussed what such vital interests are. How is, for example, the interest to regain health a vital interest?

To ensure data protection, the role of a "data controller"[49] who is responsible for the compliant processing of personal data is necessary. Data controllers assure the enforcement of protection principles (e.g., legitimate purpose, accuracy of data) to protect personal data against accidental or unlawful destruction, loss, or disclosure. In our graphical model, the controller can be easily represented by a sub-zone. The implementation of the DPD into national laws has resulted in legal heterogeneity in European member states, caused by variable interpretations of key terms, like anonymisation, informed consent and research exemption.[50] Nonetheless, the DPD provides the possibility of exemptions that may also cover exemptions to the consent principle[51] for the purposes of health research conducted in the general interest (e.g. infectious diseases), a possibility that should always be considered by researchers. Here too, as in the case of "vital interest", the term "general interest" has to be interpreted regarding the importance of public interest, for example the population of all patients with a specific disease. As a consequence, the ease of access to patient data for research purposes, the so-called secondary use of patient data, has become a debated topic in the research community.

In 2014, the European Parliament has voted on its position on the new EU proposal for a General Data Protection Regulation (GDPR), which is being negotiated among the European Parliament, the Council of the European Union and the European Commission.[52] This new European framework for privacy protection will set rules under which personal data are to be handled in all EU countries and how health and research data have to be treated. The research community in

---

[47]Art. 8 (2) (a): the data subject has given his explicit consent to the processing of those data.

[48]Art. 8 (2) ©: processing is necessary to protect the vital interests of the data subject.

[49]European Directive 95/46/EC; 1999, (18): processing is carried out under the responsibility of a controller.

[50]Marieke Verschuuren, Gérard Badeyan, Javier Carnicero, Mika Gissler, Renzo Pace Asciak, Luule Sakkeus, Magnus Sternbeck, Walter Devillé. "Working group on Confidentiality and Data Protection of the Network of competent Authorities of the Health Information and Knowledge strand of the EU Public Health Programme. 2003-08". Eur J Public Health. 18 (2008):550–551.

[51]Art. 8 (4): …Member States may, for reasons of substantial public interest, lay down exemptions in addition to those laid down in paragraph 2 either by national law or by decision of the supervisory authority.

[52]EU Commission, Data Protection Newsroom, Commission proposes a comprehensive reform of the data protection rules. January 25, 2012, accessed March 3, 2015. http://ec.europa.eu/justice/newsroom/data-protection/news/120125_en.htm.

Europe has been concerned about the consequences of the current wording of this draft regulation,[53] because, it may endanger many types of clinical research, epidemiological analysis, biobank research, and research using population-based registries. According to the draft regulation, researchers have to ask for a patient's "specific consent" every time a new research question is applied to already available data (the reuse of existing data). This would result in the cumbersome practice of asking patients for "re-consent" every time already collected data is reused for a new research purpose.

Anonymous data, in contrast, fall outside the scope of the DPD and anonymised patient data can be used for research purposes without consent. But fully anonymous data are often not suitable to answer many forms of health research questions, for example concerning research in the area of personalized medicine and research about relations between exposure, disease onset, treatment regime and life style. The strengthening of trust in the research domain, the supportive use of contractual agreements, stewardship, third party involvement and peer-based control should be considered.

3.2 The Idea of Privacy Zones Was Born

We developed a standardized graphic model to describe data privacy frameworks in primary care research and the LHS using a flexible zone model. Our zone model is based on two components: first, a number of principles extracted from privacy protection frameworks and legal requirements based on an analysis of the legal landscape that can give guidance for issues of data security and confidentiality, and second, formal descriptions for building privacy zones, an identification risk gradient and data flow structures that can represent complex relations between data flow and privacy requirements. For example, the incorporation of genomic data into personal medical records increases the risk of patient identification. In general, the sharing of genetic data requires de-identification by removing explicit identifiers (e.g., name, address, date of birth or social security number) and incorporation of sound security design principles. But all methods suffer from an absence of formal modelling of inferences from linked data sets raising the question, how current privacy and data protection frameworks can withstand advanced re-identification efforts, like genotype–phenotype inference, location–visit pattern, family structure pattern, and dictionary attack.

---

[53]LERU (League of European Research Universities); The European Parliament's position on the General Data Protection Regulation threatens EU Research. October 6, 2014, accessed March 3, 2015. http://www.leru.org/index.php/public/news/the-eps-position-on-the-general-data-protection-regulation-threatens-eu-research/.

The graphic model must accommodate the special problems arising from incorporating international research processes and cross-border data sharing into the LHS. International research with primary care data is still being hampered by a lack of a universally agreed definition of privacy.[54] Such a universally binding definition of privacy may be impossible to achieve because such a definition depends on the context of privacy protection.[55]

To describe the impact of privacy principles and privacy requirements on the data flow in the TRANSFoRm use cases, we searched for some representation and came up with the zone concept. It allows us to consider the European legal requirements issued by the DPD for privacy protection in the care zone and privacy protection in the research zone, but can be specific enough to include local and national requirements by introducing sub-zones. The concept of the zone combines several of the privacy requirements, including legal requirements, policies, responsibilities, scope of the data controller, to create environmental constraints that are displayed as zone or sub-zone.

Although, doubt has been raised if a general private versus public dichotomy is still existent when confronted with novel privacy concerns, like technical developments of person tracking and Big Data, it is obvious that privacy protection changes according to the context of use. Context is a set of interacting agents and their norms and rules.[56] For example, the roles in the LHS are doctor, patient, researcher, and investigator. Doctors must share patient health data within the scope of their treatment activities, observing specific norms and rules (see: Code of Conduct), and with the defined purpose to improve the health of the patient. In this way, the flow of information about a subject (acting in a particular role) from one actor to another actor is governed by particular transmission principles.[57] The medical confidentiality laid down in the Code of Conduct and other guidelines together with DPD Art 8 (3), precludes any disclosure. In our model the concept of privacy zones considers such contextual constraints and integrity,[58] including the set of actors, roles, norms and rules. The zone combines in a graphic way context characteristics relevant for patient care and research and applies these characteristics to the data flow. Thus, the zone concept plays a central role in our model; it ensures that the context integrity (for example the context of care) of privacy

---

[54]It should be considered that privacy protection and the protection of personal data is not the same. See: Raphaël Gellert, Serge Gutwirth, Beyond accountability, the return to privacy? In: Daniel Guagnin, Leon Hempel, Carla Ilten, Inga Kroener, David Neyland, Hector Postigo, editors, Managaing Privacy through accountability, Palgrave Macmillan 2012.

[55]Helen Nissenbaum, Privacy in context: technology, policy, and the integrity of social life. Stanford: Stanford University Press, 2010.

[56]Definition from: Nissenbaum, 2010, Ibid.

[57]Helen Nissenbaum, H., "Privacy as contextual integrity", Washington Law Review 79(1) (2004): 119–158.

[58]Helen Nissenbaum, 2004, op. cit.

Fig. 1 Generation of a risk gradient for the identification of patients. Three connected zones and their data sources are shown. *Care Zone*: directly identifiable patient data (e.g. personal data in EHR records). *Non-Care Zone*: pseudonymised data (e.g. cohort study data, clinical trial database, primary care database). *Research Zone*: (coded) anonymised data, data used in research projects. *Arrow*: indicates the risk gradient for patient identification

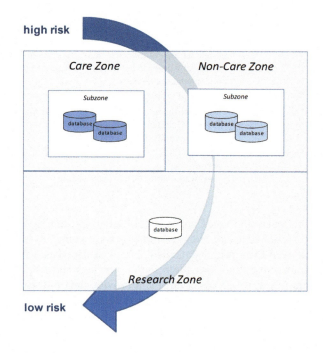

requirements is always considered.[59] Because the concept of "privacy in context"[60] is very comprehensive and encompasses social developments, we limited our use of the term to the dependency of privacy requirements on the data processing environment. When data moves into another zone the context changes accordingly. The overriding rule in our model should be: data flows from a zone with a high risk of patient identification to a zone with a lower risk (Fig. 1).

Zones are areas containing data sources controlled by similar policies/access rules and applicable regulations (Table 3). Three main zones, with a decreasing risk of patient identification, were created: Care Zone, Non-Care Zone and Research Zone. The Care Zone is the area of the treating physician and the patient and of diagnosis and treatment. Here patient identifiability constitutes an essential element of the medical treatment and of the special trust relationship between patient and physician. Therefore, in the Care Zone personal identifiable medical data is stored and used within the care context by the treating physician.

For non-care purposes, patient data is moved outside the Care Zone. Once data is outside the Care Zone and not any more protected by medical confidentiality (Code of Conduct), patient data becomes vulnerable and may be collected for non-care purposes. For example, in some countries, EHR data is regularly transferred to primary care databases. Because data are used for non-care purposes, we called

---

[59]Raphaël Gellert, 2012, op. cit.
[60]Nissenbaum, 2010, op. cit.

Table 3 The main building blocks of the graphic model to display privacy frameworks

Building blocks	Specifics
Zone	• Areas of low, middle and high risk of identification of patients • Areas of rule-based similarity in regard to purpose of data collection, applied policies and regulations
Subzone	• Comparable areas in which data can be used for the same or for a similar purpose
Data protection filter	• Tools for the anonymisation, pseudonymisation, coding and data aggregation • Privacy Enhancing technologies (PET) • Other methods to ensure privacy
Data linker	• Linkage of data sets, enables the connection within or between zones and subzone • One-way coding or two-way coding

this zone the Non-Care Zone, an area that may contain research databases (e.g., for clinical trials, cohort studies), and secondary use databases that have been derived from primary medical care data. Databases that maintain data in the Non-Care Zone often provide research services and usually make use of policies that employ strict access control and provide data in form of pseudonymous or anonymous data. In addition, access may be based on the necessity of an explicit consent (by presenting the patient with an option to agree or disagree with the collection, processing, or disclosure of personal information) or on country-specific or local regulations (e.g., exemptions to consent for research). In the case of NIVEL-PCD[61], EHR data is collected under an opt-out regime; patients can object to the collection of their data for research purposes. Such an approach may result in a problem, because once patient data is transferred to the Non-Care Zone, it is often no longer possible for database owners to identify patients that have objected to the use of their data. These preconditions have to be addressed by suitable consent forms allowing the patient to opt-out for "future use of their data", with the restriction that the data already used in a study cannot be removed retrospectively from an analysis. For the processing of health data for research not always an explicit consent (opt-in) is necessary, because the option for some kind of exemption from this rule is provided by the DPD. NIVEL primary care database[62] operates under provisions of the Dutch law to use extracts of EHR records for research purposes. These provisions are described in the Code of Conduct for health research, issued by the Dutch Federation of Biomedical Scientific Societies[63] and approved by the

---

[61]NIVEL (Nederlands instituut voor onderzoek van de gezondheidszorg), Utrecht, The Netherlands, accessed March 3, 2015. http://www.nivel.nl/.

[62]https://www.nivel.nl/en/dossier/nivel-primary-care-database.

[63]Dutch Code of Conduct for Medical Research, Commissie Regelgeving en Onderzoek, available at: www.federa.org/sites/default/files/bijlagen/coreon/code%20of%20conduct%20for%20medical%20research%201.pdf.

Dutch Data Protection Agency and the Dutch Data Protection Act,[64] requiring that researchers have no access to identifiable patient information and results cannot be traced back to individual persons, health care providers or health care organisations. Participating physicians may withdraw from NIVEL at any time, and without stating reasons.

The third zone is an area where research questions and research projects are addressed employing often de-identified and anonymous data. In the Research Zone the researcher receives data suitable for processing and analysis for specific research projects, based on approved protocols. The Research Zone can receive data from the Non-Care Zone. In the Research Zone de-identified data may be used, but each record may have its own ID (e.g., a pseudonym), which provides a link to the identifiable information stored separately in the None-Care Zone. The researcher often receives not complete databases but data extracts from primary care databases located in the None-Care Zone; these data extracts should be in a form that makes re-identification of patients practically impossible. Such a coding of patient records can be done in two ways: by "one-way" or "two-way" coding. With one-way coding, it is always possible to translate the identifier (ID) into a code number (CN), but not the other way around. Thus, one-way coding can be considered as anonymous and irreversible, because it is not possible to go back from the CN to the ID. With two-way coding, the latter is still possible.[65] For the researcher in the Research Zone it is important to receive data in a form that is suitable to answer the research question, but still coded in a form that makes breaking the anonymisation too expensive.

One should not forget that the Research Zone is not an unrestricted area, but researchers are bound by their own rules, the Code of Conduct for researchers, policies of their institutions and the control by the peer community. In addition, agreements under which data is transferred and provided for processing are used for privacy protection[66] and recently, possibilities to charge fines to prevent breaches of privacy by researchers are discussed.

3.3 The Graphic Representation of Privacy Frameworks for Compliant Data Flow

The developed privacy protection principles were incorporated into a structured representation to create possible frameworks based on elements of structured analysis and data flow diagrams. A set of formal components (zones/subzones, data

---

[64]Dutch Personal Data Protection Act (unofficial translation): http://www.coe.int/t/dghl/standardsetting/dataprotection/national%20laws/NL_DP_LAW.pdf.

[65]Evert-Ben van Veen, "Europe and tissue research: a regulatory patchwork." Diagn Histopathol 19(9) (2013): 331–336.

[66]Evert-Ben van Veen, 2013, Ibid.

sources, privacy filters/data linkers, actors and roles) were created (Table 3) to be used as building blocks with the aim to allow users a systematic and structured analysis of data flows found in common clinical research scenarios and the LHS. These graphical elements (Fig. 2) can be arranged in two dimensions, to create and test different options of the data flow (e.g. branching, conditional data flow).

A multitude of rules, policies and regulations govern the use of medical data and in the TRANSFoRm use cases of the LHS, different databases in different regions and countries have to be accessed and sometimes linked. To account for this heterogeneity, only three different zones turned out as insufficient and the element of a subzone had to be introduced. Databases in different countries or regions may operate under a different set of rules and policies. Our subzones define such differences in the type of data sources, for example differences in access rules for care databases, health insurance databases and national cancer registries. Each subzone is homogenous in terms of rules and regulations and the extent to which individuals are identifiable. Additional components of our concept of zones/subzones are the roles of informed consent, contractual agreements and database statutes that regulate the transfer of data within and between zones/subzones.

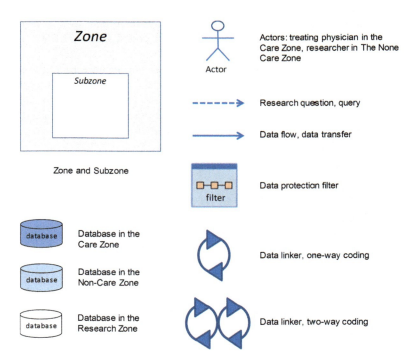

Fig. 2 Definitions of the building blocks of the graphic zone model. The building blocks and their graphic representations are shown

Two types of newly introduced symbols play an important role for enabling the representation of a proper research data flow: first, privacy filters that operate on data and second, data linkers that allow the connection of data sets within or between zones/subzones. We have not defined the technical functions of each privacy filter and linker in detail, and leave a detailed report on the functions and algorithms to an update of the model. Linking can be performed if one set of data relating to an individual is connected with the same identifier (pseudonym) as a record by that same individual in another database. Such use of irreversible pseudonyms can allow for the linkage of records of the same individual simultaneously anonymizing these records (e.g., by "fuzzy matching"). In principle, pseudonymisation is a method by which direct identifiers of a data subject (e.g. name and date of birth) are removed and replaced by a unique number, the pseudonym. Another term for pseudonymisation is "coding of data". In case of combining data from different data sources, adequate precautions must be employed to ensure that individuals are not identifiable due to any combination and linking of data available. To assess the risk of identifiability, one should consider that even a set of laboratory values may become identifiable in a specific context. Especially in the case of the linking of data to genomic data sets, the need for pseudonymisation of indirectly identifiable data must be considered.

Data controller is the natural or legal person, public authority, agency or any other body which alone or jointly with others determines the purposes and means of the processing of data concerned for research. Our term of "controller" is used in a broader meaning than the one in the DPD. In the Directive, "controller" applies only to personal data. As the Directive is not applicable to anonymous data, it lacks a term for the entity responsible for a database with anonymous data. But even for anonymous data, somebody should be responsible. This holds especially for anonymous data that may become indirectly identifiable after linking. Privacy filters are software tools to render data less identifiable or even anonymous to a subsequent controller of these data (as defined in this paper). Nonetheless, the application of pseudonymization and de-identification alone are often not sufficient for the protection against identification. This is because current frameworks tend to be narrow in their interpretation of what is inferable from genomic data as well as what information is already openly available for relating genomic data to identified data.

Once data has entered a zone, the rules and policies of this zone apply. In the LHS the physician can play a double role, as a care giver and as a researcher. As a researcher the physician receives data from many different sources. But as a researcher, the physician is not interested in the single person as such, but in patterns about persons and populations. When the physician act as a researcher and the treatment of the patient is not anymore the main focus, the physician must leave the Care Zone.

3.4 Application of the Graphic Model to Two Research Use Cases

3.4.1 Use Case Linkage of Databases

Often research aims to extract information about selected cases from linked databases. In our model, such linking may be done within a zone or subzone or between zones or subzones. We distinguish two sub-use cases, first linking of a data base in the Care Zone (e.g. EHR) with one in the Non-Care Zone and second, linking two databases in the Non-Care Zone (e.g. primary care databases, registers).[67]

Link Between Databases in the Care Zone with Non-care Zone

In this use case a database in the Care Zone (e.g. EHR database, hospital data warehouse) is linked with a database in the Non-Care Zone. A concrete research scenario may be a physician in need to enrich data of his/her patient with clinical data from a genetic database or clinical database. Another scenario is the recruitment of suitable patients for clinical trials based on data in the physician's EHR and a database (e.g. prescription database).
Data flow:

- Trigger for data linkage by researcher
- Checking the permissibility of linkage of the care database with the Non-Care Zone database by data controllers
- Authorization of data linkage (e.g. by data protection committee)
- Preparation of linkage procedure in Care and Non-Care Zone databases
- Pseudonymisation of care data inside the Care Zone
- Performance of linkage in the Non-Care Zone (e.g. using new pseudonyms)
- Linked database is coded two times (the data are re-coded using a different pseudonymisation method (different key))
- Linked database or data extracts transferred into Research Zone after additional privacy filtering according to the risk of identification through database linkage
- Linked database analyzed according to research question by researcher in Research Zone

In this use case a link is generated between a database in the Care Zone with a database in the Non-Care Zone (Fig. 3) with the purpose to identify patients suitable to participate in a clinical trial. Because in this case already pseudonymised data are linked and combined, research on this data is possible in the Non-Care Zone, although researchers in the Non-care Zone are bound by additional rules and obligations.

---

[67]Kuchinke W, 2014, op. cit.

Development Towards a Learning Healthcare System ...

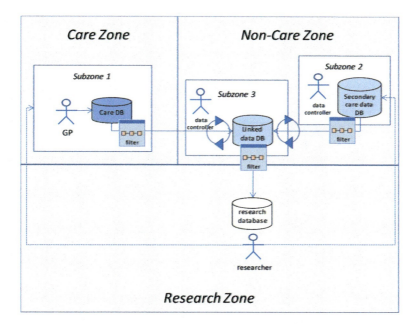

Fig. 3 Linking of databases in the Care and the Non-Care Zone by creation of a linked data database. The researcher receives anonymised data in response to his research question and authorisation, after data originating in a primary care database and a secondary care database have been linked in a separate database of linked data sets (Subzone 3)

Link Between Databases in the Non-care Zone[68]

In this use case two databases in the Non-Care Zone (e.g. primary care database, cancer database, biobank, and register) are linked (Fig. 4). A research scenario may describe a researcher who links data of a prescription database or a cancer register with data derived from a primary care database to study genetic risk factors from a cohort study on medication effects or adverse effects.

Data flow:

- Trigger of data linkage by researcher
- Checking the permissibility of linkage of the Non-Care Zone databases by data controllers
- Authorization of data linkage (e.g. ethical or data protection committee)
- Preparation of linkage procedure in both None-Care Zone databases
- Performance of linkage (e.g. use new pseudonyms)
- Linked database is re-coded (one way coding for anonymous data, two way coding for pseudonymous data)

---

[68]Kuchinke W, 2014, op. cit.

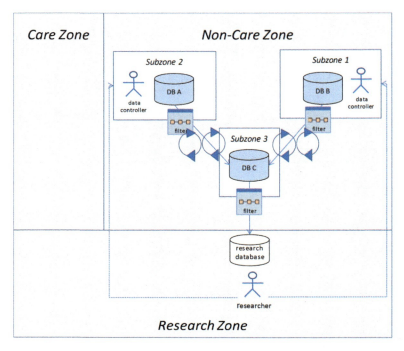

Fig. 4 Linking of two databases in the Non-Care Zone. The researcher receives pseudonymised or coded anonymised data or completely anonymised data in response to his research question and authorisation. Linking is done in Subzone 3

- Linked database or data extracts are transferred into Research Zone after privacy filtering
- Linked database analyzed according to research question by researcher in Research Zone

Because in this case, already pseudonymised data is linked and combined, research on this data is possible in the Non-Care Zone as well as in the Research Zone after an additional filter step. Records with identifiable data must be received, stored and managed in a controlled manner with transition processes of data quality, linkage, and de-identification being undertaken with minimal access to identifiable data. Safe Havens or TTPs as means of controlling access to identifiable data may facilitate the use of potentially identifiable data.[69] In some cases, it may be possible to use irreversible pseudonyms allowing the linkage of records of the same individual and at the same time ensuring effective anonymisation of records.

---

[69]Evert-Ben van Veen, Patient data for health research. A discussion paper on anonymisation procedures for the use of patient data for health research, Den Haag: MedLawConsult, 2011.

3.4.2 Application Example

In the TRANSFoRm project the large Dutch primary care database NIVEL is a partner and NIVEL's integration in the data flow may be an example of how a database service provider can be integrated into the research process. NIVEL's primary care database (NIVEL-PCD)[70] is holding general practice data of nearly 2 million patients. We used our graphic model and its notation to describe the dataflow for the TRANSFoRm Diabetes use case and discussed data provision and data protection measures with NIVEL staff. The aim was to obtain a correct graphical representation of the privacy framework satisfying NIVEL's data protection requirements and enabling research on linked data sets.

NIVEL-PCD services are based on the fact that under certain conditions, Dutch law allows the use of data from electronic health records for research purposes. According to this legislation, neither obtaining a new informed consent from patients, nor approval by a medical ethics committee, is obligatory for this type of observational studies without directly identifiable data.[71] In this agreement, the patient has always the option to opt-out. This approach has been approved by the applicable governance bodies of NIVEL-PCD.[72] This regulation forms the basis for the collection of data recorded by participating physicians by NIVEL to be processed and entered into a database. All patient data are pseudonymised already before leaving the Care Zone using a service of a Trusted Third Party (TTP) (Fig. 5). Thus, identifiable information does not leave the Care Zone. Using these generated pseudonyms, it is technically possible to link data from different health care and other databases, without having to resort to identifiable information. This can be done, because extracts from EHR data are generated in the practices, sent to NIVEL and all data processing and linking is performed in the Non-Care Zone under special rules and obligations and the use of a TTP for managing the generated pseudonyms. Data extracts are stored in a repository in the Non-Care Zone (DB A) that functions as a master database, and are protected by own pseudonyms. DB A holds no identifiable data. In addition, database B holds a patient identifier and another pseudonym. This patient identifier allows for patient identification within the context of the practice. The two pseudonyms can be linked via the TTP. Making it possible to select anonymous patients in DB A and invite them to participate in an additional study for which informed consent must be obtained. After approval by a steering committee, DB A serves as a source for new data extracts that can be made after request for specific research projects (B1, B2 and B3).

---

[70]NIVEL Primary Care Database, http://www.nivel.nl/en/dossier/nivel-primary-care-database, accessed March 3, 2015.

[71]Dutch Civil Law, Article 7:458.

[72]Dutch civil Law http://www.dutchcivillaw.com/civilcodebook077.htm, accessed March 3, 2015.

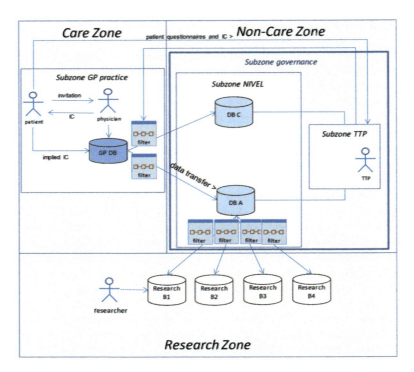

Fig. 5 NIVEL example. By the Dutch law, the use of extracts of electronic health records for research purposes is allowed, based on research exemption with implied consent by the patient. Data recorded by participating GP practices (Care Zone) are sent regularly to NIVEL. These patient data have been pseudonymised before leaving the Care Zone using a Trusted Third Party (TTP) to ensure that easily identifiable information does not leave the Care Zone. Thus, extracts from EHR data are generated in the practices, sent to the NIVEL subzone and stored in a repository in the Non-Care Zone (database A), protected by a pseudonym. From this master database, data extracts can be generated for specific research projects B1, B2, B3 and B4, following approval by a steering committee (NIVEL subzone). These research extracts are subject to a second pseudonymisation step. The researcher will never obtain direct access to NIVEL database A. Instead, to allow for the linkage of data and at the same time, to prevent unauthorised linking of data in the Research Zone, different pseudonyms are generated for each research project. These research data extracts (B1–B4) can then be analysed freely by the researcher (*DB* database, *IC* informed consent, *TTP* trusted third party)

These research extracts are subject to a number of quality and security checks and receive a second pseudonym (Fig. 5). To allow the linkage of data and at the same time, to prevent unauthorized linking of data in the Research Zone, different pseudonyms are generated for each research project.

4 Discussion

Here we reflect on our experiences with the graphical model and not on the privacy framework of TRANSFoRm, because the project is still running. The application example by NIVEL will be a part of it. With the help of the graphical model it was possible to design alternative privacy frameworks for TRANSFoRm use cases and discuss them with involved stakeholders, especially with data providers, data managers, and involved physicians. In general, the presentation of the graphic model was welcomed by data providers, data managers, and physicians and the graphic model was soon used to enrich the discussion. On the other hand, lawyers and legal experts reacted to the graphic model often by incomprehension. The model didn't help structuring their thoughts and they couldn't locate their privacy protection concepts or methods inside the model. What was criticised by legal scientists was that the graphical model doesn't suggest a concrete solution for a new privacy framework. Obviously, it must be stressed that the graphic model is a tool to design possible frameworks using two dimensions and graphs that can be helpful to create concrete privacy frameworks.

Often the hardest step in building a privacy framework is the characterisation what the system is and what is meant by privacy in each step during data processing and data sharing. In the case of privacy protection for the LHS we are potentially dealing with thousands of GP practices, hundreds of clinical trial sites, millions of patients, and several large data repositories. Without clarity on the nature of the LHS and its privacy framework, it will be difficult to create secure and workable solutions. Researchers should spend more time studying the dataflow within the context of the LHS. In general, legal requirements demand that all personal data must be processed fairly and lawfully; collected for specified, explicit and legitimate purposes; the processing of data for scientific purposes is possible by employing appropriate safeguards; data must be accurate and kept up to date; identification of data subjects should be possible for no longer than necessary; and a data controller must exist to ensure these requirements.

Representing knowledge through graphical models has become popular, because graphs can be an intuitive way of representing and visualising the relationships between variables and concepts. Especially, a graph allows abstracting conditional relationships between concepts. Graphical knowledge modelling is a way of representing knowledge structures by linking concepts, procedures and principles in a way that describes a phenomenon; and in this way, it has been used to support the transfer of expertise within organisations.[73] We wanted to use graphical modelling for representing legal requirements and use them to exchange

---

[73]Josianne Basque, Gilbert Paquette, Beatrice Pudelko, Michel Leonard. Collaborative Knowledge Modelling with a Graphical Knowledge Representation Tool: A Strategy to Support the Transfer of Expertise in Organisations. 491–517. In: Knowledge Cartography. Software Tools and Mapping Techniques. (Editors) Alexandra Okada, Simon Buckingham Shum, Tony Sherborne. Springer, London (2008).

expertise between stakeholders involved in the data flow of the TRANSFoRm framework representing the LHS. For this concept, we did not use a static representation, like a concept map,[74] but a dynamic one that considers the flexibility of different data flows for modelling. Because the use of UML activity diagrams[75] is common, we extended this approach to represent legal requirements for privacy frameworks. These privacy requirements should be despicable across both medical care and clinical research domains, considering different data sources (primary, secondary, phenotype data, genotype data), variations between countries and regions in Europe (legal, ethical, practice related) and include policies of primary care databases (e.g. GPRD, LINH), GP practices, clinical research databases (e.g. clinical trials, cohort studies) and repositories (genetic databases). The model has to be graphic, because only in this way privacy requirements can be set easily in relation to the data flow (data chain). As a result, we developed for the TRANSFoRm project a standardised graphic representation to describe legal requirements for data privacy in primary care research in the LHS using a zone model.[76] The zone model has been described in detail and was applied to some of the most common research use cases.[77]

We then used the graphic model to create frameworks that incorporate these principles. The graphic model should not be mistaken for a generic privacy compliance framework, but it employs generic privacy concepts to enable the creation of privacy frameworks that are able to facilitate the understanding of privacy rules in the settings of care and medical research. The strong point is that our graphic model allows for flexibility and choices. It is possible to design different ways of data flows that are all compliant to a set of privacy requirements. In this way, one can test different privacy frameworks to find an optimal solution. Recently, concepts like data guardians or custodians have been introduced to strengthen data against re-identification and to enable the de-pseudonymisation of patients, in case a research result is clinically relevant for an individual.[78] Such innovative components can be easily integrated in the graphic model by introducing a new sub-zone. The graphic model is constructed for global validity; though the shown use cases and especially the NIVEL example are valid for European conditions. To be applied for US conditions, differences in treatment of personal data must be

---

[74]A diagram that depicts suggested relationships between concepts (Wikipedia).

[75]Agile Modeling. UML 2 Activity Diagrams: An Agile Introduction (2014). http://agilemodeling.com/artifacts/activityDiagram.htm.

[76]Wolfgang Kuchinke, Christian Ohmann, Robert A. Verheij, Evert-Ben van Veen, Theodorus N. Arvanitis, Adel Taweel, Brendan C. Delaney, "A standardised graphic method for describing data privacy frameworks in primary care research using a flexible zone model." Int J Med Inform. 83(12) (2014): 941–957.

[77]The zone model of privacy protection and its application has been described in detail in: Kuchinke, W., 2014, op.cit.

[78]Harald Aamot, Christian Dominik Kohl, Daniela Richter and Petra Knaup-Gregori. Pseudonymization of patient identifiers for translational research. BMC Medical Informatics and Decision Making 2013, 13:75–90.

accounted for and it may be necessary to introduce a new symbol for de-identification of data according to HIPAA rules.

The TRANSFoRm project creates a European LHS and during the project, we found that our discussion with stakeholders involved in data sharing always returned to our graphic representation to change data flows and symbols to find new ways to handle privacy requirements and data protection policies. In this way, our model can be used to analyse privacy requirements of data and service providers, like NIVEL and CPRD,[79] and to contribute to discussions of researchers, physicians, database owners,[80] software developers, and others confronted with similar privacy protection problems. To work with three privacy zones is a new approach; until now different concepts were used to model differences in privacy environments and context. Data service providers are increasingly responsible for the storage, processing and integration of patient data leading to the problem of sensitive data stored in systems not under control of the entity which collected the health data. For example, this is the case for primary care databases that are no longer under the control of a physician or hospital. The increasing role of research services and their governance has long been ignored to a large degree by legal regulations.

The main simplification associated with our approach is that with the graphic model it is possible to create privacy frameworks as complicated as one likes. For each data flow, the principles and the different data access and processing policies of the involved data providers have to be observed. It is the duty of data providers, acting as data controllers, to ensure legal compliance of their data processing operations. In this context, in our NIVEL example, the Dutch laws are considered by applying the NIVEL data access and processing policies. The privacy risk assessment must be case specific and should be conducted using the graphic model of a specific data flow in a specific privacy framework. In our case, the NIVEL framework required measures against the increased risk for unauthorised linking of data. Therefore, different pseudonyms were generated for each of the research data extracts.

More and more data is collected from patients, in future even by wearable or implanted sensors, or is made available on open databases on the Internet. A large part of this data will be personal or potentially identifiable data, raising concerns about the suitability of conventional privacy requirements for this development[81, 82].

---

[79]Clinical Practice Research Datalink (CPRD) is an observational and interventional research service, accessed March 3, 2015. http://www.cprd.com/intro.asp.

[80]For example, the Non-Care Zone for CPRD is characterized by CPRD governance based on UK and European laws as well as NHS and other guidelines and the use of anonymised data and charters, PET, security measures, legal arrangements, contracts, Standard Operating Procedures (SOP), audits. See: http://www.cprd.com/governance/.

[81]Omer Tene, "Privacy - The next generations," Int Data Privacy Law 1(1) (2011): 15–27.

[82]Bradley A. Malin, Khaled El Emam, Christine M. O'Keefe, "Biomedical data privacy: problems, perspectives, and recent advances." J Am Med Inform Assoc 20 (2013): 2–6.

Any simple and comprehensive solution, like the prohibition of access to all data without explicit consent, may slow down research efforts by excluding "sensitive topics" or biasing results by omitting hard to reach patient groups. With regard to epidemiological research in public health using population-based disease registries, alternatives to the obligation to obtain consent may be necessary and the use of exemptions should be considered. More complex solutions for privacy protection involving safe havens, stewardships, data custodians and third parties may be required.

The increased re-used of patient data for research purposes and the spreading of big data in medical research has resulted in a reappraisal of existing data protection frameworks and their ability to ensure data protection and to build trust. Faced with growing concerns about the viability of privacy protection methods in the face of technological changes, improved statistical analytics and the need to collect personal and identifiable data for research, regulators seem to aim for a more restrictive use of data, in particular to restrict the use of pseudonymous data, tighten conditions for consent or broaden the interpretation of what constitutes personal information. An example for such a restrictive policy of using personal data is the current draft of the European Data Protection Regulation and its requirement for an explicit form of consent in order to strengthen confidence and individual control over personal data. Use of big data may not only increase the risk to privacy; it may even change the nature of that risk and turn the conventional form of informed consent partly meaningless. For example, it could be shown that even a small set of data, like date of birth, postal code and gender, is in the long run already sufficient to identify persons in a health information database without knowing the names or other personal information[83, 84]. It seems therefore, as if a "hardening" of the Research Zone may be required, for example by enforcing fines for privacy breaches done by researchers.

An outcome of our analysis of the legal background of privacy and data flow inside the LHS was that a privacy framework for medical research should consider that primary care data routinely stored on computers either within general practices or at national or regional databases can be linked to other healthcare datasets, like hospital admissions records, death certificates, and disease registries. Any privacy protection framework should consider these activities and should allow as much research as possible for linked data sets without endangering the privacy of patients. In this way privacy protection could act an enabler, instead of a burden to the researcher.

---

[83] Melissa Gymrek, Amy L. McGuire, David Golan, Eran Halperin, Yaniv Erlich, "Identifying Personal Genomes by Surname Inference," Science 339(6117) (2013): 321–324.

[84] Paul Ohm, "Broken Promises of Privacy: Responding to the Surprising Failure of Anonymization." UCLA Law Review 57 (2010): 1701–1711.

One problem for the implementation of the LHS is that clinical research and clinical care domains are still for the most part disconnected, because each uses different legal standards, best practices and access rules. The LHS requires data sharing between both domains to be efficient and secure. The data protection framework that is necessary for the LHS to achieve interoperability between clinical research and clinical care, must consider these different contexts. Critical data sources have been moved beyond the traditional clinical boundaries of the general practice/hospital or healthcare network; identifying genetic information is stored in databases that are not part of the care domain anymore but located outside of a hospital, or even in the internet (e.g. 23andMe is collecting and storing genetic information[85, 86]). Thus privacy governance activities will increasingly need to regard the nature, basics, and manifestations of trust and must understand that context plays an important role to privacy requirements.[87] In this environment, standard instruments (e.g. anonymisation techniques, one-way or two-way coding) and other advanced methods (obfuscation, privacy-aware linking, k-anonymity, differential privacy) are used, but the risk of re-identification of patients still exists and thus, the necessity for a risk assessment for any deployment of data (privacy impact assessment). Researches should move from a dichotomous approach (consent or anonymisation) to a risk-based approach considering any risk of re-identification. In addition, recent developments (e.g. persistent identifiers) may improve traceability and auditability of breaches of confidentiality. Proper and complete provenance tracking and means of identification of privacy breaches and their punishment should ensure that data consumers like researchers are not trying to identify persons from their data; this will be an important part of the research domain that complements the more technical privacy protection methods based on PET. This risk-based approach is already reflected in our zone model; flexibility and context sensitivity is achieved by the definition of zones/subzones that represent different categories of risks; data is moving between these zones modeled by privacy functions/filters (e.g. pseudonymisation). New concepts (e.g. direct/indirect care, safe haven) can be easily implemented in the graphic model as additions of new subzones.

---

[85]In the privacy statement 23andMe states that the company will not sell, lease, or rent the client's individual-level information (i.e., information about a single individual's genotypes, diseases or other traits/characteristics) to any third party or to a third party for research purposes without your explicit consent. See: https://www.23andme.com/about/privacy/.

[86]There is a lot of discussion about the impact of collecting and storing genetic information if one considers that relatives have very similar genomes. See: Charles Seife, 23andMe Is terrifying, but not for the reasons the FDA thinks. The genetic-testing company's real goal is to hoard your personal data. Scientific American web site (ScientificAmerican.com), November 27, 2013, http://www.scientificamerican.com/article/23andme-is-terrifying-but-not-for-reasons-fda/.

[87]Nissenbaum, H., 2004, op. cit.

The zone model allows a formal and standardized structuring and representation of confidential and data privacy requirements to further analysis and comparison of frameworks with the final aim to achieve flexibility through formalization. Nonetheless, we plan to further develop the graphic zone model, employing new specific symbols, a more comprehensive notation, and the inclusion of patient empowerment by considering differential, fine-tuned informed consents. Though, privacy requirements can be displayed sufficiently, the representation of trust needs improvements. In the LHS the need for trust has to go beyond privacy protection and informed consent.

Acknowledgments TRANSFoRm project is partially funded by the European Commission—DG INFSO (FP7 247787).

Bibliography

Aamot, Harald, Christian Dominik Kohl, Daniela Richter, and Petra Knaup-Gregori. 2013. Pseudonymization of patient identifiers for translational research. *BMC Medical Informatics and Decision Making* 13: 75–90.

APEC Secretariat. 2005. APEC privacy framework. Singapore. ISBN 981-05-4471-5.

Basque, Josianne, Gilbert Paquette, Beatrice Pudelko, and Michel Leonard. 2008. Collaborative knowledge modelling with a graphical knowledge representation tool: a strategy to support the transfer of expertise in organisations. In *Knowledge cartogrcphy. Software tools and mapping techniques*, ed. Alexandra Okada, Simon Buckingham Shum, and Tony Sherborne, 491–517. London: Springer.

Enserink, Martin, and Gilbert Chin. 2015. The end of privacy. *Science* 347(6221): 490–491.

EU Commission. 2012. Data protection newsroom. *Commission proposes a comprehensive reform of the data protection rules*, 25 Jan 2012. http://ec.europa.eu/justice/newsroom/data-protection/news/120125_en.htm. Accessed 3 March 2015.

EU Directive 95/46/EC of the European Parliament and of the Council of 24 October 1995 on the protection of individuals with regard to the processing of personal data and on the free movement of such data. Official Journal of the European Communities, 1999; No. L281/31-281/39.

Foster, Ian. 2011. Building a secure learning health system. In *Digital infrastructure for the learning health system. Workshop series summary*, ed. Claudia Grossmann, Brian Powers, and J.Michael McGinnis, 161–165. Washington, D.C.: Institute of Medicine, The National Academies Press.

Friedman, Charles P., Adam K. Wong, and David Blumenthal. 2010. Achieving a nationwide learning health system. *Science Translational Medicine* 2: 57cm29.

Gellert, Raphaël, and Serge Gutwirth. 2012. Beyond accountability, the return to privacy? In *Managing privacy through accountability*, ed. Daniel Guagnin, Leon Hempel, Carla Ilten, Inga Kroener, David Neyland, and Hector Postigo. London: Palgrave Macmillan.

Grossmann, Claudia, Brian Powers, and J. Michael McGinnis, eds. 2011. *Digital infrastructure for the learning health system*. The foundation for continuous improvement in health and health care: workshop series summary. Washington D.C.: Institute of Medicine, National Academies Press.

Gymrek, Melissa, Amy L. McGuire, David Golan, Eran Halperin, and Yaniv Erlich. 2013. Identifying Personal Genomes by Surname Inference. *Science* 339(6117): 321–324.

Health Insurance Portability and Accountability Act of 1996. *Public Law* 104-191, US 104th Congress, Washington.

ICH E6. 1996. Good clinical practice: consolidated guideline, CPMP/ICH/135/95.
Keung, Sarah N. Lim Choi, Lei Zhao, Theodoros N. Arvanitis, Vasa Curcin, Brendan Delaney, Jean-François Ethier, Anita Burgun, Mark McGilchrist, Piotr Brodka, Wlodzimierz Tuliglowicz, and Anna Andreasson. 2015. *TRANSFoRm: implementing a learning healthcare system in europe through embedding clinical research into clinical practice.* Paper presented at the 48th Hawaii international conference on system sciences, Hawaii, US, HICSS-49, 6 Jan 2015.
Kuchinke, Wolfgang, Christian Ohmann, Robert A. Verheij, Evert-Ben van Veen, Theodorus N. Arvanitis, Adel Taweel, and Brendan C. Delaney. 2014. A standardised graphic method for describing data privacy frameworks in primary care research using a flexible zone model. *International Journal of Medical Informatics* 83(12): 941–957.
LERU (League of European Research Universities). 2015. *The EP's position on the general data protection regulation threatens EU research,* 6 October 2014, accessed 3 March 2015. http://www.leru.org/index.php/public/news/the-eps-position-on-the-general-data-protection-regulation-threatens-eu-research/. Accessed 3 July 2015.
Malin, Bradley A., Khaled El Emam, and Christine M. O'Keefe. 2013. Biomedical data privacy: problems, perspectives, and recent advances. *Journal of the American Medical Informatics Association* 20: 2–6.
Nass, Sharyl J., Laura A. Levit, and Lawrence O. Gostin (eds.). 2009. *Beyond the HIPAA privacy rule: enhancing privacy, improving health through research.* Washington, D.C.: Institute of Medicine, The National Academies Press.
Nissenbaum, Helen. 2004. Privacy as contextual integrity. *Washington Law Review* 79(1): 119–158.
Nissenbaum, Helen. 2010. *Privacy in context: technology, policy, and the integrity of social life.* Stanford: Stanford University Press.
NIVEL (Nederlands instituut voor onderzoek van de gezondheidszorg), accessed 3 March 2015. http://www.nivel.nl/.
OECD. 2013. OECD privacy framework. OECD Paris, France, accessed 3 March 2015. http://www.oecd.org/sti/ieconomy/oecd_privacy_framework.pdf. Accessed 3 July 2015.
Ohm, Paul. 2010. Broken promises of privacy: responding to the surprising failure of anonymization. *UCLA Law Review* 57: 1701–1711.
Ohmann, Christian, Wolfgang Kuchinke, Helen Corley. 2011. Deliverable 3.2: *A review of the European clinical trial and data confidentiality legal frameworks.* TRANSFoRm, 31 March 2011.
Olsen, Leigh Anne, Dana Aisner, J. Michael McGinnis. The learning healthcare system: workshop summary (IOM roundtable on evidence-based medicine). Washington D.C.: Institute of Medicine, The National Academies Press (USA).
Rosenberg, Doug, and Matt Stephens. 2007. *Use case driven object modeling with UML. Theory and practice.* Berkely: APress.
Skiba, Diane J. 2011. Informatics and the learning healthcare system. *Nursing Education Perspectives* 32(5): 334–336.
Spanish Data Protection Agency. 2009. *International Standards on the Protection of Personal Data and Privacy.* Madrid: The Madrid Resolution.
Tene, Omer. 2011. Privacy—The next generations. *International Data Privacy Law* 1(1): 15–27.
The Hastings Center Report. 2015. 45(1). The Hastings Center, Garrison, NY, USA.
Unified Modelling Language (UML). 2015. Accessed 3 March 2015. http://www.uml.org. Accessed 3 July 2015.
US Federal Register. 2015. U.S.-EU safe harbor framework documents. Register, 24 July 2000, accessed 3 March 2015. http://export.gov/safeharbor/eu/eg_main_018493.asp. Accessed 3 July 2015.
Van Quathem, Kristof. 2005. Controlling personal data—The case of clinical trials. Data Protection Compliance Advisor, Covington & Burling.

van Veen, Evert-Ben. 2013. Europe and tissue research: a regulatory patchwork. *Diagnostic Histopathology* 19(9): 331–336.
van Veen, Evert-Ben. 2011. Patient data for health research. A discussion paper on anonymisation procedures for the use of patient data for health research, Den Haag, MedLawConsult.
Verschuuren, Marieke, Gérard Badeyan, Javier Carnicero, Mika Gissler, Renzo Pace Asciak, Luule Sakkeus, Magnus Sternbeck, and Walter Devillé. 2008. Working group on confidentiality and data protection of the network of competent authorities of the health information and knowledge strand of the EU public health programme, 2003-08. *The European Journal of Public Health* 18: 550–551.

Could the CE Marking Be Relevant to Enforce Privacy by Design in the Internet of Things?

Eric Lachaud

Abstract This paper aims at evaluating the relevance of using the CE marking process to enforce Data Protection by Design principles suggested by Article 23 of the proposed General Data Protection Regulation in connected devices involved in the Internet of Things. The CE marking is a conformity assessment process (A quick presentation of the basic principles of the CE marking is available on the website of the European Commission. Accessed June 14, 2015 http://europa.eu/legislation_summaries/other/l21013_en.htm. More information can be found within the recently updated guide issued by the European Commission's "Guide to the implementation of directives based on the New Approach and the Global Approach", 2014. Accessed May 21, 2015 http://ec.europa.eu/enterprise/newsroom/cf/itemdetail.cfm?item_id=7326.) designed by the European Commission during the 1980s to allow manufacturers to voluntarily demonstrate their compliance with mandatory regulations on safety, health and environment. This process offers some interesting features for the enforcement of data protection rules in products especially in the context of the globalization of trade. It promoted a co-regulation process between public and private stakeholders and contributed to the spreading of European technical standards worldwide. However, it does not fully address data protection issues raised by the IoT and it has been criticized for its lack of reliability. Moreover, this process has never been designed to include an unlimited list of requirements and adding data protection requirements could undermine it. Another option might be to transform the CE marking in an overarching European mark housing different certification schemes dedicated to the compliance of products. This option might preserve the existing process and offer the opportunity to set up a scheme arranged according a similar process but dedicated to the enforcement of Data Protection by Design principles.

Keywords Privacy by design · Internet of things · General data protection regulation · Data protection certification · CE marking

---

E. Lachaud (✉)
Tilburg Institute for Law, Technology, and Society (TILT), Tilburg University, P.O. Box 90153, 5000 LE Tilburg, The Netherlands
e-mail: eric.lachaud@outlook.com

1 Introduction

"The meaning of things lies not in the things themselves, but in our attitude towards them". A. de Saint Exupéry

Analysts[1] foresee the IoT as one of most disruptive evolutions in ICT for next decade. Some authors[2] even talk of "a new revolution of the Internet" insofar as the IoT promises to create a global network of connected devices "from the fridge in your home, to sensors in your car; even in your body".[3] Connected devices will be able to collect and share a huge amount of data regarding the behavior and bodily condition of their handlers. This evolution could certainly provide useful breakthroughs in many domains. It could also be very intrusive to the extent that it could promote a permanent monitoring of sensitive personal data.[4]

---

[1] Deloitte "Tech Trends 2014, Inspiring Disruption" (Deloitte's annual Technology Trends report 2014), 55. Accessed June 14, 2015. http://dupress.com/wp-content/uploads/2014/02/Tech-Trends-2014_FINAL-ELECTRONIC_single.2.24.pdf.

[2] "Internet of Things is a new revolution of the Internet. Objects make themselves recognizable and they obtain intelligence thanks to the fact that they can communicate information about themselves and they can access information that has been aggregated by other things" in Ovidiu Vermasen "Europe's Internet of things Strategic Research Agenda 2012" "in Internet of Things 2012" ed. by Ian G. Smith (New Horizons, 2012). "We are only in the very nascent stage of the so-called "Internet of Things," when our appliances, our vehicles and a growing set of "wearable" technologies will be able to communicate with each other" in John Podesta et al. "Big Data: Seizing Opportunities, Preserving Values", 2014 (Executive Office of the President). Accessed June 14, 2015 http://www.whitehouse.gov/sites/default/files/docs/big_data_privacy_report_may_1_2014.pdf. A study led in 2013 estimated that 4 billion objects were connected in 2010, 15 billion in 2012 and 80 billion are expected to be connected in 2020 in IDATE "Internet of things: Outlook for the top 8 vertical markets", 2013. Accessed June 14, 2015 http://www.idate.org/fr/Research-store/Collection/In-depth-market-report_23/Internet-of-Things_785.html.

[3] "The Internet of Things promises to bring smart devices everywhere, from the fridge in your home, to sensors in your car; even in your body. Those applications offer significant benefits: helping users save energy, enhance comfort, get better healthcare and increased independence: in short meaning happier, healthier lives. But they also collect huge amounts of data, raising privacy and identity issues". Foreword of Nelly Kroes in Ian G. Smith "Internet of Things", 2012 (New Horizons). See Janna Anderson and Lee Rainie "The Internet of Things Will Thrive by 2025", 2014 (Pew Internet Project report). Accessed February 21, 2015. http://www.pewinternet.org/files/2014/05/PIP_Internet-of-things_0514142.pdf.

[4] "Smart objects can accumulate a massive amount of data, simply to serve us in the best possible way. Since this typically takes place unobtrusively in the background, we can never be entirely sure whether we are being (observed) when transactions take place" in Riad Abdmeziem and Djamel Tandjaoui "Internet of Things: Concept, Building blocks, Applications and Challenges", 2014 (Cornell University Library). arXiv preprint arXiv:1401.6877. Accessed June 14, 2015. http://arxiv.org/pdf/1401.6877v1.pdf. See also: European Commission fact sheet "IoT Privacy, Data Protection, Information Security" for an overview of the different threats rose by IoT". Last accessed June 14, 2015 http://ec.europa.eu/information_society/newsroom/cf/dae/document.cfm?doc_id=1753; Federal Trade Commission. "Internet of Things: Privacy and Security in a Connected World". FTC Staff Report, 2015. Last accessed June 14, 2015. http://www.ftc.gov/system/files/documents/reports/federal-trade-commission-staff-report-november-2013-workshop-entitled-inter-

To tackle this threat, the proposed General Data Protection Regulation (*Hereinafter GDPR*) suggests that controllers and processors apply principles of *Data Protection by Design or by Default*[5] (*hereinafter DPbD*). Article 23[6] and Recital 61[7] of the proposed GDPR, require controllers and processors to apply technical and organizational measures that ensure data protection compliance throughout the lifecycle of products and services. However, the European regulation neither specifies when and how a manufacturer becomes a data controller nor does it say if DPbD applies strictly to the European manufacturers or to non-EU suppliers as well? The Working Party 29 (hereinafter WP29) argues[8] that the market destination of the device makes the manufacturer liable under the European regulation, but WP29s opinions are however non-binding. Article 4 of the proposed GDPR states that the European regulation applies when the data of European

---

Footnote 4 (continued)

net-things-privacy/150127iotrpt.pdf. See also Harald Sundmaeker et al. "Vision and Challenges for Realising the Internet of Things", 2010. (CERP-IoT—Cluster of European Research Projects on the Internet of Things). Last accessed June 14, 2015 http://bookshop.europa.eu/en/vision-and-challenges-for-realising-the-internet-of-things-pbKK3110323/.

[5]Data Protection by Design and Data Protection by default represents the European interpretation of the concept of Privacy by Design primarily elaborated by the Privacy Commissioner of Ontario at the end of the 1990s. This approach encourages controllers and processors at including data protection measures from the design stage of their products and services. Since 2009, this approach has been strongly supported by the European authorities and has been integrated into the reform of the European data protection framework in 2012. Article 23 of proposed regulation prefers talking about data protection rather than privacy to be consistent with other provisions and the European approach considering privacy as larger than data protection. Article 23 also makes a difference between Data Protection by Design and Data Protection by Default. The latter requires that the safeguards be applied without any intervention of the end user. The 7 Foundational Principles leading the implementation of Privacy by Design are presented on the dedicated website of the Commissioner of Ontario. Last accessed May 21, 2015 https://www.privacybydesign.ca/index.php/about-pbd/7-foundational-principles/. The European counterpart has recently been detailed in ENISA, 2015. Privacy and Data Protection by Design—from policy to engineering. European Union Agency for Network and Information Security. December 2014, p. iii. Last accessed May 21, 2015 https://www.enisa.europa.eu/activities/identity-and-trust/library/deliverables/privacy-and-data-protection-by-design.

[6]Article 23 of the proposed General Data Protection Regulation (hereinafter GDPR) in amended version of the European Parliament requires controllers to "implement appropriate and proportionate technical and organizational measures and procedures in such a way that the processing will meet the requirements of this Regulation and ensure the protection of the rights of the data subject" Accessed June 14, 2015 http://www.europarl.europa.eu/meetdocs/2009_2014/documents/libe/dv/comp_am_art_01-29/comp_am_art_01-29en.pdf.

[7]Recital 61 of the Parliament version of the proposed GDPR also states that "The principle of data protection by design require data protection to be embedded within the entire life cycle of the technology, from the very early design stage, right through to its ultimate deployment, use and final disposal".

[8]On the basis of Article 4 of the Directive 95/46/EC. See Article 29 Data Protection Working Party "Opinion 8/2014 on the on Recent Developments on the Internet of Things", 2014, 10.

residents is involved in a process even if the process is located outside Europe. Article 3 adds that the processes included in products must be compliant as well. However, do these provisions apply to built-in processes included in connected devices? The question is still open insofar as the regulation of products remains a topic largely unsettled in the GDPR. Moreover, stakeholders involved in data protection are increasingly demanding conformity assurance. Authorities are seeking to make businesses accountable[9] of their conformity. The public, and especially the European citizens, are increasingly concerned by the massive and unregulated collection of data[10] and request more transparency in the processing of their personal data.[11] The overhaul of the data protection general framework has created momentum to assess the possible options to address regulation issues raised by the IoT.

The CE marking is a conformity assessment process[12] designed by the European Commission during the 1980s. It offers the EU and non-EU manufacturers the capacity to voluntarily demonstrate their compliance with mandatory regulations on safety and health and environment. The CE marking approach was originally a

---

[9]Article 22 of the Parliament version of the proposed GDPR states that "the controller shall adopt appropriate policies and implement appropriate and demonstrable technical and organizational measures to ensure and be able to demonstrate in a transparent manner that the processing of personal data is performed in compliance with this Regulation".

[10]83 % of users of mobile services in Europe are concerned by collection of data and 65 % check the data collected by their smartphone's apps in GSMA "Mobile Privacy: Consumer research insights and considerations for policy makers", 2014. Accessed June 14, 2015 http://www.gsma.com/publicpolicy/wpcontent/uploads/2014/02/MOBILE_PRIVACY_Consumer_research_insights_and_considerations_for_policymakers-Final.pdf; TRUSTe "UK Consumer Confidence Privacy Report", 2014. Accessed June 14, 2015 http://info.truste.com/lp/truste/Web-Resource-HarrisConsumerResearchUK-ReportQ12014_LP.html; TRUSTe "Internet of Things Privacy Index—US Edition", 2014 underlines that 83 % of the 2000 people surveyed are concerned by the idea that personal information are being collected by smart devices. Accessed June 14, 2015 http://www.truste.com/resources/?doc=468; Sciencewise. Big Data, Public views on the collection, sharing and use of personal data by government and companies, April 2014. Assed May 21, 2015 http://www.sciencewise-erc.org.uk/cms/assets/Uploads/SocialIntelligenceBigData.pdf; Cited in Data Protection Rights: What the public want and what the public want from Data Protection Authorities. Prepared by the ICO for the European conference of Data Protection Authorities, Manchester, May 2015. Accessed May 21, 2015.

[11]The apparent failure of Google+ network and the changes suggested by Facebook's founder in his last keynote of April 30, 2014 seems at suggesting a slight inflexion—to be confirmed—in the way data are shared by people on social media. Accessed June 14, 2015 http://newsroom.fb.com/news/2014/04/f8-introducing-anonymous-login-and-an-updated-facebook-login/. The public have also developed strategies in order to avoid the full disclosure of their personal data. See Symantec—State of Privacy Report 2015 (February 2015) http://www.symantec.com/content/en/us/about/presskits/b-sta.

[12]A quick presentation of the basic principles of the CE marking is available on the website of the European Commission. Accessed June 14, 2015 http://europa.eu/legislation_summaries/other/l21013_en.htm. More information can be found in the European Commission's "Guide to the implementation of directives based on the New Approach and the Global Approach", 2014. Accessed June 14, 2015 http://ec.europa.eu/enterprise/newsroom/cf/itemdetail.cfm?item_id=7326.

regulatory arrangement aimed at tackling technical harmonization issues that endangered the achievement of the single market. It was undoubtedly a creative solution to address this urgent topic but it demanded many adjustments to achieve its full potential. Even today, its efficiency and reliability remains criticized by commentators.[13] Why then suggest in this paper the re-use of such a process in data protection? The CE marking offers interesting features to enforce the Data Protection by Design principles, especially in the context of the globalization of trade.[14] It suggested an original process of co-regulation and provides a strong incentive for manufacturers to demonstrate their conformity to the regulation. Moreover, its arrangement achieved unexpected and valuable outcomes that are worth underlining. This option presents shortcomings as well. The CE marking process only ensures manufacturer's compliance when the device is put on the market but does not address data protection issues occurring during the lifecycle of the device and its reliability is still challenged as quoted above. The CE marking has been initially designed to enforce a certain type of requirements and adding data protection requirements in the existing process could undermine its functioning. The relevance of such a solution seems dubious. Another option might be to redesign the CE marking to transform it in an overarching certification mark dedicated to the compliance of products. This option would offer the opportunity to preserve the existing process and create apart a European wide scheme dedicated to DPbD compliance.

The first section of this paper demonstrates why the CE marking may be seen as a suitable solution to enforce DPbD principles. The second shows the practical and theoretical shortcomings making this solution impracticable as such. The last evaluates the relevance of redesigning the CE marking process in order to create an overarching mark capable of housing certification schemes under the same mark.

2 Why CE Marking Could Be a Relevant Option for DPbD Enforcement?

This section defines the IoT both as connected devices and as a communication network connecting these devices. It explains why this paper intends to focus on the compliance issues of connected devices rather than on the network and discusses legal provisions applicable to manufacturers. It finally questions the possible contributions of certification schemes in the regulation of the IoT.

---

[13]ANEC "Caveat Emptor—Buyer Beware" 2012 (The European Association for the Co-ordination of Consumer Representation in Standardization). Accessed June 14, 2015 http://www.anec.eu/attachments/ANEC-SC-2012-G-026final.pdf. See also Consumer Research Associates Ltd. "Certification and Marks in Europe", 2008 (Study commissioned by EFTA), 11. Accessed June 14, 2015 http://www.efta.int/sites/default/files/publications/study-certification-marks/executive-summary.pdf.

[14]"Regulation framework of IoT has to be global because IoT has no border especially with globalization" in Rolf. H Weber "Internet of Things—New security and privacy challenges", Computer Law & Security Review, Volume 26, Issue 1, January 2010, pp. 23–30.

2.1 The Twofold Nature of the IoT

The IoT is generally defined as a communication protocol by analogy with the Internet of services. R.H. Weber[15] presents the IoT as an "information architecture facilitating the exchange of goods and services in global supply chain networks". The most well-known example of the IoT network, he adds, are the RFID networks "based on an Electronic Product Code (EPC)". The European funded project CASAGRAS[16] similarly defines the IoT as "a global network infrastructure, linking physical and virtual objects through the exploitation of data capture and communication capabilities". However, the IoT does not encompass only a communication protocol. Sundmaeker et al.[17] stress that "In the IoT, "things" are expected to become active participants in business, information and social processes where they are enabled to interact and communicate amongst themselves and with the environment by exchanging data and information "sensed" about the environment, while reacting autonomously to "real/physical world" events and influencing them by running processes that trigger reactions and create services with or without direct human intervention". Weber[18] argues that the things are physical objects carrying RFID tags with a unique EPC. CASAGRAS supplements the definition by stating that IoT "will be characterized by a high degree of autonomous data capture, event transfer, network connectivity and interoperability. WP29 adds[19] that the exchange can be done directly "machine to machine" (M2M) or with human intervention in between (B2B).

Therefore, the IoT may be defined as a communication infrastructure set to connect a full array of devices with autonomous capabilities to collect process and transfer data. Both of these features define the twofold nature of the IoT, in which a connected device collect the data and the communication infrastructure and transfers it to third party systems. Smartphones are undoubtedly the most illustrative example of the twofold nature of the IoT. They incorporate different types of sensors by design offering the capability to collect data concerning localization,

---

[15]Rolf. H Weber "Internet of Things—Need for a New Legal Environment", 2009, Computer Law & Security Review, Volume 25, Issue 6, November 2009: 522–527.

[16]CASAGRAS—Coordination and support action for global RFID-related activities and standardisation. European Internet of things Initiative. Accessed June 14, 2015 http://www.iot-i.eu/iot-database/all/organizations/internet-of-things-initiative/fines-future-internet-enterprise-systems/casagras.

[17]Harald Sundmaeker et al. "Vision and challenges for realising the Internet of Things", 2010. CERP-IoT—Cluster of European Research Projects on the Internet of Things European Commission—Information Society and Media DG–EUR-OP, 2010, 43. Accessed June 14, 2015. http://www.theinternetofthings.eu/sites/default/files/Rob%20van%20Kranenburg/Clusterbook%202009_0.pdf.

[18]Rolf. H. Weber "Internet of Things—New security and privacy challenges" 2010, 23.

[19]Article 29 Data Protection Working Party "Opinion 8/2014 on the on Recent Developments on the Internet of Things." 2014, 5.

behavior and bodily condition of their handler. They are also able to transmit these data to third parties through a bunch of communication protocols[20] most of the time operated by telecommunication companies.

The communication side of the IoT has been voluntarily excluded from the scope of this paper. It focuses on data protection issues raised by built-in mechanisms embedded by design in connected devices. This choice has been made to be in line with the very nature of the CE marking process that focuses on the compliance of manufactured products. Moreover, data protection compliance issue of communication infrastructures, especially those related to cloud computing, represents one of the trickiest issues of data protection and requires be treated independently.

2.2 Challenges for the Regulation of the IoT

The data protection regulation of products remains unclear so far. Nothing in Directive 95/46 EC and in the proposed GPDR addresses this issue. Only *data processing activities* enter into the scope of the regulation according Article 3.2 (a).[21] Does this wording mean that the scope of the regulation would be limited to the collection and communication processes without any monitoring of built-in collection mechanisms embedded in devices? Does it mean conversely that processes included by design in the IoT devices should be considered as data processing, as the WP29 suggested in its recent opinion about the IoT[22]? However, what should the status of devices communicating personal information within a domestic network be, without any professional and commercial purposes? Would they fall under the domestic use exemption set in Article 2.2 (d) and Recital 15 of the proposed regulation? What should then be the status of the same device when built-in collection mechanisms are used later or simultaneously for commercial purposes? Would it mean that a same collection mechanism would have different legal statuses depending of its purposes and the period of time? The final use of the device remains uncertain when it is put on the market and legal provisions applicable by default are questionable. It is too early to bet on the interpretations that will be provided by the European authorities about the notion of processes in connected devices, but it promises to be a tricky issue. One can nevertheless assume that a broad interpretation of Article 3 would include connected device manufacturers in the scope of the regulation.

---

[20]GSM, Bluetooth, Wifi 802.11, NFC to cite only the most known.

[21]Article 3.2 (a) of the Parliament version of the GPDR states: "This Regulation applies to the processing of personal data of data subjects residing in the Union by a controller or processor not established in the Union, where the processing activities are related to: (a) the offering of goods or services, irrespective of whether a payment of the data subject is required, to such data subjects in the Union".

[22]Working Party "Opinion 8/2014, 11" see Footnote 17 for details.

The relevance of applying DPbD principles to IoT devices is also questionable. Why use these particular principles rather than general principles of data protection set in Article 5 of the GDPR? Both principles are complementary and Article 23 even states that DPbD principles should be implemented in order to "ensure the protection of the rights of the data subject, in particular with regard to the principles laid out in Article 5[23]". However, privacy by design principles as suggested in Article 23 could be more suited to the regulation of products to the extent that they are performance requirements.[24] Performance requirements are generally more suited for technical issues and they are recommended in the drafting of technical standards. Furthermore, this type of devices is generally produced in large series, making modifications much more difficult once they have been designed. Thus, implementing data protection requirements at the early stage of the design process could be easier and more efficient for manufacturers.

Article 23 suggests a very limited set of principles and the European Parliament removed, in its proposal, the right for European Commission to "*adopt delegated act*" to supplement principles presented as examples[25] in text. The European Parliament version of the regulation may hinder the efficiency of DPbD by limiting the possibility to extend the set of DPbD principles. How can the lawmaker be sure that DPbD principles defined in the regulation will be sufficient to address the future technological evolutions? Is there, for instance, any principle suited to the collection and processing of genetic data? Will this sensitive issue merely be a matter of minimization and pseudonymization?

The GDPR relies essentially on a deterrence model with a higher level of financial sanctions in case of infringement. Will it be enough to ensure that DPbD principles be correctly implemented in the absence of enforcement? The growing complexity of technical systems make these tasks harder for Data Protection Authorities (hereinafter DPAs), who do not always have the financial resources to hire laboratories to perform assessment tasks.

The proposed GDPR offers through Article 22 and 39[25] the capacity for the authorities to transfer to businesses the burden of demonstrating conformity with the regulation. The European Parliament suggests in Recital 77 that certification

---

[23]Article 23.1 and Recital 61 of the draft GDPR.

[24]Article 23.1 states that "Data protection by design shall have particular regard to the entire lifecycle management of personal data from collection to processing to deletion, systematically focusing on comprehensive procedural safeguards regarding the accuracy, confidentiality, integrity, physical security and deletion of personal data." Article 23.2 states that "The controller shall ensure implement mechanisms for ensuring that, by default, only those personal data are processed which are necessary for each specific purpose of the processing and are especially not collected, or retained or disseminated beyond the minimum necessary for those purposes." Recital 61 states that "the principle of data protection by default requires privacy settings on services and products which should by default comply with the general principles of data protection, such as data minimization and purpose limitation".

[25]Article 23.1 and Recital 61 of the Draft GDPR.

[26]Article 22 requires controllers to be accountable of their compliance. Article 39 encourages the set up of certification schemes dedicated to data protection.

could be a means to "export European data protection standards by allowing non-European companies to more easily enter European markets by being certified". The WP 29[27] encourages the authorities to explore the use of certification schemes to regulate privacy and security issues in this domain.

The regulation of the IoT needs innovative solutions. Paradoxically, the innovation may lie in a regulation instrument set up thirty years ago. The next section explores the possible contributions that the CE marking process could offer to enforce the DPbD principles in IoT devices.

2.3 Possible Contributions of the CE Marking

The CE marking process could be attractive insofar as it offers a ready to use conformity assessment process applicable to non-EU manufacturers. The CE marking process is able to translate high profile legal provisions into technical requirements. It sets different processes of assessment depending on the risk carried out by devices and manages the issuance and maintenance of a public sign fully recognized by stakeholders. The original features of the CE marking could offer suitable solutions to some issues that data protection authorities are seeking to address. For instance, it could offer the opportunity to share the burden of regulation and participate in the realization of a single market of data that remains a high profile objective for the authorities.[28] It could also spread European data protection standards worldwide and thus participate in the achievement of extraterritorial ambitions enshrined in Article 3.2[29] of the proposed regulation.

2.4 Innovative Division of Regulation Work

A *New Approach policy* has been suggested in 1985[30] by the European Commission to reorganize the lawmaking process set to ensure technical harmonization because the current one was burdensome and endangered the achievement of European single market.

---

[27]Working Party "Opinion 8/2014, 24" see Footnote 17 for details.

[28]Speech of Viviane Reding, former Vice-President of the European Commission, EU Commissioner for Justice "Data protection reform: restoring trust and building the digital single market- European Commission" 2013 (SPEECH/13/720-17/09/2013). Accessed June 14, 2015 http://europa.eu/rapid/press-release_SPEECH-13-720_en.htm.

[29]Article 3.2 (a) of the Parliament version of the GDPR states "This Regulation applies to the processing of personal data of data subjects in the Union by a controller or processor not established in the Union".

[30]The New Approach policy has been adopted in Europe by the Council Resolution 85/C136/01 of 7 May 1985. This policy has been set up to speed-up the harmonization of EU requirements for product safety and reduce the technical barriers between member states in order to realize

This new policy suggested that *essential requirements* should be enacted by legislators and included in clear and concise provisions[31] in the annexes of New Approach Directives dedicated to a range of product. Essential requirements could be translated, at the request of the legislator, in technical standards by European standardization bodies.[32] Once the standard was agreed between the stakeholders involved in its drafting, it would become mandatory, harmonized in the wording of the European Commission, and replacing former standards issued on the same subjects and already in force in the standards' library of member states. As a result, the New Approach policy succeeds in speeding up the law making process. More importantly, it offered the stakeholders innovative divisions of the regulation work in which general principles are defined by the legislator while technical requirements necessary to supplement the legal principles are entrusted to standardization bodies who became a new actor in the European legislative process.

Footnote 30 (continued)

the single market before 1992. The "New Approach" Legislative Commission has defined four main principles: (i) The products must at least comply with the principles laid down in directives before to be introduced on the market; (ii) These principles are defined in the Directives so-called "New Approach". They are available at the request of the legislator in technical standards by the European standardization bodies. These standards are technical specifications designed to facilitate compliance with the principles set out in the Directives "New Approach". These standards called harmonized standards are mandatory in all member states. Member states must repeal that all texts that contradict these harmonized standards; (iii) The application of standards remain voluntary; (iv) The products that comply with the standards benefit of a «presumption of conformity» with the principles set out in the Guidelines. They can be distributed in all the Member states. In Mark. R. Barron. "Creating Consumer Confidence or Confusion? The Role of Product Certification in the Market Today", 2007 (Marquette Intellectual Properties Maw review, Volume 11 Issue 2), 427. A full presentation of the foundations of the CE marking process can be found in the 'Blue Guide' on the implementation of EU product rules issued by "the European Commission" 2014, 6. Accessed June 14, 2015 http://ec.europa.eu/enterprise/newsroom/cf/itemdetail.cfm?item_id=7326. See also Jacques Pelkmans "The New Approach to Technical Harmonization and Standardization". Journal of Common Market Studies, XXV, No 3, 3 March 1987. Accessed June 14, 2015. https://courses.washington.edu/eulaw09/supplemental_readings/Pelkmans_New_Approach_Harmonization.pdf.

[31] An interesting example of wording of essential requirements can be found into the annex of the Directive 2006/95/EC Low voltage. Accessed June 14, 2015. http://eur-lex.europa.eu/LexUriServ/LexUriServ.do?uri=OJ:L:2006:374:0010:0019:EN:PDF.

[32] The European Committee for Standardization called Comité Européen de Normalisation (CEN) has been created in 1961 in order to harmonize technical standards drafted in Europe. The CEN is headquartered in Brussels. He is composed of the 28 members of the European Union and the EFTA countries (Iceland, Norway and Switzerland). The CEN like the ISO is working with two sectorial partners: The Comité de Normalisation Electrotechnique (CENELEC) who is composed of the National Electrotechnical Committees of 30 European countries. The CENELEC is responsible for developing standards in electrotechnical area on behalf of the CEN. The European Telecommunications Standards Institute (ETSI) is responsible for developing standards in telecommunications. This process involves over 600 companies and institutions from 55 European countries. For instance, the ETSI is at the origin of the DECT and GSM standards.

Data Protection regulation faces rapid technological changes including a growing technical complexity in many domains. To be relevant over time, the regulation must be technology neutral and drafted in general principles. However, technical issues like the compliance of build-in mechanisms cannot be efficiently enforced through general principles and the proposed framework does not suggest any process to translate general principles of DPbD into workable technical requirements. The European Parliament version of the regulation even denies the right for the Commission to supplement the legislative work. How then to breach the gap between general principles and technical requirements? The process suggested by the New Approach policy could be helpful to complete this task and to adopt such an approach could also promote a new way of regulation in data protection that could be used to address other technical issues.

2.5 Co-regulation

The CE marking promoted an original co-regulation arrangement between the stakeholders involved in this process. Lawmakers, standardization bodies, manufacturers and private conformity assessment bodies (hereinafter CAB) must collaborate but remain responsible for their part of the process. The lawmaker issues high-level requirements in new approach Directives, while standardization bodies supplement them with technical requirements. Manufacturers or Private conformity assessment bodies (hereinafter CAB) audit and attest the conformity with the technical standards and national authorities monitor CABs[33] and manufacturers on their own market.

Certification processes are today at the intersection of two main idea streams. The first seeks at encouraging businesses to demonstrate their compliance through mechanisms of accountability.[34] The second seeks at involving regulated bodies into their own regulation process in order to improve their efficiency and acceptance by the regulated bodies. Co-regulation, also called collaborative governance,[35] aims at meeting these challenges required by modern regulation and the original regulatory arrangements suggested by the CE marking process could help to achieve these goals.

---

[33]Regulation 765/2008/EC of July, 9 2008 setting out the requirements for accreditation and market surveillance relating to the marketing of products and repealing Regulation (EEC) No 339/93. Accessed June 14, 2015. http://eur-lex.europa.eu/LexUriServ/LexUriServ.do?uri=OJ:L:2008:218:0030:0047:en:PDF. See also the accreditation process in the Blue Guide on the implementation of EU product rules p. 73.

[34]Colin Bennett. "International Privacy Standards: Can Accountability Be Adequate", 2010 (Privacy Laws and Business International), 3.

[35]Dennis D. Hirsch. "The Law and Policy of Online Privacy: Regulation, Self-Regulation or Co-Regulation?" 2010 (ExpressO), 7. Accessed June 14, 2015 http://works.bepress.com/dennis_hirsch/1.

2.6 Accountability

The CE marking promoted original accountability processes in which only the conformity to essential requirements is mandatory. The manufacturer remains free to choose the method to demonstrate its conformity and the requirements on which it intends to make its demonstration.[36] Before affixing the CE mark on its product, it must issue technical documentation describing the procedures followed to ensure the conformity with essential requirements. In order to encourage conformity with harmonized standards, the CE marking process offers suppliers who voluntarily comply with harmonized standards, a presumption of conformity allowing the manufacturers to market their product in all member states without any further administrative procedure. The presumption of conformity offers a powerful incentive for manufacturers to comply with harmonized standards. It also gives some flexibility to manufacturers to demonstrate their conformity by different ways, and it ensures the traceability of the assessment through the obligation to document this process. Therefore, using a regulatory approach mixing obligations, flexibility and practical benefits may certainly encourage manufacturers to implement DPbD in connected devices. The absence of such an incentive within the current drafting of Article 39 limits the attractiveness of data protection certification schemes.

Moreover, data protection authorities have sought for some time to promote a general principle of accountability from businesses, now recognized by Article 22 of the proposed regulation. Businesses themselves might be interested in accountability procedures in order to demonstrate their good will to customers and, eventually, to the court. The public, who discovers new infringements in data protection on a daily basis, is increasingly reluctant to share data without any assurance of transparency in the processing of their data. It sounds very unlikely that authorities could respond alone to the growing demand of conformity assurance insofar as they do not have enough resources and competences to assess the overwhelming flow of connected devices. Therefore, encouraging businesses to voluntarily demonstrate their conformity through self or third party assessment processes might be a valuable solution to involve stakeholders in this quest for conformity assurance.

2.7 Risk Based Approach

The assessment requirements set in the CE marking process are closely related to the risk presented by the product. The *Global Approach* enacted in 1993

---

[36]Conformity assessment under the CE marking could be carried out, at the manufacturer's discretion, with respect to the harmonized standards or directly against essential requirements included into the Directive. The provisions of the Directive can also serve as requirements of substitution in case of absence of standards in this arena. In order to be assessable, new approach Directives must be written in such way that they can be easily audited by the certification bodies. In Section III of the Council Resolution of May 7, 1985 states "They (the Directives) should be so formulated as to enable the certification bodies straight away to certify products as being in conformity, having regard to those requirements in the absence of standard".

establishes eight levels of conformity assessment[37] depending of the level of risk. A self-assessment process done by the supplier is deemed sufficient for products presenting low risks while the highest module requires manufacturers to set up a *Total Quality Management* system (TQM)[38] in which an external monitoring must be regularly performed by a third party certification body on each product unit.

The one size fits all approach prevailing so far in Directive 95/46 EC is commonly denounced as red tape because it requires from businesses burdensome and unnecessary procedures even for processes presenting low risks. Some commentators[39] argue that this situation undermines business opportunities and the achievement of the single market of data. The Data Protection Commissioner[40] and scholars[41] in UK actively support the idea of including a risk-based approach in data protection regulation in order to improve the efficiency of the proposed regulation. This proposal could mitigate, they argue, the effort of conformity assessment required from businesses according the risk presented by products, thus allowing the focusing of monitoring effort on risky processing. This approach has been partially endorsed in the proposal made by the European Council[42] but remains criticized[43] and have to be confirmed during the tripartite negotiation.

---

[37]Council Decision 93/465/EEC July 22, 1993 concerning the modules for the various phases of the conformity assessment procedures and the rules for the affixing and use of the CE conformity marking, which are intended to be used in the technical harmonization directives. Accessed June 14, 2015 http://eur-lex.europa.eu/LexUriServ/LexUriServ.do?uri=OJ:L:1993:220:0023:0039:EN:PDF.

[38]The different levels of assessment are called assessment modules in the European regulation. Module (A) requires the manufacturer to conduct itself a conformity assessment of its product in order to establish a Self-Declaration of Conformity (SdoC). At the other end, module (H) requires the manufacturer to set up a Total Quality Management (TQM) system in which a third party body certifies each unit of product. For a detailed presentation of the Global Approach and associated assessment modules, see European Commission "Guide to the Implementation of Directives Based on the New Approach and the Global Approach" 2014, 28. Accessed June 5, 2015. http://ec.europa.eu/enterprise/policies/single-market-products/files/blue-guide/guidepublic_en.pdf.

[39]John Wagley "EU Privacy Proposal Criticized", 2013 (Security Management website magazine).

[40]ICO "Comparative analysis of the European Commission text and the European Parliament's LIBE (civil liberties) Committee amendments of Proposed draft EU General Data Protection Regulation and 'law enforcement' Directive", 2013, 2.

[41]Neil Robinson et Al. "Review of the European data protection directive" 2009. (Cambridge: RAND), X. Accessed June 5, 2015 http://www.rand.org/pubs/technical_reports/TR710.html.

[42]Privacy & Information Security Law Blog "Council of the European Union Proposes Risk-Based Approach to Compliance Obligations" Posted on February 2, 2015. Accessed June 5, 2015 https://www.huntonprivacyblog.com/2014/10/29/council-european-union-proposes-risk-based-approach-compliance-obligations/. See also European Delegations' comments regarding risk based approach. European Council Accessed June 5, 2015 http://register.consilium.europa.eu/doc/srv?l=EN&f=ST%2012267%202014%20REV%202.

[43]Article 29 data protection working party, 2014 'Statement on the role of a risk-based approach in data protection legal frameworks' WP 218 Adopted on 30 May 2014 http://ec.europa.eu/justice/data-protection/article-29/documentation/opinion-recommendation/files/2014/wp218_en.pdf.

2.8 Widespread European Standards

The long-awaited GDPR promises to harmonize the legal framework within the Member States but its influence on sub-regulatory rules like standards and private codes of conduct remain hypothetical and definitely slower. Using the CE marking and especially the mechanism of *harmonized standards*[44] could offer the opportunity to streamline technical regulations in all member states and complete data protection harmonization.

The CE marking contributed to what some authors call the *ratcheting up effect* of European regulation.[45] Manufacturers who were interested to market their products on the highly profitable European market have been encouraged to comply with European standards in order to benefit from the presumption of conformity. Thereby, European standards have had a discreet but deep influence on manufacturing processes in foreign countries, promoting the European viewpoint on safety and health worldwide. The CE marking process influenced international standardization processes through a series of agreements[46] signed during the 90s in order to organize cross collaboration between European and international standardization bodies to prevent standard duplication. European standards also benefit from the behaviour of multinational companies that commonly align their internal processes on the most stringent rules in order to prevent further needs for adjustment in the different regions in which they make business.[47] Therefore, using the same process for data protection could have the same effects and including DPbD requirements in the CE marking might be an interesting way to spread European data protection[48] worldwide.

---

[44]See Footnote 28 section (ii).

[45]The ratcheting effect can be defined as the influence played by the regulation of one region or country on some others. One observes a ratcheting effect when "businesses adopt a uniform set of data practices that satisfy the rules of the most protective jurisdiction". To use words of economics, there is a ratcheting effect when "regulations in one jurisdiction create positive externalities in another jurisdiction" in Mark Rotenberg and Daniel Jacobs "Updating the Law of Information Privacy: The New framework of The European Union", 2012 Harvard Journal of Law & Public Policy, Vol. 36, 637–641.

[46]The Vienna agreement has been signed in 1991 by the International Standardization Organisation (ISO) and the Comité Européen de Normalisation (CEN) and renewed in 2001. The Dresden Agreement has been signed in 1996 between the International Electrotechnical Committee (IEC) and the Comité de Normalisation Electrique (CENELEC). These agreements allow international standards to become European standards and vice versa when relevant. An interesting abstract of the content of these agreements has been published by the American standardization body ANSI. Accessed June 14, 2015 http://publicaa.ansi.org/sites/apdl/.../ISO-CEN-Vienna.doc.

[47]Streamlining procedure reminded by Scott Taylor, representing Hewlett Packard during "Accountable organisations deserve benefits from regulators" panel at CPDP 2015. Brussels, January 22, 2015.

[48]The influence of European data protection regulation on the other frameworks is already underway underlined Mark Rotenberg and Daniel Jacobs in "Updating the Law of Information Privacy: The New framework of The European Union", 2012 Harvard Journal of Law & Public Policy, Vol. 36, 637–641.

Finally, the CE marking contributed to the realization of the single market without penalizing international free trade to the extent that non-EU manufacturers have benefited from the presumption of conformity.[49] Hence, the CE marking played a role of *one stop shop* approach for non-EU suppliers that surely prevented any accusation of protectionism through technical barriers that could have been claimed by EU partners against the requirement to apply only European standards.

The CE marking process is now a well-known procedure especially by manufacturers outside EU. Using it to enforce data protection requirements would prevent the adding of new administrative layers with the possible issues related to the introduction of a new procedure. Some authors even stress that it could be useful "to establish a governing body similar to the W3C for the IoT to oversee the standardization and certification processes".[50]

The CE marking process offers two valuable contributions to the data protection regulation. Firstly, it suggests a workable process of co-regulation, relevantly combining contributions of the European lawmaker, standardization bodies, private assessment bodies and manufacturers. Secondly, it provides, with the presumption of conformity, a strong incentive for manufacturers to demonstrate their conformity with the regulation. These two features could be beneficial to data protection regulation when the authorities seek to encourage co-regulation and accountability within the future framework. However, such a procedure only addresses conformity issues of built-in processes but does not solve those raised by processes implemented during the lifecycle of the device. The major role played by self-assessment in the CE marking process still receives criticisms for its reliability.[51] Furthermore, the CE marking has not been initially designed for an elaborate future extension of the number of requirements assessable through this process. The introduction of data protection requirements could undermine the whole system of the CE marking. The next section reviews why the CE marking could not be a suitable solution to enforce DPbD in connected devices.

3 Shortcomings of This Solution

Using a signaling approach to attest the conformity of IoT devices does not guarantee the conformity lasting beyond the moment the device is marketed. Affixing a certification mark on the product could even be misleading for consumers to the

---

[49]The supplier self declares the conformity of its product with the requirements of applicable legislation without any mandatory third party intervention. In European Commission DG Trade "European Commission submission to the WTO about 'Supplier's Self Declaration of Conformity", 2003, 1.

[50]Charith Perera et al. "Privacy of Big Data in the Internet of Things Era" 2015. (IEEE IT Special Issue Internet of Anything), 6. Accessed May 27, 2015. http://arxiv.org/abs/1412.8339.

[51]95 % of the declaration of conformity in the CE marking process result from self-assessment processes in Consumer Research Associates Ltd. "Certification and Marks in Europe", 2008 (A Study commissioned by EFTA), 11. See also ANEC. "Caveat Emptor—Buyer Beware", 2012 (The European Association for the Co-ordination of Consumer Representation in Standardization. Accessed June 14, 2015 http://www.anec.eu/attachments/ANEC-SC-2012-G-026final.pdf.

extent that data processing might change during the lifecycle of the device. Third party businesses could implement non-compliant features at any time. Moreover, the relevance of this option remains challenged by the weaknesses of the CE marking itself and by the unforeseeable consequences that the introduction of DPbD requirements might cause to the balance of this process.

3.1 Limits of a Seal Policy for the IoT

The visual demonstration of compliance, a masterpiece of the CE marking process, may be questionable for certain connectible devices. Would this solution still be relevant on nanometric devices on which the sign would not be viewable[52]? How could the manufacturer affix a conformity mark on devices like cameras or webcams which are not submitted to the same regulation requirements when sold alone or as part of a CCTV system?

The issuance of a seal attests to the conformity when the assessment is made and it attests to the conformity in specified conditions of use. The compliance cannot be challenged if the product is used under different conditions than those for which it has been assessed. But conditions of use for IoT devices are not always precisely defined when the device is marketed. A camera may have many different final uses and uses can change during the lifecycle of the device. The same camera that has been previously used for domestic purposes can be sold to become part of a CCTV system. Therefore, the guarantee of conformity offered by the CE marking may be limited in time and scope.

3.2 Misleading

Affixing a CE marking on a product could be misleading by falsely inferring the compliance of the whole system while only one part has been assessed. The presence of the CE mark on the product does not mean that external third party bodies audited the product. It only signifies that the manufacturer realized the necessary tests in order to comply with the European regulation before marketing its product.[53]

---

[52]The American authorities recently authorize manufacturers to remove certification sign from marketed devices and display these signs on accompanying documentation. "Obama signs E-Label Act, allows manufacturers to remove rear logos", 2014, Electronista.com website. Accessed June 14, 2015 http://www.electronista.com/articles/14/11/27/regulatory.symbols.on.devices.can.be.removed.shown.in.software.instead/#ixzz3QW0GCzju.

[53]This definition is confirmed in Article 2.20 of Regulation 765/2008 of July,9 2008 stating that the CE marking is "a marking by which the manufacturer indicates that the product is in conformity with the applicable requirements set out in Community harmonization legislation providing for its affixing".

Thus, the CE marking presents a basic ambiguity related to its process of issuance. Depending on the assessment process required in the Directive, it could be a mark of self-certification[54] or a third party certification mark. The only viewable difference for the consumer lies in the obligation for the manufacturer to mention, under the mark, the name of the certification body that delivered the third party certification. This ambiguity explains why some consumer associations[55] denounce the misleading effect of CE marking to the extent that it lets the public falsely believe "*products that have been tested as to their safety by an independent party, or even by a public authority*" they argue.[56]

The development of *Quality Management Standards* (hereinafter QMS) further to the introduction of the ISO 9000 standards[57] in 1987 and the requirement set by certain modules to organize a Total Quality Management process has also contributed to blurring boundaries between safety and quality in the CE marking rationales.[58] The misunderstanding in the CE marking purposes is periodically highlighted by the different surveys led by authorities[59] and scholars.[60] These studies show that certification marks are generally recognized[61] by consumers but their purposes are not very well understood.[62] The most common confusion lies in the belief that the CE marking is a certificate of European origin or a European quality

---

[54]See Footnote 50.

[55]The European Consumer Consultative Group (ECCG). "ECCG, Opinion on CE Marking", 2008. Accessed June 14, 2015 http://ec.europa.eu/consumers/cons_org/associations/committ/opinions/eccg_op_02022008_en.pdf.

[56]The European Consumer Consultative Group (ECCG), 1.

[57]ISO 9000 standard series offers to manage the quality of production systems rather than the quality of the products. For this reason, some authors call them "metastandards". Mr Uzumeri for instance defines the metastandards as "lists of design rules to guide the creation of entire classes of management systems. Since systems theorists use the term metasystem for lists of this type, it follows that this type of management standard should be referred to as a metastandard". Mustata Uzumeri "ISO 9000 and Other Metastandards: Principles for Management Practice?", 1997 Academy of Management Executive, 11(1): 21–36.

[58]This blurring has certainly also participated to the confusion of the public in the actual purposes of the CE marking.

[59]Commission Staff Working Document on Knowledge-Enhancing Aspects of Consumer Empowerment 2012–2014, "Consumer attention and understanding of labels and logos", 2012 (SWD, Final, 19.7.2012 4.1), 26.

[60]P.T. van der Zeijden et al. "Keurmerken, erkenningsregelingen en certificaten; klare wijn of rookgordijn? Zoetermeer: EIM Onderzoek voor Bedrijf en Beleid", 2002.

[61]NF mark (FR) is recognised by 64, 5 % of the people interviewed. The Kitemark (UK) by 44, 7 %. The KEMAKEUR (NL) by 39, 4 %. The GS Mark (Germany) by 28, 2 % in "Eurobarometer Europeans and EC logos", 2000 (INRA for The Directorate-General for Health and Consumer Protection). Accessed June 14, 2015 http://ec.europa.eu/public_opinion/archives/ebs/ebs_137_en.pdf.

[62]Consumer Research Associates Ltd. "Certification and Marks in Europe", 40.

mark.[63] The ambiguous meaning of the CE abbreviation does not help.[64] Indeed, few are aware whether CE means *"Conformité Européenne"* (European conformity) or *"Communauté Européenne" (European Community)*" or something else?[65]

3.3 Unreliable

National authorities are in charge to ensure the permanent monitoring of the market and to sanction infringements by the withdrawal[66] from the market of non-compliant products. Studies led on the CE marking monitoring[67] concluded with the insufficient monitoring and the lack of deterring sanctions in case of infringement. The default of monitoring entailed a low confidence in the CE marking and encouraged manufacturers to voluntarily affix additional certification marks[68] to demonstrate their conformity.[69] This situation created an accumulation of certification marks on the products contributing to the public's[70] confusion. Some commentators[71] concluded in a slightly provocative manner that CE marking should be deemed as a *caveat emptor (buyer beware)* mark[72] rather than a quality mark.

---

[63]Commission Staff Working Document on Knowledge-Enhancing Aspects of Consumer Empowerment 2012–2014, SWD (2012) Final, 19.7.2012 cited in ANEC. "Caveat Emptor—Buyer Beware".

[64]"What does the acronym "CE" represent? Although no explanation is provided in Regulation 765/2008, it is thought to mean "Conformité Européenne". The absence of clear explanation as to its exact meaning contributes to the confusion around what CE Marking is." in ANEC "Caveat Emptor—Buyer Beware", 5.

[65]The article dedicated to the CE marking in the English edition of Wikipedia underlines that "in former German legislation, the CE marking was called "EG-Zeichen" meaning "European Community mark".

[66]The principle of withdrawal is defined by Article 21b of Directive 93/68/EEC.

[67]«The results of the study research conducted on behalf of Teknikföretagen, the Association of Swedish Engineering Industries shows that a lack of efficient market surveillance on the Internal Market is undermining confidence in CE marking» in Consumer Research Associates "Certification and Marks in Europe", 43.

[68]The study conducted by Teknikföretagen confirmed the demand of additional marks because of a lack of confidence in CE marking. In Consumer Research Associates "Certification and Marks in Europe", 18.

[69]The German GS mark. A co-regulated certification mark monitored by German ministry of Industry has a growing success in Europe for these reasons. In Consumer Research Associates "Certification and Marks in Europe", 43.

[70]"The proliferation of labels may create confusion rather than facilitate purchasing. Organisations, surveys and studies point to a risk of information overload and the need for clearer and more reliable labels". In Commission Staff Working Document on Knowledge-Enhancing Aspects of Consumer Empowerment 2012–2014, SWD (2012) Final, 27.

[71]The ANEC is the European consumer association involved in standardization. A presentation of its action is available on its website. Accessed June 14, 2015 http://www.anec.eu.

[72]ANEC "Caveat Emptor—Buyer Beware". See Footnote 46.

Although a bit exaggerated, this opinion gives a good idea of the type of criticisms still at stake against the CE marking process.

Some theoretical remarks about the very nature of certification schemes are worth referring to at this stage of the discussion. First, the lack of monitoring may have negative consequences on the reliability of certification marks. A certification scheme is a management system in which every component is closely related to the others. Each one participates in the balance of the whole. If one of them does not work properly, the balance of the scheme could be quickly compromised and the confidence in the scheme severely undermined. Second, the notion of confidence is a key component in the balance of a certification scheme and trust is a fragile construction. Long to build and difficult to maintain, trust can be destroyed at any time for true or false reasons. Therefore, the balance of certification schemes remains strongly dependent of external influences.

3.4 No Legal Status

The CE marking has been recently registered as a trade mark to offer some protection against counterfeiting and the opportunity for manufacturers to claim damages in case of misuse. However, The CE marking is neither a certification mark nor a collective mark[73] and although the attempts to address it,[74] the issue is still pending.

The presumption of conformity offered by the CE marking has no legal value for the manufacturers. They remains liable even if their product has obtained a third party certification.[75] Furthermore, the General Safety of Products Directive[76] allows member state authorities to remove dangerous products from the market even if they have demonstrated their compliance to the European standards. From a theoretical point of view, some scholars[77] argue that certification marks should

---

[73]Article 2.4 of the COM/2003/0240 final—Communication from the Commission to the Council and the European Parliament—Enhancing the Implementation of the New Approach Directives http://eur-lex.europa.eu/smartapi/cgi/sga_doc?smartapi!celexplus!prod!DocNumber&lg=en&type_doc=COMfinal&an_doc=2003&nu_doc=240.

[74]Christian Bock "CE Marking: What can legal metrology learn from intellectual property"—Milestone in Metrology III—Rotterdam conference 2009. Accessed June 14, 2015 http://fr.slideshare.net/cbock/ce-marking-what-can-legal-metrology-learn-from-intellectual-property.

[75]Jacques Ghestin "Normalisation et contrat", ed. "Le droit des normes professionnelles et techniques",1985, (Bruylant), 504.

[76]Article 8 of the Directive 2001/95/EC of the European Parliament and of the Council of December, 3 2001 on general product safety.

[77]Stephen Pericles Ladas. "Patents, trademarks and related rights", 1975, Vol. II, p. 1290 et seq.—Harvard: Cambridge University Press. Larry Allman "Callman on Unfair Competition, Trademarks and Monopolies" 1998 (4th ed., St Paul: West Group) Vol 3, Par. 17.18, p. 76 and R. Rozas et Al. "Impact of Certification on Innovation and The Global Market Place" 1997, 598 and N. Dawson "Certification Trade Marks Laws and Practice", 1988 (Intellectual Property Publishing, Ltd, London), 11 in Jeffrey Belson "Certification Marks", 2002. (Sweet and Maxwell—London), 73.

be regarded as a warranty for consumers to the extent that they certify that a product or a service presents certain qualities. For others,[78] the willingness to make certification a warranty is irrelevant because certification bodies are not able to verify every product and a certification body does not aim to be an insurer.[79] There is no legal relationship between the notion of quality and certification, they argue. Quality remains an expectation that cannot involve legal sanctions in its absence. Others[80] again, define certification as an indication of conformity at a given time without any legal consequences. This assertion is partially true to the extent that certification marks have legal consequences in some Member States[81] but the legal status of certification is inconsistent at the European level.[82] Data protection could inherit these shortcomings and this could seriously limit the expected results. Moreover, adding new rationales to the CE marking could introduce more confusion in the legibility of the CE marking and undermine its balance.

3.5 Limited Scope

The New Approach policy has been primarily enacted to address harmonization issues concerning safety and health but nothing in the legislation limits of the scope of the CE marking per se. Extending its rationale to include data protection requirements would be technically possible. The European legislation is regularly updated and its scope modified and broadened. The European Commission even underlines that it may cover other fundamental requirements than safety and health.[83] Essential requirements focusing on environmental protection have been already introduced in a New Approach Directive.[84] However, such an update could

---

[78] Jeffrey Belson "Certification Marks", 73.

[79] Some certification schemes like Google Trusted Stores or Trusted Shops Gmbh in Germany offer complete refund of the purchase when a buyer make the request. Google Trusted Stores "How the program works". Accessed June 14, 2015 https://support.google.com/trustedstores/answer/1669761?hl=en.

[80] "A certificate is only an indication of the situation at a given moment in time (t) at which it is checked whether a product, process or person meets the requirements. It does not give any guarantee that such a product, process or person functions that well at $t + 1$." In Meike Bokhorst "Effectiveness of certification and accreditation as a public policy instrument in the Netherlands" (Paper presented at ECPR conference in Reykjavik, 2010), 12.

[81] Germany, The Netherlands, Belgium, Luxembourg, UK, Portugal, Spain and France have established a dedicated legal framework to certification. In Astrid Cormoto Uzcategui Angulo "Las marcas de certificacion". (PhD diss., Universidad Federal de Santa Catarina—Brasil, 2006), 62.

[82] B. Brett Heavner "World-wide Certification-Mark Registration A Certifiable Nightmare", Bloomberg Law Reports, December 14, 2009.

[83] European Commission 2014. The Blue Guide on the implementation of EU product rules, 32.

[84] Directive 2000/14/EC of the European Parliament and of the Council of 8 May 2000 on the approximation of the laws of the Member States relating to the noise emission in the environment by equipment for use outdoors—OJ L 87 of 31/03/2009.

not be made straightforward. A EU-commissioned study on privacy seals made very clear that the recourse to the CE marking in data protection, although it could be valuable, should be limited to the products in which "it is possible to achieve EU-wide policy and regulatory consensus[85]". Indeed, the enactment of a European-wide regulation on data protection does not prevent divergences in national regulations dedicated to the connected devices. Furthermore, the suitability of the process itself remains questionable. For instance, would the assessment modules in force be adapted to devices containing upgradable software? Is the validity period of 10 years not too long for products in which technological components evolve rapidly? How to differentiate between a CE mark granted for safety, health, environment and one issued for data protection? By using a special mention on the mark? By creating different marks according their destination? A proliferation of marks may worsen the legibility issue already at stake.

3.6 Redundant Enforcement Tool

Article 33.1 of the proposed GDPR requires controllers and processors to conduct a *Data Protection Impact Assessment*[86] (Hereinafter DPIA) in order to evaluate and mitigate the risks existing for data protection in new processing including personal data. Controllers are also required to review and update the DPIA periodically[87] and keep its documentation available on request to DPAs.[88] The DPIA is designed, in the proposed regulation, as a self-assessment process that should be

---

[85]Rodrigues, R., Wright, D., Barnard-Wills, De Hert, P., D., Remoti, L., Damvakeraki, T., Papakonstantinou, V., Beslay, L., Dubois, N., 2014. EU privacy seals project: Challenges and possible scope of an EU privacy seal scheme: final report study deliverable 3.4, 25.

[86]"…The controller or the processor acting on the controller's behalf shall carry out an assessment of the impact of the envisaged processing operations on the rights and freedoms of the data subjects, especially their right to protection of personal data" states the Commission and the Parliament version of Article 33.1. Accessed June 14, 2015 http://www.europarl.europa.eu/meetdocs/2009_2014/documents/libe/dv/comp_am_art_01-29/comp_am_art_01-29en.pdf. "…The controller shall, prior to the processing, carry out an assessment of the impact of the envisaged processing operations on the protection of personal data." states the Council version of Article 33.1 in Council of the European Union, 2014. Proposal for a Regulation of the European Parliament and of the Council on the protection of individuals with regard to the processing of personal data and on the free movement of such data (General Data Protection Regulation) [First reading] Chapter I, 27. Accessed June 14, 2015 http://data.consilium.europa.eu/doc/document/ST-13772-2014-INIT/en/pdf.

[87]"The assessment shall be documented and lay down a schedule for regular periodic data protection compliance reviews" states Article 33.3b of Parliament version of the GDPR.

[88]"The controller and the processor and, if any, the controller's representative shall make the assessment available, on request, to the supervisory authority" states Article 33a of Parliament version of the GDPR.

done by controllers against the provisions of the data protection law. Nevertheless, a few sectorial standards[89] have already been issued to help the controllers, and a general assessment methodology is in preparation[90] within the technical committees of the ISO. 95 % of the CE marks affixed on products marketed in EU result from a self-assessment done by manufacturers.[91] Thus, one can wonder if the DPIA would not be redundant and even conflicting with the CE marking process.

However, the DPIA is so far optional and its status is pending within the proposed GDPR. The proposal of the Council seeks to water-down this requirement, making it mandatory only for certain types of processing, presenting a high risk for data protection.[92] There would be far less incentive for controllers to use it in this case. Moreover, the DPIA does not require the use of third party assessments, although they are generally considered to be more reliable. Neither does it require the assessment to be made against recognized requirements like international standards. It lets the controller make its own interpretation of the law, raising the risk to undermine the consistency of the assessment. The DPIA does not issue any formal and public attestation of conformity. It remains a purely self-regulatory instrument of which the reliability is strongly disputed. Finally, the scope of the DPIA is questionable at the stage of the reform. In the Commission and Parliament version of Article 33, the scope of the DPIA appears larger than data protection.[93] Therefore, the relationship between this process and the CE marking largely depends on the future status of the DPIA in the proposed regulation.

In sum, all of this challenges this proposal and questions the relevance of this option. The introduction of these new requirements would certainly not address the shortcomings of the CE marking and, even worse, data protection could inherit these shortcomings. Furthermore, the CE marking process has not been designed to achieve such a purpose and including data protection requirements could

---

[89]The Privacy and Data Protection Impact Assessment Framework for RFID Applications issued in 2011. Accessed June 14, 2015 http://cordis.europa.eu/fp7/ict/enet/documents/rfid-pia-framework-final.pdf. See also ISO 22307:2008—Financial services—Privacy impact assessment issued in 2008. A quick presentation of the content of the standard is available on the website of the ISO. Accessed June 14, 2015 http://www.iso.org/iso/home/news_index/news_archive/news.htm?refid=Ref1133.

[90]The ISO/IEC WD 29134—Privacy impact assessment—Methodology is still a Working Draft (WD) in the drafting process of the International Standardization Organization. Accessed June 14, 2015.

[91]Manufacturers are also required in the CE marking process to document the procedures they followed to ensure their conformity and keep this documentation available on request to the authorities.

[92]Article 33.1 of the Council version of the GDPR says "Where a type of processing in particular using new technologies, and taking into account the nature, scope or purposes of the processing, is likely to result in a high specific risks for the rights and freedoms of individuals".

[93]"...The controller or the processor acting on the controller's behalf shall carry out an assessment of the impact of the envisaged processing operations on the rights and freedoms of the data subjects, especially their right to protection of personal data" states the Commission and the Parliament version of Article 33.1. Accessed June 14, 2015 http://www.europarl.europa.eu/meetdocs/2009_2014/documents/libe/dv/comp_am_art_01-29/comp_am_art_01-29en.pdf.

undermine its functioning. Another option could be to re-organize the CE marking process in order to transform it into an overarching certification mark capable of housing multiple schemes dedicated to the conformity of products. This option would preserve the existing CE marking and allow the setting up of a *data protection CE marking*. This option would be in line with provisions included in Article 39 of the proposed GDPR, in which lawmakers intend to promote an overarching European-wide certification scheme. The section below discusses the most salient features of this possible option.

4 CE Marking as an Overarching Certification Mark for Products

This last section aims at providing an overview of how a data protection CE marking might look. It focuses on the changes to be made in the existing procedure and the innovative features that could be introduced to improve it.

4.1 *Updating the Existing Process*

Most of the components and processes in force in the existing CE marking are suitable to be used as such in a data protection CE marking. The DPbD principles specified in the Council version of Recital 61[94] provide a list of essential requirements that could be included as such in New Approach Directives dedicated to the regulation of the connected devices. The European standardization bodies recently accepted a new request from the European Commission, asking them to draft "Privacy by Design principles for security technologies".[95] A further extension of

---

[94]Recital 61 of the Council version of the GDPR introduces a list of principles to apply while the Commission and Parliament version does not provide any details on the measures to implement. Article 23.2 in all versions requires applying data minimisation and transparency in the processing of personal data. The measures suggested in the Council version of Recital 61 consist at "minimising the processing of personal data, pseudonymising personal data as soon as possible, transparency with regard to the functions and processing of personal data, enabling the data subject to monitor the data processing, enabling the controller to create and improve security features".

[95]During the plenary meeting of CEN-CENELEC JWG 8 'Privacy management in products and services' took place in Paris on March, 5 2015, the Standardization bodies jointly accepted the standard request on 'Privacy management in the design and development and in the production and service provision processes of security technologies'. "The request aims at the implementation of Privacy-by-design principles for security technologies and/or services lifecycle. The new standardization deliverables are intended to define and share best practices balancing security, transparency and privacy concerns for security technologies, manufacturers and service providers in Europe". Accessed June 14, 2015 http://www.cencenelec.eu/standards/Sectors/DefenceSecurityPrivacy/Privacy/Pages/default.aspx. See also the standardization request issued by the European Commission. Accessed June 14, 2015 ftp://ftp.cencenelec.eu/EN/EuropeanStandardization/Fields/Privacy/EN_privacy.pdf.

this mandate, in order to design DPbD requirements for connected devices, would be certainly possible. Manufacturers might use a DPIA to document their compliance for low risk devices.[96] This could offer the opportunity to streamline the process of technical documentation required in the CE marking and address the issue related to the absence of common requirements in the DPIA. Private conformity assessment bodies with competences in data protection should be accredited at the European level[97] in order to ensure a mutual recognition of the assessments realized in the different member States. All these updates seem achievable without requiring, in first analysis, fundamental changes to the legislation already in force.

4.2 Certificate Without Seal

As quoted above, the meaning of the CE mark and certification marks generally speaking remain misunderstood by the public and it could be worse if authorities came to issue a series of CE marks. However, does it really matter and what is the most important? That the product is compliant or that the public understands that the product is compliant? The CE marking certification process does not aim at differentiating products in order to provide a marketing advantage to certified one. It only attests the conformity with regulatory requirements and most of certification schemes dedicated to conformity do not issue any sign even on the consumer market.[98] Moreover, the CE marking is a certification scheme used as an *entry gate* on the European market. Only compliant products are authorized on the market. The availability of the product on the market plays as the sign of its conformity. Why not then envisage issuing a data protection CE marking without any sign. This could prevent proliferation of certification marks on products that contributes to the confusion of the

---

[96]Module (A) of the assessment modules requires the manufacturer to conduct itself a conformity assessment of its product in order to establish a Self-Declaration of Conformity (SdoC). See Footnote 38.

[97]The European Commission created an original mechanism of accreditation of Conformity Assessment Bodies in order to facilitate mutual recognition of conformity assessment within the CE marking process. Every Conformity Assessment Bodies authorized in a Member State to verify the conformity of products with essential requirements must be prior declared and recognised—notified in the European Commission language—by the European Commission. Once accepted by the Commission, conformity assessments realized by the notified bodies are recognised in all member states.

[98]Most of the certification schemes in food safety, building, housing industry and sanitary certification do not issue a seal. A recent study led by the European Commission found 464 agrifood certification schemes active in the UE in which a large majority of them do not deliver a sign. See "the Inventory of certification schemes for agricultural products and foodstuffs marketed in the EU Member States"—Study conducted by Areté for DG AGRI? Accessed June 14, 2015 http://ec.europa.eu/agriculture/quality/certification/inventory/inventory-data-aggregations_en.pdf.

Could the CE Marking Be Relevant to Enforce ...   159

public. The American Senate recently allowed US manufacturers to remove conformity seals from devices and display the seal only in the accompanying documentation.[99]

4.3 National DPAs as Market Monitoring Authorities

The leniency of national authorities in charge of market surveillance, who tend rarely to sanction the infringements and even more rarely to withdraw products from the market, have largely contributed to the low confidence in the CE marking. In order to improve monitoring capacities, it could be interesting to involve national DPAs in addition to the traditional market surveillance authorities, in the monitoring process of IoT devices. This could offer the opportunity to entrust the monitoring of devices presenting high risk and/or high complexity, to the DPAs and leave the monitoring of the others to traditional market monitoring authorities. Nothing so far in the proposed regulation[100] prohibits DPAs to play this role. Although the role and power of DPAs will be strengthened thanks to the high monetary sanctions envisaged in the future regulation, the question of the DPA's resources has not been addressed in the draft so far. Adding a new role to national DPAs without any further resources would be unrealistic and bound to fail. Some solutions could be explored. For instance, why not envision setting up a funding system managed by the Commission or the future European Data Protection Board, entrusted to grant financial resources to DPAs according the volume of goods entering into each member State.

4.4 Algorithmic Regulation

Ongoing researches, seeking to implement automatic processes to enforce data protection policies throughout algorithmic sequences,[101] offer interesting perspectives

---

[99]News on Electronista.com, November, 27 2014 "Obama signs E-Label Act, allowing manufacturers to remove rear logos". Accessed June 14, 2015 http://www.electronista.com/articles/14/11/27/regulatory.symbols.on.devices.can.be.removed.shown.in.software.instead/#ixzz3QW0GCzju. See the full text of the act on the website of the US Senate. Accessed June 14, 2015 http://www.fischer.senate.gov/public/_cache/files/4b6e357d-1414-4974-b1c7-9b0751cdd931/071014---e-label-act.pdf.

[100]Article 52 1 (a) of the Parliament version of the proposed regulation states that the role of DPAs consist to "monitor and ensure the application of this Regulation" and to "monitor relevant developments, insofar as they have an impact on the protection of personal data, in particular the development of information and communication technologies and commercial practices" adds subsection 1 (d).

[101]See the recent experiment led by Microsoft Bing team which implemented a so called Legalease meta-language in order to translate data protection requirements in encoded instructions. See Shayak Sen et Al. "Bootstrapping Privacy Compliance in Big Data Systems", 2014 (SP '14 Proceedings of the 2014 IEEE Symposium on Security and Privacy, Oakland): 327–342. See also the experiment of INRIA team which suggest a log architecture in order to implement "strong accountability" in Denis Butin et al. "Log Design for Accountability". Article presented at the 4th International Workshop on Data Usage Management, 2013. Accessed June 14, 2015 http://www.ieee-security.org/TC/SPW2013/papers/data/5017a001.pdf.

to supplement the procedural processes in force. Although this type of process remains so far unable to enforce all legal principles,[102] one can assume that they might be relevant for those that do not require interpretation. Certain technical requirements designed from legal DPbD principles could be drafted in such a way that they can be easily enforced by algorithms. Hoepman[103] already defined a library of basic privacy design strategies that could be used as a foundation to draft such *binary requirements*. Interesting experiments have also been made in the course of the Bitcoins project that provides a certain form of certification of the transactions through the *Blockchain* technology "that stands as proof of all the transactions recorded by computers participating to the network".[104] *Smart contracts*[105] in bitcoin transactions are also promising and could bring some interesting and workable solutions to this issue. In the same vein, some authors recently suggested that "certification mechanism for the IoT would be similar to the 'certificate authority model' that is used for the Internet[106]" These innovative approaches remain at the very early stage of development and leave room for improvement insofar as it does not address yet the need to issue and monitor a recognized certificate of conformity.

5 Conclusion

A data protection CE marking could certainly be useful to regulate the connected devices involved in the IoT. It does not aim at providing a one size fits all solution for the regulation of the IoT but it could be the first link in the regulatory chain, enforcing the conformity of data collection and transmission through built-in processes. Furthermore, a data protection CE marking may be an interesting means

---

[102]Bert J. Koops and Ronald Leenes "Privacy regulation cannot be hardcoded. A critical comment on the "privacy by design" provision in data protection law", 2013. International Review of Law, Computers & Technology.

[103]Jaap-Henk Hoepman "Privacy Design Strategies" Article Presented at the Privacy Law Scholars Conference (PLSC) 2013. Accessed June 14, 2015. http://arxiv.org/abs/1210.6621.

[104]The blockchain in the Bitcoin project is a "public ledger of all Bitcoin transactions that have ever been executed. It is constantly growing as 'completed' blocks are added to it with a new set of recordings. The blocks are added to the blockchain in a linear, chronological order". Investopedia entry for Blockchain, http://www.investopedia.com/terms/b/blockchain.asp.

[105]Smart contracts are "computer protocols that facilitate, verify, or enforce the negotiation or performance of a contract" says the Wikipedia entry—http://en.wikipedia.org/wiki/Smart_contract. See also the contributions of the American economist Nick Szabo on its blog. Accessed June 14, 2015 http://szabo.best.vwh.net/idea.html. See also the Ethereum project which offers an open source framework for developers to easily design smart contracts in their applications. Accessed June 14, 2015 https://www.ethereum.org.

[106]Charith Perera (2015). "Privacy of Big Data in the Internet of Things Era." 2015 (IEEE IT Special Issue Internet of Anything), 6.

to promote accountability and co-regulation and a discreet vehicle of the EU to spread European values. It may be tempting to include Data Protection by Design requirements into the existing CE marking process, but this option would be hazardous for both the CE marking and data protection. Indeed, it remains uncertain whether the current process would meet all the requirements for the regulation of connected devices and accept such changes without undermining the whole process. It may be more fruitful to transform the existing CE marking into an overarching certification mark, housing a series of schemes dedicated to the conformity of products. This option could offer the opportunity to design a data protection CE marking, dedicated to the regulation of connected devices involved in the IoT. The project seems, in first analysis, technically achievable without deep changes in the legislation already in force. Its success and sustainability depend on the issuance of workable Data Protection by Design principles in the final version of the regulation. It should also require the involvement of European standardization bodies to draft, within a reasonable period of time, technical requirements adapted to the different types of connected devices. It should rely on the incentive made to private certification bodies to develop expertise in data protection. However, this project would require further assessment to evaluate its consequences on legal and practical points of view. The legislative process to organize such a new process has not been discussed in this paper and the role of Article 39 of the draft GDPR dedicated to certification schemes must be clarified. The design of assessment modules and the subtle balance between self and third party assessments should be discussed more in depth.

Bibliography

Abdmeziem, R., and D. Tandjaoui. 2014. Internet of things: concept, building blocks, applications and challenges, Cornell University Library. arXiv preprint arXiv:1401.6877. http://arxiv.org/pdf/1401.6877v1.pdf. Accessed 14 June 2015.

Allman, L. 1998 Callman on unfair competition, trademarks and monopolies, vol. 3, 4th edn. St Paul: West Group.

Anderson, J.A., et al. 2014. The internet of things will thrive by 2025, 2014 (Pew Internet Project report). http://www.pewinternet.org/files/2014/05/PIP_Internet-of-things_0514142.pdf. Accessed 21 Feb 2015.

ANEC. 2012. Caveat emptor—buyer beware. The European Association for the Co-ordination of Consumer Representation in Standardization. http://www.anec.eu/attachments/ANEC-SC-2012-G-026final.pdf. Accessed 14 June 2015.

Barron, M. 2007. Creating consumer confidence or confusion? The role of product certification in the market today. *Marquette Intellectual Properties Maw Review* 11(2).

Belson, J. 2002. Certification marks. Certification marks. London: Sweet and Maxwell.

Bennett, C.J. 2010. International Privacy Standards: can accountability be adequate (Privacy Laws and Business International).

Bock, C. 2009. CE marking: what can legal metrology learn from intellectual property. *Milestone in Metrology III—Rotterdam conference 2009.*

Bokhors, M. 2010. Effectiveness of certification and accreditation as a public policy instrument in the Netherlands. Paper presented at ECPR conference in Reykjavik.

Consumer Research Associate Ltd. 2008. *Certification and marks in Europe*. A study commissioned by EFTA.

Dawson, N. 1988. Certification trade marks laws and practice. In: Trade marks laws and practice. London: Intellectual Property Publishing Ltd.

European Commission. 2014. The Blue Guide on the implementation of EU product rules. http://ec.europa.eu/enterprise/newsroom/cf/itemdetail.cfm?item_id=732. Accessed 22 May 2015.

Ghestin, J. 1985. Normalisation et contrat, ed. Le droit des normes professionnelles et techniques (Bruylant).

Heavner, B. 2009. World-wide certification-mark registration a certifiable nightmare. Bloomberg law reports, 14 Dec 2009.

Hirsch, D.D. 2010 The law and policy of online privacy: regulation, self-regulation or co-regulation? (ExpressO), 7. http://works.bepress.com/dennis_hirsch/1. Accessed 14 June 2015.

Hoepman, J.-H. 2013. Privacy design strategies. Article presented at the privacy law scholars conference (PLSC). http://arxiv.org/abs/1210.6621. Accessed 14 June 2015.

Jacobs, D. 2012. Updating the law of information privacy: the new framework of the European Union. *Harvard Journal of Law & Public Policy* 36.

Koops, B.-J., and R. Leenes. 2013. Privacy regulation cannot be hardcoded. A critical comment on the "privacy by design" provision in data protection law, 2013. *International Review of Law, Computers & Technology*.

Ladas, S. 1975. *Patents, trademarks and related rights*, 1975, vol. II, p. 1290 et seq. Harvard: Cambridge University Press.

Pelkmans, J. 1987. The new approach to technical harmonization and standardization. *Journal of Common Market Studies* XXV(3), 3 March 1987. https://courses.washington.edu/eulaw09/supplemental_readings/Pelkmans_New_Approach_Harmonization.pdf. Accessed 14 June 2015.

Perera C., et al. 2015. Privacy of big data in the internet of things era. *IEEE IT Special Issue Internet of Anything* 6. http://arxiv.org/abs/1412.8339. Accessed 14 June 2015.

Podesta, J., et al. 2014. Big data: seizing opportunities, preserving values (Executive Office of the President). http://www.whitehouse.gov/sites/default/files/docs/big_data_privacy_report_may_1_2014.pdf. Accessed 14 June 2015.

Robinson, N. 2009. Review of the European data protection directive. Cambridge: RAND. X. http://www.rand.org/pubs/technical_reports/TR710.html. Accessed 14 June 2015.

Rodrigues, R., D. Barnard-Wills, D. Wright, P. De Hert, L. Remoti, T. Damvakeraki, V. Papakonstantinou, L. Beslay, and N. Dubois. 2014. EU privacy seals project : challenges and possible scope of an EU privacy seal scheme: final report study deliverable 3.4. Trilateral research, Vrije Universiteit Brussel for the Institute for the Protection and Security of the Citizen (IPSC).

Rozas, R., et al. 1997. *Impact of certification on innovation and the global market place*. London: Intellectual Property Publishing Ltd.

Sundmaeker, H., et al. 2010. Vision and challenges for realising the Internet of Things. CERP-IoT—Cluster of European Research projects on the internet of things European Commission—Information Society and Media DG-EUR-OP. http://www.theinternetofthings.eu/sites/default/files/Rob%20van%20Kranenburg/Clusterbook%202009_0.pdf. Accessed 14 June 2015.

Uzcategui-Angulo, A.C. 2006. Las marcas de certificacion. PhD diss., Universidad Federal de Santa Catarina, Brasil.

Uzumeri, M. 1997. ISO 9000 and other metastandards: principles for management practice? *Academy of Management Executive* 11(1).

Van der Zeijden, P.T., et al. 2002. Keurmerken, erkenningsregelingen en certificaten; klare wijn of rookgordijn? Zoetermeer: EIM Onderzoek voor Bedrijf en Beleid.

Wagley, John. 2013. EU privacy proposal criticized (Security Management website magazine).

Weber, Rolf. H. 2009. Internet of things—Need for a new legal environment. *Computer Law & Security Review* 25(6), Nov 2009.

Weber, Rolf. H. 2010. Internet of things—New security and privacy challenges.

Visions of Technology

Big Data Lessons Understood by EU Policy Makers in Their Review of the Legal Frameworks on Intellectual Property Rights, Access to and Re-use of PSI and the Protection of Personal Data

Hans Lammerant and Paul De Hert

Abstract This article's focus is on how the advent of big data technology and practices has been understood and addressed by policy makers in the EU. We start with a reflection on of how big data affects business processes and how it contributes to the creation of a data economy. Then we look at EU policy making on big data and its understanding of the role and impact of ICT in the economy. We study 3 major legal frameworks affecting data flows and uses: intellectual property rights, access to and re-use of PSI and the protection of personal data. We explore how these frameworks affect the use of big data and how this is perceived and dealt with in the policy documents. In order to widen our perspective, we also take a comparative look at similar legal frameworks and policies in the US.

Keywords Big data · Policy · Data protection · Copyright · Open data

1 Why Understanding the Technology Visions of Policy Makers?

This article looks at EU policy developments regarding big data and, second step, see how it interacts with legal frameworks regulating data flows. Instead of taking, for instance, data protection law as a point of departure to look at big data, we

---

H. Lammerant (✉) · P. De Hert
Vrije Universiteit Brussel—Law, Science, Technology and Society (LSTS),
Pleinlaan 2, 1050 Brussels, Belgium
e-mail: hans.lammerant@vub.ac.be

P. De Hert
e-mail: paul.de.hert@vub.ac.be

want to take the inverse approach and start from the perspective of big data and the data economy resulting from it. How do legal frameworks affect the use of big data and the operation of a data economy?

A data economy is built upon establishing data value chains and dependent on the possibility to collect, aggregate and process data from diverse sources in an automated process. Legal frameworks affecting data flows have therefore an impact on big data processing and define the space for a data economy. Seen from this perspective, not only data protection, but also other legal frameworks like intellectual property rights and the regulation of access and re-use of public sector information (PSI) frame or influence the data flows on which big data operates. Although not developed specifically for big data, these frameworks regulate and condition the access to and processing of specific types of data and shape therefore the data economy.

Policy makers and legislators look for legal frameworks that allow an economy to flourish and capture the benefits of technological developments while balancing between all the values and interests at stake. To reach that ambition they try to understand the effects of technological developments and grasp what it means in terms of how an economy functions. In other words, they form a certain idea about the effect of ICT on the economy and society, and develop regulation based on that idea. The initial reaction of policy makers on technological developments often starts with adapting and patching legal frameworks, when its impact is not very clear yet or is considered not too profound. Only later more fundamental reviews will be made, dependent on the new understanding of the societal impact.

Therefore visions on how technology changes economical processes are important drivers of legal change. Policy documents are also a testimony of policy learning about the impact of technology.

This article will focus on how the advent of big data technology and practices has been understood and addressed by policy makers.[1] We look into the development of the EU big data policy and how it interacts with the legal frameworks regulating data flows. First we give a short introduction of our understanding of how big data affects business processes and how it results in a data economy (Sect. 1). Next we describe the evolution of the EU policy on big data and what it says on the role and impact of ICT in the economy (Sects. 2, 3 and 4). Following that we consider 3 major legal frameworks affecting data flows and uses: intellectual property rights (Sect. 5), the protection of personal data (Sect. 6) and access to and re-use of PSI (Sect. 7). We explore how these frameworks affect the use of big data and how this is perceived and dealt with in the policy documents. In order to widen the analysis, these sections also look at similar legal frameworks and policies in the US. In a last section we present our conclusion that recent EU policy

---

[1]This article is the result from research done as part of the BYTE project (http://byte-project.eu/). The authors are solely responsible for the opinions expressed.

documents reflect a new understanding of the data economy, but that the translation of this vision into the legal frameworks shows mixed results (Sect. 8).[2]

2 Data Economy and the Big Data Value Chain

The term *big data* is vaguely defined and partly a buzz word which came into popular use only recently. Therefore it appeared only recently in EU policy documents. As a term it is absent in the main EU policy documents such as the one on the Digital Agenda,[3] on cloud computing[4] or related documents from 2010 and 2012. It only appears in the policy documents of 2013 and 2014.[5] These policy documents bear witness of the learning process of the EU institutions concerning the economic and social impact of ICT developments.

Big data as a phenomenon is enabled by new developments in distributed computing like cloud technology, allowing to deal with very large amounts of data at much higher speed. However, big data cannot be equated with these technologies or cannot be limited to these aspects of volume and velocity. It also implies qualitative changes in terms of what can be done with this data: a variety of structured and unstructured data sources can be much easier linked with each other and analysed in new ways. New business models are built upon the capacity to capture value from data through the development of a data value chain along which data is transformed into actionable knowledge.

The concept of value chain was first introduced by Michael Porter and consists of a series of linked activities through which value is created.[6] These linkages are relationships between the performance of one activity and the cost and performance of another. The construction of a data value chain implies a new way to

---

[2] The scope of this article does not allow us to be exhaustive and limits us to exploring the subject. We focus on regulatory issues concerning the access to, linking of and (re-)use of data and the legal environment in which this takes place. Other elements of policies, like concerning investment in infrastructure or research projects, we leave outside of our consideration. Also specific regulations, e.g. on law enforcement, remain outside the remit of this article.

[3] European Commission, *A Digital Agenda for Europe*, COM(2010)245, 19 May 2010; European Commission, *The Digital Agenda for Europe – Driving European growth digitally*, COM(2012)784, 18 December 2012.

[4] European Commission, *Unleashing the Potential of Cloud Computing in Europe*, COM(2012) 529, 27 September 2012.

[5] European Commission, *Towards a thriving data-driven economy*, COM(2014) 442, 2 July 2014; European Commission, *Report on the Implementation of the Communication 'Unleashing the Potential of Cloud Computing in Europe' Accompanying the document Communication from the Commission to the European Parliament, the Council, the European Economic and Social Committee and the Committee of the Regions 'Towards a thriving data-driven economy'*, SWD/2014/0214, 2 July 2014.

[6] Porter, Michael E., *Competitive advantage: Creating and sustaining superior performance*. Free Press, New York, 1985.

optimize output or to create new products and services and a new configuration of activities and actors to do so.

This data value chain has obtained a central role in a data-driven knowledge economy and pushes organisations and administrations to open up their data sources and business processes in order to reap the benefits, resulting in a new 'data ecology' consisting of a diversity of actors providing, collecting or analysing data and acting upon the results. Old organisational barriers are penetrated by data flows. Old legal frameworks regulating such data flows come under pressure. They present barriers to this new data-driven economy or have difficulties to assure the balance between interests and values embedded in them.

The reconfiguration of activities by the construction of a data value chain also changes the role of the Internet. Where the Internet was first conceived as a separate economic space alongside the traditional economy, it evolved into a market place and distribution channel. In this vision economic actors remain units outside the Internet but meet each other through it. With the construction of a data value chain the Internet penetrates these economic units and becomes also the environment in which value creating activities take place and get linked to each other. Economic activity over the Internet broadens from an information economy, focussed on content and services for human customers, to a data economy where data mostly flows between a range of non-human actors processing this data, often in real time. The development of the Internet-of-Things will further augment this evolution. Big data practices are of course possible in contexts outside a data economy, like for data-intensive scientific uses, but a widespread commercial use is correlated with the possibility to build data value chains. This evolution of the role of the Internet is, as will be seen, reflected in recent policy documents as it brings up new regulatory questions about e.g. the space for data mining in the context of IPR,[7] and about profiling in the discussions on the upcoming data protection regulation (General Data Protection Regulation or GDPR).

Key element in the construction of data value chains is the interoperability of datasets, or assuring that datasets can be combined and analysed together. The European Interoperability Framework (EIF)[8] provides a useful conceptual model of interoperability levels: legal, organisational, semantic and technical interoperability.[9] The two first interoperability levels, the legal and organisational, leave no

---

[7]European Commission, *The Digital Agenda for Europe—Driving European growth digitally*, COM(2012)784, 18.12.2012, p. 6.

[8]European Commission, annex II of the "Communication: Towards interoperability for European public services" - COM(2010) 744 final.

[9]Technical interoperability concerns the technical aspects of linking information systems. Organisational interoperability concerns how organisations cooperate to achieve their goals. It implies aligning business processes and the related data exchanges. Legal interoperability concerns how to deal with differences in legal status. Datasets can have a different legal status and be subjected to different legal rules, what can lead to obstructions linking them or to limitations of data exchange. Semantic interoperability ensures that the precise meaning of exchanged information is understood and preserved throughout the data exchanges. It involves developing descriptions or other metadata and vocabularies concerning the exact format of information and the meaning of data elements and their relations. Growing levels of semantic interoperability make it easier to link otherwise isolated data sources.

Visions of Technology   167

doubt about the fact that interoperability of datasets is more than just *technique*. It is (also) influenced by legal frameworks, organisational structures and needs investment in data quality in order to capture potential benefits.[10]

Our focus on legal frameworks regulating data flows implies a focus on legal interoperability, but we will touch upon other aspects when useful. Legal interoperability is affected by several legal frameworks, developed in contexts where big data or interoperability of data were still unknown notions. In later sections we will consider 3 major legal frameworks affecting data flows and uses: intellectual property rights, access to and re-use of public sector data (PSI) and the protection of personal data. We will look into how these frameworks affect data flows and how policy deals with them when confronted with the new big data practices. First we will have a look at some of the basic documents in the history of EU policy making on big data.

3 Digital Agenda for Europe (2010–2012): Overview of Major Relevant Actions

The European Commission developed since the 1990s broad policy documents concerning the information society. These include a strong focus on developing a stable legal environment for commercial activities over the Internet: regulation of e-commerce, adapting intellectual property rights, and so on. But also a strong focus on economic development. In the 1990s this is linked with the liberalisation of telecommunication services. From 2000 onwards the e-Europe 2002[11] and 2005[12] action plans and the i2010 strategic framework[13] focus a lot on improving Internet access through broadband as a key enabler and the development of a rich content industry and services making full use of this potential. This includes making public services accessible over the Internet. All 3 legal frameworks we consider in this article are reconsidered and adapted to the Internet economy in this earlier period. The data protection directive 95/46/EC is adapted in its final drafting to take better account of the context of digital telecommunications networks[14] and in 2002 the E-Privacy Directive addressed specifically the electronic

---

[10]This conceptual model was developed for public services, but we use it here in a generalised meaning.

[11]European Commission, *eEurope 2002—An Information society for all—Draft Action Plan prepared by the European Commission for the European Council in Feira - 19-20 June 2000*, COM(2000)233, 24.5.2000.

[12]European Commission, *e Europe 2005: An information society for all. An Action Plan to be presented in view of the Sevilla European Council, 21/22 June 2002*, COM(2002)263, 28.5.2002.

[13]European Commission, *i2010—A European Information Society for growth and employment*, COM(2005) 229, 1.6.2005.

[14]European Commission, *Europe's Way to the Information Society. An Action Plan*, COM(94)347, 19.07.1994, p. 6.

communication sector.[15] Copyright law has been harmonised and adapted,[16] while a new regime of database protection was introduced.[17] Re-use of PSI got regulated in 2003.[18] These policies, and the directives drafted in this period, are based on the information economy vision: the Internet as market place visited by human clients (see our discussion *above*). In the later part of this article we will consider their functioning in the newer context of a data economy. First we look into how the more recent EU policy digested the advent of big data.

The 2010 *Europe2020* strategy (updated in 2012) sets out a vision on how the EU has to develop its social market economy. This vision functions as a coordinating umbrella vision for more specific policy initiatives. Part of the *Europe2020* strategy were 7 flagship initiatives, one of them being the 'Digital Agenda for Europe'. The main focus of this Digital Agenda is "a digital single market based on fast and ultra-fast Internet and interoperable applications" and builds upon the earlier action plans.

The Digital Agenda contains a comprehensive agenda concerning the digital economy. It identified a wide range of obstacles: fragmented digital markets, lack of interoperability, rising cybercrime and risk of low trust in networks, lack of investment in networks, insufficient research and innovation efforts, lack of digital literacy and skills and missed opportunities in addressing societal challenges. The actions defined in answer to these obstacles are as wide ranging.[19]

A first relevant action in the Digital Agenda, within the aim to create a digital single market, concerns the opening up of access to content. The main problem is that the European digital market is still very fragmented, both concerning private and public data or content. Action points identified are simplifying copyright clearance, management and cross-border licensing.[20] Part of this has been the review of the PSI Directive in 2013 and the adoption of Directive 2014/26/EU on collective rights management and multi-territorial licensing, but also the ongoing review of the data protection framework with the proposed General Data Protection Regulation (GDPR) and e-commerce related legislation.[21] The Commission planned continued action on e-commerce related issues and

---

[15]European Parliament and the Council, Directive 2002/58/EC of 12 July 2002 concerning the processing of personal data and the protection of privacy in the electronic communications sector (Directive on privacy and electronic communications).

[16]European Parliament and the Council, Directive 2001/29/EC of 22 May 2001 on the harmonisation of certain aspects of copyright and related rights in the information society.

[17]European Parliament and the Council, Directive 96/9/EC of 11 March 1996 on the legal protection of databases.

[18]European Parliament and the Council, Directive 2003/98/EC of 17 November 2003 on the re-use of public sector information (PSI-directive).

[19]We will focus on the actions and resulting policy initiatives that concern the access, linking and use of data.

[20]European Commission, *A Digital Agenda for Europe*, COM(2010)245, 19.5.2010, p. 9.

[21]European Commission, *The Digital Agenda for Europe—Driving European growth digitally*, COM(2012)784, 18.12.2012, p. 5.

intellectual property rights. These plans were repeated in the intention in the 2012 update of the Digital Agenda to make proposals to strengthen the European data industry, specifically on "issues such as common licensing conditions and the implementation of charging rules to enable public data to fuel the development of online content".[22] Also problems concerning text and data mining were mentioned by announcing the structured stakeholder dialogue *Licences for Europe* held in 2013, which addressed cross-border portability of content, user-generated content (UGC), data and text mining, access to audiovisual works and cultural heritage institutions. At this point the attention was still limited to text and data mining for scientific research purposes.[23] As part of this effort on content the Commission also focused on public data. It presented its policy in the Communication on 'Open data. An engine for innovation, growth and transparent governance'.[24] Public sector information is seen as a resource. With an active open data-policy this resource is made available for the European economy.

A second important action area linked to data policies in the Digital Agenda is the focus on interoperability and standards. This concerns a wide range of hardware, software, IT services and it can also concern data. Standardization has always been an important instrument in the single market and it also plays a key role in creating a functioning data economy. When content remains locked up in incompatible formats, licenses, etc., the data economy remains very fragmented. The focus on making data sources more interoperable through standardisation is mostly present in the effort to enhance the interoperability between public administrations.

Big data was not mentioned in the Digital Agenda, but the agenda nevertheless contained attention for cloud computing as part of its innovation strategy, with the development of "an EU strategy for cloud computing notably for government and science" as specific action. Cloud computing is an important enabling infrastructure for big data processing, and attention for cloud computing is the only element specific to big data in this Digital Agenda. The Commission next outlined a specific policy agenda on cloud computing in its 2012 *Communication Unleashing the Potential of Cloud Computing in Europe.*[25] It presented 3 key actions: enhance standards and certification, establishing safe and fair contract terms and conditions (through model contracts and contractual clauses, and a code of conduct for cloud computing providers) and the launch of the European Cloud Partnership. Especially the action on contracts has an important effect on the access and use of data, even when it concerns infrastructure rather than big data processing itself. It

---

[22]Ibid., p. 6.

[23]European Commission, *On content in the Digital Single Market*, COM(2012)789, 18.12.2012; Results can be found on http://ec.europa.eu/licences-for-europe-dialogue/en.

[24]European Commission, *Open data. An engine for innovation, growth and transparent governance*, COM(2011)882, 12.12.2011.

[25]European Commission, *Unleashing the Potential of Cloud Computing in Europe*, COM(2012) 529, 27 September 2012.

can create a more predictable and safe environment in terms of data security and data protection and prohibit cloud providers to abuse data on their servers for other purposes. The state of work and results are presented in the 2014 *Report on Implementation accompanying the Communication on a data-driven economy.*[26]

The Digital Agenda for Europe (2010–2012) did not contain yet a fully developed vision on the data economy. It continued to consider the Internet as a market place where consumers and e-commerce enterprises meet. Most attention therefore went to ensuring this market functions properly both for enterprises and consumers and to integrate national markets into a single digital market.

Big data or a data economy as such does not enter the picture yet. Cloud computing does but mainly as an infrastructure delivering more flexible IT resources to companies. Attention therefore goes to market conditions between cloud providers and enterprises buying cloud services. Interoperability concerns mostly hardware and software, but not so much data apart from specific applications. Data protection is still mostly seen as an element to establish trust for consumers in e-commerce, which remains in the older vision of an information economy, while specific big data-related concerns are not considered.

Notwithstanding this several important elements linked to a data economy are already present in the Agenda. IPR-related problems for data and text mining appear in 2012 as a distinct issue.[27] PSI and open data are present as policy issues. The e-government policy and the policies on research data and geospatial data (both are distinct areas of policy planning) appear to be driving areas from which new practices putting data central are developed.

4 Towards a Thriving Data-Driven Economy (2014): Four Regulatory Issues

The European Commission presented an updated version of its vision on the data economy in its 2014 *Communication on a data-driven economy.*[28] It builds upon the ideas first formulated by Commission Vice-President Neelie Kroes in a strategic initiative on the data value chain in November 2013,[29] in response to the European Council's conclusions of its meeting on 24–25 October 2013, where big

---

[26]European Commission, *Report on the Implementation of the Communication 'Unleashing the Potential of Cloud Computing in Europe' Accompanying the document Communication from the Commission to the European Parliament, the Council, the European Economic and Social Committee and the Committee of the Regions 'Towards a thriving data-driven economy'*, SWD/2014/0214, 2 July 2014.

[27]European Commission, *On content in the Digital Single Market*, COM(2012)789, 18.12.2012.

[28]European Commission, *Towards a thriving data-driven economy*, COM(2014) 442, 2 July 2014.

[29]European Commission, *A European strategy on the data value chain*, November 2013.

data as a concept appeared on the EU policy agenda.[30] This policy agenda aims to "provide the right framework conditions for a single market for big data and cloud computing". It puts data forward as the central element in the future knowledge economy. Data-driven innovation is defined as "the capacity of businesses and public sector bodies to make use of information from improved data analytics to develop improved services and goods". Improved data analytics are seen as key to more efficient business and production processes. With this communication a more profound understanding of the impact of big data on business processes appears in EU policy documents.

The Communication further points to the slow embracing of this 'data revolution' in Europe compared to the US. Among the causes the "complexity of the current legal environment" is mentioned. To reverse this the EU must "make sure that the relevant legal framework and the policies, such as on interoperability, data protection, security and IPR are data-friendly". Other needs include an accelerated digitisation of public services and sharing and developing its public data resources. In order to develop a data-driven economy good quality, reliable and interoperable datasets, backed by an enabling infrastructure have to be present, as well as an adequate skills base and close cooperation between public and private partners. The action plan announces several initiatives to make progress towards such data-driven economy, including the development of open data policies and standards and several regulatory issues.

The first regulatory issue concerns personal data protection and consumer protection. After the adoption of the GDPR the Commission plans to work on guidance for issues important in the big data context, like data anonymization, data minimization and privacy by design. Further regulatory work concerns ensuring the application of consumer law on big data technologies. The second issue raised is data mining and its relation to the copyright framework. Thirdly, the Commission plans to explore the security risks related to big data technologies and propose risk management and mitigation measures. Finally issues concerning data ownership and data transfer will be considered. Mentioned are data location requirements, presenting a barrier for cloud computing and big data, and data ownership and liability in the context of the Internet of Things.

In the 2014 Communication the European Commission develops a new vision on the data economy and puts forward the central role of (big) data in the knowledge economy. Regulatory issues raised concern similar areas as before, like IPR and data protection, but now with attention to their impact on the data value chain. Similarly, the attention for open data includes more attention for interoperability and an investment in semantic interoperability.

However, its proposed actions clearly build on the earlier initiatives. The earlier importance of the Internet as a market place does not disappear, but gets supplemented with attention for specific issues linked to data value chains.

---

[30]European Council, *Conclusions – 24–25 October 2013*, EUCO 169/13, 25 October 2013, §3.

5 A Digital Single Market Strategy for Europe (2015): Three Pillars of Reform

The Communication *A Digital Single Market Strategy for Europe* of 6 May 2015 is the first major policy document on the digital economy of the newly installed Commission led by Jean-Claude Juncker.[31] Although still focussing a lot on the Internet as a market place, an in-depth vision on the data economy is now clearly integrated.

The first pillar focuses on market integration by the removal of obstacles for cross-border trade and of the differences between online and offline trade. Under this pillar the Commission envisions to review the copyright framework and make legislative proposals before end 2015. Again mostly focussed on audio-visual content, it also foresees creating "greater legal certainty for the cross-border use of content for specific purposes… through harmonized exceptions". Purposes mentioned are research and text and data mining.

The second pillar aims at reform of the telecommunications and the media sector to enhance market integration and competition. The Commission also plans before end 2015 a comprehensive assessment of the role of platforms like search engines, social media, e-commerce platforms, …. These platforms have been innovators and early adopters in the creation of a data value chain and building new business models around it. The success of some platforms has now led to concerns over their growing market power. This assessment will therefore consider issues like transparency (e.g. in search results), how the platforms use the information they collect, the ability of individuals and business to switch platforms, and other issues connected to the bargaining power of the platforms. Further under this pillar the Commission foresees measures to improve trust and security in digital services and protection of personal data. This involves the continuation of earlier initiatives like the GDPR and the Network and Information Security Directive, which are proceeding through the legislative process, and a review of the ePrivacy Directive after the adoption of the GDPR.

The third pillar is more focussed on the data economy. It will propose in 2016 a European 'Free flow to data' initiative. Where the GDPR prevents member states to restrict the flow of personal data within the EU, with this initiative the Commission wants to tackle other restrictions to data flows and on the location of data for storage and processing (e.g. for security reasons). In this context it wants to address issues like ownership, interoperability, access to data and data portability. The Commission also plans a European Cloud Initiative involving issues like cloud services certification, contracts, liability, switching of providers and so on.

Further the Commission wants to put extra effort into interoperability and standardisation. The focus of the standardisation effort is now broadened from

---

[31]European Commission, *A Digital Single Market Strategy for Europe*, COM(2015)192, 6 May 2015.

hardware and software to the data component. The Commission points to the need to define standards "essential for supporting the digitisation of our industrial and services sectors (e.g. Internet of Things, cyber security, big data and cloud computing)". This effort also has a e-government component. The Commission plans to review the European Interoperability Framework and further focuses on achieving cross-border interoperability. It will present a new e-Government Action Plan 2016–2020 with several initiatives to extend national e-government services across borders. These initiatives include the interconnection of business registers, an initiative to pilot the 'Once-Only' principle cross-border, extending and integrating European and national portals towards a 'Single Digital Gateway' and accelerating the transition of member states towards full e-procurement and interoperable e-signatures.

The 2015 Communication takes up unfinished initiatives from the former Commission, like the review of copyright law and the GDPR, and builds further upon earlier work. The data economy gets more attention alongside the digital market perspective. Together with the earlier communication on a data-driven economy this communication presents a clear policy agenda on big data.

In the following parts we take a closer look at the legal outcomes of these policy initiatives. We study three major EU legal frameworks that affect or have the potential to affect data value chains. None of them has been developed specifically for big data. To broaden the perspective we compare the EU and US legal frameworks, explore how they apply on data and limit its use. We further look at how the advent of big data was received in this context. The first legal framework we consider is the one of intellectual property rights (IPR).

6 The Intellectual Property Rights Framework (Framework 1): Adequate for Big Data?

6.1 The Application of Copyright on Datasets Is not Straightforward

IPR protect intellectual creations and reserve certain exclusive rights concerning their use and distribution to their creators or those to whom these rights have been transferred. Each regime defines what falls under its protection. Certain regimes can apply to data and datasets. Most relevant are copyright and database protection.

Protection by IPR of datasets is a major obstacle for access, linking and use of data and therefore also for big data processing. These limitations can be legitimate, but the framework is not well-adapted for a situation where data flows in large amounts between a broad range of actors and gets processed in real time. When the data is protected by copyright or database protection, authorization of the right holders is required. This can lead to large transaction costs or delays, and is practically impossible in certain use cases, e.g. text mining on thousands of articles or webpages. Solutions can be developed in licensing schemes specifically adapted to data mining practices, but such solutions remain limited to right holders applying them. A more radical solution would be limiting the protection and allowing the specific data use without requiring authorization. This can be done on several levels: the subject matter of the protection, the extent of the reserved usage, the exceptions on these reserved rights.

Copyright can exist over the individual data as well as over the database as a whole. Copyright protection for databases results from the copyright for collections. The protection concerns the organisation and structuring of the data but does not extend to the individual data items itself. Copyright of individual data items grants exclusive rights on the individual item, but is independent from the copyright over the database structure as a collection. Both have to be checked separately and can belong to different right holders.

General principle of copyright is that it protects expressions, but not ideas in itself, nor procedures, methods or mathematical concepts.[32] Aim is to protect products of human intellect and creativity. Trigger for the protection is therefore some sort of originality. Originality implies originating from an author, but also being the result of some intellectual or creative effort.[33] Novelty is not required, but the mere investment of effort in copying information does not reach the threshold for copyright protection. Also, purely factual information is not protected under copyright. Basic idea is that facts are discovered and not the result of creativity. Copyright protection given to the expression does not extend to the underlying facts. This factual information can be used by others, as long as they do not reproduce it in the protected expression.

The application of copyright on datasets is therefore not straightforward: not all data is protected by copyright, but only those that meet the originality-requirement. For instance, maps have been subject to copyright controversies, as the factual geographical information as such lacks the element of creativity.[34] The *Infopaq*-decision of 16 July 2009 the European Court of Justice (EUCJ) concerned the application of copyright law on a search engine of newspaper articles, providing summaries of articles. It stated that the protection by copyright applies only when the data "is original in the sense that it is its author's own intellectual creation".[35] This originality requirement also needs to be checked when reproduction in part is concerned. In this case a string of 5 words before and after the keyword were stored. The Court considered a word in isolation not to be the intellectual

---

[32]WIPO Copyright Treaty, art. 2; TRIPS, art. 9 §2.

[33]Paul Goldstein, *International Copyright. Principles, Law, and Practice*, Oxford University Press, New York, 2001, p. 161–164.

[34]Janssen, Katleen, and Jos Dumortier, "The Protection of Maps and Spatial Databases in Europe and the United States by Copyright and the Sui Generis Right", *J. Marshall J. Computer & Info. L.*, Vol. 24 No.195, 2006, pp. 207–211.

[35]EUCJ, C-5/08, *Infopaq International A/S v Danske Dagblades Forening*, 16 July 2009, §37.

creation of the author, but that such creation could be achieved "through the choice, sequence and combination of those words".[36] Isolated words were therefore not covered by protection, but strings of 11 words could be and this needed to be checked by the national court. This decision clarified the originality requirement upon which copyright protection in the EU is based and made clear that the specific technical characteristics of text and data mining methods are legally relevant in the context of copyright law. Methods based on 'bag-of-words' sets, making a frequency distribution of words in a text and thereby taking all words out of their context, can avoid the applicability of copyright protection, but not those methods using longer strings.

The lack of clarity concerning the application of copyright on data has been resolved in divergent ways. Based on a similar economical reasoning the US courts refuse to extend copyright protection to claims purely based on investment, while in the EU the policy maker has generally chosen to strengthen the protection. The resulting legal environment strongly affects big data practices.

6.2 Striking Differences with the US Copyright Framework

The US courts have seen a lot of legal battles on what can be protected by copyright and what not, including cases concerning several sorts of data and compilations of data.[37] Main legal precedent is *Feist*, in which the Supreme Court made clear that effort or investment is as such not protected by copyright. It took distance from court decisions which granted protection to 'sweat of the brow' or 'industrious collection', through which courts had earlier developed a protection for factual collections. Instead it reaffirmed that originality was an essential requirement, grounded on the objectives of copyright protection listed in the Constitution "to promote the Progress of Science and useful Arts". Copyright also needs to allow others to build upon the ideas and information contained in a work, which is the rationale for only granting protection to the expression but not to facts. The case concerned the white pages of a telephone directory, consisting of an alphabetically ordered lists of names with their town and telephone number. The Court considered that such lists of facts lacked any originality and were not protected by copyright. Factual data in databases or other works are available for reproduction or extraction, even when this extraction is substantial.

The EU has on the contrary resolved the issue by introducing an extra legal protection on databases with a sui generis right. Database protection is provided by directive 96/9/EC of 11 March 1996 on the legal protection of databases. This

---

[36]Ibid., §45.

[37]An overview of case law can be found in Leslie C. Ruiter and Gerald van Belle, Data Extraction: Beyond the Sweat of the Brow, http://www.stokeslaw.com/uploads/pdf/data_and_the_law-gerald_van_belle_and_leslie_ruiter.pdf.

directive contains 2 forms of protection for databases, one as copyright, another as a sui generis right. These protections can coincide.

The copyright on databases protects databases where "the selection or arrangement of their contents" is a result of "the author's own intellectual creation".[38] This protection does not extend to the contents. This remains similar to the copyright protection on databases in the US, derived by the courts from the protection of collections.

The sui generis-right protects the maker of a database who has made a substantial investment in the creation of a database. Only the costs associated with "obtaining, verification or presentation of the contents" as a whole are taken into account, not the cost associated with obtaining, creating or updating individual data items.[39] No originality is required, protection is based on the investment. The maker of the database is given the right to prevent extraction and re-utilization of the whole or of a substantial part of the contents of that database. This right does not prevent lawful use, consisting of extracting or re-utilizing insubstantial parts of database contents. The substantiality can be assessed both quantitatively and qualitatively. Further may this use not conflict with the normal exploitation of the database or unreasonably prejudice the legitimate interests of the maker of the database. Result is that any big data processing involving a substantial part of a database will need permission from the right holder during the 15 years term of protection.

The EU introduced the sui generis protection based on the assumption that property rights attract investments and therefore stimulate the economy. In the US protection was refused on a similar economical reasoning. Ian Hargreaves pointed in his review of the intellectual property framework to the evaluation in 2006 by the European Commission of the Database directive. This evaluation shows less investment instead of growth, while the US market kept growing without such protection.[40] Hargreaves sees this as an example of policy development inconsistent with the available evidence.[41] The European Commission has kept the directive unchanged seen the large support of the concerned industry for the directive. It can be questioned if such large support shows the economic value of the directive in general or if it shows the value for a specific interest group. Would a data economy be better off with less protection of databases through IPR?

A second difference between the EU and US frameworks can be found in the exceptions to the reserved rights provided in these frameworks. The EU

---

[38]European Parliament and the Council, Directive 96/9/EC of 11 March 1996 on the legal protection of databases, art. 3§1.

[39]Maarten Truyens & Patrick Van Eecke, "Legal aspects of text mining", *Computer law & security review* 30 (2014), 160.

[40]European Commission, *First evaluation of Directive 96/9/EC on the legal protection of databases*, DG Internal Market and Services Working Paper, 12 December 2005, pp. 22–23.

[41]Hargreaves, Ian, *Digital Opportunity: Review of Intellectual Property and Growth*, May 2011, p.19.

harmonised copyright and adapted it to the digital environment in the Information Society or InfoSoc directive.[42] The directive includes a set of quite precise exceptions, which are mostly optional. All these exceptions are limited by the 'three-step test': they can "only be applied in certain special cases which do not conflict with a normal exploitation of the work or other subject-matter and do not unreasonably prejudice the legitimate interests of the rights holder".[43] Relevant is the exception on the right of reproduction for temporary acts of reproduction which are transient or incidental, are an essential part of technological processes like transmission or other lawful uses and have no independent economic significance.[44] This exception was meant for caching and temporary storage during digital communication. In the big data context the question is if this can also be used for text and data mining. In *Infopaq* the EUCJ stated that the copies made of the newspaper for the search for keywords could be considered a temporary and transient act of reproduction that fell under the exception if those copies were indeed automatically deleted at the end of the process. This exception could not apply for the further storage or printing of the strings of 11 words, when these fell under protection. In a second decision in the same case the EUCJ broadened the exception by using the 3-step test.[45] The exception for temporary storage did draw attention also in other decisions,[46] but in general the scope for data mining remains quite narrow.

The US Copyright Act of 1976 does not contain a long list of specific exceptions, but grants an exception to the fair use of a copyrighted work.[47] What constitutes fair use is illustrated with the purposes of "criticism, comment, news reporting, teaching (including multiple copies for classroom use), scholarship, or research". Further the law provides four factors to consider in order to determine what is fair use. The 'four factor-test' involves the "purpose and character of the use", the "nature of the copyrighted work", the "amount and substantiality of the portion used" and the "effect of the use upon the potential market for or value of the copyrighted work". The evaluation is made globally and no factor is more important than another, although the economic impact has in practice got more importance. As part of the evaluation of the purpose courts have looked to the transformative nature of the new use. The more a new use is distant from the earlier use and the less it can be conceived as a mere re-packaging and copying, the

---

[42]European Parliament and the Council, Directive 2001/29/EC of 22 May 2001 on the harmonisation of certain aspects of copyright and related rights in the information society.

[43]Ibid., art.5 §5.

[44]European Parliament and the Council, Directive 2001/29/EC of 22 May 2001 on the harmonisation of certain aspects of copyright and related rights in the information society, art. 5 §1.

[45]EUCJ, C-302/10, *Infopaq International A/S v Danske Dagblades Forening*, 17 January 2012.

[46]EUCJ, C-360/13, *Public Relations Consultants Association Ltd v. Newspaper Licensing Agency Ltd and Others*, 5 June 2014 (aka the Meltwater decision); EUCJ, C-403/08 and C-429/08, *Football Association Premier League Ltd*, 4 October 2011.

[47]17 U.S.C. §107.

more chance it makes to be considered fair.[48] The case law on this fair use-exception is very extensive and concerns a wide range of Internet-related practices like hyperlinks, copying or reproducing of images and text by search engines, ... The fair use framework proved able to flexibly deal with new technical developments.

This comparison between the EU and US copyright framework showed some striking differences. First, the US has limited the protection of databases to copyright and has never extended IPR protection purely on grounds of investment. Secondly, its fair use regime proved to be much more technology neutral and adaptable to new technological developments. These 2 differences in the IPR regime lead to a large difference in playing field for big data processing.

Several countries did notice the problem the IPR regime poses for text and data mining and adapted the exceptions or are discussing a change. The UK added an exception for computational analysis for the purpose of non-commercial research. In its international strategy on IPR the UK government further included the aim "to secure further flexibilities at EU level that enable greater adaptability to new technologies".[49] Japan updated its copyright law with a new exception giving space for information analysis.[50] The Australian Law Reform Commission recommended after a public consultation to adopt a general 'fair use'-exception like in the US as a flexible and technology-neutral solution.[51] A similar review took place in Ireland, leading to the recommendation to add exceptions for 'content-mining' for purposes of education, research or private study to both copyright and the protection of databases as well as a fair use-exception.[52] Both recent reports have not led yet to legislative action.

IPR policy in the EU has been focused at strengthening the protection of right holders, motivated by a concern to develop a strong content industry. The sui generis protection of databases is an early witness, but recent policy documents have kept this focus. The recent shift in perception towards a data economy created the space to raise the problems a strict IPR framework poses for text and data mining. This issue entered the agenda in 2012[53] and the Juncker Commission plans to adapt the exceptions regime for text and data mining.[54] On the other hand, no discussion did arise on the usefulness of the sui generis-protection of databases.

---

[48]Netanel, Neil. (2011). Making Sense of Fair Use. UCLA: UCLA School of Law.

[49]Intellectual Property Office, *The UK's International Strategy for Intellectual Property*, 11 August 2011, p. 13.

[50]Triaille, Jean-Paul, Jérôme de Meeûs d'Argenteuil and Amélie de Francquen, *Study on the legal framework of text and data mining (TDM)*, March 2014, pp. 10–11.

[51]Australian Law Reform Commission, *Copyright and the Digital Economy. Final Report*, ALRC Report 122, 30 November 2013, p. 13.

[52]Copyright Review Committee, *Modernising Copyright. The Report of the Copyright Review Committee for the Department of Jobs, Enterprise and Innovation*, Dublin, 2013.

[53]European Commission, *On content in the Digital Single Market*, COM(2012)789, 18.12.2012.

[54]European Commission, *A Digital Single Market Strategy for Europe*, COM(2015)192, 6 May 2015.

Visions of Technology 179

We can conclude that the EU still focuses on strengthening protection with IPR from information economy perspective, but has now more attention to its fine-tuning in the context of a data economy. If this suffices for an adequate big data policy can be questioned. The legal interoperability of data remains lower in the EU due to IPR protection.

In the US no such debate on IPR can be found. The US IPR framework proved to be open and adaptable for new legal developments thanks to its fair use-regime. This does not mean no conflicts between new technological applications and IPR did arise. But these have not been subject of policy making, but of legal disputes and court decisions.

7 Protection of Personal Data (Framework 2)

7.1 A Comparison Between the EU and US Legal Regime

The legal frameworks dealing with personal data are very different in their foundations and grounded in different constitutional cultures. This results in a quite different environment to deal with big data.

First we would like to clarify the difference between privacy and data protection and show how both get a very different place in the EU and the US. Protection of personal data is based on the fundamental right to privacy, but has evolved into a framework of rights and duties which exceeds the right to privacy and has acquired the status of an autonomous fundamental right in itself. Both rights do partially overlap, but function with a different logic.[55] Both set of tools are used in very different ways in the EU and the US.

The general European data protection framework is provided by directive 95/46/EC,[56] but it is rooted in earlier instruments like *Convention for the Protection of Individuals with regard to Automatic Processing of Personal data* (also known as Convention 108), adopted by the Council of Europe in 1981. The data protection framework had a profound impact on the fundamental right jurisprudence concerning the right to privacy. The right to protection of personal data developed into a fundamental right in itself, distinct from the right to privacy.

This data protection framework provides that all processing of personal data requires a legal ground. In other words, all processing of personal data is regulated and subject to a set of rules guaranteeing the accountability of the processor and the transparency of the processing. The European data protection framework

---

[55]Gutwirth, S., De Hert, P., Regulating Profiling in a Democratic Constitutional State, in Hildebrandt, M. and Gutwirth, S. (eds.), *Profiling the European Citizen: Cross-Disciplinary Perspectives*, Springer Science + Business Media B.V. 2008, 271–293.

[56]European Parliament and the Council, Directive 95/46/EC on the protection of individuals with regard to the processing of personal data and on the free movement of such data, 24.10.1995.

applies to all processing of personal data. Personal data is defined very broadly as "any information relating to an identified or identifiable natural person". Also the range of activities to which the directive applies is very broad. Processing is defined as "any operation or set of operations that is performed upon personal data, whether or not by automatic means". This means that whenever data in a big data-context contains information linked to an identifiable natural person, the processing has to be according to the data protection principles and mechanisms have to be implemented to allow data subjects to exercise their rights. The only possibility to escape this framework is by anonymisation of the data.

Directive 95/46 provides principles to which any processing of personal data has to conform, like legitimacy (several grounds for legitimate processing are foreseen, including the consent of the data subject), finality, proportionality and relevance, accuracy, transparency, data subject participation and control, data security. The directive further provides the rights of data subjects, like the right to information about the data processing, to access the data, to object and to rectification. It also specifies the obligations of data controllers, like assuring the confidentiality and the security of the personal data and notifying or prior checking of automated processing to the supervisory authority. The directive foresees a control mechanism through the establishment of independent supervisory authorities. These data protection authorities have powers to investigate, to intervene and to start legal proceedings against violations of the data protection laws. The Commission proposed a new General Data Protection Regulation (GDPR)[57] in 2012 and the review is still ongoing. The draft versions contains generally the same principles, but provide more detailed implementations.

The US framework does not subject all processing of personal data to legal rules guaranteeing more transparency and control for data subjects. Personal data can be freely used unless it is forbidden. The basic structuring of the legal framework is based on opacity tools.

The 4th Amendment to the US Constitution protects people "in their persons, houses, papers, and effects" against the government. Searches are only allowed with a warrant and upon probable cause. This 4th Amendment protection only applies towards the government and not towards private actors. Outside this limited area processing of personal data is in principle allowed, except when specific laws forbid it or subject it to certain rules. Privacy law between private actors was first established through tort law. Four privacy tort actions are recognized in the Second Restatement of Torts and can be considered as opacity tools between private actors, but these have no practical relevance for big data.

This comparison shows a fundamentally different situation in which big data processing using personal data can take place. This US constitutional framework gives free space for such big data processing, as long as no other specific law

---

[57]European Commission, Proposal for a Regulation of the European Parliament and of the Council on the protection of individuals with regard to the processing of personal data and on the free movement of such data (General Data Protection Regulation), COM(2012)11, 5 April 2012.

provides constraints. The EU framework does only allow unconstrained big data processing with anonymized data. When using personal data, big data processing has to be able to fulfill the requirements of data protection law.

This does not mean that data protection has no place in US law. The growing use of computers and the surveillance scandals from the Nixon and FBI director Hoover-era led to the formulation of Fair Information Practice Principles (FIPP):[58]

- There must be no personal data record-keeping systems whose very existence is secret.
- There must be a way for a person to find out what information about the person is in a record and how it is used.
- There must be a way for a person to prevent information about the person that was obtained for one purpose from being used or made available for other purposes without the person's consent.
- There must be a way for a person to correct or amend a record of identifiable information about the person.
- Any organization creating, maintaining, using, or disseminating records of identifiable personal data must assure the reliability of the data for their intended use and must take precautions to prevent misuses of the data.

The first 4 FIPP have to do with transparency, while the last one sets an accountability standard. These FIPP are similar to the principles underlying data protection in Europe, but have only been put into law in specific areas. The Privacy Act of 1974 implements the FIPP in the government and is the main legal framework concerning the treatment of personal information by the federal government. It regulates and restricts the collection, retention and disclosure of personal data. Further does it grant individuals a right of information, access and amendment or correction.[59] Since the 1970s a range of laws containing privacy protection for specific sectors have been established. These laws implement the FIPPs fully or partially and make them applicable on big data practices with data regulated by these laws.

The Federal Trade Commission (FTC) plays an important role in regulating privacy in the private sector. It regulates and supervises market practices and has the authority to enforce trade law through investigatory and litigation powers. Basic consumer protection is provided by the FTC Act, which forbids "unfair or deceptive acts or practices in or affecting commerce", while the FTC also has the authority to enforce other specific consumer protection laws, like the FCRA or COPPA, and the EU-US Safe Harbor Framework. The FTC has taken up the role

---

[58]US Department of Health, Education & Welfare, *Records, Computers and the Rights of Citizens, Report of the Secretary's Advisory Committee on Automated Personal Data Systems*, July 1973.

[59]Chris Jay Hoofnagle, "Big Brother's Little Helpers: How ChoicePoint and Other Commercial Data Brokers Collect, Process, and Package Your Data for Law Enforcement", *N.C.J. Int'l L. & Com. Reg.*, Vol. 29, No. 595, Summer 2004.

of the de facto data protection authority by enforcing privacy policies of companies. The legal status of privacy policies has been ambiguous and enforcement under contract law failed in practice. The FTC has treated violations by a company of its published privacy policy as such a deceptive and in several occasions unfair act. It developed through settlements a common law-like jurisprudence establishing norms concerning transparency, data collection and use, and data security. This jurisprudence evolved towards treating the disrespect of industry standards on these issues as a form of deceptive act. FTC settlements and opinions have therefore become an important source of law.[60] This FTC practice has widely broadened the areas where processing of personal data is subjected to constraints.

7.2 Impact of Data Protection on Big Data Economics

The rules contained in the EU data protection framework as well as in the US FIPP have received heavy criticism from industry and a range of scholars for not being suited for big data. These critics consider it to be an obstacle for technical development and the scientific and economic advantages a wider implementation of big data can bring,[61] or consider it broken and not effective any more to protect privacy in the age of big data.[62] Criticism has been levelled at the notions of personal data versus anonymous data, principles like purpose limitation, data minimisation, and consent as base for legitimate processing of personal data. On the other hand, a range of scholars and the WP29 defend the application of the data protection framework in the big data context and refuse to see enough ground in the fruits of progress arguments in terms of economy, security or science to lower the protection of privacy given by the data protection framework.

The GDPR drafts show some attempts to limit the application of the data protection framework or to lower the obligations in certain circumstances, like the inclusion of pseudonymous data. These contested attempts for legal fine-tuning embody the plea by the critics of the current data protection framework to move

---

[60]Solove, Daniel J. and Hartzog, Woodrow, "The FTC and the New Common Law of Privacy", *Columbia Law Review*, Vol. 114, No. 583, 2014.

[61]Tene, Omer and Jules Polonetsky, Big Data for All: Privacy and User Control in the Age of Analytics, 11 *Nw. J. Tech. & Intell. Prop.* 239 (2013); Ira S. Rubinstein, Big Data: The End of Privacy or a New Beginning?, NYU School of Law, Public Law Research Paper No. 12-56; Lokke Moerel, Big Data Protection: How to Make the Draft EU Regulation on Data Protection Future Proof, Tilburg University, 2014.

[62]Ohm, Paul, Broken Promises of Privacy: Responding to the Surprising Failure of Anonymization, 57 *UCLA Law Review* 1701 (2010), 1701–1777; Schwartz, Paul M. and Solove, Daniel J., The PII Problem: Privacy and a New Concept of Personally Identifiable Information, *New York University Law Review*, December 2011, 1814–1894; Alessandro Mantelero, Defining a new paradigm for data protection in the world of Big Data analytics-2014 ASE BIGDATA-SOCIALCOM-CYBERSECURITY Conference, Stanford University, May 27–31, 2014.

the attention from data collection to a risk-based approach based on the actual use of personal data. These proposals involve a scaled approach through which the application of data protection principles gets modulated.[63] The WP29 has reacted to this plea with its statement on a risk-based approach[64] and other recent recommendations. It points to the risk-based elements present in the data protection framework, while making re-interpretations of data protection principles like purpose limitation which are more compatible with this approach.

The Obama administration has taken the initiative to remedy the piecemeal privacy law by an overall consumer privacy regulation, called the Consumer Privacy Bill of Rights.[65] This Consumer Privacy Bill of Rights gives a wider implementation of the FIPP in the digital economy.

These principles will be further developed through multistakeholder processes in order to develop enforceable Codes of Conduct. The FTC would enforce this Bill. This can happen through a new authority provided by law or through its authority to prohibit deceptive and unfair practices. The Obama administration takes a double approach towards the further development. It prefers to enact the Consumer Privacy Bill of Rights through legislation in order to increase legal certainty, but if Congress does not want to vote this proposal into law, the implementation can anyway go on through the development of codes of conduct.

The law proposal itself has not seen a lot of action the last 2 years in Congress. The privacy multistakeholder processes have resulted in a Code of Conduct for transparency in mobile apps,[66] while such a process is ongoing concerning the commercial use of facial recognition technology.[67] Although both affect specific big data practices, the results of this initiative remain limited. Where a Consumer Privacy Bill of Rights would subject commercial big data practices to the FIPP, the situation remains one of piecemeal regulation in distinct laws and FTC enforcement of privacy statements.

---

[63]Tene, Omer and Jules Polonetsky, Big Data for All: Privacy and User Control in the Age of Analytics, 11 *Nw. J. Tech. & Intell. Prop.* 239 (2013), 258–259; Paul Ohm, Broken Promises of Privacy: Responding to the Surprising Failure of Anonymization, 57 *UCLA Law Review* 1701 (2010), 1759–1777; Schwartz, Paul M. and Solove, Daniel J., The PII Problem: Privacy and a New Concept of Personally Identifiable Information, *New York University Law Review*, December 2011, 1879–1894.

[64]WP29, *Statement of the WP29 on the role of a risk-based approach in data protection legal frameworks*, 30 May 2014.

[65]White House, *Consumer Data Privacy in a Networked World: A Framework for Protecting Privacy and Promoting Innovation in the Global Digital Economy*, 23 February 2012.

[66]NTIA, "Privacy Multistakeholder Process: Mobile Application Transparency", 12 Nov 2013. www.ntia.doc.gov/other-publication/2013/privacy-multistakeholder-process-mobile-application-transparency.

[67]NTIA, "Privacy Multistakeholder Process: Facial Recognition Technology", 11 June 2015. http://www.ntia.doc.gov/other-publication/2015/privacy-multistakeholder-process-facial-recognition-technology.

The last years a lot of policy debate has taken place concerning big data and privacy, reflected in several important reports. One focus were data brokers, the other was specifically on big data and privacy. The Government Accountability Office (GAO),[68] the Committee on Commerce, Science and Transportation in the Senate[69] and the FTC[70] have investigated the data broker industry and the problems it poses concerning privacy. The reports all conclude that consumers can not exercise rights foreseen in FIPPs towards this industry. The FTC recommends to subject the different branches of this industry to legislation similar to FCRA and to assure transparency, access and amendment for consumers.

Further president Obama launched a Big data review, focused on big data and privacy. It resulted in 2 reports. The first report[71] was made by a working group of senior Administration officials led by John Podesta and resulted from a broad process with stakeholder consultations and academic workshops. This report of the Big Data and privacy working group gives an overview of big data practices in the public and private sector, and points to both the positive gains as the dangers involved. It notes several areas where big data presents challenges like the marketplace, schools, the danger of new forms of discrimination and using data as a public resource. The report makes recommendations like: advance the Consumer Privacy Bill of Rights, pass national data breach legislation, extend privacy protections to non-US persons, ensure data collected on students in school is used for educational purposes, expand technical expertise to stop discrimination at the lead civil rights and consumer protection agencies and amend the Electronic Communications Privacy Act.

Parallel the President's Council of Advisors for Science and Technology (PCAST) conducted a study of the technological trends underpinning big data, in order to assess the technical feasibilities of different policy approaches.[72] Also this report start with a broad sketch of uses of big data and the possible tradeoffs between privacy, security and convenience. PCAST states that a policy focusing on limiting data collection is not a broadly applicable or scalable strategy. Also because a lot of privacy problems arise after the collection with the fusion of data sources. It argues that the use of data is the place where consequences are produced and the technically most feasible place for protections. Further, some

---

[68]United States Government Accountability Office, *Information Resellers. Consumer Privacy Framework Needs to Reflect Changes in Technology and the Marketplace*, GAO-13-663, September 2013.

[69]Senate Committee on Commerce, Science, and Transportation, *A Review of the Data Broker Industry: Collection, Use, and Sale of Consumer Data for Marketing Purposes*, 18 December 2013.

[70]Federal Trade Commission, *Data Brokers. A Call for Transparency and Accountability*, May 2014.

[71]White House, *Big Data: Seizing Opportunities, Preserving Values*, 1 May 2014.

[72]President's Council of Advisors on Science and Technology, *Big Data: A Technological Perspective*, White House, 1 May 2014.

techniques for privacy protection used in the past do not seem robust anymore in the context of big data, like anonymization, data deletion (as old data sources can later prove useful in combination with others) or distinguishing the treatment of data from metadata (as metadata can be as much a risk to privacy as the data itself). Also the notice and consent framework is considered unworkable. This framework places the burden of privacy at the individual, while this individual is placed in an unequal position in relation with the provider. The responsibility for using the personal data in accordance with the preferences of the data subject should better be shifted to the provider.

This assessment puts doubt to the robustness of FIPPs, underlying the proposed Consumer Privacy Bill of Rights and other privacy regulations. PCAST still endorses these principles as sound, but states that big data puts effective operationalisation at risk. It suggests several adaptations in line with its recommendation to focus on use of data instead of data collection. Concerning rights meant as consumer empowerments, PCAST recommends to recast these empowerments as obligations of the entity using the data whenever such empowerment has become practically impossible to exercise in a meaningful way.

On both sides of the Atlantic protection of personal data and its relation to big data has become a policy issue. The different starting situations influences the debate on both sides. Also for personal data the EU framework is more restrictive for big data than the US privacy laws. Except in certain sectors, personal data can be freely used in the US where in the EU this is only the case for anonymized data. Result has been the development of strong data economy in the US, with the development of specialised actors in the data value chain, like data brokers, and the development of such data value chains based on personal data, like targeted advertising. The lack of a general protection of personal data in the US incited the FTC to become implicitly a rule maker based on consumer law and the reasonable expectations of the consumer. The Obama administration has made attempts to make privacy protection more general as part of consumer law, but without result till now.

The policy debate shows a similar struggle with the practical implementation of data protection principles or FIPP on big data processing. The underlying question is if the legal mechanisms to ensure transparency about what data processors do with personal data, developed for the information market, are still effective mechanisms in a data economy and if they allow building data value chains. Although the space given to data protection principles vary much, policy makers on both sides of the Atlantic do not question the underlying principles of data protection, but are looking for a more 'data-friendly' implementation. This is not directly reflected in the EU policy documents discussed earlier, but rather in the WP29 statements and in the legislative process of the GDPR. The policy documents generally present data protection as an important tool to build consumer trust. However, in the Communication on a data-driven economy the Commission announces that after the adoption of the GDPR it will work on guidance concerning big data-related problems like on such as data anonymisation and pseudonymisation, data minimisation.[73]

Of all 3 legal frameworks the tension between data protection and the realisation of economic opportunities with a data economy remains the most difficult to resolve. In the US the status quo remains the most 'data-friendly' solution from a commercial perspective and the federal government proves to be a too weak actor to force change. In the EU the outcome is less clear, but a clear demand from big data companies exists to soften the protection.

8 Public Sector Information and Open Data (Framework 3)

The evolution of the policies concerning public sector information shows more similarities on both sides of the Atlantic. The official policy motivation in the EU tends to be more integrated in the general economic motivation to develop an information society, while the Obama administration mainly stresses governmental transparency. But a look at the situation early 2000 shows it was the US being the forerunner in creating a market in PSI,[74] while much more barriers remained in the EU.[75]

The underlying logic is on both sides the same. Older frameworks of passive transparency, or access to documents on request, get augmented with active transparency and open data policies. Where the passive transparency procedures were tools to enlarge governmental transparency, the active transparency policies are more economically motivated. They look at PSI from a market perspective and try to avoid that public bodies have a distorting effect. Open data policies present a shift to a data economy perspective and include more attention to data quality.

Main regulatory focus of EU policy on public sector information (PSI) has been the review of the directive on the re-use of public sector information or PSI-directive 2003/98/EC, which was realised in 2013. The directive concerns PSI held by public sector bodies in member states. EU-law differentiates access from re-use of PSI, as it has no competence to regulate access to PSI in member states, except

---

[73]European Commission, *Towards a thriving data-driven economy*, COM(2014) 442, 2 July 2014.

[74]Gelmann, Robert, "The Foundations of United States Government Information Dissemination Policy", in Aichholzer, Georg and Herbert Burkert, *Public Sector Information in the Digital Age*, Edward Elgar Publishing Ltd., Cheltenham, 2004, 123–136.

[75]Volman, Yvo, "Exploitation of Public Sector Information in the Context of the *eEurope* Action Plan", in Aichholzer, Georg and Herbert Burkert, *Public Sector Information in the Digital Age*, Edward Elgar Publishing Ltd., Cheltenham, 2004, 93–107; Weiss, Peter N., "Borders in Cyberspace: Conflicting Public Sector Information Policies and their Economic Impacts", in Aichholzer, Georg and Herbert Burkert, *Public Sector Information in the Digital Age*, Edward Elgar Publishing Ltd., Cheltenham, 2004, 137–159.

on environmental information. The right of access consists of a right to see and to take knowledge of the content of documents, but does not imply automatically that this information can be used without restrictions. Re-use is defined as: "the use by persons or legal entities of documents held by public sector bodies, for commercial or non-commercial purposes other than the initial purpose within the public task for which the documents were produced".[76] In other words, re-use concerns the further use of the information, after having received knowledge of it. When public authorities make a further use of information outside their public task, it is also considered re-use. E.g. the commercialisation of certain data in order to recuperate costs, like publishing maps.[77] Specific sectoral rules exist on access and re-use, like the Inspire-directive.

Aim of the 2003 PSI directive was to create an internal market of PSI. It wants to assure that all private actors can use PSI in an equal manner. The PSI has to be available for re-use both for commercial and non-commercial purposes under the conditions stipulated by the directive. States are not obliged to give access or to allow re-use, but once the permission for re-use is given it must be done under equal conditions for all players and in a transparent manner. The conditions linked to the re-use of documents have to be non-discriminatory for comparable categories of re-use.

The PSI directive also fitted in the vision of the European Commission on economic development of the information society. PSI had to fuel a market of rich content and therefore become available for such content producers. Therefore charges for the re-use of PSI have to be limited to the marginal costs incurred for their reproduction, provision and dissemination and may not include costs linked to the original collection of data. The directive provides an exception when public sector bodies are required to generate revenue to cover a substantial part of their costs and for libraries. Secondly, monopolies by public sector bodies have to be prevented. Non-discriminatory access and re-use concerns also public sector bodies for activities outside their public tasks. Commercial activities by public sector bodies outside their public task have to take place under the same market conditions as for private actors.

Where the 2003 PSI-directive fits in the vision of the Internet as content market, the revision in 2013 shows a shift towards attention for the data value chain. The revised PSI-directive provides that when possible PSI is made available "in open and machine-readable format together with their metadata. Both the format and the metadata should, in so far as possible, comply with formal open standards".[78] This improves semantic interoperability and facilitates the use of the data

---

[76]European Parliament and the Council, Directive 2003/98/EC on the re-use of public sector information (PSI-directive), 17 November 2003, art. 2(4).

[77]Janssen, Katleen, *The EC Legal Framework for the Availability of Public Sector Spatial Data. An examination of the criteria for applying the directive on access to environmental information, the PSI directive and the INSPIRE directive*, ICRI, Leuven, 4 December 2009, p. 65.

[78]European Parliament and the Council, Directive 2003/98/EC on the re-use of public sector information (PSI-directive), 17 November 2003, art. 5(1).

in automated and aggregated ways. Other recent policy initiatives concerning open data and e-government also represent this shift. On the one hand by making data more accessible through data portals. The Commission planned in its Communication on Open data[79] in 2011 to set up 2 data portals: the European Union Open Data Portal[80] to make available its own data resources and a pan-European data portal with data from the Commission, member states and public sector bodies.[81] Further it shifted more attention to data quality by supporting projects to enhance semantic interoperability.

The open data policy is also driven by the Commission's effort to enhance interoperability between public administrations as part of its e-government policy. Through cross-border exchanges of information between member state and EU public administrations it tries to enable European public services. Objective is to aggregate 'basic' public services and to make them Europe-wide accessible in cross-border services. The European eGovernment action plan 2011–2015 did set the objectives to have by 2015 a number of key cross-border services available online.[82] Such increased interoperability between public administrations would not only lead to more efficient and effective public administrations, but also have strong impact on the data economy. Open data policies have limited effect when data cannot be linked easily and remains locked in incompatible formats. Interoperability between open data sources turns these sources into big data.

The European Commission developed a European Interoperability Strategy (EIS)[83] and a European Interoperability Framework (EIF),[84] and promotes now the adoption of national interoperability frameworks by member states in line with the EIF.[85] The EIS combines a top-down approach through European policy development and coordination with a bottom-up, sectoral approach through projects. The practical implementation of this sectoral approach is found in the program on Interoperability Solutions for European Public Administrations (the ISA program),[86] supporting activities to facilitate cross-border digital collaboration between public administrations from member states and EU institutions. The top-down approach is further developed in the EIF, which defines an agreed approach to interoperability. It sets principles of and a conceptual model for European

---

[79]European Commission, *Open data. An engine for innovation, growth and transparent governance*, COM(2011) 882, 12.12.2011.

[80]https://open-data.europa.eu.

[81]http://publicdata.eu.

[82]European Commission, *The European eGovernment Action Plan 2011-2015. Harnessing ICT to promote smart, sustainable & innovative Government*, COM(2010)743, 15.12.2010, p. 4.

[83]European Commission, *Towards interoperability for European public services,* COM(2010) 744, annex I, 16.12.2010.

[84]European Commission, *Towards interoperability for European public services*, COM(2010) 744, annex II, 16.12.2010.

[85]An overview of the progress can be found on http://www.daeimplementation.eu/dae_actions.php?action_n=26.

[86]http://ec.europa.eu/isa/.

public services and describes interoperability levels, interoperability agreements and governance.

The revision of the PSI directive figured already in the 2010 Digital Agenda, but the shift in attention towards the use of PSI as resource for a data economy became more visible first in the Communication on Open Data in 2011. The attention for interoperability in the e-government initiatives was also an early sign of attention for data value chains. It shows that e-government initiatives also function as tools for policy learning.

In the US we see similar attention shifts in the PSI policy. The Obama administration has from the start in 2009 given a strong impulse for enlarging the availability and access to public sector information, building upon pre-existing legislation for passive and active transparency.

Passive transparency, the giving access to information on request, is provided by the Freedom of Information Act. Active transparency, the providing of information on the initiative of the government, is regulated by the E-Government Act of 2002 and the Paperwork Reduction Act. The Paperwork Reduction Act dates from 1980, but was strongly revised in 1995 and was also the convenient place to include the framework for an information management and dissemination policy. It prevents agencies to restrict dissemination by using exclusive distribution arrangements, to restrict use, resale or redissemination or to make it subject to fees or royalties, and to ask user fees exceeding the cost of dissemination.[87] Another important element is that copyright protection is not available for the US government.[88] This legal framework led an early foundation for private sector use of PSI and the development of an information market.

The Obama administration added to this legislative framework a policy initiative by the executive branch to give a stronger implementation of open government policy.[89] Rationale behind this policy is on the one hand strengthening democracy by enhancing accountability towards the public and participation of the public. On the other hand the objective is to make the government more effective by strengthening cooperation within the government and with private actors.

This resulted in the Open Government Directive, presenting a policy road map for the implementation of open government by executive departments and agencies.[90] It instructed agencies to make more government information available online in open formats. When deciding about publishing information, the presumption should be in favor of openness, that is "to the extent permitted by law and subject to privacy, confidentiality, security, or other valid restrictions". The publication of information should preferably be in open formats. An open format

---

[87]Gelmann, Robert, "The Foundations of United States Government Information Dissemination Policy", in Aichholzer, Georg and Herbert Burkert, *Public Sector Information in the Digital Age*, Edward Elgar Publishing Ltd., Cheltenham, 2004, 130.

[88]Ibid., 126.

[89]White House, *Memorandum on Transparency and Open Government,* 21 January 2009.

[90]Office of Management and Budget, Open Government Directive, M-10-06, 8 December 2009.

is defined as "one that is platform independent, machine readable, and made available to the public without restrictions that would impede the re-use of that information." This definition contains elements of legal (no restrictions of re-use) and of technical interoperability (platform independent, machine readable). Objective of the open format is that the information can be "retrieved, downloaded, indexed, and searched by commonly used web search applications". This focus on open formats shift the attention from information markets to making data useful for more developed data value chains.

A second policy initiative focused on digital government.[91] It lists 4 main principles: an information-centric approach, a shared platform approach, a customer-centric approach and a platform ensuring security and privacy. All point again to making PSI useful in data value chains. The information-centric approach introduces an attention for semantic interoperability. It promotes a shift in thinking about digital information, away from the old approach focused primarily on presentation. An information-centric approach should focus on making data and content accurate, available and secure. It needs to turn unstructured content into structured data and to associate this structured data with valid metadata. Providing this data through web APIs enhances interoperability and makes the data assets widely available. It also supports device-agnostic security and privacy controls, shifting the focus from securing devices to securing data.

The evolution of policies on public sector information are similar in the EU and the US. Both developed PSI as a resource for information markets, but more recently focussed more on making PSI useful for data value chains. Their open data policies evolved from setting up data portals and a focus on quantity of datasets to a focus on quality of data in terms of interoperability. Non-discriminatory frameworks and licenses improve legal interoperability, attention for open and standard formats and descriptions improve technical and semantic interoperability. The aim to improve public services is present on both sides, but e-government efforts differ due to the specific EU attention for cross-border interoperability.

We can conclude that both e-government and open data policies are important elements of big data policies in the EU and the US. They also present important areas of policy learning, especially on interoperability and on what the construction of data value chains involves.

9 Conclusions: Adapting Legal Frameworks to a Data Economy Remains Unfinished Business

In this article we looked at how policy makers digested big data in their ICT- or Internet-related policies. Big data practices depend on the possibility to build data value chains and are therefore very much affected by legal frameworks regulating

---

[91]White House, *Digital Government: Building a 21st Century Platform to Better Serve the American People*, 23 May 2012.

access, linking to and use of data. Such data value chains also change the role of the Internet from a market place where content and service providers meet human customers into a space where a lot of non-human actors exchange and process data in real time. We called this a change from an information economy to a data economy, as it changes the level of interactions over the Internet. This change puts stress on existing legal frameworks and can change views on the objectives which these legal frameworks have to reach.

We first looked at how the EU has adapted its policies to the advent of big data. On policy level the EU has developed in its 2014 *Communication on a data-driven economy*[92] a profound vision on a data economy, overcoming and deepening the earlier focus on the Internet as marketplace. It has more attention to the role of data and the need to make data interoperable for the creation of data value chain as part of a data economy. This change in focus also raises the question if the legal frameworks affecting such data value chains are still adequate and if this change in policy vision is also translated into new objectives concerning the legal frameworks regulating data flows. In this respect we found a more mixed picture.

We looked at 3 legal frameworks: intellectual property rights, the protection of personal data and the regulation of public sector information. To give a broader perspective we also compared these legal frameworks and policy responses in the US.

The copyright framework in the US has less problems with big data. Comparing the European IPR framework with the US shows some striking differences. First, the US has limited the protection of databases to copyright and has never extended IPR protection purely on grounds of investment. Secondly, its fair use regime proved to be much more technology neutral and adaptable to new technological developments. Both improve legal interoperability of data sources.

Result is that copyright raised no policy debate in the US, while the courts are the main actors in dealing with new technologies in this context. In other countries we see reviews of IPR policies and recommendations to adapt the copyright frameworks with new exceptions. This debate also started in the EU, but remains embedded in and limited by a focus on strengthening the IPR framework. Active policy attention for a data economy by the EU has not led yet to a thorough revision of the IPR framework and more legal interoperability. A revision of the exceptions in the copyright framework to improve the space for data mining is planned, but the database directive remains outside the policy focus.

On both sides of the Atlantic protection of personal data and its relation to big data has become a policy issue. The different starting situations influences the debate on both sides. Also for personal data the EU framework is more restrictive for big data than the US privacy laws. Except in certain sectors, personal data can be freely used in the US where in the EU this is only the case for anonymized data. Result has been the development of strong data economy in the US, with the development of specialised actors in the data value chain, like data brokers, and

---

[92]European Commission, *Towards a thriving data-driven economy*, COM(2014) 442, 2 July 2014.

the development of such data value chains based on personal data, like targeted advertising. The lack of a general protection of personal data in the US incited the FTC to become implicitly a rule maker based on consumer law and the reasonable expectations of the consumer. The Obama administration has made attempts to make privacy protection more general as part of consumer law, but without result till now.

The policy debate shows a similar struggle with the practical implementation of data protection principles or FIPP on big data processing. The underlying question is if the legal mechanisms to ensure transparency about what data processors do with personal data, developed for the information market, are still effective mechanisms in a data economy and if they allow building data value chains. Although the space given to data protection principles varies a lot, policy makers on both sides of the Atlantic do not question the underlying principles of data protection, but are looking for a more 'data-friendly' implementation. Of all 3 legal frameworks the tension between data protection and the realisation of economic opportunities with a data economy remains the most difficult to resolve.

The evolution of policies on PSI are similar in the EU and the US. Both developed PSI as a resource for information markets, but recently focussed more on making PSI useful for data value chains. Their open data policies evolved from setting up data portals and a focus on the quantity of datasets to a focus on quality of data in terms of interoperability. Non-discriminatory frameworks and licenses improve legal interoperability, attention for open and standard formats and descriptions improve technical and semantic interoperability. The aim to improve public services is present on both sides, but e-government efforts differ due to the specific EU attention for cross-border interoperability. We can conclude that both e-government and open data policies are important elements of big data policies in the EU and the US. They also present important areas of policy learning, especially on interoperability and on what the construction of data value chains involves.

On the whole we can conclude that the recent EU policy documents reflect an improved understanding of big data and a data economy, but the translation of this vision into new objectives for legal frameworks dealing with data flows proves to be more difficult and shows mixed results.

Bibliography

Alessandro Mantelero. 2014. Defining a new paradigm for data protection in the world of Big Data analytics-2014 ASE BIGDATA-SOCIALCOM-CYBERSECURITY Conference. Stanford University, 27–31 May 2014.

Article 29 Data Protection Working Party. 2014. *Statement on the role of a risk-based approach in data protection legal frameworks*, WP218, 30 May 2014.

Australian Law Reform Commission. 2013. *Copyright and the Digital Economy. Final Report*, ALRC Report 122, 30 Nov 2013.

Copyright Review Committee. 2013. *Modernising Copyright. The Report of the Copyright Review Committee for the Department of Jobs, Enterprise and Innovation*, Dublin.

European Commission. 1994. *Europe's way to the information society. An action plan*, COM(94)347, 19 July 1994.
European Commission. 2000. *eEurope 2002—An Information society for all—Draft Action Plan prepared by the European Commission for the European Council in Feira—19–20 June 2000*, COM(2000)233, 24 May 2000.
European Commission. 2002. *e Europe 2005: An information society for all. An Action Plan to be presented in view of the Sevilla European Council, 21/22 June 2002*, COM(2002)263, 28 May 2002.
European Commission. 2005. *i2010—A European Information Society for growth and employment*, COM(2005) 229, 1 June 2005.
European Commission. 2005. *First evaluation of Directive 96/9/EC on the legal protection of databases*, DG Internal Market and Services Working Paper, 12 Dec 2005.
European Commission, *A Digital Agenda for Europe*, COM(2010)245, 19 May 2010.
European Commission, *The European eGovernment Action Plan 2011–2015. Harnessing ICT to promote smart, sustainable and innovative Government*, COM(2010)743, 15 Dec 2010.
European Commission, *Towards interoperability for European public services,* COM(2010) 744, annex I & II, 16 Dec 2010.
European Commission, *Open data. An engine for innovation, growth and transparent governance*, COM(2011) 882, 12 Dec 2011.
European Commission, *Unleashing the potential of cloud computing in Europe*, COM(2012)529, 27 Sept 2012.
European Commission, *The Digital Agenda for Europe—Driving European growth digitally*, COM(2012)784, 18 Dec 2012.
European Commission, *On content in the Digital Single Market*, COM(2012)789, 18 Dec 2012.
European Commission, *A European strategy on the data value chain*, November 2013.
European Commission, *Towards a thriving data-driven economy*, COM(2014) 442, 2 July 2014.
European Commission, *Report on the Implementation of the Communication 'Unleashing the Potential of Cloud Computing in Europe' Accompanying the document Communication from the Commission to the European Parliament, the Council, the European Economic and Social Committee and the Committee of the Regions 'Towards a thriving data-driven economy'*, SWD/2014/0214, 2 July 2014.
European Commission, *A Digital Single Market Strategy for Europe*, COM(2015)192, 6 May 2015.
European Council. 2013. Conclusions. EUCO 169/13, 24–25 Oct 2013.
Federal Trade Commission. *Data Brokers. A Call for Transparency and Accountability*, May 2014.
Gelmann, Robert. 2004. The foundations of United States Government Information Dissemination Policy. In *Public sector information in the digital age* eds. Aichholzer, Georg and Herbert Burkert, 123–136. Cheltenham: Edward Elgar Publishing Ltd.
Goldstein, Paul. 2001. *International copyright principles, law, and practice*. New York: Oxford University Press.
Gutwirth, Serge and Paul De Hert. (2008) Regulating profiling in a democratic constitutional state. In *Profiling the European Citizen: Cross-disciplinary perspectives* eds. Hildebrandt, M. and Gutwirth, S, 271–293. Berlin: Springer.
Hargreaves, Ian. 2011. *Digital Opportunity: Review of Intellectual Property and Growth*, May 2011.
Hoofnagle, Chris Jay. 2004. Big brother's little helpers: How choice point and other commercial data brokers collect, process, and package your data for law enforcement. *N.C.J. Int'l L. & Com. Reg* 29: 595 (Summer 2004).
Intellectual Property Office. 2011. *The UK's International Strategy for Intellectual Property*, 11 August 2011.
Ira S. Rubinstein, *Big Data: The End of Privacy or a New Beginning?*, NYU School of Law, Public Law Research Paper No. 12–56.

Janssen, Katleen, *The EC Legal Framework for the Availability of Public Sector Spatial Data. An examination of the criteria for applying the directive on access to environmental information, the PSI directive and the INSPIRE directive*, ICRI, Leuven, 4 December 2009.

Janssen, Katleen, and Jos Dumortier. 2006. The protection of maps and spatial databases in Europe and the United States by copyright and the Sui Generis Right. *Journal of Marshall Journal Computer & Information Law* 24(2): 195–225.

Leslie C. Ruiter and Gerald van Belle. Data extraction: Beyond the Sweat of the Brow. http://www.stokeslaw.com/uploads/pdf/data_and_the_law-gerald_van_belle_and_leslie_ruiter.pdf.

Lokke Moerel. 2014. *Big data protection: How to make the draft EU regulation on data protection future proof.* Tilburg University.

Netanel, Neil. 2011. Making sense of fair use. *Lewis & Clark Law Review* 15(2011): 715–771.

Office of Management and Budget. 2009. Open Government Directive, M-10-06, 8 Dec 2009.

Ohm, Paul. 2010. Broken Promises of privacy: Responding to the surprising failure of anonymization, 57. *UCLA Law Review* 1701: 1701–1777.

Porter, Michael E. 1985. *Competitive advantage: Creating and sustaining superior performance.* New York: Free Press.

President's Council of Advisors on Science and Technology. 2014. *Big Data: A Technological Perspective*, White House, 1 May 2014.

Schwartz, Paul M. and Daniel J. Solove. 2011. The PII problem: Privacy and a new concept of personally identifiable information, *New York University Law Review*, 1814–1894.

Solove, Daniel J. and Woodrow Hartzog. 2014. The FTC and the new common law of privacy. *Columbia Law Review* 114: 583.

Tene, Omer and Jules Polonetsky. 2013. Big data for all: Privacy and user control in the age of analytics. *Northwestern Journal of Technology and Intellectual Property* 11: 239.

Triaille, Jean-Paul. 2014. Jérôme de Meeûs d'Argenteuil and Amélie de Francquen. *Study on the legal framework of text and data mining (TDM)*.

Truyens, Maarten, and Patrick Van Eecke. 2014. Legal aspects of text mining. *Computer law & security review* 30: 153–170.

US Department of Health, Education & Welfare. 1973. *Records, Computers and the Rights of Citizens, Report of the Secretary's Advisory Committee on Automated Personal Data Systems*.

US Government Accountability Office. 2013. *Information Resellers. Consumer Privacy Framework Needs to Reflect Changes in Technology and the Marketplace*, GAO-13-663.

US Senate Committee on Commerce, Science, and Transportation. 2013. *A Review of the Data Broker Industry: Collection, Use, and Sale of Consumer Data for Marketing Purposes*, 18 December 2013.

Volman, Yvo. 2004. Exploitation of public sector information in the context of the *eEurope* action plan. In *Public sector information in the digital age* eds. Aichholzer, Georg and Herbert Burkert, 93–107. Cheltenham: Edward Elgar Publishing Ltd.

Privacy and Innovation: From Disruption to Opportunities

Marc van Lieshout

Abstract In this chapter I present an approach of privacy from the perspective of innovation theory. I bring two conceptual approaches together. First, I disentangle privacy in three interconnected concepts: information security, data protection and the private sphere. Each of these concepts has its own dynamics and refers to a specific logic: technology in case of information security, regulation in case of data protection and society in case of the private sphere. By interconnecting them, a more nuanced perspective on the innovative incentives stemming from privacy considerations arises. Second, innovation is considered to be hampered by market and system imperfections. These imperfections reduce the efficiency of the innovation system. Analysing which imperfections exist helps in overcoming them by identifying adequate counter-strategies. I will use a policy study that has been performed for the Dutch Ministry of Economic Affairs to elaborate the relation between privacy and innovation in more detail. The resulting tone is optimistic: during the study several indications for a more privacy respecting approach by firms were found. Still, the challenges to be addressed are huge.

Keywords Privacy · Innovation theory · Market and systems imperfections · Information security · Data protection · Privacy and innovation

1 Introduction

Many organisations perceive privacy as an 'innovation killer'. Innovative applications which make use of personal data are constrained because they need to meet legal obligations. Data mining and data analytics are at the forefront of today's

---

M. van Lieshout (✉)
TNO Strategy & Policy, Van Mourik Broekmanweg 6, 2628 XE Delft, The Netherlands
e-mail: marc.vanlieshout@tno.nl

information and communication technologies (ICTs).[1] For data analytics, 'purpose specification'—that needs to be articulated before data collection and processing are to take place—seems to be a relic of past times. In a similar way the 'Right to be Forgotten'[2] is considered to complicate business processes of organisations that collect vast amounts of personal data. Google, being brought to court by a Spanish citizen and convicted in a ruling by the European Court of Justice (May 13, 2014), faces an on-going stream of requests to remove specific links from the results of a search query.[3] According to the court ruling, justified requests must refer to removal of data being "inaccurate, inadequate, irrelevant, or excessive".[4] So, not all personal data will be removed on request from search queries but only data that are inaccurate, inadequate, irrelevant or excessive, categories that are hard to define with precision. Google received more than 220,000 requests in the months following upon the publication of the rulings. As a consequence, Google stated that it will look to specifics within the legal framework of the European Union in order to stay ahead of new requirements that need to be met.[5] It established an advisory network that will try to discover common terms and approaches to deal with the various requests.[6]

This illustration is interesting: though it seems to show that regulation may stifle innovation (Larry Page, Google's CEO, warned that the ruling "could damage the next generation of Internet start-ups and strengthen the hand of repressive governments inclined to restrict online communication"[7]), the ruling could force Google to be creative and think of novel ways to deal with the thousands of requests that flood its offices. Up till now, Google allegedly uses man power to deal with the requests, and has not introduced other more innovative approaches.[8]

---

[1] World Economic Forum, Unlocking the value of personal data: from collection to usage (World Economic Forum 2013).

[2] European Court of Justice (2014). Factsheet on the 'Right to be Forgotten' Ruling (C-131/12), http://ec.europa.eu/justice/data-protection/files/factsheets/factsheet_data_protection_en.pdf (visited March 4, 2015).

[3] http://www.economist.com/news/international/21621804-google-grapples-consequences-controversial-ruling-boundary-between (visited March 4, 2015).

[4] European Court of Justice (2014, p. 2).

[5] http://www.theguardian.com/technology/2015/feb/19/google-acknowledges-some-people-want-right-to-be-forgotten (visited March 4, 2015).

[6] https://www.google.com/advisorycouncil/ (visited March 12, 2015).

[7] http://www.washingtonpost.com/news/morning-mix/wp/2014/05/30/google-ceo-warns-right-to-be-forgotten-could-stifle-innovation-and-empower-repressive-regimes/ (visited March 4, 2015).

[8] http://www.pcworld.idg.com.au/article/560060/how-google-dealing-right-forgotten-requests/ (vistied March 4, 2015).

A Dutch example shows how organisations may retreat to a combination of organizational and technical innovations to deal with potential issues of privacy invasion. The Dutch railway organization, NS, had acquired negative attention some years ago with the introduction of the 'OV-chip'card, a contactless RFID-based public transport card. The Dutch Radboud University showed in 2008 that the OV-chip could easily be hacked.[9] The chip in use was an old and basically outdated version of the MiFare Classic chip, with a modest level of protection. While this vulnerability could only indirectly be attributed to the Dutch railway organization (it was NXP[10] that sold the relatively unsecure chips to TransLink Systems, the organisation that introduced the OV-chip into the Dutch public transport system), the NS still was publicly held responsible. TransLink Systems had been warned as early as 2005 by the Dutch Data Protection Authority that it should refine its procedures and guidelines under which data collected through the OV-chip would be used for business and client purposes. In 2010, the Dutch DPA warned TransLink Systems, the Dutch NS and two other public transport providers (of two major Dutch cities) that they did not provide sufficient detail on how they would use collected data of students travelling with the student version of the public transport card.[11] As a result of the negative experiences and the negative public image, NS appointed a privacy officer who is able to halt projects and activities that could be invasive to customer privacy, and who has the responsibility to safeguard the privacy of travellers in NS activities.[12] Since NS has adopted several policies in which it consciously incorporates privacy considerations. One example is the monitoring of passenger movements on railway stations, a relevant activity for both the spatial organisation of railway stations and for determining economic hotspots within the railway station. NS used a system in which infrared detection of passengers was combined with using MAC addresses of Bluetooth and WiFi connections that were used by passengers. To prevent identification by MAC addresses, these addresses were complemented with information about the day on which the monitoring took place, and the resulting data were subsequently one-way hashed. An encompassing information policy that also included procedures for removal of data that were not needed anymore and campaigns to raise awareness by the employees who had access to the data, complemented the NS approach.[13]

This creative and innovative approach shows that privacy may have interesting innovation consequences as well. At the same time, NS does not publicly convey this image of respecting privacy in its activities. In our research we have met with

---

[9] https://ovchip.cs.ru.nl/Main_Page (visited March 4, 2015). The hack took place in 2008.

[10] The MiFare chip was originally a product produced by Philips, but at the time the hack became public, the chips were made and sold by NXP, the successor of Philips.

[11] https://cbpweb.nl/nl/nieuws/ov-bedrijven-bewaren-reisgegevens-studenten-ov-chipkaart-strijd-met-de-wet (visited March 4, 2015).

[12] Arnold Roosendaal et al., Actieplan Privacy (Delft: TNO-report R11603, 2014), 24 ff.

[13] Roosendaal, 24 ff.

a number of organisations that all express the intention to respect consumer privacy, but which all are hesitant to advertise this fact.[14]

Privacy and innovation thus do not really seem to merge in an easy and convenient manner. In this chapter I will start with presenting a pragmatic perspective on privacy, by disentangling it into three intersecting circles: information security, data protection and the private sphere. Then I will present some studies that researched the relationship between privacy and innovation. The next section introduces innovation theory and especially the existence of market and system imperfections as a conceptual approach to understanding how privacy and innovation practices can be related. This will be followed by a discussion of the results of a study the PI.lab performed for the Dutch ministry of Economic Affairs on privacy and innovation. The PI.lab is an expertise centre, formed by the Dutch organisations, Radboud University, SIDN, Tilburg University and TNO.[15] The study was dedicated to studying how Dutch businesses dealt with privacy requirements in their practices and policies. The concluding section will add some perspectives on research in this field.

2 The Concept of Privacy

Many authors have presented views on privacy.[16] In this paper, I will explore a different kind of approach, one that will start by distinguishing three main pillars. The European Charter of Fundamental Rights distinguishes between a right to privacy (article 7) and a right to the protection of personal data (article 8).[17] While the right to privacy relates to fundamental notions of the integrity of the body, the intimacy of the family, the sacrosanct place of the house and the right to confidential communications, the right to data protection refers to fundamental principles to be obeyed when personal data are at stake. These principles are articulated in the present EU directive on data protection.[18] One of these principle is the principle that data controllers and processors should take appropriate technical and

---

[14]We did not publish on this issue, but we encountered this attitude at a number of Dutch organisations. A few of these will be mentioned in this article.

[15]See http://pilab.nl/ (visited March 12, 2015).

[16]See for instance Rachel Finn, David Wright, and Michael Friedewald, "Seven types of privacy", in Serge Gutwirth, Yves Poullet et al. (eds.), *European Data Protection: Coming of Age* (Dordrecht: Springer, 2013) who present a challenging sevenfold dimensioning of privacy, based on previous work by amongst others Roger Clark.

[17]EC, *Charter of Fundamental Rights of the European Union*, (Brussels: Official Journal of the European Communities, C 364/1, 2000).

[18]For the FIP, see Robert Gellman, Fair Information Practices: A Basic History, http://bobgellman.com/rg-docs/rg-FIPShistory.pdf (visited March 9, 2015). The title of the EU Data Protection Directive 95/46/EC states: "[O]n the protection of individuals with regard to the processing of personal data and on the free movement of such data". See: http://eur-lex.europa.eu/LexUriServ/LexUriServ.do?uri=CELEX:31995L0046:en:HTML (visited March 9, 2015).

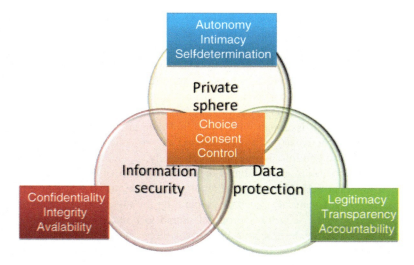

Fig. 1 Conceptual relations between private sphere, data protection and information security

organisational measures for safeguarding the security of the systems that store and process personal data. Together with the right to privacy and the right to data protection this leads to the following scheme (see Fig. 1).

The figure presents three intersecting circles with distinct orientations:

1. Information security focuses on technical requirements stemming from the treatment of personal data and the basic principles used in information security: confidentiality of data, integrity of data and availability of data.[19]
2. Data protection focuses on legal and regulatory issues stemming from laws and regulations, such as the EU directive on protecting personal data, 95/46/EU[20] and the ePrivacy directive (2002/58/EU), in line with the OECD Fair Information Principles and Practices.[21]
3. The Private sphere focuses on issues of individual privacy, the autonomy of the individual, the safeguarding against interference by others and the ability to determine one's life.

Choice, control and consent are the three basic principles that relate to each of the circles. By subdividing the discourse on privacy in these three domains I try to overcome the following barriers:

1. The discussion on privacy issues is often obfuscated by the dominance of technical issues (encryption as the key to all privacy problems) over societal

---

[19]Mark Stamp, *Information Security—Principles and practice* (Hoboken, New Jersey: John Wiley & Sons, 2006): p. 2.

[20]With the 95/46/EC directive to be replaced by the General Data Protection Regulation in due time.

[21]See http://oecdprivacy.org/ (visited March 5, 2015).

and regulatory ones.[22] Information security clearly plays a role, but it is obvious that good technical solutions cannot fix everything.

2. Gutwirth and Gellert argue that data protection is based on the definition of a number of procedures on how to deal with personal data, while privacy deals with social phenomena such as autonomy and the right to self-fulfilment. These latter issues always need to be assessed in a specific context. No general rules or procedures are available to decide if and in what sense privacy is infringed.[23] When dealing with issues in which personal data and privacy play a role we need both perspectives (the procedural and the so-called substantive one). In the scheme I use I address both aspects separately.

3. Data protection is often considered to be the domain of lawyers and legally trained professionals. However, the legal perspective alone is not at all sufficient to cover relevant DP issues. By separating the DP-approach from the private sphere and the security approach I intend to both emphasize the relative relevance of the legal (procedural) issues which are brought forward by the data protection legislative frameworks while keeping a strict eye on the technical and societal dimensions which are explicit parts of these frameworks as well.[24]

The figure presents and reconciles the various dimensions one has to deal with when personal data are at stake. Innovation processes play a role in each of the three domains. Innovation in security processes relates to new encryption techniques, such as homomorphic encryption. These encryption techniques can be part of the 'appropriate organisational and technical measures' which are requested in the data protection regulations. Data protection impact assessments and data protection audits are examples of organisational innovations. And new approaches which combine technical, organisational and user-related dimensions, such as information processes that use data vaults, are an example of a more complex and multidimensional innovation process, directed at enhancing autonomy, choice and control by the data subjects.[25]

---

[22]One telling example is the response of Phil Zimmerman during the panel on privacy innovations who responded to a question on whether privacy was more than securing data, that, indeed, in the end it all comes down to using encryption for safeguarding data. In my view, which I also introduced during the panel, this perspective falls short to capture on what privacy is about.

[23]Raphael Gellert and Serge Gutwirth. "The legal construction of privacy and data protection" *Computer Law and Security Review (CLSR)* 29 (2013): 522–530.

[24]One interesting issue in this respect is the capabilities data protection officers need to have. In the Directive and the Regulation it is emphasized that DPOs should have sufficient legal and technical knowledge. Given the need for additional DPOs (triggered by the new Regulation) one would expect multidisciplinary vocational courses to emerge that teach basic and advanced legal and technical insights.

[25]Examples of these innovations will be provided in the next section. One example relates to the opportunity to organize one's CV in a data vault, thereby anticipating on the increasing number of self-employed professionals who need to convey their professional details to (potential) clients.

3 Privacy and Innovation—A Conceptual Framework

According to the OECD Manual on innovation, innovation is either 'something new to the firm, something new to the market or something new to the world'.[26] Over the years, the OECD has expanded its definition of innovation in order to capture a broader array of activities: in addition to technological innovations, organisational innovations have become part of the definition. New production methods, new service distribution models, new ways of organising the collection and distribution of data within an organisation are all examples of innovation. Theoretical and conceptual approaches on the differences between the service oriented character of many business practices today and the older industry-led production model have led to variations on the traditional 'innovation-diffusion' model, but the main elements of this model are still in place.[27]

Innovation systems face different kind of imperfections. One such imperfection is, for instance, a regulatory framework that is not up to date and that blocks innovative activities because it forbids specific services that are part of a novel approach of businesses.[28] Modelling imperfections from an innovation systems perspective has led to the identification of market and system imperfections.[29] The model I use identifies five categories of imperfections that can arise in the innovation system and four categories of imperfections that can arise in the functioning of the market (Table 1).

3.1 Market Imperfections

Market imperfections refer to imperfections in the functioning of a market. These imperfections may have consequences for individuals, an example being the exclusion from specific products of services.

---

[26]OECD Oslo Manual (1997). The measurement of Scientific and Technological activities—Proposed guidelines for collecting and interpreting technological innovation data. http://www.oecd.org/science/inno/2367580.pdf (visited March 4, 2015).

[27]Richard Barras, "Towards a theory of innovation theory in services", *Research policy* 15 (4) (2000): 161–173. Everett M. Rogers, *Diffusion of Innovations* (5th edition) (New York: Free Press, 2003).

[28]This illustration could be applied to the example I provided before on purpose specification in data analytics situations. Purpose specification as such may not be sufficient to block an innovation in the field of data analytics, but combined with other regulatory requirements it may hinder innovative practices.

[29]Martijn Poel, *The impact of the policy mix on service innovation—The formative and growth phases of the sectoral innovation system for internet video services in the Netherlands*, (Enschede: GildeprintDrukkerijen 2013). Poel discusses these imperfections as market and systems failures. In a study project I have been part of in recent years, some participants urged to use the less intrusive vocabulary of 'imperfections' instead of 'failures'. I will follow this approach in this contribution.

Table 1 Categories of market and system imperfections

Market imperfections	Imperfections in the innovation system
Externalities/spill-overs	Imperfections in infrastructural provision and investment
Public goods	Lock-in or path dependency
Information a-symmetry	Institutional imperfections
Market power	Interaction failures
	Capabilities failures

Poel (2013, pp. 55–56)

Externalities and spill-overs refer to the situation in which the activities of one party have consequences, or spill over to other parties. These spill-overs can be of different kinds: knowledge spill-overs, market spill-overs or network spill-overs.[30] A well-known example of a positive spill-over relates to the so-called network externalities: in a service which relies on the exchange between its participants (a social media app for instance), each participant profits from the addition of a new participant.

Public goods are goods that embody public values, such as knowledge that could become available to everyone who would like to use it. Non-exclusivity however could be a barrier for innovation. When no party is able to capture the competitive advantages of exclusive knowledge no one is willing to invest in realising this knowledge. But exclusive availability of knowledge may hinder innovation as well since only one party may capture the benefits. Open innovation approaches, in which knowledge is shared in order to enhance the benefits for all, have been shown to offer advantages, especially in the domain of information and communication technologies where network effects are important.[31]

Information asymmetry refers to differences in access to relevant information. App developers, for instance, are usually small firms[32] which cannot afford to invest in coping with the peculiarities of privacy regulations. They cannot compete with larger organisations who can afford to hire a privacy officer information asymmetry may also refer to the relationship between firms and customers, in which a customer usually lacks detailed insights on what a firm knows and can do with information collected over the individual.

---

[30]James Medhurst et al., *An economic analysis of spill overs from programmes of technological innovation support*, (Report prepared for ICF GHK 2014). https://www.gov.uk/government/uploads/system/uploads/attachment_data/file/288110/bis-14-653-economic-analysis-of-spillovers-from-programmes-of-technological-innovation-support.pdf.

[31]Eric von Hippel and Georg van Krogt, "Open Source Software and the "Private-Collective" Innovation Model: Issues for Organization Science." (MIT Sloan School of Management Working Paper 4739-09, 2009).

[32]See for instance the EC Green Paper on Mobile Health (Com(2014) 219 final, that indicates that 64 % of mobile app have less than 10 employees (p. 7).

Market power refers to market dominance. Facebook is a clear example. Facebook has captured over one billion users. The mere presence of so many 'peers' on Facebook makes Facebook an interesting medium. Privacy-respecting alternatives to Facebook, such as Diaspora,[33] face the problem of not offering the same level of customer spread as Facebook does. Competing with the market dominance of Facebook is not an easy challenge.

3.2 Systems Imperfections

Imperfections in the innovation system refer more exclusively to arrangements between parties (firms, governments and customers) that block the process of innovation. These imperfections can be of various kinds as well.

Imperfections in infrastructural investments relate to those provisions that create opportunities to offer new services and products. The roll-out of broadband and 4G telecommunications networks is one such provision that is deemed essential to keep the innovation fabric running. Up till the nineties of the past century these infrastructures were considered public goods. Since then, market forces determine the creation of new infrastructures. Public intervention may be necessary to guarantee availability of new infrastructure in locations which are hardly interesting from a commercial perspective. Another example of such interventions is the case concerning network neutrality. The US Federal Communications Commission decided on February 26, 2015 that firms had to obey the principle of network neutrality.[34] No price competition on network bandwidth is allowed. In the European Union, a similar debate is going on, with the Commission leaning towards favouring net neutrality but as yet no clear decision has been made.[35]

Lock-in or *path dependencies* relate to the restriction of choice once a choice for a system has been made. The dominance of Microsoft in previous decades with its Windows Operating System and the dominance of Apple with its closed platform are examples. Lock-in creates fixed avenues of innovation. For customers it means that switching costs are high (having to replace all Apple related equipment and services comes at a high price), while new services need to fit in existing paths to be interesting to these customers.[36]

Institutional imperfections refer to failure in the institutional domain to enhance innovative practices. As a 'rule', the regulatory framework lags behind business practices.[37] 'Purpose specification' for instance, is deemed obsolete, given the

---

[33]See https://diasporafoundation.org/.

[34]http://www.zdnet.com/article/net-neutrality-becomes-the-law-of-the-land/.

[35]http://chrismarsden.blogspot.nl/2015/03/access-on-wolf-in-sheeps-clothing.html.

[36]http://www.macgasm.net/2012/02/09/state-apples-ecosystem-lockin/.

[37]Technology neutral regulatory frameworks are presented as alternative to this lagging behind, but—as the example of network neutrality shows—they are difficult to maintain.

changes in collecting and processing personal data. However, the regulatory framework still requires purposes to be defined as legitimate basis for data processing. Other imperfections could relate to a failing supervisory authority, for instance, one that lacks sufficient manpower to exercise all its responsibilities.

Imperfections in the interaction between the dominant players within an innovation network may result in sub-optimal solutions. These could lead to missing out opportunities because of groupthink among the most dominant actors.[38]

Capabilities imperfections refer to a sector's lack of skills and competences to fully capture the benefits of an innovation. Again, the size of the average app firm (64 % fewer than 10 employees) may lead to problems in capturing relevant developments taking place in the app market. Such developments relate to privacy as well.

While this set of market and system imperfections relate to innovation systems in general, they can also be related to issues concerning privacy as well. One such issue is *Privacy by design*. Promoted by the Canadian Information and Privacy Commissioner Ann Cavoukian[39] and adopted by the International data protection authorities in its 31st international conference in Madrid,[40] privacy by design is one of the placeholders in the new Regulation.[41] Considering privacy by design as an innovative practice enables analysing the impact of potential market and systems imperfections on the rise and spread of privacy by design.

4 Privacy and Innovation: It Takes Two to Tango?

An oft-heard statement is that privacy has a stifling effect on innovation. In a report that formed the basis of a statement for the US Government, Lenard and Rubin concluded that "the 'familiar solutions' associated with the Fair Information Principles and Practices are a potentially serious barrier to much of the innovation we hope to see from the big data revolution."[42] However, empirical evidence on the relation between privacy and innovation is scarce. One article which empirically studied the impact of privacy regulations on business processes, concluded that the overall consequences of having to deal with privacy are negative.[43] The authors

---

[38]Martijn Poel, *The impact of the policy mix on service innovation—The formative and growth phases of the sectoral innovation system for internet video services in the Netherlands*, (Enschede: Gildeprint Drukkerijen, 2013), 56.

[39]Ann Cavoukian, *Privacy by design—Take the challenge*, (Ontario 2009).

[40]http://thepublicvoice.org/TheMadridPrivacyDeclaration.pdf; https://www.priv.gc.ca/information/conf2013/res_06_openness_e.asp.

[41]Article 23 of the proposed General Data Protection regulation deals with data protection by design and by default.

[42]Thomas M. Lenard and Paul H. Rubin, *The Big Data revolution—Privacy Considerations*, (Washington: Technology Policy Institute, 2013), 3.

[43]Avi Goldfarb, and Catherine Tucker, "Privacy and innovation", In: Josh Lerner and Scott Stern (eds.), *Innovation Policy and the Economy*. (Chicago: University of Chicago Press), 65–89.

Privacy and Innovation: From Disruption to Opportunities

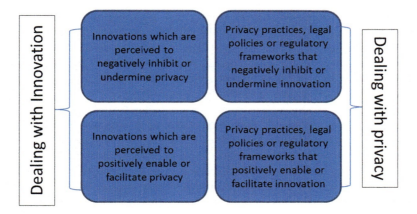

Fig. 2 Relation between privacy and innovation (Cave et al. 2011, p. 10)

studied the consequences of privacy regulations on the adoption of Electronic Medical Records (EMRs) in the United States. They were able to show that adopting privacy regulations had a negative impact (compared to having no regulation) on the adoption of EMRs, which was subsequently shown to have detrimental effects on the quality of care delivered. The study focused on neonatal mortality rates. The research showed a decrease in the number of incidences when a hospital had access to EMRs, which enabled exchange of patient information in critical situations. Doctors in hospitals that did not utilize EMRs were not able to consult all available information on a patient's health situation (in this case of new-born babies), which could have detrimental effects on the patients. The authors concluded that privacy regulations explained about 5 % of the variation in EMR adoption.[44] The authors also studied on-line advertisements and showed that privacy considerations have a significant effect on the efficiency of online advertisements and thus on online advertisement revenues. Targeted advertisements were 65 % more effective than advertisements that could not use targeting information to address dedicated groups of customers. The authors conclude that "privacy protection will likely limit the scope of the advertising-supported internet" and that "without targeting, it may be the case that publishers and advertisers switch to more intentionally disruptive, intrusive, and larger adds."[45] A final conclusion is that "ultimately privacy policy is interlinked with innovation policy and consequently has potential consequences for innovation and economic growth."[46]

In another study, performed for the European Parliament, the relation between privacy and innovation was split in four different segments (see Fig. 2).[47]

---

[44]Avi Goldfarb, and Catherine Tucker, 81.

[45]Avi Goldfarb, and Catherine Tucker, p. 77.

[46]Avi Goldfarb, and Catherine Tucker, p. 85.

[47]Jonathan Cave et al., *Does it help or does it hinder? Promotion of innovation on Internet and citizen's right to privacy,* (Brussels: European Parliament, 2011).

The fourfold relationship between privacy and innovation was investigated in a number of case studies (biometrics, cloud computing, online behavioural advertisement, RFID and location based services). The overall conclusions of the study are that innovation practices hardly take notice of privacy concerns and that the dominant logic within these practices promotes innovation at the expense of privacy. The conceptual approach adopted in the study for innovation enabled the study to differentiate between various aspects of innovation (technological dimension, organisational dimension, regulatory dimension and user perspective). Emergent new technologies are based upon opportunities to collect sensitive personal data (gene technologies, biometrics) and to collect an ever broader array of personal data (RFID sensor data and location based services). Awareness of these practices and developments within organisations and user constituencies is low or absent. The study recommends distinguishing between normative dimensions of privacy and an economic dimension of privacy.[48] Policy interventions should relate to a number of issues such as clarifying consent, offering more fine-grained privacy rights and checking for possibilities to reconcile privacy and economic regulations.[49]

5 Action Plan Privacy—The Dutch Situation

The preceding section explored a number of perspectives related to privacy and innovation. Some of the studies I presented, show that adherence to privacy demands blocks innovation, and may have detrimental impacts on relevant societal practices such as health care. According to these studies, privacy blocks innovation, or stated the other way around, adopting innovation practices means giving up on privacy. In order to better unravel the processes of innovation that are at stake I introduced a conceptual approach towards privacy that distinguishes between the technical (emphasized by information security), legal/regulatory (emphasized by data protection) and societal (emphasized by the private sphere) aspects. This conceptualisation enables us to classify between technical innovations, institutionally oriented innovations and societal innovations.[50] We used these distinctions to examine innovative privacy practices in the Netherlands in a study, commissioned by the Dutch Ministry of Economic Affairs. The study took a rather optimistic point of departure in presuming that

(a) It is possible to identify innovative practices that promote privacy.
(b) These practices may have a positive economic impact, while safeguarding privacy as well.

---

[48]Jonathan Cave et al., p. 97.

[49]Jonathan Cave et al., pp. 98–100.

[50]This does not assume that data protection for instance *only* deals with regulatory innovations. As the example in the text indicate what is manifest in the cross cutting of data protection with information security, and the private sphere, data protection deals with technical and societal innovations as well. The distinctions should help in pinpointing and focusing, reducing complexity.

We adopted as starting point that one can identify a certain willingness to engage with privacy as an agent of change. A recent report by Deloitte concisely phrases this in its title that says: "Having it all—Protecting privacy in the age of analytics."[51] It is not the only expression of a changing mood. In consultancy projects we are engaged with, several organisations indicated a willingness to include privacy and data protection in customer oriented services, but were reluctant to 'go public' with this approach.

The Action Plan Privacy was based on three subsequent steps:

1. Inventorying best practices and best technologies that could support practices to respect privacy.
2. Identifying organisations that had already implemented privacy respecting approaches or that offered privacy respecting services.
3. Analysing these practices from an innovation policy point of view, and arriving at a set of recommendations to the client, the Dutch Ministry of Economic Affairs.

In the first step, the assumption was that many more privacy tools are available than is generally presupposed. However, these tools are hardly known and hardly implemented. The inventory identified the following three categories of privacy tools:

1. Tools and technologies that are directed at safeguarding privacy within a service; these tools relate to privacy by design approaches (strategies and patterns), the use of anonymous credentials and anonymisation and pseudonymisation techniques, and standards for information security. Many of these tools relate to the technical pillar of our approach to privacy, which deals with information security. Organisational tools in this category relate to Privacy Impact Assessments (PIAs), Privacy Officers, and the use of a Privacy Maturity Model to identify the level of privacy awareness and privacy actions within an organisation. These tools relate to the data protection pillar as they take the regulatory framework as starting point.
2. Tools that are directed towards privacy respecting information architectures and networks. These are technical (inserting a digital vault for instance) and organisational (sticky policies, development and implementation of context aware privacy policies).
3. Tools that are directed towards enhancing the position of the data subject. These tools underscore the private sphere pillar of our approach. Examples of these tools are privacy dashboards, informed consent, private browsing, Do not track and the use of TOR networks and encryption are examples. They cover technological, organisational and regulatory dimensions.

---

[51]Deloitte. *Having it all—Protecting privacy in the age of analytics.* http://www2.deloitte.com/content/dam/Deloitte/ca/Documents/Analytics/ca-en-analytics-ipc-big-data.pdf (visited March 5, 2015).

The study was not able to identify the use of these tools in practice. It identified available technologies and tools, some of them still within the academic world, some of them already available as a commercial product. Trusted third parties for instance, are well-known as an approach to cope with sensitive data. And privacy impact assessments (or: data protection impact assessments) are already introduced in a variety of settings.

Within the second step, some anecdotal evidence was collected on organisations that had embedded privacy tools and techniques in their organisation. I have already mentioned the NS. Another organisation that based its primary product on a privacy respecting approach is CV-OK. CV-OK developed a data vault that individuals can use for storage of accredited documents such as diplomas and other reference documents. With the rise of flex contracts, in which employees change jobs more frequently and with a rising number of self-employed individuals the need for such a data vault is growing. This organisation decided to develop a secure and privacy respecting data vault that could be used by individuals to store and forward documents they need when soliciting for a job or a task. Their approach embodies an attitude that respects privacy, with the user in control, obeying data protection regulation and using security techniques to realise secure storage and handling of personal data.

The Action Plan Privacy also discussed the role of privacy/data protection officers and the role of branch organisations in promoting awareness and reflection on business processes and services that respect privacy. Large organisations in which personal data is processed need awareness campaigns to raise overall awareness for how to deal with these data and organisational rules concerning access, use and management of personal data.

Within the study we identified three sort of privacy approaches. *Privacy as service enabler* refers to firms that adopt approaches that respect privacy in the services they offer to their clients. The NS is an example of such a firm. These firms go beyond the mere need for compliance with the data protection regulations and try to build in user control, choice, and autonomy in their approach. *Privacy as a niche market* refers to firms that bring new and innovative systems and products for respecting or maintaining privacy on the market. Qiy is an example of such an approach, where the data subject is able to determine which data are released to which party for which purposes.[52] We concluded that an important challenge for these niche firms is to turn niche products into mainstream products. Finally, *privacy as compliance* refers to those firms that adopt a pragmatic approach towards privacy and seek to comply with the regulatory framework. This could reduce privacy awareness to a so-called tick box approach, in which minimal effort is invested in complying with the necessary regulations.

The final part of the Action Plan Privacy was identifying the market and systems imperfections and the presentation of policy recommendations in order to fix these imperfections. On the market side, firms do not know what kind of tools and

---

[52]See https://www.qiy.nl/en/ (visited March 9, 2015).

practices are available (information asymmetry). They may experience triggers to search for privacy respecting approaches, for instance due to regulatory requirements. An example is the EU 'Recommendation on privacy and data protection principles when using RFID applications' that promotes the use of PIA when a firm develops an RFID application. This Recommendation has however not led to the widespread adoption of practices to respect privacy when offering RFID applications.[53] Organisations are not (sufficiently) aware of the principles they should obey, and supervision by supervisory authorities is not strict enough to act as a trigger.

This last aspect is a manifestation of institutional imperfections. Imperfections in market dominance play a role as well. Most large system integrators are rather reluctant to position themselves as offering privacy respecting architectures, networks and services. They hardly advertise their measures to maintain privacy. Information and cyber security is a relevant market window, but data protection and privacy still is treated with caution.

The most prominent system imperfections are the institutional imperfections, the capability imperfections and the interaction imperfections. Supervisory authorities do not have the capacity to exert real pressure on the market to obey data protection regulations. In the Netherlands, a complaint voiced during a consultation workshop was that the Dutch DPA is not willing to give advice beforehand. Firms would appreciate the DPA offering a helping hand on which kind of practices are allowed but the Dutch DPA refrains from providing that service. The branch organisations indicated that many firms feel they are missing the capabilities to respond to the regulatory requirements. With the advent of the General Data Protection Regulation, branch organisations feel that the regulatory requirements impose larger pressures on data processing organisations without offering sufficient support to cope with these requirements.

A positive outcome of this systems imperfection is the emergence of a maturing juridical consultancy market that develops new services to help small firms that deal with personal data (such as app developers). Unfortunately, it is very problematic to insert truly new approaches to privacy into the market (sophisticated trusted architecture and key encryption schemes, for instance). Turning the innovations in a commercially interesting proposition is difficult. One such initiative is Qiy. Qiy set itself the objective to realise a structure of secured exchange of information between various parties such that these parties can share minimal sets of information in a trusted environment in a manner that respects privacy. Over the past five years, Qiy is trying to create a business case for this approach. It needs consensus with many stakeholders to make the solution it offers attractive (network externalities). At this moment (March 2015) it is not clear whether it will succeed in its mission to realise such a secured infrastructure with sufficient support of all relevant stakeholders.

---

[53]See EC, *DG CONNECT INTERNAL REPORT on the implementation of the Commission Recommendation on the implementation of privacy and data protection principles in applications supported by radio-frequency identification*, (Brussel 2014).

The examples of Qiy and CV-OK demonstrate that market and system imperfections need to be addressed to realise a functioning market of privacy respecting technologies and services. In the Action Plan Privacy we presented a number of recommendations that are meant to solve or to overcome the experienced market and system imperfections. Information asymmetry requires awareness campaigns. Branch and interest organisations play a role in establishing awareness and promoting practical approaches to privacy respecting solutions. The branch and interest organisations that were consulted indicated willingness to play such a role. They indicated that the implementation of the General Data Protection Regulation, which is now expected to be realised at the end of 2015, forms an important trigger for informing their customers on what needs to be done.[54] Institutional imperfections are more difficult to address. The implementation of the GDPR is a trigger for firms to check whether their approach is still privacy compliant or needs to be attuned. The requirement to have a data protection officer appointed will lead to the need for more skilled and trained data protection officers. From the perspective I sketched in this chapter such a data protection officer should have capabilities on the technical, the legal and the organisational domain. This is also the way the capabilities are phrased in the GDPR (and in the current data protection directive).

6 Conclusions

The awareness for the societal role of privacy is growing. Firms start to realise that privacy itself can be an innovative agent of change. By inserting privacy principles in the innovation equation, innovative systems can be implemented that realise public and economic value by making use of personal data and that meet privacy expectations. In this chapter I used the approach of market and systems imperfections to address innovation. Overcoming identified market and system imperfections is a way to realise innovative capacity. Privacy was addressed in terms of three interconnected spheres of influence: the private sphere, data protection and information security. Each of these spheres is characterised by a dominant logic: technological principles in the case of information security, regulatory principles in the case of data protection and societal principles in the case of the private sphere. By having this split, it is possible to have a separate look at what is needed from a technical perspective, a regulatory perspective and a societal perspective. Issues dealing with privacy and innovation should look for innovation in each of the spheres.

The Dutch Action Plan Privacy, commissioned by the Dutch Ministry of Economic Affairs and performed by the PI.lab (in which TNO participates), was used to discuss the innovative capacities of privacy. The Action Plan Privacy

---

[54]The full implementation period of the GDPR will last for two years. Starting at the end of 2015 thus implies that the GDPR will be fully effective at the end of 2017.

concluded that three strategies can be utilized by firms to become more respectful of privacy: a firm could decide to embed privacy in their service activities (privacy as service enabling), a firm could develop new niche products that help protect privacy (privacy as a niche market), and a firm could decide to restrict itself to being compliant (privacy as compliance). Examples of firms using the first strategy are privacy sensitive firms that deal with personal data as a by-product. Examples of firms using the second strategy are innovative firms offering privacy respecting services and systems. Examples of firms using the third strategy are firms that take the regulatory framework as starting point and seek the easiest way to be compliant.

Overall, the Action Plan has an optimistic tone with respect to the opportunities privacy offers as an innovation strategy. The upcoming General Data Protection Regulation already influences privacy behaviour of firms. Firms realise they might have to strengthen their privacy profile to keep on track with the requirements of the new GDPR. A second important motive is that firms realise that negative incidents have a considerable impact on their reputation. In a number of situations, a direct link can be made between how a firm treats privacy matters and a confrontation with an incident with a severe impact on that firm. Thirdly, emerging technical and organisational solutions help avoiding the 'all or nothing' approach that seems to hinder privacy innovations. System integrators start to implement privacy solutions that can be tuned to the specific requirements of a firm. Privacy by design strategies and patterns help in fine-tuning solutions to the specific systems in use. Internal Data Protection Officers and awareness campaigns promote privacy respecting attitudes in organisations. Instruments such as PIA become standardised. Consultancy firms help to implement these tools and check for compliance of existing data processing approaches. These activities help in overcoming identified market and systems imperfections and in embedding an approach that respects privacy as part of a competitive firm strategy. Government intervention is necessary to help organise a business climate that respects privacy.[55]

However, we need to balance this optimistic tone with the following observations. Firstly, the emergence of personal data markets will continue to put pressure on protecting privacy. Secondly, the continuing development of an 'app-economy', in which many small firms whose business model is almost exclusively based on collecting, processing and disseminating personal data, will pose serious problems in controlling whether appropriate data protection strategies are implemented and secured.

An active approach by public and private organisations (governmental organisations included) is prerequisite to have the best of both world: innovative practices, creating economic and public value, and new services that truly respect the privacy of its customers.

---

[55]The Ministry of Economic Affairs published a policy letter on Big Data and Privacy in which it underscores the relevance of a privacy respecting approach towards big data and in which it stated that the recommendations of the Action Plan Privacy should be implemented by a working group that the Ministry will establish on Big Data and Privacy.

Acknowledgments I would like to thank Arnold Roosendaal (TNO, PI.lab) and the anonymous reviewers for their constructive comments on earlier versions of this chapter.

Bibliography

Barras, Richard. 2000. Towards a theory of innovation theory in services. *Research Policy* 15(4): 161–173.
Cave, Jonathan, Marc van Lieshout, Neil Robinson, Rebecca Schindler, Gabriela Bodea, and Linda Kool. 2011. *Does it help or does it hinder? Promotion of innovation on Internet and citizen's right to privacy*. Brussels: European Parliament.
Cavoukian, Anne. 2009. *Privacy by design—Take the challenge*. Ontario: Information and Privacy Commissioners Office.
Edquist, Charles. 1997. *Systems of innovation: Technologies*. Institutions and Organizations: Pinter.
European Commission. 2000. *Charter of fundamental rights of the European Union*. Brussels: Official Journal of the European Communities C 364/1 Brussels.
European Commission. 2014. *DG CONNECT internal report on the implementation of the Commission Recommendation on the implementation of privacy and data protection principles in applications supported by radio-frequency identification*. Brussel.
European Court of Justice. 2014. Factsheet on the 'right to be forgotten' ruling (C-131/12). http://ec.europa.eu/justice/data-protection/files/factsheets/factsheet_data_protection_en.pdf.
Finn, Rachel, David Wright, and Michael Friedewald. 2013. Seven types of privacy. In *European data protection: Coming of age*, ed. Serge Gutwirth, Yves Poullet, et al. Dordrecht: Springer.
Gellert, Raphael, and Serge Gutwirth. 2013. The legal construction of privacy and data protection. *Computer Law and Security Review (CLSR)* 29: 522–530.
Goldfarb, Avi, and Catherine Tucker. 2012. Privacy and innovation. In *Innovation policy and the economy*, ed. Josh Lerner, and Scott Stern, 65–89. Chicago: University of Chicago Press.
Hippel, Eric von, Georg van Krogt. 2009. *Open source software and the "Private-Collective" innovation model: Issues for organization science*. MIT Sloan School of Management Working Paper 4739-09.
Lenard, Thomas M., and Paul H. Rubin. 2013. *The big data revolution: Privacy considerations*. Washington: Technology Policy Institute.
Medhurst James, Joel Marsden, Angina Jugnauth, Mark Peacock, Jonathan Lonsdal. 2014. *An economic analysis of spillovers from programmes of technological innovation support*. Report prepared for ICF GHK.
Nelson, Richard R. 1993. *National innovation systems: A comparative analysis*. Urbana: University of Illinois.
Poel, Martijn. 2013. *The impact of the policy mix on service innovation—The formative and growth phases of the sectoral innovation system for internet video services in the Netherlands*. Ph.D thesis, Technical University Delft, Delft.
Rogers, Everett M. 2003. *Diffusion of innovations*, 5th ed. New York: Free Press.
Roosendaal, Arnold, Marc van Lieshout, Colette Cuijpers, Ronald, Leenes. 2014. *Actieplan Privacy*. Delft: TNO-report R11603.
Solove, Daniel. 2002. Conceptualizing privacy. *California Law Review* 90(4): 1087–1155.
Stamp, Mark. 2006. *Information security: Principles and practice*. Hoboken, New Jersey: Wiley.
World Economic Forum. 2013. *Unlocking the value of personal data: From collection to usage*. World Economic Forum.

Behavioural Advertising and the New 'EU Cookie Law' as a Victim of Business Resistance and a Lack of Official Determination

Christina Markou

Abstract This paper looks into Article 5(3) of the ePrivacy Directive on cookies and more specifically, on the practical effect of its 2009 amendment which changed the legal approach towards the use of cookies to opt-in. The new rule had minor practical effect as except that notice about cookie use has overall been improved, behavioural advertising, which is the privacy-invasive commercial practice that the recent amendment of the rule mainly intended to tackle, is still conducted without prior real user consent. The paper inquires into the reasons behind the failure of the rule and finds them in the logical, yet unfounded, business resistance, the rule's negative publicity as well as in the (misguided) scepticism of EU officials and a lack of enthusiasm or determination at national level. The latter is translated into relaxed national implementations and absence of official guidance for compliance 'moments' before the rule was to enter into force. The final blow to the rule was given by a change in the approach of the UK ICO, which essentially aligned the law to the practice of implied consent adopted by many online businesses and by the reaction of the DPWP, which, far from strongly opposing this 'back off', confusingly moved closer to the updated stance of the UK ICO. The paper finally suggests that if they really want to, data protection authorities can restore a strict opt-in approach towards behavioural advertising and insist in business compliance with it.

Keywords Cookies · Opt-in rule · E-privacy · Cookie law · Directive 2009/136/EC

C. Markou (✉)
8 Kassandras Street, Strovolos, Nicosia 2021, Cyprus
e-mail: c.markou@euc.ac.cy

1 Introduction

When I was looking into behavioural advertising and inevitably into cookies, which comprise the main technology enabling it, I came across literature criticizing Article 5(3) of the e-Privacy Directive[1] for affording insufficient protection against the risks involved. Article 5(3) is often referred to as 'the EU cookie law'[2] due to dictating limits to the freedom of online businesses to use cookies (and other similar tracking technologies). Back then, the particular provision was only subjecting cookies to an opt-out approach, which meant that businesses could freely use cookies provided that they informed users about this use and the latter did *not* object. That was in 2009 however and Directive 2009/136/EC[3] had already passed into law. The particular measure amended Article 5(3) of the ePrivacy Directive introducing a decisive switch to an opt-in scheme, which meant that under the *new* EU cookie law, the use of cookies would not be allowed without prior user consent. Given that cookies are also used for non-controversial purposes such as user customization,[4] the main reason behind the introduction of stricter regulation has most probably been the need to improve user protection against the multiple risks inherent in behavioural advertising. I thought therefore, that the criticisms had become dated and that the concerns surrounding the unfettered use of tracking technologies for the collection of detailed information about users had been addressed. After all, it has been stated that "the impact of this new 'consent' rule is enormous and affects the entire ecosystem".[5] Only if I knew, that the brave change of the law would not be followed by an equally brave change in practice. It has been three years since the new rule came into force and yet, businesses still adhere to an opt-out approach, albeit slightly modified. User tracking and behavioural advertising are being conducted (almost) as freely as before and yet, it is not entirely certain that businesses will face enforcement action for breaking the law.

This paper investigates what went wrong and looks for the reasons behind this full circle back to an opt-out approach towards cookies and behavioural advertising. It starts off with a brief explanation of cookies and behavioural advertising as

---

[1]Directive 2002/58/EC of the European Parliament and of the Council of 12 July 2002 concerning the processing of personal data and the protection of privacy in the electronic communications sector.

[2]See for example, "How to comply with EU cookie law", *ComputerWeekly*, accessed November 9, 2014, http://www.computerweekly.com/guides/How-to-comply-with-the-EU-cookie-law.

[3]Directive 2009/136/EC of the European Parliament and of the Council of 25 November 2009 amending Directive 2002/22/EC on universal service and users' rights relating to electronic communications networks and services, Directive 2002/58/EC concerning the processing of personal data and the protection of privacy in the electronic communications sector and Regulation (EC) No 2006/2004 on cooperation between national authorities responsible for the enforcement of consumer protection laws.

[4]See infra p. 10.

[5]Phil Lee, P. "The impact of cookie 'consent' on targeted adverts," *Journal of Database Marketing & Customer Strategy Management* 18.3 (2011): 205.

well as of the recognized risks posed to the user. As this paper essentially comprises a discussion on cookies and their regulation, it has to proceed on the understanding that such discussions are useful and topical, rather than obsolete as is often suggested. Thus, the paper proceeds with defending its relevance by explaining why the relevant topic is by no means anachronistic. It then offers a description of the law governing cookies before and after 2009 mainly to flag up the drastic change in the legal approach towards cookies and therefore, to behavioural advertising. It also points out the recognized weaknesses of the old approach, which the new regulation supposedly came to address and places much emphasis on the interpretation of the new rule by the Data Protection Working Party (DPWP) to confirm that the change was indeed intended to be dramatic. It should be noted that the DPWP mainly consists of the leaders of the 28 national data protection authorities and is entrusted with interpreting and advising on data protection legislation.[6] The paper describes the business implementations of the new rule, thereby mirroring the sharp contrast between them and the intended content of the rule. It then looks for the reasons behind this failure mainly in the businesses' reaction to it and in the general handling of the rule by national and EU officials. It shows that the new cookie law may have been set up for failure and also, failed by those who were supposed to defend and insist in compliance with it. Finally, it suggests that the failure of the new rule to limit behavioural advertising and/or its privacy-invasiveness may not be irreversible but this mainly depends on enforcers, namely national data protection authorities which must take action against behavioural advertisers who, openly engage in behavioural advertising and yet, do not seek prior consumer consent.

2 Cookies and Behavioural Advertising

Cookies are invariably described as small text files that websites send on the computer of their users to store their behaviour exhibited in the form of clicks, page views, product searches and purchases. A cookie can alternatively serve as a unique identifier linking the computer containing it to user behaviour that is stored on website servers. For the purposes of behavioural (or targeted) advertising, the cookie-sending website reads the cookie (or the information stored in it) every time the cookie-containing computer access that website, which, as a result, adjusts its advertising or commercial content to the information contained or linked to the cookie.[7] Of course, the cookie-collected data is raw and thus, has to

---

[6]See Articles 29 and 30, Directive 95/46/EC of the European Parliament and of the Council of 24 October 1995 on the protection of individuals with regard to the processing of personal data and on the free movement of such data.

[7]For straightforward information on cookies, their evolution and purposes, see "What is a cookie?", *YouTube*, accessed November 14, 2014, https://www.youtube.com/watch?v=I01XMRo2ESg&feature=player_embedded.

undergo analysis aiming at the extraction of meaningful information about the personal circumstances, preferences and characteristics of users. These processes often referred to as web or data mining,[8] lead to the construction of individual profiles and are widely known as online profiling. In practice, a user who has been reviewing offerings of books on diabetes may be shown advertisements (or other commercial content) referring to blood glucose starter kits, for example. Another who has spent time on web pages with fashion magazines in Arabic may be shown expensive bags or even books on therapeutic herbs if the website holds evidence suggesting that a percentage of Arabs interested in fashion prefer alternative (rather than traditional) medical treatments. Indeed, group profiling (or the segmentation of individuals into various groups) is often involved in online profiling[9] and can lead to the production of wholly unforeseeable information about users that cannot expressly (or directly) be justified by their behaviour and which may or may not be true. Hildebrandt explains it as follows: "A group profile identifies and represents a group (community or category), of which it describes a set of attributes… Imagine if a person is included in the group of people with blue eyes and red hair and imagine that it is the case that a group profile is constructed for this category that indicates 88 % probability of a specific type of skin disease. This does not mean that this particular person has an 88 % chance to have this disease, because this may depend on other factors (like age, sunlight…)".[10]

Cookies sent and read by a business on its *own* website are called first-party cookies. However, following an agreement with a number of websites, a third party (often, a network advertising agency, such as DoubleClick[11]) serves cookies and relevant advertisements on those websites. These third-party cookies track users across a number of websites and collect information on their behaviour in multiple domains. As a result, they enable the construction of particularly detailed user profiles, thus posing greater risks than first-party ones and yet, they are more heavily used than first-party ones.[12]

A lot has been written on the risks involved in behavioural advertising and the use of cookies that enable it. As the 'personal data' processing involved is often conducted without user knowledge, let alone consent, behavioural advertising entails a violation of informational privacy, which conceptualizes the right of

---

[8]Van Well, L. and Royakkers, L., "Ethical Issues in Web Data Mining." *Ethics and Information Technology* 6.2 (2004): 129.

[9]Ibid at p. 133.

[10]Mireille Hildebrandt, "Profiling: From Data to Knowledge, The challenges of a crucial Technology," *Datenschutz und Datensicherheit 30* (2006): 549, accessed 9 November 2014, http://www.fidis.net/fileadmin/fidis/publications/2006/DuD09_2006_543.pdf.

[11]"Google Inc.", http://www.google.com/doubleclick/, accessed November 10, 2014.

[12]A 'cookie sweep' recently conducted in eight Member States has found that 70 % of all recorded cookies were third-party ones, see Article 29 Data Protection Working Party, "Cookie sweep combined analysis—Report" WP 229: 2, accessed March 13, 2015, http://ec.europa.eu/justice/data-protection/article-29/documentation/opinion-recommendation/files/2015/wp229_en.pdf.

individuals to control the collection and use of their personal data.[13] Indeed, the user tracking involved has even resulted in the invention of a new term, namely "dataveilance",[14] which refers to "a new form of surveillance, a method of watching not through the eye or the camera, but by collecting facts and data."[15] Given that, "Put in terms perhaps more appropriate for the information society, [privacy] might be classed as the right not to be subject to surveillance",[16] the informational privacy issues inherent in behavioural advertising (and cookies) are self-evident. Additionally, behavioural advertising presupposes the existence of detailed personal information stored on servers while as Bernal states "... wherever data exists, it is vulnerable..."[17] Unsurprisingly therefore, commentators have pointed towards the existence of risks to reputation, marriage and employment. These cannot wholly be excluded given the frequent data exchanges and the not so rare security breaches that may result in the data falling into the hands of parties who may (adversely) affect individuals.[18] There is also the risk of users inadvertently finding out private facts relating to other computer users specifically by visiting a website displaying behavioural advertisements based on the profile of the latter. For example, a husband who logs on Amazon and comes across a homepage displaying advertisements referring to books on depression may infer that his wife, whom he knows to be a frequent Amazon user, suffers from the particular condition.[19] Indeed, commercial websites are often explicit about what a user has been doing, specifically, through the display of text such as 'you looked at X product' and 'you might want to check Y product'. Ohm also refers to an example given by security researcher, Ross Anderson, of a woman who keeps her pregnancy secret while considering termination and sees the computer which she shares with her boyfriend suddenly starting to receive baby-related advertisements.[20]

Price and quality discrimination are additional risks associated with behavioural advertising. These involve offering different prices or quality to different users in accordance with their personal circumstances or characteristics as known

---

[13] Van Well and Royakkers, supra n. 8, pp. 130–131.

[14] Roger Clarke in Clarke, R., "Information Technology and Dataveillance," *Communications of the ACM* 31.5 (1987): 498.

[15] Daniel J. Solove, "Privacy and power: Computer databases and metaphors for information privacy," *Stanford Law Review* (2001): 1417.

[16] Ian J. Lloyd, *Information technology law*, (3rd edition, Butterworths, 2000), 33.

[17] Paul Alexander Bernal, "A right to delete?" *European Journal of Law and Technology* 2(2) (2011): Sect. 2.4, accessed March 5, 2015, http://ejlt.org/article/view/75/144.

[18] Jerry Kang, "Information privacy in cyberspace transactions," *Stanford Law Review* (1998): 1240, quoting Gary Marx; Lilian Edwards, "Consumer Privacy, Online Business and the Internet: Looking for Privacy in All the Wrong Places," *International Journal of Law and Information Technology* (2003): 231–232 and Solove, supra n. 15, pp. 1434, 1453.

[19] Inadvertent disclosure of private facts to other computer users is a risk emphasized by Cranor, F. L., "'I Didn't Buy it for Myself': Privacy and Ecommerce Personalization," WPES (2003): 111–117, accessed November 9, 2014, http://lorrie.cranor.org/pubs/personalization-privacy.pdf.

[20] Paul Ohm, "The Rise and Fall of Invasive ISP Surveillance," *University of Illinois Law Review* (2009): 41, accessed November 9, 2014, SSRN: http://ssrn.com/abstract=1261344.

by the business. Thus, a user known to the business as a frequent buyer of expensive products may be offered higher prices than the ones offered to others.[21] As the European Commission has recently acknowledged, not all users will benefit from such practices.[22] The fact that those who will lose, will do so as a result of secret processing of their personal data renders the practice privacy-invasive and also unfair: "Price discrimination may be a good thing in a free market economy, but the fairness again depends on consumers' awareness of the way they are categorized".[23] This holds true also of quality discrimination, which means that "the person is denied an opportunity of purchasing products/services that are made available to others."[24] The European Consumer Commissioner explains this risk, which she labels as 'commercial discrimination' clearly:

> ... personal and behavioural information can also reveal how much you are actually willing to pay for a service. It can reveal the risks you are likely to incur, be it in late payments, illnesses, or even the likelihood you will return the goods you buy. If this personal information is used to extract the maximum price possible from you or to block your access to some services altogether, then commercial discrimination can damage the confidence in digital trade and services. People may resent a world where they would have to systematically pay for who they are or the risks they personally incur.[25]

There are also issues with decisional privacy (or autonomy). These have heavily been discussed by American scholars who have pointed to the chilling effect that knowledge of the fact that one is being tracked may have on behaviour. Solove for example, wrote that "The mere knowledge that one's behavior is being monitored and recorded certainly can lead to self-censorship and inhibition".[26] Froomkin acknowledged that "For some, just knowing that their activities are being recorded may have a chilling effect on conduct, speech, and reading".[27] Indeed, a user who

---

[21]Taylor, R. C., "Consumer Privacy and the Market for Customer Information," *The RAND Journal of Economics 35(4)* (2004): 631.

[22]European Commission, 'Data Collection, Targeting and Profiling of Consumers for Commercial Purposes in the Online Environment' (Background Paper 2009): 10–11; European Commission, "Report on cross-border e-commerce in the EU", Commission Staff Working Document, SEC 283 final (2009): 14, 18.

[23]Mireille Hildebrandt, "Profiling into the future: An assessment of profiling technologies in the context of Ambient Intelligence," *FIDIS Journal* 1 (2007): 10, accessed November 9, 2014, http://www.fidis.net/fileadmin/journal/issues/1-2007/Profiling_into_the_future.pdf.

[24]Lee Bygrave, "Minding the Machine: Art15 of the EC Data Protection Directive and Automated Profiling", *Privacy Law and Policy Reporter* 40 (2000): Sect. 4.2, accessed 15 November 2014, http://folk.uio.no/lee/oldpage/articles/Minding_machine.pdf.

[25]European Commission Speech 09/156, Meglena Kuneva, "Roundtable on Online Data Collection, Targeting and Profiling", (speech given at Roundtable on Online Data Collection, Targeting and Profiling, Brussels, March 31, 2009), accessed November 9, 2014, http://europa.eu/rapid/press-release_SPEECH-09-156_en.htm.

[26]Supra n. 15, p. 1418.

[27]Michael A. Froomkin, "The death of privacy?" *Stanford Law Review* (2000): 1470. See also Kang, supra n. 18, pp. 1260–1261.

knows about the workings of behavioural advertising may adjust its online behaviour so that he prevents businesses from drawing inferences about himself. He may therefore avoid looking at luxury items, 'sensitive' book titles or non-mainstream products for fear of price discrimination, a security breach or unwanted disclosure of related private facts concerning him. Finally, Zarsky concentrates on the imposition of a different kind of 'autonomy' restriction made possible through user tracking and behavioural advertising. More specifically, he says that advertisers can prevent individuals from achieving personal objectives such as quit smoking by filling in their shopping experience with tobacco offerings, thus narrowing down their options and weakening resistance.[28]

3 Cookie Regulation Topical

Given that behavioural advertising is mainly facilitated by cookies, the relevant practice can be regulated through regulating the use of cookie technology. The discussion around cookies and their regulation is not futile or even anachronistic. Though the rise of other technologies such as java script and fingerprinting has caused some to assert that "the web cookie is dying",[29] the EU cookie law remains pertinent. Some of these newer technologies operate similarly to cookies in the sense that they involve access to user terminals and therefore, to the private sphere of users, which comprises the rationale behind Article 5(3).[30] Indeed, Hoofnagle et al. who emphasize that "advertisers use new, relatively unknown technologies to track people, specifically because consumers have not heard of these techniques"[31] list five relevant techniques, namely ETags, Flash cookies, HTML5 local storage, Evercookies and fingerprinting, all of which, with the exception fingerprinting, relying "upon writing files to the user's computer".[32] What is more, the DPWP has

---

[28]Tal Zarsky, "Mine your own business: making the case for the implications of the data mining of personal information in the forum of public opinion," *Yale JL & Tech.* 5 (2002): 38–39.

[29]"The Web Cookie is dying. Here's the creepier Technology that comes next," *Forbes*, (2013), accessed November 9, 2014, http://www.forbes.com/sites/adamtanner/2013/06/17/the-web-cookie-is-dying-heres-the-creepier-technology-that-comes-next/.

[30]See Recital 24, ePrivacy Directive.

[31]Chris Jay Hoofnagle et al., "Behavioral Advertising: The Offer You Cannot Refuse," 6 Harvard Law & Policy Review 273 (2012): 273, accessed November 25, 2014, SSRN: http://ssrn.com/abstract=2137601. See also "Javascript: Advantages and Disadvantages", *Jcscripters.com*, accessed November 25, 2014, http://www.jscripters.com/javascript-advantages-and-disadvantages/.

[32]Hoofnagle et al., ibid at p. 281. For further discussion on existing tracking technologies, see Omer Tene and Jules Polonetsky, "To Track or 'Do Not Track': Advancing Transparency and Individual Control in Online Behavioral Advertising," *Minnesota Journal of Law, Science & Technology*, 13.1 (2012): 14–19, accessed November 25, 2014, SSRN: http://ssrn.com/abstract=1920505.

recently dealt with device fingerprinting, that is, the collection and processing of a combination of various information elements regarding the device used by users including IP addresses, "which is sufficiently unique…to act as a unique fingerprint for the device…",[33] thereby enabling covert user tracking without the need for cookies.[34] Importantly, as the DPWP has confirmed, even fingerprinting can under certain circumstances involve the storage or access to information stored on the user's terminal equipment to which Article 5(3) applies.[35] Additionally, as the DPWP also confirmed, the application of the said provision is not confined to cookies but extends to all similar tracking technologies,[36] cookies thus being used as an umbrella term for all comparable tracking technologies including web beacons and web bugs. In any event, the development of mechanisms such as evercookies or zombie cookies, ensuring persistent cookies which are difficult to reject through browser settings or which can be re-created following deletion[37] serves as proof that cookies remain the underpinning tracking technology. Moreover, at least for the time being, Internet giants like Amazon, eBay and DoubleClick admit to using cookies extensively[38] meaning that businesses do not prefer other technologies to cookies. Finally, as it is elsewhere pointed out,[39] studies confirm that cookie usage is on the rise. All in all, cookies are probably here to stay and in any event, the EU cookie law is by no means just about cookies.

---

[33]Article 29 Data Protection Working Party, "Opinion 9/2014 on the application of Directive 2002/58/EC to device fingerprinting" WP 224: 6, accessed November 25, 2014, http://ec.europa.eu/justice/data-protection/article-29/documfentation/opinion-recommendation/files/2014/wp224_en.pdf.

[34]Ibid.

[35]Ibid at p. 11.

[36]Article 29 Data Protection Working Party, "Opinion 1/2009 on the proposals amending Directive 2002/58/EC on privacy and electronic communications (e-Privacy Directive)" WP 159: 10, accessed November 2015, http://ec.europa.eu/justice/policies/privacy/docs/wpdocs/2009/wp159_en.pdf: "…the chosen wording is not limited to the current issue of cookies, but implies any other new technology that could be used to track the users' behaviour using their browser".

[37]"How a new type of "evercookie" tracks you online," *The Economist*, (2014), accessed November 25, 2014, http://www.economist.com/blogs/economist-explains/2014/08/economist-explains-3; Woody Leonhard, "Zombie cookies won't die: Microsoft admits use, HTML5 looms as new vector," *InfoWorld*, (2011), accessed November 25, 2014, http://www.infoworld.com/article/2620781/internet-privacy/-zombie-cookies--won-t-die--microsoft-admits-use--html5-looms-as-new-vector.html.

[38]See for example, "Cookies and Internet Advertising", *Amazon.co.uk*, (2012), accessed November 25, 2014, http://www.amazon.co.uk/gp/help/customer/display.html?ie=UTF8&nodeId=201149560&ref_=gw_cookie_uk.

[39]Christine Riefa and Christiana Markou, "Online Marketing: Advertisers Know You are a Dog on the Internet," in Savin, A., Trzaskowski, J. eds, *Research Handbook on EU Internet Law* (Denmark: Edward Elgar, 2014): 397. See also Julia Angwin, "The Web's New Gold Mine: Your Secrets," *The Wall Street Journal*, (2010), accessed November 15, 2014, http://online.wsj.com/articles/SB10001424052748703940904575395073512989404.

4 The Law on Cookies Before and After 2009

Before 2009, Article 5(3) of the ePrivacy Directive permitted 'the use of electronic communications networks to store information or to gain access to information stored in the terminal equipment of a subscriber or user' only if the latter were offered clear information about the relevant processing and the right to refuse it. As already explained,[40] it therefore subjected cookie usage to an opt-out scheme, which did not necessitate obtaining user consent before installing a cookie on user terminals. Anyone with basic knowledge about cookies and their use by websites could confirm that the common way of compliance with Article 5(3) and its opt-out approach almost invariably consisted of information about cookies being hidden in (often) long and technical privacy policies themselves hidden behind tiny links at the very bottom of websites. Those privacy policies often also contained information on how to block cookies by setting browsers to do so and/or details about industry initiatives which users could utilize asking not to be sent third-party cookies, specifically by clicking on an opt-out button. Unsurprisingly, user knowledge about cookies was limited as was the number of users actually exercising the right to refuse them. Indeed, a study conducted by the UK government in 2011, i.e., two years after the 2009 amendment of the ePrivacy Directive, which placed cookies on the spotlight inevitably improving user awareness, still found "limited knowledge and understanding of internet cookies".[41] Additionally, it concluded that even though most users care about internet privacy, "…few users adjust default privacy settings actively"[42] and that 85 % were unaware of existing opt-out solutions.[43]

Commentators thus rightly to criticized Article 5(3) for affording insufficient protection against the risks inherent in the use of cookies:

> …how will this information and opt-out opportunity be supplied? Will a hyperlink to a privacy policy be sufficient? What if the privacy policy is unintelligible? …What if (as seems anecdotally to be the case) consumers never read privacy policies anyway? What if a tick box is supplied, already ticked, which gives permission to set cookies, unobtrusively tucked away at the bottom of the page? Or a box whose rubric reads 'Tick this box if you don't want us to set cookies', so putting the onus on the unsuspecting consumer? *Neither of these would surely have been acceptable under a requirement of explicit prior consent, but may well be in an opt-out regime.*[44]

It is true that such practices could not be considered as inconsistent with Article 5(3) and indeed, in 2003, the UK Information Commissioner (UK ICO) gave

---

[40]Supra p. 1.

[41]"Research into consumer understanding and management of internet cookies and the potential impact of the EU Electronic Communications Framework," *Department for Culture, Media & Sport (DCMS)* (2011): 24, accessed November 9, 2014, https://www.gov.uk/government/uploads/system/uploads/attachment_data/file/77641/PwC_Internet_Cookies_final.pdf.

[42]Ibid at p. 21.

[43]Ibid at p. 27.

[44]Edwards, supra n. 18, p. 239, emphasis added.

guidance for compliance, which commentators said it "would... boil down to explaining in the privacy policy how the web browser can be used to refuse cookies".[45] Later in 2007, the UK ICO appeared somewhat stricter requiring a notice displayed to all visitors about the use of cookies but still accepting the inclusion of information about cookies and the right to refuse in a privacy policy.[46] Notably, this guidance as well as an Opinion of the DPWP stating that information about the use of cookies may have to be displayed more prominently than "simply being part of a search engine's privacy policy"[47] was given by reference to the pre-2009 Article 5(3) and its mere opt-out approach, even though as already stated, businesses were not complying with this approach either.

With the adoption of the Directive 2009/136/EC in 2009, Article 5(3) of the ePrivacy Directive has changed to provide as follows:

> Member States shall ensure that the storing of information, or the gaining of access to information already stored, in the terminal equipment of a subscriber or user *is only allowed on condition that the subscriber or user concerned has given his or her consent*, having been provided with clear and comprehensive information, in accordance with Directive 95/46/EC, inter alia, about the purposes of the processing. This shall not prevent any technical storage or access for the sole purpose of carrying out the transmission of a communication over an electronic communications network, or as strictly necessary in order for the provider of an information society service explicitly requested by the subscriber or user to provide the service.[48]

Obviously, according to the new provision,[49] simply offering users the right to refuse cookies normally installed by default is not permissible. Websites should not use cookies unless users consent to their use having been provided with clear relevant information. The change in the wording (and the approach) is evidently dramatic and was naturally expected to have a profound impact on how websites are engineered to deal with their visitors, as they would have to seek and obtain user consent. Sure enough when the new cookie law came into force, it hit the headlines. Numerous news articles and blog posts were capturing this sense of a

---

[45]Frederic Debusseré, "The EU E-Privacy Directive: A Monstrous Attempt to Starve the Cookie Monster?" *Int'l JL & Info. Tech.* 13 (2005): 91, referring to UK Information Commissioner, 'Guidance to the Privacy and Electronic Communications (EC Directive) Regulations 2003—Part 2: Security, Confidentiality, Traffic and Location Data, Itemised Billing, CLI and Directories' (2003), version 1, November 2003: 5, accessed November 9, 2014, http://gov.gg/ccm/cms-service/download/asset/?asset_id=36034.

[46]"Data Protection Good Practice Note, Collecting Personal Information using websites" (2007): 2–3, accessed November 9, 2014, http://webarchive.nationalarchives.gov.uk/20100402134332/, http://ico.gov.uk/upload/documents/library/data_protection/practical_application/collecting_personal_information_from_websites_v1.0.pdf.

[47]Article 29 Data Protection Working Party, "Opinion 1/2008 on data protection issues related to search engines" WP 148: 20, accessed 9 November 2014, http://ec.europa.eu/justice/policies/privacy/docs/wpdocs/2008/wp148_en.pdf.

[48]Emphasis added.

[49]For a thorough analysis of the amended provision, see Eleni Kosta, "Peeking into the cookie jar: the European approach towards the regulation of cookies", *International Journal of Law and Information Technology* 23(4) (2013): 380.

breakthrough that the new law was believed to achieve. Some headlines are characteristic: "Cookie law makes most UK websites illegal: what you need to know",[50] "Why your site is now illegal in Europe",[51] "A simple guide to cookies and how to comply with EU cookie law".[52] Non-official guides often compiled by non-lawyers seem to have gotten it right: "The definition of consent is open to interpretation, but must involve some form of communication where the individual knowingly indicates their acceptance. This may involve clicking an icon, dismissing a banner, sending an email or subscribing to a service".[53] After all, it was plain common sense that the switch to the *exact opposite* (opt-in) approach would have to require some active involvement of the user in indicating acceptance of cookies which was not previously required and was therefore missing.

Importantly, most of those guides were just echoing official interpretations and indeed, it was *officially* confirmed that that was the correct approach under the new Article 5(3). More specifically, the DPWP has issued four different Opinions relating to cookies during the years 2010, 2011, 2012. In 2010, the DPWP has expressly clarified that the then prevailing practice of including information on how to opt-out through browser settings within privacy policies is not compliant with the new Article 5(3) rule.[54] It called for prior opt-in mechanisms, which require *an affirmative data subject's action* to indicate consent *before* the cookie is sent to the data subject[55] and drew a clear distinction between compliant opt-in consent and *non-acceptable* passive, opt-out or *implied* consent:

> ...consent means active participation of the data subject prior to the collection and processing of data. The opt-out mechanism often refers to a 'non' reaction of the data subject after such processing has already started. *Furthermore, under opt-out mechanism there is no active participation; simply the will of the data subject is assumed or implied. This does not meet the requirements for legally effective consent.*[56]

A year later in 2011, the DPWP issued an Opinion[57] on the requirement of consent defined in Article 2(h) of the Data Protection Directive (DPD) as "a freely

---

[50]Jessica Chambers, "Cookie law makes most UK websites illegal: what you need to know," (2011), accessed November 9, 2014, http://blog.silktide.com/2011/05/cookie-law-makes-most-uk-websites-illegal-what-you-need-to-know/.

[51]Craig Buckler, "Why your site is now illegal in Europe," *Sitepoint*, (2012), accessed November 9, 2014, http://www.sitepoint.com/europe-website-cookie-privacy-law/.

[52]Olivia Solon, "A simple guide to cookies and how to comply with EU cookie law," *wired.co.uk*, (2012), accessed 9 November 2014, http://www.wired.co.uk/news/archive/2012-05/25/cookies-made-simple.

[53]Ibid.

[54]Article 29 Data Protection Working Party, "Opinion 2/2010 on online behavioral advertising" WP 171: 13–16, accessed November 9, 2014, http://ec.europa.eu/justice/policies/privacy/docs/wpdocs/2010/wp171_en.pdf.

[55]Ibid at p. 16.

[56]Ibid at pp. 15–16, emphasis added.

[57]Article 29 Data Protection Working Party, "Opinion 15/2011 on the definition of consent" WP 187, accessed November 9, 2014, http://ec.europa.eu/justice/policies/privacy/docs/wpdocs/2011/wp187_en.pdf.

given, specific and informed indication of one's wishes by which the data subject signifies his agreement to personal data relating to him being processed".[58] According to the DPWP, consent is a *positive act* that *excludes* any system giving a right to object or refuse *after* the processing has taken place.[59] In relation to the 'specific' ingredient, it stated that "...blanket consent without specifying the exact purpose of the processing is not acceptable... It cannot apply to an open-ended set of processing activities".[60] Finally, it was adamant that making access or membership to social networks conditional upon acceptance of cookies for behavioural advertising without offering the possibility of separate acceptance or rejection is not 'free and specific consent'.[61]

A few months later, the DPWP issued an Opinion discussing the EASA/IAB code on online behavioural advertising drafted by those engaging in the particular practice as represented by the European Advertising Standards Alliance (EASA) and the Internet Adverting Bureau Europe (IAB).[62] This Opinion has been more explicit on the practical implementation of the new rule, as the DPWP has specified ways in which opt-in consent can be obtained. These are "a static information banner on top of a website requesting the user's consent to set some cookies", "splash screen on entering the website explaining what cookies will be set by what parties if the user consents", "a default setting prohibiting the transfer of data to external parties, requiring a user click to indicate consent for tracking purposes" and "a default setting in browsers that would prevent the collection of behavioural data".[63] All of them appear consistent with a requirement of *active* participation by the user *specifically in response to a notice regarding cookie use*. In the same Opinion, the DPWP emphasized the need for the provision of information 'directly on screen interactively through layered notices' on the parties who set cookies and the fact of monitoring across different websites for behavioural advertising purposes.[64]

Finally, in 2012, the DPWP issued an Opinion on the exemptions from the requirement of consent provided for in the new Article 5(3). These refer to cookies used "for the sole purpose of carrying out the transmission of a communication over an electronic communications network, or as strictly necessary in order for the provider of an information society service explicitly requested by the subscriber or

---

[58]By virtue of Article 2(f), ePrivacy Directive, this definition is applicable also in the context of the ePrivacy Directive and thus, in relation to the cookie rule.

[59]Ibid at p. 10.

[60]Ibid at p. 17.

[61]Ibid at pp. 18–19.

[62]Article 29 Data Protection Working Party, "Opinion 16/2011 EASA/IAB Best Practice Recommendation on Online Behavioural Advertising" WP 188, accessed November 9, 2014, http://ec.europa.eu/justice/data-protection/article-29/documentation/opinion-recommendation/files/2011/wp188_en.pdf.

[63]Ibid at pp. 9–10.

[64]Ibid at p. 5.

user to provide the service".[65] It clearly arises from this Opinion that those cookies that are vital to the operation of the Internet are temporary-session cookies (*not* persistent-tracking ones) and are exempted from the 'consent' requirement. The DPWP specifically illustrated that authentication cookies, security cookies, shopping cart cookies, multimedia player cookies, customization cookies (remembering user language, for example) and social plug in cookies for commenting or sharing content by logged-on social network members are *all* exempted from consent.[66] It also clarified that tracking and behavioural advertising do *not* fall within any of the exemptions[67] and again spoke of 'banners and consent *requests*', thus insisting in some active participation towards cookie acceptance.[68]

And it was not just the DPWP. In 2011, the UK ICO published similar guidance on the new rule stating that it required active expression of consent through, for example, splash screens or sign-up processes.[69]

5 Business Implementations

Despite official and unofficial guidance to the effect that the new Article 5(3) *really* meant a dramatic change of approach towards *real* (opt-in) consent and away from its implied (opt-out) counterpart, a look at how major online businesses now comply with the rule reveals three main implementations, none of which is consistent with those that were being described by the DWPW or the UK ICO until 2012. The first involves a certainly more-prominent-than-before (and sufficiently highlighted) notice placed on top or at the bottom of webpages informing users that cookies are used and that by using the website, the user consents to them. The notice is accompanied by a link leading to a page with information on the various cookies used and instructions on how to opt-out or disable cookies through browsers. The opportunity to accept some cookies and reject others such as behavioural advertising ones is not available.[70] The second involves the display of a notice stating that cookies are used accompanied by a link to learn more, which, if followed, leads to information on how to reject cookies through browsers and other opt-out initiatives. This notice however is in small black fonts (not in any way highlighted) placed at a top place within the homepage intermingling

---

[65]Article 29 Data Protection Working Party, "Opinion 4/2012 on Cookie Consent Exemption" WP 194, accessed November 9, 2014, http://ec.europa.eu/justice/data-protection/article-29/documentation/opinion-recommendation/files/2012/wp194_en.pdf.

[66]Ibid at p. 11.

[67]Ibid pp. 6, 9–10.

[68]Ibid at p. 6.

[69]"Changes to the rules on using cookies and similar technologies for storing information," *UK Information Commissioner's Office (ICO)*, (2011): 6–7, accessed November 9, 2014, https://www.huntonprivacyblog.com/wp-content/files/2011/12/Initial-guidance.pdf.

[70]See for example, *Google UK*, http://www.google.co.uk and *PriceGrabber UK*, http://www.pricegrabber.co.uk.

with (and overshadowed by) other flashy commercial content.[71] The third involves the addition of a microscopic link at the very bottom of pages next to the links 'terms of use' and 'privacy policy' reading 'Cookies' or 'Ad targeting policy', which contains information about cookies and the ways to avoid them.[72]

It is not difficult to see that these implementations comprise tiny and reluctant steps away from the pre-2009 business approach towards cookies. Admittedly, notice has been improved in the sense that the fact of cookie use is no longer hidden within privacy policies and in some cases it is sufficiently prominent and also provided directly on screen as prescribed by the DPWP. Recall however that official guidance on the *old* Article 5(3) spoke about notice that is more prominent than simply being part of privacy policies.[73] It seems therefore that in response to the new (opt-in) rule, businesses took measures to comply with the old (opt-out) one! Recall also that Article 5(3) is not merely a 'sufficient notice' rule but a 'consent' one. All three implementations only incorporate an opt-out approach in ways (such as information on how to set browsers to reject cookies) which were expressly rejected in the aforementioned Opinions of the DPWP. Even the first one, which is the only one referring to cookie acceptance (or consent) directly on screen, infers consent from '*general website*' use, which it makes conditional upon consent to cookies. Again, this is an approach that the DPWP explicitly rejected for not allowing for 'free and specific' consent.[74] Notably, despite arguments to the effect that such an approach may only be incompatible with EU privacy law in the case of monopolistic providers,[75] the DWPD has not limited rejection of the relevant approach to monopolies[76] and has

---

[71] See, for example, *Amazon UK*, http://www.amazon.co.uk.

[72] It looks like this: "'PriceRunner UK—Compare UK Prices and Find Deals Online Copyright © 1999–2014 PriceRunner|Terms & Conditions|Privacy Policy|Cookie Policy'", *PriceRunner UK*, http://www.pricerunner.co.uk, accessed on November 15, 2014. See also, *eBay UK*, http://www.ebay.co.uk, accessed on November 15, 2014.

[73] Supra n. 47 and associated text.

[74] See supra at p. 9.

[75] Borgesius, supra n. 101, citing the Dutch Data Protection Authority Letter to the State Secretary of Education, Culture and Science, on answers to parliamentary questions about cookie policy, (2013), www.cbpweb.nl/downloads_med/med_20130205-cookies-npo.pdf, accessed on February 24, 2015 and Eleni Kosta, Consent in European Data Protection Law (PhD diss., University of Leuven, 2013): 256, 312.

[76] This is a right approach for two reasons. First, even if there are alternative service providers to which users could theoretically switch, transaction costs associated with such a switch may in some cases 'force' them to stay with the current provider, see Borgesius, supra n. 101, p. 33. Secondly, if *all* competing service providers employ the same (or a comparable) practice of gaining consent, the existence of competition and the consequent possibility of a switch does not obviously lead to user choice. Cohen made this argument in a slightly different context: "… to the extent that individuals need or want the goods or services and cannot obtain them elsewhere—to the extent, that is, that all vendors serving a given market believe collecting consumer data is a competitive necessity—one suspects that individuals may simply concede, and convince themselves that the loss of privacy associated with this particular transaction is not too great"; Julie E. Cohen, "Examined Lives: Informational Privacy and the Subject as Object," *Stanford Law Review* 52 (2000): 1397, accessed November 9, 2014, http://scholarship.law.georgetown.edu/cgi/viewcontent.cgi?article=1819&context=facpub.

further clarified that Recital 25 of the e-Privacy Directive[77] allows *conditional* access only to *specific* and not *general* content: "The emphasis on "specific website content" clarifies that websites should not make conditional "general access" to the site on acceptance of all cookies but can only limit certain content if the user does not consent to cookies (e.g.: for e-commerce websites, whose main purpose is to sell products, not accepting (non-functional) cookies should not prevent a user from buying products on this website)".[78] Given that it is this Recital that mainly casted doubt on the acceptability of the approach of inferring consent from website navigation essentially disallowing website use without cookie acceptance,[79] it should have been considered settled that the relevant approach does *not* lead to valid consent.

The lax business implementations of the rule evidently suggest that unless enforcement actions are imminent, the rule has failed. The failure of the rule is confirmed by studies inquiring into how websites have sought to implement it even in Member States such as the Netherlands which have adopted a strict implementation of Article 5(3). For example, according to a 2014 survey conducted by Leenes and Kosta amongst 96 active websites only six were implementing the rule consistently with the DPWP Opinions discussed above.[80] Why?

6 Searching for the Reasons Behind the 'Cookie Law' Failure

It may be that the new Article 5(3) was set up to fail. What is more, it was failed by those who were supposed to defend and enforce it, namely EU officials, national governments and data protection authorities.

Even though the rule was only borne in 2009, its history can be traced back to 2000 and the preparatory stages of the 2002 ePrivacy Directive. Kierkegaard described the strong opposition of the advertising industry to an opt-in approach towards cookies and explained how its arguments about the benefits of cookies and the negative consequences of an opt-in approach on the use of the Internet have

---

[77]Recital 25 of the e-Privacy Directive provides as follows: "... The methods for giving information, offering a right to refuse or requesting consent should be made as user-friendly as possible. Access to specific website content may still be made conditional on the well-informed acceptance of a cookie or similar device, if it is used for a legitimate purpose".

[78]Article 29 Data Protection Working Party, "Opinion 02/2013 providing guidance on obtaining consent for cookies" WP 208: 5, accessed November 9, 2014 http://ec.europa.eu/justice/data-protection/article-29/documentation/opinion-recommendation/files/2013/wp208_en.pdf.

[79]Discussing whether the said approach amounts to valid consent under EU law, Kosta writes that "... the explicit reference on the conditionality of access in Recital 25 complicates the situation", Kosta, supra n. 75, p. 321. See also DPWP, ibid.

[80]Ronald Leenes and Eleni Kosta, "Taming the cookie monster with Dutch law—A tale of regulatory failure," *Computer Law & Security Review: The International Journal of Technology Law and Practice* 31.1 (2015), doi:10.1016/j.clsr.2015.01.004.

convinced the European Commission eventually resulting in the pre-2009 opt-out provision.[81] As she observed, "The explanatory memorandum for Article 5 echoes the position advanced by the IAB in its arguments against an opt-in requirement".[82]

Given the mere opt-out approach of the original Article 5(3), the years that followed have naturally seen unfettered user tracking and behavioural advertising. The spread of the practice must have heightened concerns about the risks inherent in it and relevant pressure by privacy activists eventually resulted in the new Article 5(3). These concerns surrounding are vividly captured in a speech given by the European Consumer Commissioner two months before the introduction of the (amending) Directive 2009/136/EC:

> Currently, consumers have little awareness of what data is being collected, how and when it is being collected and what it is used for. And they are also not able to control this process. The current opt-out systems are partial, sometimes nowhere to be found, they are difficult or cumbersome and most of all, they are unstable. Avoiding tracking is currently technically difficult if not impossible…Behavioural targeting on the internet will become increasingly pervasive and consumers understandably feel uncomfortable. Today I want to send one very clear message to those involved in all aspects of the digital world—*Consumer rights must adapt to technology, not be crushed by it*. The current situation with regard to privacy, profiling and targeting is not satisfactory.[83]

The new rule has however strongly been opposed by the advertising industry insisting (again) that the new rule can be harmful to businesses, consumers and the economy in general.[84] There have also been videos referring to it as "the stupid cookie law"[85] which "should die"[86] and many journalists (or technology columnists) would seem to side with opponents often using very strong language:

> The EU's arrogance in presuming to legislate for a global world wide web is matched only by its hilarious technological incompetence: cookies have dozens of uses besides the advertising and tracking purposes that this directive is aimed at "protecting" against, most of which enable key features of web pages that users will be severely inconvenienced without. Cookies are a core component of how today's internet works.[87]

---

[81] Kierkegaard, Sylvia Mercado, "How the cookies (almost) crumbled: Privacy & lobbyism," *Computer Law & Security Review: The International Journal of Technology Law and Practice* 21.4 (2005): 310–322.

[82] Ibid at p. (emphasis added).

[83] Supra n. 25, emphasis added.

[84] Kathleen Hall, "EC cookie privacy laws threatens UK's digital economy," *ComputerWeekly.com*, (2011), accessed November 9, 2014, http://www.computerweekly.com/news/1280095377/EC-cookie-privacy-laws-threaten-UKs-digital-economy. See also, Mike Butcher, "Stupid EU cookie law will hand the advantage to the US, kill our startups stone dead," *techcrunch.com*, accessed November 9, 2014, http://techcrunch.com/2011/03/09/stupid-eu-cookie-law-will-hand-the-advantage-to-the-us-kill-our-startups-stone-dead/.

[85] "The stupid EU cookies law (and why it should die)", *YouTube*, accessed 9 November 2014, https://www.youtube.com/watch?v=9hLmX9FX2KA.

[86] Ibid.

[87] Milo Yiannopoulos, "Guest Opinion: The EU's legal war on cookies is barking mad," *wired.co.uk*, (2011), accessed November 9, 2014, http://www.wired.co.uk/news/archive/2011-05/11/cookies-regulations?p=2. See also Mike Butcher, supra n. 84.

This negativity has inevitably been passed to users. Consumer limited knowledge about cookies and also dissatisfaction with the rule are said to be reasons behind the rule's failure[88] but the argument can also be framed differently: user limited knowledge about cookies meant that businesses could easily turn users against the rule. Leenes and Kosta report that Dutch users preferred annoying consent-requesting pop-ups and overlays to disappear.[89] Though consumer complaints and survey results depicting consumer opinion should always be relied upon with caution,[90] consumer negative reaction to the rule may just be the other side of the (same) 'business resistance' coin. The negative publicity of the rule mainly created by the industry coupled by the unwilling and thus, user-unfriendly business implementations of it were bound to result to consumer dissatisfaction towards the rule (even though as already explained, that was a misleading depiction of its content). Cofone seems to confirm this view by referring to consumer associations complaining about the 'cookie wall'[91] that essentially prevented access to website content if cookies were not accepted. Leenes and Kosta make the same point sophisticatedly: "Website owners were thus able to create an unusual alliance with the targets (victims) of profiling against their protectors (the regulator)."[92]

This hostile environment within which the new rule found itself is perfectly logical. Consumer personal data is so important and valuable for online businesses that it is now accepted to be "the new oil of the internet and the new currency of the digital world".[93] Years of unfettered user tracking and behavioural advertising by default have caused businesses to invest in relevant data collection practices aiming at becoming able to acquire even deeper knowledge about their users. Indeed, seven years ago, Google's CEO has been blatantly honest:

> We are very early in the total information we have within Google. The algorithms will get better and we will get better at personalisation...The goal is to enable Google users to be

---

[88]Leenes and Kosta, supra n. 80.

[89]Ibid.

[90]There is almost always at least one survey evincing a conflicting stance. For example, a study conducted by the Pricewaterhouse and commissioned by the UK Department for Media, Culture and Sport found that most respondents expressed preference towards the opt-in approach and this was so despite the fact that the said approach was described to them as requiring "...repeated pop-up windows or other virtual labels on every web page visited by a user where internet cookies are in use"; "Research into consumer understanding and management of internet cookies and the potential impact of the EU Electronic Communications Framework," *Department for Culture, Media & Sport (DCMS)* (2011): 1,3, accessed November 9,2014, https://www.gov.uk/government/uploads/system/uploads/attachment_data/file/72837/PwC_DCMS_Internet_Cookies_Summary_and_Conclusions.pdf. To be fair however, many respondents admitted to possess limited a priori knowledge of cookies. Moreover, unlike Dutch users, UK users did not get to see what repeated pop-ups mean in practice.

[91]Ignacio Cofone, "The Way the Cookie Crumbles: Online Tracking Meets Behavioral Economics," (2014): 12, accessed, February 27, 2015, SSRN: http://ssrn.com/abstract=2541215.

[92]Leenes and Kosta, supra n. 80.

[93]European Consumer Commissioner, supra n. 25. See also "Understanding the Personal Data Bargain", *InternetSociety*, (2013), accessed November 9, 2014., http://www.internetsociety.org/blog/2013/02/understanding-personal-data-bargain explaining the online bargain involving free services in exchange of personal data, the latter being referred to as "info-currency".

able to ask the question such as 'What shall I do tomorrow?' and 'What job shall I take?'…We cannot even answer the most basic questions because we don't know enough about you. That is the most important aspect of Google's expansion.[94]

Given that the new Article 5(3) sought to change the established 'free surveillance' default and restrict the freedom of businesses to collect personal data, the business reaction (and opposition) has been only natural. Actually, businesses would negatively react against *any* tracking-restrictive rule and indeed, a Californian Do-not-Track Bill, which merely introduced an opt-out, rather than an opt-in approach resulted in opposition letters sent to the Senate by Google and other major businesses.[95]

Though perfectly logical however, this reaction against the rule was (and is) largely unfounded. The 'shopping cart' argument, i.e., the fact that cookies are behind basic functions of the Internet, which will simply not work (or work in the same nice way) without them is a central argument against the rule, yet as already mentioned, *all* functional cookies such as authentication, customization or shopping cart cookies are exempted from the 'consent' requirement.[96] What is more is that it is doubtful that an opt-in mechanism was impossible to implement. For this reason, the new rule was never a threat to the very operation of websites and it may have been a matter of re-programming websites so as to stop them from installing certain 'unnecessary' or non-operational cookies, a task that has not convincingly been proven excessively burdensome or costly.[97] The rule does not presuppose multiple and intrusive consent requests either as consent can be obtained for multiple cookies at a *single* 'consent request' point and what is more, consent given once can cover all subsequent connections[98] and even different websites on which a given behavioural advertising provider serves cookies.[99] Finally, it is wrong to treat the opt-in approach of Article 5(3) as if it were an outright ban. Behavioural advertising will naturally be conducted at a lesser scale but the rule, by no means, means the end of it. Advertisers remain free to seek and obtain user

---

[94]Caroline Daniel and Maija Palmer, Google's Goal: to Organize your Daily Life, *FT.com*, (2007), accessed November 9, 2014, http://www.ft.com/cms/s/2/c3e49548-088e-11dc-b11e-000b5df10621.html#axzz3HcfhNcV2.

[95]Letter, Subject: SB 761 (Lowenthal)—OPPOSITION, (2011), accessed November 9, 2014, http://regmedia.co.uk/2011/05/05/dnt_opposition_letter.pdf.

[96]See supra, p. 10.

[97]It would be interesting for computer scientists to compare this kind of re-programming with the one necessary to stop standard web logs from registering IP addresses and the URL of the requested content, which is part of how the Internet through the HTTP protocol works. Kang stated that the re-programming for the latter purpose would be overly burdensome, see Kang, supra n. 18, p. 1276, n. 328.

[98]Recital 25, e-Privacy Directive, supra n. 77.

[99]See DPWP, Article 29 Data Protection Working Party, supra n. 54 at p. 16; DPWP, Article 29 Data Protection Working Party, "Opinion 16/2011 EASA/IAB Best Practice Recommendation on Online Behavioural Advertising" WP 188: 10–11, accessed 9 November 2014, http://ec.europa.eu/justice/data-protection/article-29/documentation/opinion-recommendation/files/2011/wp188_en.pdf; DPWP, Article 29 Data Protection Working Party, supra n. 65 at p. 6.

consent and *ideally*, the more respectful of user rights and interests advertisers are, the more accepting of their practices users may become. Admittedly, users may not be well-equipped to take privacy decisions based on such sophisticated assessments (yet) and as other others report, half of the respondents surveyed in the Netherlands "always click 'OK'".[100] Yet, even results of this kind could be taken to disprove business arguments to the effect that opt-in would significantly curtail behavioural advertising and hence, many beneficial free services.

In fact, this 'advertising' argument is central in the fight of the industry against the opt-in approach towards cookies. Yet, just because behavioural advertising is used to finance useful services and thus, has a beneficial aim, does not necessarily (or automatically) mean that the particular practice leads to "a net benefit for society".[101] As Borgesius states, "Neither economic theory nor empirical economic research has provided a definite answer to the question of whether behavioural targeting… leads to more or less social welfare in the aggregate".[102] In any event, behavioural advertising is not the *only* online advertising type and thus, the *sole* source of support for the various admittedly useful services that are currently available to users free of charge. Apart from untargeted advertisements which "… simply target the broad Web audience in general"[103] and which, though 'old-fashioned' are unlikely to cease to produce some revenue, IAB UK admits the importance of contextual advertising; this tailors advertising to the content of the page viewed (and/or the keywords searched for by users) and does not rely on the collection of personal data or the use of cookies at all.[104] Companies are certainly investing in technology supporting this type of advertising[105] and in fact, the largest part of the revenue of Google which, in 2011, consisted of more than thirty billion dollars, comes from keyword[106] and hence, contextual advertising. There is

---

[100]Leenes and Kosta, supra n. 80 (referring to a survey conducted by the Dutch Consumer Union, http://www.consumentenbond.nl/test/elektronica-communicatie/).

[101]Zuiderveen Frederik J. Borgesius, "Consent to Behavioural Targeting in European Law—What are the Policy Implications of Insights from Behavioural Economics?", *Amsterdam Law School Research Paper* 43 (2013): 24, accessed November 9, 2014, SSRN: http://ssrn.com/abstract=2300969 or 10.2139/ssrn.2300969.

[102]Ibid.

[103]Langheinrich et al., "Unintrusive Customization Techniques for Web Advertising," *Computer Networks: The International Journal of Computer and Telecommunications Networking* 31 (1999): 1260.

[104]Internet Advertising Bureau, "Office of Fair Trading (OFT) Online Targeting of Advertising and Prices Market Study: Response by the Internet Advertising Bureau," *Internet Advertising Bureau*, (2012): 3, accessed November 9, 2014, http://www.iabuk.net/sites/default/files/IABresponsetoOFTmarketstudyintoOnlineTargetingofAdvertisingandPrices_6012_0.pdf.

[105]See for example, "Contextual ad leader vibrant signs new premium publishers," *Vibrant*, (2011), accessed November 9, 2014, http://www.vibrantmedia.com/press/press.asp?section=press_releases&id=182.

[106]See "What industries Contributed the Most to Google's Earnings?" *WordStream*, accessed November 9, 2014, http://www.wordstream.com/articles/google-earnings.

also the "targeted (filtered)" advertising, which is essentially contextual advertising that however involves some filtering on the basis of individual parameters such as the country of the user or his browser software, thus further improving advert relevance.[107] This seems similar to demographic advertising to which IBA UK also refers and which appears to be far less intrusive than behavioural advertising.[108] All in all, the Internet does not sit on the shoulders of behavioural advertising and in any event, as already stated, the opt-in approach does *not* ban or otherwise, stop behavioural advertising.

The existing potential of alternative advertising types (or systems) that could strike a better balance between user privacy interests and the interests of businesses seems very relevant to the question posed by Chester: "Can the digital marketing "ecosystem," as online advertisers have called it, be transformed so it balances the interests and rights of consumers while it also expands its data collection capabilities?"[109] As it has just been illustrated, privacy-friendlier advertising systems that do not necessarily fall back to the now primitive totally untargeted and largely irrelevant static advertising of the early Internet days are a technological possibility which can be given a decisive push possibly leading to miraculous results if the industry, which is in the best position to explore the relevant possibility, is forced to do so. The Article 5(3) opt-in rule, if insisted upon by regulators and enforcers, could obviously give this push, yet if regulators and enforcers are taking steps back, (as they do),[110] the industry will not take even the tiniest step forward.

Despite being largely unfounded (or at least, insufficiently justified), these arguments seem to have been effective. Indeed, the Digital Agenda Commissioner (and Vice-President of the EU Commission) gave a very industry-friendly speech at a time, specifically shortly after the introduction of the new rule, when one would expect the Commissioner to emphasize its privacy-improving properties and more generally, encourage compliance with it. Instead however, the Commissioner spoke about the major contribution of online and behavioural advertising in the availability of free content and services[111] and drew a very positive portrait of behavioural advertising passing it as one that is widely-accepted:

> *Like anyone* I can feel bored or annoyed when faced with…ads I am not interested in. So the idea of only seeing ads that are likely to interest me is an appealing one.[112]

---

[107]Langheinrich et al., supra n. 103, pp. 1260–1261.

[108]Supra n. 104.

[109]Jeff Chester, "Cookie wars: how new data profiling and targeting techniques threaten citizens and consumers in the "big data" era," in *European Data Protection: In Good Health?* ed. Gutwirth S. et al. (Netherlands: Springer, 2012), 53–77.

[110]See infra at pp. 16–20.

[111]European Commission Speech 10/452, Neelie Kroes, "European Roundtable on the Benefits of Online Advertising for Consumers" (speech given at European Roundtable on the Benefits of Online Advertising for Consumers, Brussels, September 17, 2010), accessed November 15, 2014, http://europa.eu/rapid/press-release_SPEECH-10-452_en.htm?locale=FR.

[112]Ibid.

This position sounds a lot like the "sanitized fairy-tale version" [113] of behavioural advertising which, Chester says, is presented to users by the advertising industry in both Europe and the US[114] and indeed, it is at least debatable. Research conducted by the University of Pennsylvania has showed that 66 % of US users do not like the idea of tailored ads and that this percentage increases dramatically when users are informed of the tracking involved.[115] As Borgesius illustrates, recent Eurobarometers and national surveys in EU Member States such as the UK and the Netherlands produce similar results, thus evincing a large majority of EU users also worrying about or not wanting behavioural advertising.[116]

Elsewhere in the same speech, the advertising industry arguments about the supposed impracticalities of an opt-in rule are very quickly accepted and even worse, the new Article 5(3) rule seems to be reduced to a mere 'sufficient notice' rule:

> Obviously we want to avoid solutions which would have a negative impact on the user experience. On that basis it would be prudent to avoid options such as recurring pop-up windows. On the other hand, it will not be sufficient to bury the necessary information deep in a website's privacy policies. We need to find a middle way.[117]

The current business implementation of prominent notice but 'imposed' consent, (in the form of consent given with 'general website' use) coupled with the offer of an opportunity to opt-out in the form of instructions on how to disable cookies, would seem to sit well between the two extremes as described by the Digital Agenda Commissioner and yet, it is not consistent with the cookie rule as described by the DPWP during 2010–2012.[118]

A month later, the Communications Committee of the European Commission issued a Working Document in which, as Kosta rightly observes, it did not emphasize the need for the implementation of a strict opt-in approach and regarded the development of technical solutions of compliance as an issue to be resolved by self-regulation.[119] Kosta writes that "This has been considered by the industry as a 'softening' of the European Commission on the consent requirement for Article 5(3)".[120] The message had thus started to be conveyed that the new cookie rule may not actually require the adoption of real opt-in mechanisms.

---

[113]Chester, supra n. 109, p. 70.

[114]Ibid.

[115]Joseph Turow et al., "Contrary to what marketers say, Americans reject tailored advertising and three activities that enable it," (2009), accessed November 9, 2014, http://graphics8.nytimes.com/packages/pdf/business/20090929-Tailored_Advertising.pdf.

[116]Zuiderveen Frederik J. Borgesius, supra n. 101, pp. 8–10.

[117]Supra n. 111.

[118]Supra pp. 9–10.

[119]Communications Committee (European Commission—Information Society and Mediate Directorate General), 'Working Document on the implementation of the revised Framework—Article 5(3) of the ePrivacy Directive' COCOM10-34, Brussels, 20 October 2010: 6.

[120]Kosta, supra n. 49, p. 401.

Actually, the fate of the new rule had begun to be written even earlier than that and before it was even adopted. More specifically, while it is trite knowledge that the Recitals to a Directive describe and expand upon its various rules, the amending 2009 Directive, which has changed the Article 5(3) to opt-in, keeps the text of Recital 25 of the original e-Privacy Directive, (albeit as Recital 66), which refers to a right to refuse cookies consistently with the opt-out approach of the pre-amendment Article 5(3). Some responsibility for the failure of the new rule thus lies with its drafters and indeed, thirteen Member States have interpreted Recital 66 as meaning that "…amended Article 5(3) is not intended to alter the existing requirement that such consent be exercised as a right to refuse the use of cookies…"[121] This statement was made while the Proposal for the amending Directive was being read by the Council and had been regarded by commentators as creating "further confusion".[122]

Sure enough, at national level, things were not easy for the new rule either. It has been reported that two months before the rule were to come into force none of the Member States had issued guidance for compliance, even though the rule was introduced in 2009 and was being discussed even before that.[123] In the UK, enforcement of the rule had even been postponed for one year to give websites more time to take compliance measures.[124] Yet, it was what happened just before the expiration of that one-year 'grace period' that pulled the rug from under the new rule. More specifically, in May 2012, the UK ICO changed its 2011 guidance, which spoke about a requirement of explicit (opt-in) consent consistently with the DPWP Opinions[125] and published an updated version,[126] which opened the door to implied consent and clearly mirrored a softer stance towards cookies:

> While explicit consent…might be the most appropriate way to comply in some circumstances this does not mean that implied consent cannot be compliant…For implied

---

[121] Adoption of the Proposal for a Directive of the European Parliament and of the Council amending Directives 2002/21/EC on a common regulatory framework for electronic communications networks and services, 2002/19/EC on access to, and interconnection of electronic communications networks and services, and 2002/20/EC on the authorisation of electronic communications networks and services (LA+S) (third reading)—Statements, 15864/09 ADD 1 REV 1 COR 1, http://register.consilium.europa.eu/doc/srv?l=EN&t=PDF&f=ST+15864+2009+ADD+1+REV+1+COR+1. See also Leenes and Kosta supra n. 80.

[122] N van Eijk, et al., "Online tracking: questioning the power of informed consent," *Emerald Group Publishing Limited* 14.5 (2012): 59, accessed on April 16, 2014, http://dare.uva.nl/document/2/121980.

[123] "Governments 'not ready' for new European Privacy law", *BBC News Technology* (2010), accessed November 9, 2014, http://www.bbc.com/news/technology-12677534.

[124] Chirstopher Graham, "ICO Blog: half term report on cookies compliance," *UK Information Commissioner's Office (ICO)*, accessed November 9, 2014, http://ico.org.uk/news/blog/2011/half-term-report-on-cookies-compliance.

[125] See supra pp. 9–10.

[126] "Guidance on the rules on use of cookies and similar technologies", *UK Information Commissioner's Office (ICO)*, (2012), accessed November 9, 2014, http://ico.org.uk/news/blog/2011/~/media/documents/library/Privacy_and_electronic/Practical_application/cookies_guidance_v3.pdf.

consent to work there has to be some action taken by the consenting individual from which their consent can be inferred. *This might for example be visiting a website, moving from one page to another or clicking on a particular button.*[127]

Clearly, websites were given the green light to seek to comply with the rule by making general website use conditional upon acceptance of cookies. Thus, the one-year grace period in the UK has not resulted in businesses adapting their practices to the UK ICO guidance as it was initially the intention but in the UK ICO adapting its guidance (and the law) to the business-chosen implementations of the rule. Indeed, businesses have consistently been resisting a switch to an opt-in scheme in the form initially described by the DPWP and the UK ICO: it has been reported that a month before the expiration of the one-year grace period, 95 % of UK websites were found *not* to comply with the new rule.[128] Thus, it is difficult *not* to think of the change of approach of the UK ICO as an example of online private ordering, albeit not through contracts, but through concerted business resistance to legal rules.

Notably, the updated guidance of the UK ICO not only avoids exempting behavioural advertising from the relaxed approach of implied consent but also discusses third party advertising cookies almost as vaguely as the advertising industry itself. Indeed, it refers to information and 'choice' (not consent),[129] an approach also adopted in the (industry) EASA/IAB Code and rejected by the DPWP as not complying with the new rule:

> … instead of seeking users consent, claims to provide for a way of exercising "choice". In fact it is a choice to opt out, as it offers the user the possibility to object to having his/her data collected and further processed for OBA.[130]

Also, a UK ICO 12-minute video of May 2012 summarizing the required approach of compliance contains absolutely no reference to 'consent' which is the very innovation of the new rule. Moreover, behavioural advertising is not discussed at all and the word 'choice' is heard once without being expanded upon.[131] This video is displayed on an ICO web page dedicated to cookies[132] together with the ICO updated guidance of May 2012, which has unsurprisingly been celebrated as "the death of the stupid cookie law".[133] Interestingly, a practical guide issued

---

[127]Ibid, pp. 6–7, emphasis added.

[128]Anh Nguyen, "95 % of UK organisations 'do not comply with EU cookie law," *Computerworld UK* (2012), accessed November 9, 2014, http://www.computerworlduk.com/news/it-business/3350059/95-of-uk-organisations-do-not-comply-with-eu-cookie-law/.

[129]Supra n. 95, p. 23.

[130]Opinion 16/2011, supra n. 62, p. 6.

[131]"Cookies FAQs, May 2012—ICO", *YouTube*, accessed 9 November 2014 https://www.youtube.com/watch?v=V0M8MYiGkQw.

[132]"The EU cookie law (e-Privacy Directive)," *UK Information Commissioner's Office (ICO)*, accessed November 9, 2014, http://ico.org.uk/for_organisations/privacy_and_electronic_communications/the_guide/cookies.

[133]Simon Gibbs, "The stupid cookie law is dead at last," *Libertarian Home* (2013), accessed 9 November 2014, http://libertarianhome.co.uk/2013/01/the-stupid-cookie-law-is-dead-at-last/.

by the ICO in November 2012, where it is stated that "The use of implied consent for …Targeting and Advertising cookies is unlikely to be acceptable"[134] is nowhere to be found on that page.

Regarding its own website, the UK ICO has gone from initially adopting a real opt-in approach to changing to a *business-like* opt-out approach,[135] which it admits in a relevant announcement, dated 31 January 2013.[136] The UK Department of Culture, Media and Sport has also admitted that "… the UK implementation of the revised e-privacy Directive, particularly with regard to Article 5(3) on cookies, is *light touch, business friendly* and sets a benchmark in Europe".[137] Unsurprisingly therefore, commentators have criticized the UK approach as being inconsistent with the opt-in approach of the amended Article 5(3) rule[138] and more generally, as not adhering to "either common sense or philosophical conceptions of consent."[139] Even 'softer' descriptions of the UK implementation readily reveal its strong 'opt-out' flavour: "… the British regulation… regulates it as close to an opt-out system as the wording of the amendment allows".[140]

Outside the UK, a table compiled by Fisher Field Waterhouse in May 2013 containing the various national implementations of the rule reveals that the vast majority of the Member States does not require opt-in consent and/or considers implied consent as acceptable.[141] One of the very few Member States stated to adopt an explicit opt-in scheme, namely the Netherlands, adopted an opt-in approach even before the 2009 amendment of the ePrivacy Directive, yet researchers reported in 2012 that the majority of Dutch websites were ignoring the rule.[142]

All in all, by 2013, there was enough evidence to suggest that the dramatic switch to an opt-in scheme introduced by the 2009 amendment of Article 5(3) was to remain dramatic only in theory (and 'empty' in practice), unless (perhaps) the

---

[134]"ICC UK Cookie guide," *International Chamber of Commerce (ICC) UK* (2012): 13, accessed November 9, 2014, http://www.cookielaw.org/media/1096/icc_uk_cookiesguide_revnov.pdf.

[135]See also Riefa and Markou, supra n. 39, pp. 405–406.

[136]"Changes to cookies on our website," *UK Information Commissioner's Office (ICO)*, (2013), accessed November 9, 2014, https://ico.org.uk/about-the-ico/news-and-events/current-topics/changes-to-cookies-on-our-website/.

[137]Ed Vaizey, "Open letter on the UK implementation of Article 5(3) of the e-Privacy Directive on cookies," *DCMS*, (2011): 1, accessed February 27, 2015, https://www.gov.uk/government/uploads/system/uploads/attachment_data/file/77638/cookies_open_letter.pdf, emphasis added.

[138]Andrew McStay, "I consent: An analysis of the Cookie Directive and its implications for UK behavioral advertising," *New Media & Society* 15.4 (2012): 596–611.

[139]Ibid at p. 609.

[140]Cofone, supra n. 91, p. 16.

[141]"Cookie 'consent' rule: EEA implementation," *Field Fisher Waterhouse*, http://www.fieldfisher.com/pdf/cookie-consent-tracking-table.pdf. For another table of national implementations, see Cofone, supra n. 91, pp. 8–10.

[142]N van Eijk, et al., supra n. 122, p. 71.

DPWP intervened with a new Opinion on the various (relaxed) national approaches noting that they were not in line with its earlier Opinions. Yet, as already mentioned, the DPWP consists of the leaders of the national data protection authorities and is, therefore, bound to be affected by the legal and business climate in the various Member States. Sure enough, the final blow to the rule was given by the DPWP, which never returned with a 'rule-saving' Opinion. Instead, in October 2013, the DPWP published a Working Document through which it took a significant step *away from* the 'strict opt-in'-favouring stance repeatedly expressed in its earlier Opinions and *closer to* the softer approach of the UK ICO and to current business implementations.[143] The DPWP admits to have had the opportunity to observe the various national legal and business implementations of the rule[144] and it is therefore difficult not to think that the relevant 'step back' has been influenced and even 'forced' by the failure of its previous interventions to change the relevant status quo.

As already illustrated, the previous DPWP Opinions were *categorical* in describing the new rule as requiring nothing less than a strict opt-in approach. Unsurprisingly therefore, it was not easy for the DPWP to back off. Probably for this reason, the 2013 Working Document mirrors a somewhat vague and a difficult-to-understand DPWP trying to balance between the different approaches and reconcile its new guidance with the previous ones. Thus, while it repeats that making website use conditional upon acceptance of cookies is not acceptable and that users should be able to reject non-functional cookies and still use the website such as for buying products,[145] in the same Document, the DPWP is liable to be taken as suggesting the opposite:

> Tools to obtain consent may include splash screens, banners, modal dialog boxes, browser settings...the users may signify their consent, either by clicking on a button or link or by ticking a box in or close to the space where information is presented...or *by any other active behaviour* from which a website operator can unambiguously conclude it means specific and informed consent...active behaviour means an action the user may take, typically one that is based on a traceable user-client request towards the website, *such as clicking on a link, image or other content on the entry webpage*...If the user enters the website where he/she has been shown information on the use of cookies, *and does not initiate an active behaviour, such as described above, but rather just stays on the entry page without any further active behaviour*, it is difficult to argue that consent has been given unambiguously.[146]

It is difficult not to take this passage as saying that a click on a link or image *unrelated* to cookies such as some product offering, can be a sufficient indication of consent and that it is only if the user stays inactive on the homepage, i.e., he does not use the website at all, that unambiguous consent *cannot* be inferred. This differs little, if any, from the practice of making website use conditional upon

---

[143] Article 29 Data Protection Working Party, supra n. 78.

[144] Ibid at p. 2.

[145] Ibid at p. 5.

[146] ibid at pp. 4–5, emphasis added.

acceptance of cookies which the DPWP rejects even in the same Working Document by saying that "websites should not make conditional "general access" to the site on acceptance of all cookies".[147]

Obviously, the relevant DPWP Working Document 'muddies the waters' regarding whether the said method of compliance, which is now followed by several major businesses, is acceptable or not and can back up arguments to the effect that it actually is. This is so despite the fact that the practice, at least in the form now employed, only *infers* consent and additionally applies to an *open-ended set of purposes* (as it covers everything from performance cookies to advertising cookies). Recall that in its previous Opinions, the DPWP has unequivocally stated that consent *cannot* just be inferred and that it should refer to specific purposes, rather than to an open-ended set of processing activities.[148] It would seem that the UK ICO and the DPWP have followed the tactic also employed by the Dutch Data Protection Authority by reference to which Leenes and Kosta observe: "The regulation is being clarified, amended and *watered down* when there is too much opposition…"[149] The Italian regulator also, has issued relevant guidance requiring that the consent request be displayed on the homepage but considering consent expressed through any click on the page acceptable.[150] Along similar lines, the guidance of the Spanish data protection authority states that implied consent can be valid.[151]

It is therefore unsurprising that, in seeking to obtain user consent (as they say), businesses do not really behave significantly differently from before the 2009 amendment and the new rule is, at best, treated as a mere 'sufficient notice' rule rather than as a (real) 'consent' one. Equally unsurprising is the fact that the new rule comes widely to be accepted as an opt-out rule, just like its predecessor. Indeed, one of the oldest and most respectable industry-based privacy watchdogs, TrustE warns EU websites about the European Cookie Sweep Initiative of 2014 and advises them to give clear information and an opportunity to *opt-out*.[152]

There can be little doubt that the new rule has largely failed to achieve its purpose of limiting user tracking and the privacy-invasive nature of behavioural advertising. The reasons for this failure could be summarized to lie with a combination of (i) concerted business resistance to the new rule, (ii) press hostility, (iii) official skepticism

---

[147]ibid at p. 5.

[148]Supra p. 9.

[149]Leenes and Kosta, supra n. 80, emphasis added.

[150]"Simplified Arrangements to Provide Information and Obtain Consent Regarding Cookies," *Garante*, (2014), accessed February 27, 2015, http://www.garanteprivacy.it/web/guest/home/docweb/-/docweb-display/docweb/3167654.

[151]Pablo Rivas, "Spanish Data Protection Agency Releases Guidance on Cookies Regulation," *Hogan Lovells*, (2013), accessed February 27, 2015, http://www.hldataprotection.com/2013/06/articles/consumer-privacy/the-spanish-data-protection-agency-finally-releases-its-guidance-on-cookies/.

[152]Eleanor Treharne-Jones, "European Cookie Sweep Initiative: Are you Compliant?" *TRUSTe Blog*, (2014), accessed November 9, 2014, http://www.truste.com/blog/2014/07/17/european-cookie-sweep-initiative-are-you-compliant/.

and un-readiness both at EU and national level and ultimately, (iv) an official step backwards to opening the door to implied consent and effectively, to the pre-2009 opt-out approach. Given that the online advertising industry is now a multi-billion one and that many online business models rely on data collection in one way or another, businesses had every reason to oppose the rule and resist its application. The question why the DPWP, data protection authorities and other officials did not defend the rule rigorously enough to avoid its collapse is more difficult to answer but it may be that they felt that the resistance coming from the 'business-press-user' alliance was too powerful to fight. After all, it is widely accepted that lawyers and policymakers cannot understand technology better than those who constantly use, study and develop it for their business purposes. It is therefore not unlikely that regulators were at least influenced, if not convinced, by the arguments of those who 'know better'.

7 Irreversible?

With major businesses now implementing an opt-out approach towards cookies, the question arises whether the status quo is irreversible. The very purpose of the 2009 amendment of Article 5(3) was to limit the privacy-invasive nature of behavioural advertising and protect users against its multiple risks. Is the Internet ever going to see a real opt-in approach in relation to behavioural advertising cookies? This will depend on national data protection authorities and the enforcement action they will be willing to take. For example, is Amazon, which admits to using behavioural advertising cookies, going to face enforcement action for not asking its visitors to click to consent to the relevant practice? TRUSTe and Fieldfisher report that, "Between 2009 and 2013, there was no meaningful enforcement of the EU's new cookie consent law".[153] It seems that so far, one of the very few directly relevant enforcement actions was taken in 2014 in the Netherlands against a behavioural advertising agency which only offered an opportunity to opt-out from receiving behavioural ads and did not seek user consent to installing relevant cookies.[154] It has been reported that the Spanish regulator has imposed the first fine for incompliance with the law on cookies in early 2014, yet that was not for the failure of the website to secure opt-in consent but to provide adequate information.[155]

---

[153]"Cookie audits—are you ready?" *TRUSTe and Fieldfisher*, accessed November 10, 2014, http://webcasts.acc.com/handouts/Whitepaper-_EU_Cookie_Audits_Are_you_Ready.pdf.

[154]Ibid at p. 7.

[155]Nuria Pastor, "History in the making: the first 'cookie rule' fines in Europe," *Fielsfisher*, (2014), accessed November 10, 2014, http://privacylawblog.fieldfisher.com/2014/history-in-the-making-the-first-cookie-rule-fines-in-europe; Cynthia O' Donoghue, "Spain: First European Cookie Fine Issued by Spanish Data Protection Authority," *Mondaq*, (2014), accessed November 10, 2014, http://www.mondaq.com/x/296196/Data+Protection+Privacy/First+European+Cookie+Fine+Issued+By+Spanish+Data+Protection+Authority and "Spain: AEPD issues first European cookie fine," *DataGuidance*, (2014), accessed November 10, 2014, http://www.dataguidance.com/dataguidance_privacy_this_week.asp?id=2203.

Leenes and Kosta observe that after the publication of the investigation findings of the Dutch Data Protection Authority, "...the number of websites implementing differentiated opt-in mechanisms... is growing",[156] something that apparently confirms that data protection authorities can, through rigorous enforcement, revive the rule. Obviously however, the fact of a *single* enforcement action taken in one of the very few Member States which adopted a strict opt-in approach[157] does not portray a very encouraging future for the EU cookie rule.

Regulators should perhaps start by flagging the fact that implied consent is *not* acceptable in relation to behavioural cookies. For example, there seems to be no reason why the November 2012 guide of the UK ICO which expressly says so[158] should be 'hidden' within its website and not displayed on its cookie-dedicated page where the rest of the guidance on cookies is displayed. Similarly, the DPWP should perhaps be more explicit than simply referring the issue of whether users are given real choice to national data protection authorities without offering any guidance. As already illustrated, even the 2013 Working Document of the DPWP contains statements, such as those rejecting the method of inferring consent from website use,[159] on which national data protection authorities can base a relevant enforcement action, if they are willing to do so. Especially in the UK, the November 2012 guidance almost expressly disallows implied consent for behavioural cookies. There is therefore little hindering the taking of enforcement action against behavioural trackers that not obtain sufficient user consent. This is so especially given that behavioural cookies are widely accepted to be highly intrusive while (even) the UK ICO says to businesses that, "the more privacy intrusive [their] activity, the more priority [they] will need to give to getting meaningful consent".[160]

8 Conclusion

This paper has looked into Article 5(3) of the ePrivacy Directive on cookies and more specifically, on the practical effect of its 2009 amendment which changed the legal approach towards the use of cookies to opt-in. It showed that it only had minor practical effect in the sense that except that notice about cookie use is now prominently displayed in some cases, behavioural advertising, which is exactly the practice that the new rule intended to tackle, is still conducted without prior real user consent. The paper searched for the reasons behind the failure of the rule and found them in the logical, yet unfounded, business resistance to its application

---

[156]Leenes and Kosta, supra n. 80.
[157]Supra p. 20.
[158]See supra at p. 19.
[159]See supra at p. 21.
[160]See supra n. 126, p. 16.

and in the rule's negative publicity which created a hostile environment around the new rule. This environment was further cultivated by the (misguided) scepticism of EU officials and a lack of enthusiasm or determination at national level translated into relaxed national implementations of the rule and a lack of official guidance of compliance 'moments' before the rule was to enter into force. The 2012 change in the guidance of the UK ICO, which essentially aligned the law to the practice of implied consent adopted by many online businesses and the reaction of the DPWP, which, far from opposing this 'back off', it resorted to a similar change of approach moving closer to the updated stance of the UK ICO meant that the rule was too 'weak' to comprise a real 'threat' to the slightly-changed status quo created by online businesses.

The situation regarding the practical impact of the new cookie rule is not irreversible. Regulators *do* have the tools to break free from the new, yet still largely unsatisfactory status quo and demand strict opt-in consent to behavioural advertising cookies, which are widely accepted to be highly intrusive and privacy-invasive. Of course, they will have to want to use them, thus taking relevant enforcement action. Unfortunately, enforcement has so far been sparse. Whether national data protection authorities are, in fact, equipped with the necessary willingness (or determination) is thus uncertain and there is also the issue of the inadequacy of resources, which is a notorious problem in the particular context.

It is of course true that consent, even when properly sought, has been under some sort of attack regarding its effectiveness as a tool of privacy protection; myopia, bounded rationality and other factors tend to render (even) opt-in consent, a manifestation only of *weak* or *insufficient* user control[161] and thus, lead to the exploration of alternative approaches of (more drastic) protection such as the imposition of a prohibition on behavioural tracking in certain domains such as on news websites.[162] The idea however, is rarely to abandon consent altogether[163] and there are also those who reject the scepticism over the 'consent' approach towards privacy protection.[164] The truth is that privacy is so intrinsically connected with consent[165] that the latter could never be abandoned as a principal tool of privacy protection. What is more is that any alternative approaches of protection including prohibitions

---

[161]Borgesius, supra n. 101, pp. 29–37; Tene and Polonetsky, supra n. 32, pp. 39–54; Mantelero, Alessandro. "The future of consumer data protection in the EU Re-thinking the "notice and consent" paradigm in the new era of predictive analytics," *Computer Law & Security Review* 30.6 (2014): 643–660.

[162]Borgesius, supra n. 101, pp. 51–56.

[163]Tene and Polonetsky for example, call for "dimming the highlight on user choice while focusing on businesses' obligations under the FIPs", supra n. 32, p. 48. Obviously, this is a call to focus on other tools of protection *as well* rather than to abandon the 'consent' approach.

[164]Ryan M. Calo, "Against Notice Skepticism in Privacy (and Elsewhere)," *Notre Dame Law Review 87(3)* (2012): 1027, accessed 5 March 2015, https://cyberlaw.stanford.edu/files/publication/files/ssrn-id1790144.pdf.

[165]Tene and Polonetsky, supra n. 32, p. 41.

may not have a much better luck as regards their enforcement than consent. Indeed, one cannot but wonder whether the law could get certain businesses stop behavioural tracking altogether (consistently with a relevant prohibition) given that it has failed to convince or force them merely to display a clear 'consent' request.[166]

Hoofnagle et al. observe that "Behavioral advertising—and the tracking that goes with it—is the offer you cannot refuse, not necessarily because you are tempted by it, but because sophisticated, market-dominant actors control the very platforms you use to access the web".[167] The opt-in approach of the new Article 5(3) as initially interpreted by the DPWP and the UK ICO was a way to shift this control to users who would thus be given the chance to let us know (in practice) whether they actually care about behavioural advertising and its privacy risks or not. Scepticism over consent is often expressed through questions such as the following: 'what does it mean to be "for privacy" or "against tracking", and at the same time unwilling to check a box…to preserve one's rights?'[168] Yet, statements of this kind overlook the fact that at least in most Member States users did *not* get to be offered that checkbox in the first place. As the Article 5(3) opt-in approach has never consistently been implemented by (major) online businesses, EU officials, governments, data protection authorities and the advertising industry did not get to know and will never do, unless appropriate enforcement action is taken forcing its implementation. But do they want to know? Maybe not![169]

Bibliography

Angwin, Julia. 2010. The web's new gold mine: Your secrets. *The Wall Street Journal*. http://online.wsj.com/articles/SB10001424052748703940904575395073512989404. Accessed 9 Nov 2014.

Article 29 Data Protection Working Party. 2015. Cookie sweep combined analysis. http://ec.europa.eu/justice/data-protection/article-29/documentation/opinion-recommendation/files/2015/wp229_en.pdf. Accessed 13 Mar 2015.

Article 29 Data Protection Working Party. Opinion 02/2013 providing guidance on obtaining consent for cookies. WP 208. http://ec.europa.eu/justice/data-protection/article-29/documentation/opinion-recommendation/files/2013/wp208_en.pdf. Accessed 9 Nov 2014.

Article 29 Data Protection Working Party. Opinion 1/2008 on data protection issues related to search engines. WP 148. http://ec.europa.eu/justice/policies/privacy/docs/wpdocs/2008/wp148_en.pdf. Accessed 9 Nov 2014.

---

[166]Arguably, enforcement could in this case be somewhat improved as a result of the limited and hence, more manageable number of the addressees of the prohibition but then again as the recent official cookie sweep has found that more than half of third-party (advertising) cookies are set by just 25 advertising businesses, see Data Protection Working Party, supra n. 12, pp. 2, 15.

[167]Supra n. 31, p. 278.

[168]Tene and Polonetsky, supra n. 32, p. 38.

[169]McStay reports that based on the findings of a study they commissioned, the UK Department for Media, Culture and Sport has remarked that if an opt-in were implemented, nearly 50 % of Internet Users would *not* accept third party (advertising) cookies, supra n. 138, p. 607.

Article 29 Data Protection Working Party. Opinion 1/2009 on the proposals amending Directive 2002/58/EC on privacy and electronic communications (e-Privacy Directive). WP 159: 10. http://ec.europa.eu/justice/policies/privacy/docs/wpdocs/2009/wp159_en.pdf. Accessed 9 Nov 2014.

Article 29 Data Protection Working Party. Opinion 15/2011 on the definition of consent. WP 187. http://ec.europa.eu/justice/policies/privacy/docs/wpdocs/2011/wp187_en.pdf. Accessed 9 Nov 2014.

Article 29 Data Protection Working Party. Opinion 16/2011 EASA/IAB best practice recommendation on online behavioural advertising. WP 188. http://ec.europa.eu/justice/data-protection/article-29/documentation/opinion-recommendation/files/2011/wp188_en.pdf. Accessed 9 Nov 2014.

Article 29 Data Protection Working Party. Opinion 2/2010 on online behavioral advertising. WP 171. http://ec.europa.eu/justice/policies/privacy/docs/wpdocs/2010/wp171_en.pdf. Accessed 9 Nov 2014.

Article 29 Data Protection Working Party. Opinion 4/2012 on cookie consent exemption. WP 194. http://ec.europa.eu/justice/data-protection/article-29/documentation/opinion-recommendation/files/2012/wp194_en.pdf. Accessed 9 Nov 2014.

Article 29 Data Protection Working Party. Opinion 9/2014 on the application of Directive 2002/58/EC to device fingerprinting. WP 224. http://ec.europa.eu/justice/data-protection/article-29/documfentation/opinion-recommendation/files/2014/wp224_en.pdf. Accessed 25 Nov 2014.

BBC News Technology. 2010. Governments 'not ready' for new European Privacy law. http://www.bbc.com/news/technology-12677534. Accessed 9 Nov 2014.

Bernal, Paul Alexander. 2011. A right to delete. *European Journal of Law and Technology* 2.2. http://ejlt.org/article/view/75/144. Accessed 5 Mar 2015.

Borgesius, Zuiderveen Frederik J. 2013. Consent to behavioural targeting in European law—What are the policy implications of insights from behavioural economics? *Amsterdam Law School Research Paper* 43. SSRN: http://ssrn.com/abstract=2300969. Accessed 9 Nov 2014.

Buckler, Craig. 2012. Why your site in now illegal in Europe. *Sitepoint*. http://www.sitepoint.com/europe-website-cookie-privacy-law/. Accessed 9 Nov 2014.

Butcher, Mike. Stupid EU cookie law will hand the advantage to the US, kill our startups stone dead. techcrunch.com, http://techcrunch.com/2011/03/09/stupid-eu-cookie-law-will-hand-the-advantage-to-the-us-kill-our-startups-stone-dead/. Accessed 9 Nov 2014.

Bygrave, Lee. 2000. Minding the machine: Art15 of the EC data protection directive and automated profiling. *Privacy Law and Policy Reporter* 40: 67–76. http://folk.uio.no/lee/oldpage/articles/Minding_machine.pdf. Accessed 9 Nov 2014.

Calo, Ryan M. 2012. Against notice skepticism in privacy (and elsewhere). *Notre Dame Law Review* 87(3): 1027–1072.SSRN. http://papers.ssrn.com/sol3/papers.cfm?abstract_id=1790144. Accessed 10 Mar 2015.

Chambers, Jessica. 2011. Cookie law makes most UK websites illegal: What you need to know. http://blog.silktide.com/2011/05/cookie-law-makes-most-uk-websites-illegal-what-you-need-to-know/. Accessed 9 Nov 2014.

Chester, Jeff. 2012. Cookie wars: How new data profiling and targeting techniques threaten citizens and consumers in the "big data" era. in *European Data Protection: In Good Health?* ed. Gutwirth S. et al. 53–77. Netherlands: Springer.

Clarke, Roger. 1987. Information technology and dataveillance. *Communications of the ACM* 31.5: 498–512. http://www.rogerclarke.com/DV/CACM88.html. Accessed 9 Nov 2014.

Cofone, Ignacio. 2014. The way the cookie crumbles: Online tracking meets behavioral economics. *Rotterdam Institute of Law and Economics (RILE), Working Paper Series, Paper No. 2014/14*. SSRN: http://ssrn.com/abstract=2541215. Accessed 27 Feb 2015.

Cohen, Julie E. 2000. Examined lives: Informational privacy and the subject as object. *Stanford Law Review* 52: 1373–1437. http://scholarship.law.georgetown.edu/cgi/viewcontent.cgi?article=1819&context=facpub. Accessed 9 Nov 2014.

Communications Committee (European Commission—Information Society and Mediate Directorate General). 2010. Working document on the implementation of the revised Framework—Article 5(3) of the ePrivacy Directive. COCOM10–34, Brussels, 20 Oct 2010. http://itek.di.dk/SiteCollectionDocuments/Zire/Sikkerhedsnyhedsbrev/COCOM10-34%20Guidance%20Art%205%283%29%20eprivacy%20Dir.pdf. Accessed 20 Nov 2014.

ComputerWekkly. How to comply with EU cookie law. http://www.computerweekly.com/guides/How-to-comply-with-the-EU-cookie-law. Accessed 9 Nov 2014.

Cookies and Internet Advertising. 2012. Amazon.co.uk, http://www.amazon.co.uk/gp/help/customer/display.html?ie=UTF8&nodeId=201149560&ref_=gw_cookie_uk. Accessed 25 Nov 2014.

Cranor, Lorrie F. 2003. 'I didn't buy it for myself': Privacy and ecommerce personalization. *WPES*, 111–117. http://lorrie.cranor.org/pubs/personalization-privacy.pdf. Accessed 9 Nov 2014.

Daniel, Caroline, and Palmer, Maija. 2007. Google's goal: To organize your daily life. FT.com, http://www.ft.com/cms/s/2/c3e49548-088e-11dc-b11e-000b5df10621.html#axzz3HcfhNcV2. Accessed 9 Nov 2014.

Debusseré, Frederic. 2005. The EU E-privacy directive: A monstrous attempt to starve the cookie monster? 13(1): 70–97. *International Journal of Law and Information Technology*. http://gov.gg/ccm/cms-service/download/asset/?asset_id=36034. Accessed 9 Nov 2014.

Department for Culture, Media & Sport (DCMS). 2011a. Research into consumer understanding and management of internet cookies and the potential impact of the EU Electronic Communications Framework. https://www.gov.uk/government/uploads/system/uploads/attachment_data/file/77641/PwC_Internet_Cookies_final.pdf. Accessed 9 Nov 2014.

Department for Culture, Media & Sport (DCMS). 2011b. Research into consumer understanding and management of internet cookies and the potential impact of the EU Electronic Communications Framework. https://www.gov.uk/government/uploads/system/uploads/attachment_data/file/72837/PwC_DCMS_Internet_Cookies_Summary_and_Conclusions.pdf. Accessed 9 Nov 2014.

Edwards, Lilian. 2003. Consumer privacy, on-line business and the internet: Looking for privacy in all the wrong places. *International Journal of Law and Information Technology* 11(3): 226–250.

European Commission. 2009. Report on cross-border e-commerce in the EU. Commission Staff Working Document, SEC 283 final (2009). http://ec.europa.eu/consumers/archive/strategy/docs/com_staff_wp2009_en.pdf. Accessed 9 Nov 2014.

European Commission. 2009. Data collection, targeting and profiling of consumers for commercial purposes in the online environment. *Background Paper*.

Field Fisher Waterhouse. Cookie 'consent' rule: EEA implementation. http://www.fieldfisher.com/pdf/cookie-consent-tracking-table.pdf. Accessed 9 Nov 2014.

Forbes. 2013. The Web Cookie is dying. Here's the creepier Technology that comes next. http://www.forbes.com/sites/adamtanner/2013/06/17/the-web-cookie-is-dying-heres-the-creepier-technology-that-comes-next/. Accessed 9 Nov 2014.

Froomkin, Michael A. 2000. The death of privacy? *Stanford Law Review*: 1461–1543. http://osaka.law.miami.edu/~froomkin/articles/privacy-deathof.pdf. Accessed 9 Nov 2014.

Gibbs, Simon. 2013. The stupid cookie law is dead at last. *Libertarian Home*. http://libertarianhome.co.uk/2013/01/the-stupid-cookie-law-is-dead-at-last/. Accessed 9 Nov 2014.

Google Inc. Welcome to DoubleClick. http://www.google.com/doubleclick/. Accessed 10 Nov 2014.

Graham, Chirstopher. ICO blog: Half term report on cookies compliance. *UK Information Commissioner's Office (ICO)*. http://ico.org.uk/news/blog/2011/half-term-report-on-cookies-compliance. Accessed 9 Nov 2014.

Hall, Kathleen. 2011. EC cookie privacy laws threatens UK's digital economy. ComputerWeekly.com. http://www.computerweekly.com/news/1230095377/EC-cookie-privacy-laws-threaten-UKs-digital-economy. Accessed 9 Nov 2014.

Hildebrandt, Mireille. 2006. Profiling: From data to knowledge, the challenges of a crucial technology. *Datenschutz und Datensicherheit 30*: 549–552. http://www.fidis.net/fileadmin/fidis/publications/2006/DuD09_2006_548.pdf. Accessed 9 Nov 2014.

Hoofnagle, Chris Jay, et al. 2012. Behavioral advertising: The offer you cannot refuse. *Harvard Law & Policy Review* 6: 273–296. SSRN: http://scholarship.law.berkeley.edu/facpubs/2086. Accessed 25 Nov 2014.

ICC UK Cookie guide. 2012. *International Chamber of Commerce (ICC) UK*. http://www.cookielaw.org/media/1096/icc_uk_cookiesguide_revnov.pdf. Accessed 9 Nov 2014.

InternetSociety. 2013. Understanding the personal data bargain. http://www.internetsociety.org/blog/2013/02/understanding-personal-data-bargain. Accessed 9 Nov 2014.

Javascript: Advantages and disadvantages. Jcscripters.com. http://www.jscripters.com/javascript-advantages-and-disadvantages/. Accessed 25 Nov 2014.

Kang, Jerry. 1998. Information privacy in cyberspace transactions. *Stanford Law Review*: 1193–1294.

Kierkegaard, Mercado Sylvia. 2005. How the cookies (almost) crumbled: Privacy & lobbyism. *Computer Law & Security Review* 21(4): 310–322.

Kosta, Eleni. 2013. Peeking into the cookie jar: The European approach towards the regulation of cookies. *International Journal of Law and Information Technology* 23(4): 380–406.

Kosta, Eleni. 2013. Consent in European data protection law. PhD dissertation, University of Leuven.

Kroes, Neelie. 2010. European Commission Speech 10/452. European roundtable on the benefits of online advertising for consumers (speech given at European Roundtable on the Benefits of Online Advertising for Consumers, Brussels, September 17, 2010). http://europa.eu/rapid/press-release_SPEECH-10-452_en.htm?locale=FR. Accessed 15 Nov 2014.

Kuneva, Meglena. 2009. European Commission Speech 09/156. Roundtable on online data collection, targeting and profiling (speech given at roundtable on online data collection, targeting and profiling, Brussels, March 31, 2009). http://europa.eu/rapid/press-release_SPEECH-09-156_en.htm. Accessed 9 Nov 2014.

Langheinrich et al. 1999. Unintrusive customization techniques for web advertising. *Computer Networks: The International Journal of Computer and Telecommunications Networking* 31: 181–194. http://www.ra.ethz.ch/cdstore/www8/data/2159/pdf/pd1.pdf. Accessed 9 Nov 2014.

Lee, Phil P. 2011. The impact of cookie 'consent' on targeted adverts. *Journal of Database Marketing & Customer Strategy Management* 18(3): 205–209. http://www.palgrave-journals.com/dbm/journal/v18/n3/pdf/dbm201120a.pdf. Accessed 9 Nov 2014.

Letter. Subject: SB 761(Lowenthal)—OPPOSITION. (2011). http://regmedia.co.uk/2011/05/05/dnt_opposition_letter.pdf. Accessed 9 Nov 2014.

Lloyd, Ian J. 2000. *Information technology law* (3rd edn.). Butterworths.

Mantelero, Alessandro. 2014. The future of consumer data protection in the EU Re-thinking the "notice and consent" paradigm in the new era of predictive analytics. *Computer Law & Security Review* 30(6): 643–660.

McStay, Andrew. 2012. I consent: An analysis of the cookie directive and its implications for UK behavioral advertising. *New Media & Society* 15(4): 596–611.

Nguyen, Anh. 2012. 95 % of UK organisations 'do not comply with EU cookie law. *Computerworld UK*. http://www.computerworlduk.com/news/it-business/3350059/95-of-uk-organisations-do-not-comply-with-eu-cookie-law/. Accessed 9 November 2014.

Office of Fair Trading (OFT). 2014. Online targeting of advertising and prices market study. *Internet Advertising Bureau*. http://www.iabuk.net/sites/default/files/IABresponsetoOFTmarketstudyintoOnlineTargetingofAdvertisingandPrices_6012_0.pdf. Accessed 9 Nov 2014.

Ohm, Paul. 2009. The rise and fall of invasive ISP surveillance. *University of Illinois Law Review*: 41. SSRN: http://ssrn.com/abstract=1261344. Accessed 9 Nov 2014.

Riefa, Christine, and Markou, Christiana. 2014. Online marketing: Advertisers know you are a dog on the internet. In *Research Handbook on EU Internet Law*, eds. Savin, A., Trzaskowski, J. Denmark: Edward Elgar.

Ronald, Leenes, and Kosta Eleni. 2015. Taming the cookie monster with Dutch law—A tale of regulatory failure. *Computer Law & Security Review: The International Journal of Technology Law and Practice*. doi:10.1016/j.clsr.2015.01.004.

Solon, Olivia. 2012. A simple guide to cookies and how to comply with EU cookie law. wired. co.uk, http://www.wired.co.uk/news/archive/2012-05/25/cookies-made-simple. Accessed 9 Nov 2014.

Solove, Daniel J. 2001. Privacy and power: Computer databases and metaphors for information privacy. *Stanford Law Review* 53: 1393–1462. http://scholarship.law.gwu.edu/cgi/viewcontent.cgi?article=2077&context=faculty_publications. Accessed 9 Nov 2014.

TRUSTe and Fieldfisher. Cookie audits—Are you ready? http://webcasts.acc.com/handouts/Whitepaper-_EU_Cookie_Audits_Are_you_Ready.pdf. Accessed 10 Nov 2014.

Taylor, Curtis R. 2004. Consumer privacy and the market for customer information. *The RAND Journal of Economics* 35(4): 631–650. http://www.jstor.org/discover/10.2307/1593765?sid=21105660558101&uid=3737848&uid=2&uid=4. Accessed 9 Nov 2014.

Tene, Omer, and Polonetsky, Jules. 2012. To track or 'do not track': Advancing transparency and individual control in online behavioral advertising. *Minnesota Journal of Law, Science & Technology*, 13(1). SSRN: http://ssrn.com/abstract=1920505 cr 10.2139/ssrn.1920505. Accessed 15 Nov 2014.

The Economist. 2014. How a new type of "evercookie" tracks you online. http://www.economist.com/blogs/economist-explains/2014/08/economist-explains-3. Accessed 25 Nov 2014.

Treharne-Jones, Eleanor. 2014. European cookie sweep initiative: Are you compliant?" *TRUSTe Blog*. http://www.truste.com/blog/2014/07/17/european-cookie-sweep-initiative-are-you-compliant/. Accessed 9 Nov 2014.

Turow, Joseph, et al. 2009. Contrary to what marketers say, Americans reject tailored advertising and three activities that enable it. http://graphics8.nytimes.com/packages/pdf/business/20090929-Tailored_Advertising.pdf. Accessed 9 Nov 2014.

UK Information Commissioner's Office (ICO). 2011. Changes to the rules on using cookies and similar technologies for storing information. https://www.huntonprivacyblog.com/wp-content/files/2011/12/Initial-guidance.pdf. Accessed 9 Nov 2014.

UK Information Commissioner's Office (ICO). 2013. Changes to cookies on our website. https://ico.org.uk/about-the-ico/news-and-events/current-topics/changes-to-cookies-on-our-website/. Accessed 9 Nov 2014.

UK Information Commissioner's Office (ICO). Data protection good practice note, collecting personal information using websites. http://webcache.googleusercontent.com/search?q=cache:93a44GXghAYJ:webarchive.nationalarchives.gov.uk/20100402134332, http://ico.gov.uk/upload/documents/library/data_protection/practical_application/collecting_personal_information_from_websites_v1.0.pdf+&cd=1&hl=en&ct=clnk&gl=cy. Accessed 9 Nov 2014.

UK Information Commissioner's Office (ICO). 2012. Guidance on the rules on use of cookies and similar technologies. http://ico.org.uk/news/blog/2011/~/media/documents/library/Privacy_and_electronic/Practical_application/cookies_guidance_v3.pdf. Accessed 9 Nov 2014.

UK Information Commissioner' s Office (ICO). 2014. The EU cookie law (e-Privacy Directive). http://ico.org.uk/for_organisations/privacy_and_electronic_communications/the_guide/cookies. Accessed 9 Nov 2014.

Vaizey, Ed. 2011. Open letter on the UK implementation of Article 5(3) of the e-Privacy Directive on cookies. *Department of Culture, Media and Sport*. https://www.gov.uk/government/uploads/system/uploads/attachment_data/file/77638/cookies_open_letter.pdf. Accessed 9 Nov 2015.

van Eijk N, et al. 2012. Online tracking: Questioning the power of informed consent. *Emerald Group Publishing Limited* 14(5): 57–73. http://dare.uva.nl/document/2/121980. Accessed 14 Nov 2014.

Van Well, Lita and Royakkers, Lambèr. 2004. Ethical issues in web data mining. *Ethics and Information Technology* 6(2): 129–140. http://alexandria.tue.nl/repository/freearticles/612259.pdf. Accessed 9 Nov 2014.

Vibrant. 2011. Contextual ad leader vibrant signs new premium publishers. http://www.vibrantmedia.com/press/press.asp?section=press_releases&id=182. Accessed 9 Nov 2014.

Woody, Leonhard. 2011. Zombie cookies won't die: Microsoft admits use, HTML5 looms as new vector." *InfoWorld*. http://www.infoworld.com/article/2620781/internet-privacy/-zombie-cookies–won-t-die–microsoft-admits-use–html5-looms-as-new-vector.html. Accessed 25 Nov 2014.

WordStream. What industries contributed the most to Google's earnings? http://www.wordstream.com/articles/google-earnings. Accessed 9 Nov 2014.

Yiannopoulos, Milo. 2011. Guest opinion: The EU's legal war on cookies is barking mad." wired.co.uk, http://www.wired.co.uk/news/archive/2011-05/11/cookies-regulations?p=2. Accessed 9 Nov 2014.

YouTube. 2012. Cookies FAQs. May 2012—ICO. https://www.youtube.com/watch?v=V0M8MYiGkQw. Accessed 9 Nov 2014.

YouTube. The stupid EU cookies law (and why it should die). https://www.youtube.com/watch?v=9hLmX9FX2KA. Accessed 9 Nov 2014.

YouTube. What is a cookie? https://www.youtube.com/watch?v=I01XMRo2ESg&feature=player_embedded. Accessed 14 Nov 2014.

Zarsky, Tal Z. 2002–2003. Mine your own business: making the case for the implications of the data mining of personal information in the forum of public opinion. *Yale JL & Tech.* 5: 1–56. http://yjolt.research.yale.edu/files/zarsky-5-YJOLT-1.pdf. Accessed 9 Nov 2014.

Forget About Being Forgotten

From the Right to Oblivion to the Right of Reply

Yod-Samuel Martin and Jose M. del Alamo

Abstract User concerns about the dissemination and impact of their digital identity have led to the spread of privacy-enhancing and reputation-management technologies. The former allow data subjects to decide the personal information they want to disclose in the limited scope of a particular transaction. The latter help them adjust the visibility of the information disclosed and tune how other people perceive it. However, as the sources of personal information are growing without their control, and both original and deceptive data coexist, it is increasingly difficult for data subjects to govern the impact of their personal information and for the information consumers to grasp its trustworthiness so as to separate the wheat from the chaff. This paper analyses the aforementioned risks and presents a protocol, an architecture, and a business approach supporting both data subjects in replying to the information linked to them and consumers in gaining opportune access to these replies.

Keywords Personal information · Secondary disclosure · Right to be Forgotten · Right of reply

---

Y.-S. Martin (✉)
Universidad Politécnica de Madrid, E.T.S.I. Telecomunicación,
B-204.1, Av. Complutense 30–Ciudad Universitaria, 28040 Madrid, Spain
e-mail: samuelm@dit.upm.es

J.M. del Alamo
Departamento de Ingeniería de Sistemas Telemáticos, Universidad Politécnica de Madrid,
E.T.S.I. Telecomunicación, B-204.1, Av. Complutense 30–Ciudad Universitaria,
28040 Madrid, Spain
e-mail: jmdela@dit.upm.es

1 Introduction

The Web is a universal and open space of information with no central control or authority. This allows any user to produce and share a personal profile, photographs, personal comments and opinions, and even publish their social relationships. Others can also publish any new piece of information, either true or deceptive, and associate it with the aforementioned user. As a result, individuals find it difficult to control the dissemination of information about them, due to the huge amount and diversity of its sources. In addition, the consumers of this information lack any clue about its accuracy, validity, and trustworthiness. As this information relates to real-world people, the problem gets even worse: It can damage their reputation and jeopardize their privacy.

Existing solutions have failed to solve these issues. First, the Web does not forget: Once a piece of information has been published, it is easily copied and re-introduced anywhere else, becoming nearly impossible to erase. Even if some particular instances of that information can be deleted from one site, copies may survive in other places.[1] Besides, it has now become easier and cheaper to preserve information than it is to erase it, and long-lost personal data may unexpectedly resurface in the future.[2] Second, a single user cannot control the whole Web: The lack of a central authority makes it difficult to control the information available in independent administrative domains such as social networking sites or blogs, which are governed by different policies or national legislations.

Even so, we consider that data subjects must have some rights over the information that shapes how other people perceive them, so as to defend themselves against public criticism or provide clarifications. This right has long been recognized in the realm of journalism, where it is known as the Right of Reply. Consequently, we have devised a solution that translates this Right of Reply to the online domain.

The technical approach we present allows data subjects to reply to the information linked to their digital identity, and to distribute these replies to the information consumers. We do not propose modifying the original information at the source, but rather enriching it at the destination with additional metadata that describes whether or not it has been endorsed by the data subjects, together with their own explanations. As a result, each site keeps control over the information it hosts, but the consumers of that information can still inspect additional parameters to assess its accuracy and validity. In addition, our contributions put individuals back in control of all the information about them after it has been disclosed—by either themselves or others.

---

[1] As pointed by Jennifer Stoddart (Canada's Privacy Commissioner) in Office of the Privacy Commissioner of Canada, "Protect Your Personal Information Because the Internet Never Forgets, Privacy Commissioner of Canada Says—January 27, 2011."

[2] Some famous, traditional examples include Stacy Snyder's and Andrew Feldmar's cases, gathered in Mayer-Schönberger, *Delete: The Virtue of Forgetting in the Digital Age*, chap. 1 Failing to Forget the "Drunken Pirate."

The remaining of the chapter is organized as follows. The following section further introduces the new sources of personal information on the Web, which do not just arise from the data subjects themselves anymore. Then, the next section discusses the related work that may help to partially mitigate the problems described, and the drawbacks of these solutions. After that, we describe the technical approach of our solution, including the entities involved and the protocol used to communicate among them, plus the implementation details in two different scenarios. Next, we introduce potential business models for the exploitation of our solution, discussing the motivations and benefits for the different roles involved. Finally, we provide a critique of our proposal, discussing some apparent and actual shortcomings and flaws that might prevent the solution from becoming useful and valuable; and we conclude with a sketch of some promising, future lines of action.

2 Secondary Disclosures as a Source of Personal Information

Traditional privacy research has focused on managing self-disclosed information—i.e. information about individuals disclosed by themselves. Unfortunately, the general public have not understood the long-term consequences of their disclosures on the Web. Every now and then the mass media report privacy and reputation damages due to information spinning out of their owners' control. An ancient Latin proverb reads *verba volant scripta manent*, which means that spoken words fly away while written ones remain. Likewise, digital information on the Web is easy to copy and reproduce in a different time and context where consumers may misunderstand it if they do not get any further explanations.

However, Web users keep on uploading and sharing private photographs, and posting religious, sexually oriented, or political comments and opinions on their online profiles in the belief that they will remain private forever. A recent study[3] has shown that nearly two thirds of youngsters are not concerned about their use of social media potentially harming their future career. Yet, the same study reveals that 10 % have been rejected for a job because of their social media profile, and seven out of ten recruiters have rejected job candidates based on the information referring to them found on the Web.

Even worse is the information about an individual disclosed by others, i.e. secondary disclosures,[4] which is quickly leading to new privacy and reputation issues on the Web, since relating a piece of information to another user's identity is as easy as annotating it with that user's identifier, effectively linking the information to his or her profile. The best-known example of secondary disclosures is the "tagging" feature of several websites, where a user can identify (tag) another user in a

---

[3]Quinn, "Facebook Costing 16-34s Jobs in Tough Economic Climate."

[4]Wisniewski, Lipford, and Wilson, "Fighting for My Space."

photograph or a post. Another relevant example is the "user mention" feature, which allows the users of a website to post a text online and connect what is there stated to others, by mentioning their names alongside some specific marks (e.g. "@username"). These features have been rapidly adopted by major social networking services.

Previous research provides evidence of the remarkable privacy concerns some users feel, caused by disclosure practices of others: Facebook users are concerned at the impression they may project when they are tagged on pictures[5] or when their friends post information about them;[6] romantic partners exercise interpersonal electronic surveillance by inspecting the secondary disclosures of their significant other's friends;[7] Twitter users may even be exposed to the gaze of marketers and would-be robbers due to the information disclosed by their contacts;[8] etc. Research has also shown how apparently innocuous, secondary disclosed information allows inferring sensitive information about the data subject.[9]

As a result, the current situation of the Web depicts a landscape of information about individuals, arising from both the data subject and other users as well, which mixes up genuine, deceitful, current, and outdated data, with hardly any clues about its accuracy or validity.

2.1 Roles in the Disclosure Process

There are at least three roles related to the information published about an individual: the discloser, the consumer, and the data subject (Fig. 1). The **discloser** uploads and shares a piece of information e.g. a tagged photograph. The **data subject** is the user referred by the information e.g. the user depicted in the photograph. The **consumer** finds the information and retrieves it. For self-disclosed information, both the discloser and the data subject are the same person, while they may differ in the case of secondary disclosures. Figure 1 illustrates this process by depicting a suited-up, respectable data subject, who appears outraged at the disclosers saying that he is instead a foolish jester; meanwhile, the information consumers are confused by the contradicting information they find.

Both the disclosers and the data subject may be registered at the same administrative domain e.g. a social networking site, where they are subject to common rules that protect their privacy and reputation. For example, Facebook users can

---

[5]Besmer and Lipford, "Moving beyond Untagging: Photo Privacy in a Tagged World."

[6]Bornoe and Barkhuus, "Privacy Management in a Connected World: Students' Perception of Facebook Privacy Settings."

[7]Tokunaga, "Social Networking Site or Social Surveillance Site? Understanding the Use of Interpersonal Electronic Surveillance in Romantic Relationships."

[8]Humphreys, Gill, and Krishnamurthy, "How Much Is Too Much? Privacy Issues on Twitter."

[9]Pesce et al., "Privacy Attacks in Social Media Using Photo Tagging Networks."

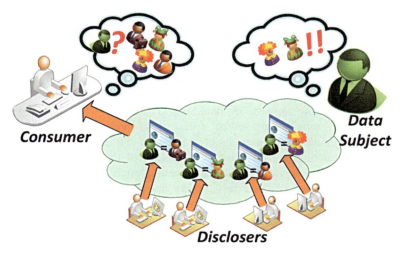

Fig. 1 Roles related to the information disclosed about an individual

tag other Facebook users in photographs, but the tagged users can delete the tag if they want—though not the photograph.

However, it is more often the case that each is registered at a different provider, which abides by its own policies and legislation. Actually, mainstream social networking sites also allow their users to link their identities to a limited set of external contents. In turn, independent sites leverage the identifiers of the users of mainstream social networks to associate pieces of information with them. For instance, it is quite easy for a discloser to tag a Twitter user on a photograph hosted at Flickr, and to publish this information, either within Twitter or anywhere else on the Web. Since each provider usually works independently with no underlying trust relationships set up with one another, the data subjects have little chance, if any, of having the tag or the information easily removed from a domain where they are not registered.

Going further, several services have begun linking doubtful or damaging information to user profiles on social networking sites. For example, in recent years, several sites[10] quickly became a place for anonymously slipping alleged sexual and drug-taking behaviours of students and classmates, referring to them by their Facebook or Twitter identifiers. More recently, non-consensual pornography sites have started doing the same with lewd pictures they host and link to Facebook or Twitter profiles of the depicted persons, in what the press has dubbed "identity

---

[10]E.g.: abouteveryone.com, Topix, or the so-called "confessions board" sites which specifically target college students such as JuicyCampus, CollegeACB, LittleGossip, etc. For a review of the evolution of these sites and a discussion from the perspective of US defamation laws, see Schorr, "Malicious Content on the Internet: Narrowing Immunity Under the Communications Decency Act."

porn".[11] In spite of the relatively limited life span of these services, several embodiments resurrect the same concept once and again with varying success.[12]

On top of these issues, the consumers usually find the information via services that neither produce nor host the information, but just point to it without checking its trustworthiness or validity. Search engines are the most widespread example, and currently there is an intense debate about to what extent they should filter this information, honouring requests by data subjects as some court decisions have ruled. In addition, some "people search" social sites[13] have come about which specialize in aggregating information about individuals extracted from other sites,[14] effectively constructing and publishing biographies of the data subjects without their knowledge.[15] While some of the search results might indeed refer to the person being sought, many of these services are not able to disambiguate between namesakes or even persons with similar names. Thus they often return results completely unrelated to the sought after individual, instead providing inaccurate and misleading information.

Summarizing, the amount of information referring to individuals is rapidly increasing on the Web; its sources are varied, distributed in space, administered by unrelated organizations, and governed by different policies and legislations. Consequently, it is nearly impossible to govern all the pieces of information referring to an individual by controlling the source of the information. So, what can users do to keep control over the information that defines how others perceive them?

3 Related Work

The political debate in the privacy realm has recently focused on the "Right to Erasure" or "Right to be Forgotten"[16] (also known as "Right to Oblivion"): upon individuals' request, data controllers should erase any data about them from their

---

[11]"Identity porn" is related, but slightly different, to the better-known concept of "revenge porn". Both refer to non-consensual dissemination of pornographic material depicting the data subject, but the latter emphasizes a specific motivation of a discloser who had been previously involved in a sexual relationship with the data subject; while the former focuses on scenarios where these contents published are also linked to the data subject's profile and identity attributes (whoever the disclosers and whichever their motivations might be). See Stroud, "The Dark Side of the Online Self: A Pragmatist Critique of the Growing Plague of Revenge Porn."

[12]Examples of existing of defunct sites that either foster or are frequently used for these practices include Is Anyone Up?, UGotPosted, Is Anybody Down?, Pink Meth, Texxxan.com, and MyEx.com.

[13]E.g.: Zoominfo, Pipl, 123people, Yasni, AnyWho, peekyou.

[14]Brennecke, Mandl, and Womser-Hacker, "The Development and Application of an Evaluation Methodology for Person Search Engines."

[15]Werbin, "Auto-Biography: On the Immanent Commodification of Personal Information."

[16]Mantelero, "The EU Proposal for a General Data Protection Regulation and the Roots of the 'Right to be Forgotten.'"

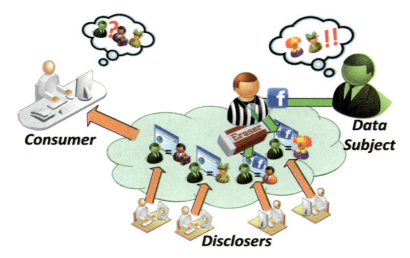

Fig. 2 Content moderation to implement the right to erasure in centralized realms

systems (unless legitimate reasons to preserve it can be alleged).[17] Maybe the simplest mechanism is the moderation facility offered by many online sites (Fig. 2), which allow data subjects to report inappropriate content to moderators (which usually rely on a mixture of human and semi-automatic tools), so as to have it erased. However, this only works if we are dealing with a closed realm where the moderators may act, and this often requires that both the disclosers and the data subject be registered at the same domain.

When different stakeholders (i.e. the discloser and the data subject) are involved in the control of some information, multiparty authorization models,[18,19] provide a solution: Each party sets its policies and a trusted evaluation process resolves conflicts. While this solution might work within trustworthy, centralized environments, it can hardly be enforced on the open Web.

Some state-of-the-art, technical solutions have been proposed to implement the right to erasure based on setting a secure channel between the data subject and the consumer, mainly relying on cryptographic means and a trusted infrastructure (Fig. 3). These solutions purportedly allow the data subject to grant or revoke permissions and hopefully control who can see the contents they disclose. For example, Privacy Rights Management[20] is the term used to describe the application of

---

[17]For a comprehensive review of the Right to be Forgotten, see Mayer-Schönberger, *Delete: The Virtue of Forgetting in the Digital Age*.

[18]Squicciarini, Shehab, and Paci, "Collective Privacy Management in Social Networks."

[19]Squicciarini, Shehab, and Wede, "Privacy Policies for Shared Content in Social Network Sites."

[20]Kenny and Korba, "Applying Digital Rights Management Systems to Privacy Rights Management."

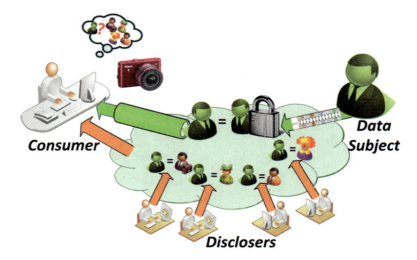

Fig. 3 Cryptographic approaches for the right to erasure, vulnerable to the analogue loophole

Digital Rights Management (DRM) techniques to personal information, in order to allow safely releasing that information beyond a single administrative domain. The dissemination is only allowed in an encrypted form, so that access controls can still be enforced when the information is being consumed, thus preventing any unauthorized usage. Other solutions replace the concept of DRM with that of 'data control',[21,22] which assumes that personal information will be always stored and cached at hosts controlled by the data subjects. Quite in the same line, some authors have proposed to combine cryptographic means with policy management in order to define sticky policies;[23,24] usage policies that are cryptographically bound to the data to which they pertain, and are enforced through a trusted, local, usage control platform. This control platform transfers the policies and (by monitoring the access and usage of the data) ensures that its constraints are enforced properly. However, all these solutions focus on self-disclosed information (secondary-disclosed information may still be out there in the wild), and require dedicated software and trusted endpoints which behave according to a set of pre-agreed rules in order to ensure that the sensitive information does not escape from the trusted domain and the control of the data subject.

---

[21]Castelluccia and Kaafar, "Owner-Centric Networking (OCN): Toward a Data Pollution-Free Internet."

[22]Sarrouh et al., "Defamation-Free Networks through User-Centered Data Control."

[23]Kelbert and Pretschner, "Towards a Policy Enforcement Infrastructure for Distributed Usage Control."

[24]Pearson and Casassa-Mont, "Sticky Policies: An Approach for Managing Privacy across Multiple Parties."

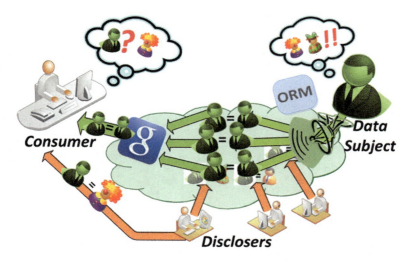

Fig. 4 Online reputation management (ORM) and search engine optimization (SEO) techniques, and their limitations

Even in trusted domains, little can be done to skip the so called analogue loophole[25] i.e. when a content can be ultimately reproduced using analogue means, it is easy to digitally recapture that analogue reproduction without restrictions. This has recently become a mainstream issue when users of private distribution services such as SnapChat realized their ephemeral photographs could be indeed easily captured by something as rudimentary as a screenshot.[26]

Lately, the European Network and Information Security Agency (ENISA) has recognized[27] the difficulty of applying that right to erasure, concluding that "*for any reasonable interpretation of the Right to be Forgotten, a purely technical and comprehensive solution to enforce the right in the open Internet is generally impossible*". Furthermore, it has been suggested that oblivion is indeed a misleading metaphor for the right to data erasure, precisely because of the radical difference between computer and human memories.[28]

As inconvenient information cannot be always erased, it would be enough if it could be hidden from the consumers' view. That is the approach followed by Online Reputation Management (ORM) techniques (Fig. 4). Their solutions are based on applying Search Engine Optimization (SEO) techniques to personal information: They advise to build positive content about oneself across a variety of sites, display one's name prominently, and link to them as much as possible "*as a*

---

[25]Diehl and Furon, "© Watermark: Closing the Analog Hole."

[26]Shein, "Ephemeral Data."

[27]Druschel, Backes, and Tirtea, "The Right to be Forgotten—between Expectations and Practice—ENISA."

[28]Markou, "The 'Right to be Forgotten': Ten Reasons Why It Should Be Forgotten."

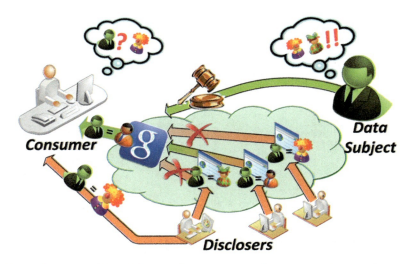

Fig. 5 Implementation of the Right to be Forgotten by search engines, and its limitations

buffer against misleading or negative search results".[29] Hopefully, this positive content will appear among the first results on search engines, above the content to hide.

This very same idea—that putting result links out of sight from search engines shifts the personal information out of mind at the consumers' side as well—, has also infected the legal realm (Fig. 5). In fact, recent court decisions in both the EU[30] and Japan[31] have established that, in certain cases, search engine administrators are obliged to remove, from the page of search results obtained when querying about the data subject's name, any links to third-party-hosted pages with personal information about them. Nonetheless, this concept has been subject to contention by legal experts[32] and there is an intense debate on how this ruling should be implemented.

Both the ORM and the judicial approaches targeting search engine results implicitly recognize that the original information sources can seldom be tackled. However, both have also taken for granted that the only way that consumers access

---

[29]Thompson and Fertik, *Wild West 2.0: How to Protect and Restore Your Online Reputation on the Untamed Social Frontier*.

[30]Ilešič, JUDGMENT OF THE COURT (Grand Chamber) In Case C-131/12, REQUEST for a preliminary ruling under Article 267 TFEU from the Audiencia Nacional (Spain), made by decision of 27 February 2012, received at the Court on 9 March 2012, in the proceedings Google Spain SL, Google Inc. v Agencia Española de Protección de Datos (AEPD), Mario Costeja González, (2014).

[31]Fujikawa, "Google Suffers New Privacy Setback in Japan."

[32]Court of Justice of the European Union, Opinion of Advocate General Jääskinen delivered on 25 June 2013 Case C-131/12 Google Spain SL Google Inc. v Agencia Española de Protección de Datos (AEPD), Mario Costeja González (2013).

information about the data subject is through search engines, and do not account for any information, for instance, directly shared by their contacts through social media, or simply accessed by random browsing.

While it is impossible in general to remove or hide data from the Internet once it was published, it might be possible to provide allegations from the affected user to help consumers in understanding its meaning. We are aware of the overheated rebuff surrounding some proposals to legislate for a compulsory Right of Reply on the digital arena,[33,34] and recognize the specificities of online media which have hindered the implementation of a traditional Right of Reply.[35] Notwithstanding that, we do not put the burden of the Right of Reply on the host of the information. As an alternative, we introduce a right-of-reply scheme where data subjects may provide clarifications to existing contents referring to them, and consuming agents can check these replies and any additional explanation, without altering the original source at all. That way, we separate the mere access to the information from the reply by the data subject.

4 Description of a Right-of-Reply Based Solution

Next we present a technological solution that allows the data subjects to provide their replies to information referring to them, and the information consumers to directly check it in turn against the data subjects. When new information about data subjects appears online, they are notified of it so that they can provide their replies and have them stored in our system. Later, when information consumers encounter that information, a consumer tool retrieves and presents the data subject replies as well. This enables the former to adjust how they are perceived by the latter.

4.1 Architecture and Protocols

The solution we propose consists of three logical entities, namely a notifier, a guarantor, and a verifier (Fig. 6). The **notifier** detects information linked to an individual and notifies the data subject's guarantor about it. The **guarantor** allows data subjects to issue replies to the information linked to them. It acts as an authoritative entity to which third parties may resort in order to ascertain the data

---

[33]Tiffen, "Finkelstein Report: Volume of Media Vitriol in Inverse Proportion to Amount of Evidence."

[34]Cornwall, "It Was the First Strike of Bloggers Ever: An Examination of Article 10 of the European Convention on Human Rights as Italian Bloggers Take a Stand against the Alfano Decree."

[35]Werkers, Lievens, and Valcke, "A Critical Analysis of the Right of Reply in Online Media."

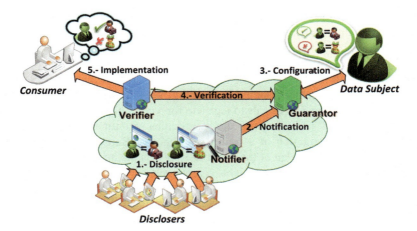

Fig. 6 Entities and messages involved in the right-of-reply solution

subject's viewpoint. Finally, the **verifier** helps the consumer obtain the reply from the data subject. Next we introduce the technical details followed in the procedure.

The whole process begins when a discloser releases a new piece of information regarding a data subject (**step 1, Disclosure**). The disclosers may publish firsthand, fresh information; but they may also propagate information already existing in another context (e.g. 'retweet' a Twitter message); or even just link a piece of information to the identity of the data subjects (e.g. tag them on a picture).

After that, at some point in time, a notifier encounters the information and sends the guarantor a message including this information or the URL address where it can be found (**step 2, Notification**).

Some service providers such as social networks, news sites, or blogs, can easily play the role of a notifier, given that they have access to any new information published there. Alternatively, a notifier can monitor information changes from the outside. For example, some social networking services provide real-time information streams (e.g. Twitter API or Facebook Open Stream), and major search engines allow subscribing to any changes in the contents they have indexed, so that a notifier can be timely alerted for new information items that might sprout in the wild. If the information is modified, deleted, etc., subsequent notifications will be triggered as well.

Once the notifier finds that personal information, first it resolves the identity of the data subject whom the information refers to, then it discovers the guarantor working on her behalf, and finally it notifies the new piece of information. The first step is usually easy to cope with, as the major social networking services semantically annotate the information they publish with the identity of the user it refers to (e.g. Google+ and Twitter mentions, or Facebook tags). However, it may be also the case that the data subject is not unambiguously identified in the information, for instance, because her real name is used instead of a unique identifier.

In this case, external mechanisms are used to filter the information and disambiguate the data subject identity (e.g. by applying clustering,[36,37] and entity matching techniques,[38,39] to natural text, or mechanisms that combine textual and social information for identity resolution[40,41]). And if, after all, any information which dubiously refers to that individual is also notified, then the mistaken data subject gets a chance to reject the information and avoid anyone else wrongfully relating it to her.

After the identity of the data subject is ascertained, the notifier discovers the guarantor working on her behalf. When the identity takes the form of an email address, then WebFinger[42] enables this step, as it provides a standardized and decentralized way of retrieving the location of service endpoints associated to an email address—i.e. the guarantor service endpoint. Major email providers already provide a WebFinger service e.g. Google enables it for all Gmail accounts with public Google+ profiles, and it is endorsed by Microsoft and Cisco as well. Besides, for email addresses whose domain does not natively support WebFinger, a functional extension called WebFist[43] comes to help. This extension allows the owner of an email address to establish a delegation to an external WebFinger provider, as long as the original domain supports DomainKeys Identified Mail (DKIM)[44]—which practically all major e-mail providers now do. Moreover, other identifiers such as Twitter or Facebook IDs can be translated into email-like addresses to be used as WebFinger identifiers. When the service endpoint is resolved, the notifier delivers the new piece of information.

Upon receiving the notification, the guarantor fetches the resource containing the information (if the notifier has not directly supplied it) and presents it to the data subject, who sets the reply (**step 3, Configuration**). The guarantor must be controlled by the data subjects on whose behalf they work, and be linked to their public identity in order to be discoverable through WebFinger. Typically, data subjects will lease the guarantor's service to an information service provider which includes this among its service offering. It might be a dedicated provider, but it will more usually be bundled up with other services. For instance, the guarantor can be offered by either the social networking service where the data subjects have their social profile published, or by their identity management provider of choice.

---

[36]Balog, Azzopardi, and de Rijke, "Resolving Person Names in Web People Search."

[37]Long and Shi, "Web Person Name Disambiguation by Relevance Weighting of Extended Feature Sets."

[38]Köpcke and Rahm, "Frameworks for Entity Matching: A Comparison."

[39]Jonas, "Threat and Fraud Intelligence, Las Vegas Style."

[40]Li, Wang, and Chen, "Identity Matching Using Personal and Social Identity Features."

[41]Berendsen et al., "Result Disambiguation in Web People Search."

[42]Jones et al., "WebFinger."

[43]Slatkin, "Bootstrapping WebFinger Decentralized Discovery with WebFist."

[44]Hansen and Hallam-Baker, "DomainKeys Identified Mail (DKIM) Service Overview."

During the configuration step, data subjects can formulate one reply to each piece of information. However, the guarantor also supports more sophisticated, data-subject-configured, automatic reply policies that may account for different attributes of the information or the context. That way, guarantors are able to infer whether the data subjects would likely endorse or deny some information without having it directly checked with the latter, therefore relieving data subjects from the overload of explicitly replying to each and every piece of information potentially related to them. For example, a data subject may provide a policy to acknowledge by default all the information that can be traced back to a domain under his or her control. In any case, the notifier can be re-configured at any time in the future, thus enabling the data subjects to regret or change their mind, and assert different replies to the same information at different occasions.

Later on, when a verifier detects a piece of information referring to an individual, first it discovers the guarantor that serves her by following the same steps described for the notifier, and then it requests the data subject's reply from the guarantor (**step 4, Verification**). The verifier must be deployed on a node trusted by the information consumer, and linked to an entity that generates the presentation of the contents to be rendered to that consumer, so as to intercept the contents before their rendering and check with the guarantors the potential replies of the data subjects. This allows arranging the verifier at either the web client side (i.e. integrated with a web browser) or the server side (e.g. within the presentation layer of a web service provider) with little distinction. Besides, we would remark the option to constrain the verification step down to the identities of the members of the consumer's social graph, as this represents a much more restricted user range whose personal information is indeed especially appreciated by the consumer. This possibility allows creating simple verifiers that just pay attention to the information disclosed about a few data subjects, which in turn encapsulate most of the value for the information consumer.

The verifier eventually receives the data subject's reply from the guarantor, and delivers it to its client, which may exhibit different behaviours (**step 5, Implementation**). Depending on the response, the verifier modifies the information that is going to be rendered by e.g. decorating it with added information (such as a tick, a cross, a question mark, an exclamation point, or a text bubble with the detailed clarification by the data subject), or even by filtering it out. This feature is somehow similar to that provided by anti-virus software which checks links and mark them as safe or dangerous to the consumer.[45]

The protocols require applying **integrity checks** to avoid Cross-Site Content-Forgery (XSCF) attacks i.e. serving different contents to different clients for a particular resource, to make it appear as something else. Upon notification, the guarantor applies a cryptographic hash function to the originally notified information. Later on, the verifier applies the same function to the version of the information it retrieves, and both results are compared during the verification step. The

---

[45]Ligouras, "Protecting the Social Graph: Client–side Mitigation of Cross–Site Content Forgery Attacks," 27–29.

Forget About Being Forgotten

Fig. 7 Verifier screenshots for the One Social Web (*top*) and the Chrome extension (*bottom*) prototypes

guarantor only delivers the reply from the data subject if both hashes match, that is, if both contents do not differ. That way, a malicious entity is barred from inducing a verifier to mistakenly apply a reply to a piece of information the data subject did not intend to do so.

4.2 Implementation Details

The three entities presented above can be mapped to different deployments, as long as some basic rules are observed. Implementations of this architecture have been deployed in two different scenarios (Fig. 7). The first one integrates with an open-source, decentralized/peer-to-peer social networking architecture, namely One Social Web[46] (OSW), which allows anyone to run their own server and join the network. OSW users can socialize with other users seamlessly, regardless of the server where these other users have registered their identity and profile. OSW builds upon the Extensible Messaging and Presence Protocol (XMPP) and includes

---
[46]https://github.com/onesocialweb.

extensions to cover all the usual social networking use cases such as user profiles, relationships, activity streams, and third party applications.

The development of our solution has included the design and implementation of a set of XMPP extensions, and an instance of each of the three entities involved: a notifier, a guarantor, and a verifier. These are respectively deployed in our facilities as add-ons to three independent OSW provider instances, which have their own users registered and are connected to the global OSW network.

In this deployment, the notifier associated to each OSW provider informs the guarantors associated to other providers about the information disclosed by their own users. The domain part of the data subject's username (user@domain) determines the guarantor to which the information must be notified. The User Interface (UI) of the guarantor informs the data subjects about the new information disclosed, and allows them to set Boolean replies in the form of agreement with or denial of the information. Accordingly, the presentation layer of the verifier shows the consumers the replies of the data subject, by decorating the original information with ticks (information agreed with), crosses (denied), and warning signs (not replied) (Fig. 7, top).

The second scenario includes two notifiers, a standalone guarantor, and a web browser extension as a verifier. In this deployment, we constrain the configuration of the notifiers and the guarantor to a restricted and highly cohesive user range: the members of our research group. One notifier is set up to receive alerts from Google Alerts service whenever Google indexes a new document containing any variants of their full names, their phone numbers, their email addresses, or their nicknames. In addition, another notifier is configured to gather tweets mentioning their Twitter nicknames, from the Twitter stream API (Application Programming Interface). The configuration of the notifiers relies on the information automatically extracted from the data subjects' electronic business cards (vCard).

Here, one standalone guarantor serves all the data subjects and thus no guarantor resolution is required. The data subjects are allowed to include a free text and a URL address as their reply, in addition to the agreement or denial flag. The guarantor also includes a rule engine to store advanced, data-subject-configured reply policies: e.g. provide a positive agreement to all the information coming from any domain the data subject declares to have control of; and apply access control lists so as to restrict the delivery of replies to blacklisted verifiers not authorized to access specific areas of the data subject's life. In cases where no rule can be applied, an "unknown" reply is returned. In that case, a notification process is also triggered, as the guarantor becomes aware of some previously unknown information, which could be used in the responses to further verifications.

Finally, the verifier has been developed as a Greasemonkey script. Greasemonkey is a browser extension framework[47] that allows installing user scripts in the browser which modify the rendered contents of a web page. Our script seeks the information available in the vCards of the consumer's contacts

---

[47]Greasemonkey for Mozilla Firefox (http://www.greasespot.net/) or Tampermonkey for Google Chrome, Opera Next and Safari (https://tampermonkey.net/).

(specifically, the members of one of his Google+ circles). Whenever such information is detected on any of the web pages the consumer is browsing through, the script triggers the verification step. If any reply has been previously configured at the guarantor, it is returned in response to the verifier, whose script modifies the original page to show an inline icon running next to the mention to the data subject. When the consumer mouse hovers the icon, the page pops up the reply (Fig. 7, bottom).

5 Potential Business Exploitation Models

Verifying the information found about an individual obviously benefits both data subjects and information consumers: the former distribute their replies to the information spread on the Web; while the latter gain an easier access to the data subject's replies, which may be used as a quality hint regarding the original information, reinforced or rebutted by the reply. But what about the motivations and benefits for the providers operating the guarantor, the notifier, and the verifier? (Fig. 8).

The **guarantor** supplies all the features and infrastructure that enable receiving notifications and responding to requests for verification of personal information, as well as the configuration of replies by the data subject. Typically, data subjects could lease the guarantor's service to an information service provider which had it included it among its service offering. It could be a dedicated provider, but it would more usually be bundled up with other identity- and reputation-related

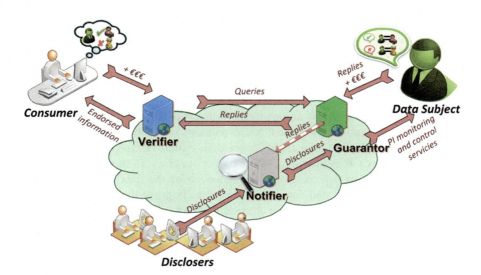

Fig. 8 Value network and flows between the entities and roles involved in the business model

services. For instance, the guarantor could be offered by the social networking service where the data subject has her social profile published, by her online identity management provider, or by her ORM.

As a matter of fact, ORMs already charge their users for allowing them to control their online reputation with solutions such as SEO or information removal services. However, not all the information can be hidden or removed, and thus providing adequate replies associated to the original information turns out to be a valuable add-on to reputation management.

Notifiers detect information linked to an individual and notify the data subject's guarantor of it. The main motivations for third party notifiers to join the system is that they can benefit in return by gaining a better knowledge about the accuracy and validity of the information they host, thus enhancing their own services and data. In the business information arena, some hospitality review sites already enrich the guests' opinions (disclosed information) with the replies from the hotel management (data subject) so as to enhance the information available to the site visitors (information consumers).[48]

Alternatively, if the notifier belongs to the guarantor provider itself, they may charge their customers (data subjects) for this service. For example, major ORM providers already provide their users with identity and reputation monitoring and alert services. Search engines also provide these services to data subjects;[49] however, they do not charge them for that, but provide it instead as a free, value-added service that increases the engagement and trust of their users.

Verifiers check the quality of the information found on the Web about a data subject. Again, consumers would preferably lease this service to a provider that bundles it up with other services. For example, antivirus companies include link scanners among their offering i.e. web browser add-ons that analyse the potential risk of links' destinations, warning their users when a suspicious one is detected. Thus, they might easily add the verification feature to their offering. Likewise, search engines nowadays sometimes decorate search results linking to suspicious contents with warnings to information consumers—a feature that could be expanded to encompass the warnings issued by our verifiers, too.

Same as in any other network-economy business, the value of the system comes from having a large number of participants that take advantage of its features and a high degree of connectedness among them—otherwise the actors may be reluctant to invest in deploying this new idea. Niche domains with a tightly cohesive, large base of users such as employment agencies, dating sites, or credit bureaus already show these features, thus providing lower barriers to entry. In these domains, consumers are willing to get accurate information about the data subjects (and are already paying for that), and data subjects want to improve their situation if their replies are able to amend misleading or inaccurate information.

---

[48]O'Connor, "Managing a Hotel's Image on TripAdvisor."
[49]"Me on the Web" by Google https://www.google.com/settings/me.

For example, employment websites charge for their services to both consumers of personal information (recruiters) and data subjects (candidates). Recruiters usually surf the Web and are already paying for background check services, looking for information that may shed some light on whether to accept or reject a candidate. On the other hand, job seekers usually pay for premium services that allow them to increase their chances of being selected by managing the profile that will be presented to recruiters. In this domain, the employment website can enhance its traditional business by bridging consumers and data subjects, who in turn benefit from using the verification and reply services.

Finally, it should be noted that search engines would benefit from implementing this approach and encouraging data subjects to exercise a right of reply rather than their right to oblivion (data subjects cannot relinquish their rights, but they can elect which one they want to enforce on each occasion). It is in the best interest of search engines to provide access to the most amount of information available, and the Right of Reply goes in line with that, by enriching currently indexed contents with the replies by the data subjects; while the Right to be Forgotten goes in the opposite direction, by removing contents from the query results.

6 Critical Discussion

We have shown a systems-based solution that allows data subjects to cope with secondary-disclosed personal information online, by providing their own replies and directing these to the information consumers. Our proposed solution directly connects the data subjects and the information consumers, without relying on the cooperation of the disclosers; thus it works in unmoderated, untrustworthy, distributed domains where other approaches had failed before. This makes it especially useful in the current Web landscape, where social media have become a channel to access information as relevant as search engines used to be,[50] despite the original publishing sources are beyond the control of the social networking sites themselves. Our solution also recognizes the dynamism inherent to the online media, by allowing for shifting information and changing replies.

However, this solution is still subject to some drawbacks and limitations. Firstly, we do not vouch for the Right of Reply as a one-size-fits-all solution. It may be a superior alternative to the application of the Right to be Forgotten when the data subject cannot control the information source nor its distribution channel, and these are in another jurisdiction, different from that of the data subject. It may also mitigate privacy and reputational impacts when conflicting rights prevent the application of the Right to be Forgotten e.g. when journalistic, historical, or public interests might be at stake. The recent experience of Google's application of the Right to be Forgotten has demonstrated that data subjects wish to enforce this right

---

[50]Weinberg, "An Introduction to Social Media Marketing."

more often than possible,[51] and their personal data remains at the original sources anyway.[52] Yet the Right of Reply may not be the optimal solution for data subjects who wish to begin with a clean state (e.g. after a sex change), deal with stories they want to remain completely hidden, or with embarrassing images depicting the data subjects which they cannot deny. Nonetheless, a reply needs not automatically be a denial, and data subjects can also find it useful to provide their reply to pictures in many circumstances: manipulated photomontages, mugshots of individuals who were eventually released as innocent (or not even trialled), etc.

Secondly, even if the solution is useful, it needs to become adopted so as to be valuable. When there are few verifiers deployed, data subjects are not incentivized to provide their replies to our system, as these will seldom reach the consumers. And if no (or very few) replies are made available through our solution, then information consumers will do not find any added value to use the verifier. This chicken-and-egg problem is frequently modelled as an "empty village" that no one wants to visit because it is empty and boring—and thus it remains the same. The bootstrapping might come from the integration of our solution by an appropriate service provider that offered it at once to a large, cohesive, user base, with little adoption effort on their side. Above we have presented some niches where data subjects and consumers share a common ground (e.g. e-recruitment, e-dating, credit scoring), and which could be a place to start, as the data subjects (i.e. job applicants, date seekers, borrowers) would already know that their replies would not be diluted but reach the targeted information consumers (recruiters, other date seekers, creditors). But maybe the most illuminating case would be that of some huge web companies that currently run their operations through the whole value chain of information.

For instance, let us take Google as an example (Microsoft might represent a similar case as well). From a data protection perspective, Google qualifies as a data controller; as it collects the browsing habits of its users, stores their e-mail messages, etc. Considering the complexity and multiple nuances implied by managing the privacy of multiple kinds of personal data, Google has provided data subjects with a tool (Google Dashboard) to manage all the personal information that it stores about its users.[53] However, as recently pointed out by the infamous decision of the Court of Justice of the EU, Google is also a data controller for keeping an index of the World Wide Web that contains loads of personal information. Hence it would seem straightforward for Google to expand the dashboard so that it would already cover their search engine indices, serving as the embryo for a guarantor. But, indeed, Google had long admitted this rationale and provides such a service, albeit only the part to receive the notifications (Me on the Web, based on

---

[51]During the first year of the application of the EU data-protection rules to Google search results, only 41.5 % of the removal requests were honored. See Laursen, "Google's Year of Forgetting [News]."

[52]Ibid.

[53]Ortlieb, "The Anthropologist's View on Privacy."

Google Alerts).⁵⁴ Should the replies be leveraged by Google, they could be employed to enrich the search results as suggested in our examples, therefore providing an infrastructure to smoothly distribute the replies. Moreover, Google could even demote or promote the information within the search results, allowing for the data subject replies as another input parameter in its ranking algorithm. These replies would represent more additional information that perfectly fits within Google business model, which is based on amassing data, rather than erasing it. Furthermore, Google already operates some of the most visited user-generated web content hosting services (Youtube, Blogger, Picasa, Google+) where the notifiers could be deployed, and they develop the most used web browser (Chrome) and mobile operating system (Android) to which the verifier functionality can be added.

In this regard, it should be noted that different business models to exploit the solution are possible, and not all of them entail monetary payments. Our goal has been to propose a balanced, self-sustainable model that provides the suitable incentives for every party to join, and which can flexibly adapt to different industries. But we vouch for everyone to be able to exercise their Right of Reply, without creating unnecessary divides that prevent the less wealthy from issuing them. We propose non-monetary incentives for the entities operating the guarantor (e.g. legal compliance, increased traffic, trustworthiness, added information) to ensure that they operate the service without introducing economic barriers to data subjects.

Thirdly, even a valuable solution could be rendered worthless if it is abused by dishonest users. For the system to prove advantageous, it would entail that information consumers cannot easily tell apart trustworthy from untrustworthy personal information on their own, and that the data subject replies be honest so as to be helpful. However, data subjects may instead be tempted to clean up or even deny inconvenient yet fair information about them. We regard this as a feature of our solution: we do not suggest that the agreement of the data subject be the only indicator that information consumers should employ when judging the trustworthiness of personal information. Rather, replies are just a further mechanism to empower them in deciding what information they trust. In many cases, they will not directly dismiss some information they find just because of a negative reply, but if it is more precise than a mere denial, they will get to know both sides of the story and possibly have their perspective adjusted.

Yet the data subjects themselves are incentivized as well to be honest in their replies: if they cry wolf by consistently denying fair information, they can make the consumers distrust them, which would then ignore legit disagreements with libellous information. We are currently working on a probabilistic model to show how our solution discourages both data subjects from issuing dishonest replies and, what is more, disclosers from releasing untrustworthy information. To put it simply, it is in the interest of the data subject to provide helpful and honest

---

⁵⁴Ibid.

replies, so that they can convincingly deny untrustworthy information, especially when they often need to defend from this. Besides, the user replies may also have a chilling effect on the disclosers aiming to disseminate untruthful information about the data subject. Malicious disclosers may be interested in sneaking as much deceitful information as they can; however, if they lie too often, they will be caught by consumers who face the replies from the data subjects, and their credibility will be damaged—achieving the opposite of what they wanted. In turn, if the data subjects do not want to let lies about them spread, they would behave properly. Eventually, these mutual incentives would reduce untrustworthy information until a Nash equilibrium is reached (where only white lies would survive).

A complementary issue is that of the information overload that both the data subjects and the information consumers would be facing, when respectively providing and receiving the replies to each piece of personal information that might appear online. We have tried to mitigate that issue at both ends of the system. The configuration of replies at the data subjects' side may be done in batches, by establishing automatic reply policies for the most common cases. And the verification results are rendered using lightweight, non-intrusive user interface hints such as the icons that allow a quick assessment by the information consumers, who can then inspect the whole reply if they wish.

Finally, some other apparent shortcomings are already tackled by our solution. One is the problem of disambiguating between namesakes, which impacts the decision on who should be notified (and whose reply should be inquired). This is not a problem when the identity of the referred to data subject is semantically annotated (as it happens in social networking services regarding e.g. photo tags or user mentions). When only plain text, natural language is present, we need to resort to person disambiguation techniques, as above mentioned. But even when this is not feasible, it should not be seen as a disadvantage, but as an opportunity to reply "this refers to someone else, but not me" to the information consumers who would anyway encounter that information without being able to disambiguate the identity of the data subject either.

A related issue is the association of an online service user identifier with a real world person identity. This is something that has long be tackled by other online services, and which currently admits several mature solutions which are based on delegating the identity check on other agents: PKI (Public Key Infrastructure) certification authorities, issuers of electronic payment instruments, or even social networking services themselves.[55]

Conversely, when the replies from the data subjects can be definitely linked to a real-world strong identity, the data subjects may start facing privacy risks derived from the disclosures associated to their own replies. If they acknowledge or deny whatever is published about them, and even provide supporting details, then they will be indeed releasing personal information—and even more might be inferred. A sophisticate attacker might even issue multiple queries to the data subject's

---

[55]E.g. Twitter verified accounts https://support.twitter.com/articles/119135-faqs-about-verified-accounts.

Forget About Being Forgotten

controller, asking for any personal details one might come up with, to which the courteous controller would irremissibly reply (on behalf of the data subject), eventually revealing all kinds of personal information. It might seem this is the price of being able to defend oneself in the public space. However, there are measures that can be taken to mitigate this theoretical risk: from the use of multiple, partial, and pseudonymous identities that prevent linkability,[56] to artificial response delays that are usually applied against brute-force attacks, and to cloaking techniques employed to mitigate similar multi-query attacks in location-based services.[57]

Another threat we deal with is the potential information forgery that would arise from the disclosers sending different versions of the same content to the notifiers and to the information consumers. This could result in the data subjects unknowingly accepting information they did not intend to acknowledge. To avoid that, during the configuration step, the guarantor stores a tamper-proof summary of the information it received (to which the data subject replied) computed by applying a cryptographic hash function to the information. The same processing is later applied by the verifier to the information it receives, and both summaries are then compared to ensure the integrity of the information.

To conclude this section, we would like to emphasize what the Right of Reply entails and what it does not. Sometimes, replies can be a plain "yes/no" answer, but more often they may come with rich contextual data that allows clarifying non-obvious pieces of information (e.g. the aforementioned mugshots of innocent persons). Likewise, the reaction of a verifier when facing a denial needs not imply hiding the offending information: depending on the situation, it may be more suitable to render the original content to the information consumers, together with the reply from the data subject, so that they can make their own judgements. This is especially applicable to published opinions, criticism, and even factual information about public persons, all of which should never be censored at all, but may be put in their place if the consumers can also gain access to replies that provide valuable details and contextualize the original information—this is the very same goal as that of the Right of Reply in the realm of journalism, where it originated.

7 Conclusions

We have introduced a novel protocol and an architecture to effectively translate the Right of Reply into the Web. Three entities assist users in, respectively, being aware of the information existing about them, providing their reply to such information, and getting the replies to the information they encounter online. Our solution allows the data subjects referred to by some outdated, inaccurate, or misleading information to provide further clarifications, and helps the consumers of

---

[56]Hansen et al., "Privacy-Enhancing Identity Management."

[57]Talukder and Ahamed, "Preventing Multi-Query Attack in Location-Based Services."

that information discern its trustworthiness. As a result, it contributes to solving the problems that many individuals are facing, as the sources of information about them are increasing, the accuracy and validity of such information is doubtful, and having the information deleted or hidden from the potential consumers seems nearly impossible. This is a workable approach as demonstrated by the proof-of-concept implementations carried out, which integrate with mainstream social networking sites and major web browsers. Besides, the solutions proposed could be seamlessly incorporated within the current business models of several industries—maybe the most obvious being the case of search engines.

As more and more information about data subjects is encountered, they will need to provide more replies, for which we are currently working on enhancing the usability, efficiency and scalability of the system. Additionally, we are developing pilots tailored to e-dating and e-recruitment service providers, which currently intermediate large amounts of information of their cohesive user base. All in all, we expect these developments to mitigate the negative impact that deceptive information may have on both the consumers and the data subjects, and ultimately incentivize the spread of honest sources.

Bibliography

Balog, Krisztian, Leif Azzopardi, and Maarten de Rijke. 2009. Resolving person names in web people search. In *Weaving Services and People on the World Wide Webeaving Services and People on the World Wide Web*, ed. Irwin King and Ricardo Baeza-Yates, 301–23. Berlin, Heidelberg: Springer Berlin Heidelberg, 2009. doi:10.1007/978-3-642-00570-1.

Berendsen, Richard, Bogomil Kovachev, Maarten Weerkamp Nastou, Evangelia-Paraskevi de Rijke, and Wouter. 2012. Result disambiguation in web people search. In *Advances in Information Retrieval*, ed. Ricardo Baeza-Yates, Arjen P. de Vries, Hugo Zaragoza, B. Barla Cambazoglu, Vanessa Murdock, Ronny Lempel, and Fabrizio Silvestri, 7224:146–57. Lecture Notes in Computer Science. Springer, Berlin, Heidelberg. doi:10.1007/978-3-642-28997-2.

Besmer, Andrew, and Heather Richter Lipford. 2010. Moving beyond untagging: Photo privacy in a tagged world. In *Proceedings of the 28th International Conference on Human Factors in Computing Systems—CHI '10*, 1563–1572. doi:10.1145/1753326.1753560.

Bornoe, Nis, and Louise Barkhuus. 2011. Privacy management in a connected world: Students' perception of facebook privacy settings. In *Workshop on Collaborative Privacy Practices in Social Media*, 19–23. Hangzhou, China. http://bornoe.org/papers/CSCW2011-Privacy-Workshop-Bornoe.pdf.

Brennecke, Roland, Thomas Mandl, and Christa Womser-Hacker. 2011. The development and application of an evaluation methodology for person search engines. In *EuroHCIR 2011 European Workshop on Human–Computer Interaction and Information Retrieval, CEUR Workshop Proceedings*, 42–45, 2011. http://ceur-ws.org/Vol-763/.

Castelluccia, Claude, and Mohamed Ali Kaafar. 2009. Owner-centric networking (OCN): Toward a data pollution-free internet. In *2009 Ninth Annual International Symposium on Applications and the Internet*, 169–72. IEEE. doi:10.1109/SAINT.2009.34.

Cornwall, Janelle L. 2011. It was the first strike of bloggers ever: An examination of article 10 of the European convention on human rights as Italian bloggers take a stand against the Alfano Decree. *Emory International Law Review* 25(1): 499–538. http://connection.ebscohost.com/c/articles/64295112/was-first-strike-bloggers-ever-examination-article-10-european-convention-human-rights-as-italian-bloggers-take-stand-against-alfano-decree.

Court of Justice of the European Union. 2013. Opinion of Advocate general Jääskinen delivered on 25 June 2013 Case C-131/12 Google Spain SL Google Inc. v Agencia Española de Protección de Datos (AEPD), Mario Costeja González.

Diehl, E., and T. Furon. 2014. © Watermark: Closing the analog hole. In *2003 IEEE International Conference on Consumer Electronics, 2003. ICCE.*, 52–53. IEEE. Accessed 24 Oct 2014. doi:10.1109/ICCE.2003.1218801.

Druschel, Peter, Michael Backes, and Rodica Tirtea. 2011. The right to be forgotten—between expectations and practice—ENISA. Heraklion, Greece. http://www.enisa.europa.eu/activities/identity-and-trust/library/deliverables/the-right-to-be-forgotten.

Fujikawa, Megumi. 2014. Google suffers new privacy setback in Japan. *The Wall Street Journal*. http://online.wsj.com/articles/google-suffers-new-privacy-setback-in-japan-1412933523.

Hansen, Marit, Peter Berlich, Jan Camenisch, Sebastian Clauß, Andreas Pfitzmann, and Michael Waidner. 2004. Privacy-enhancing identity management. *Information security technical report* 9(1): 35–44. doi:10.1016/S1363-4127(04)00014-7.

Hansen, Tony, and Phillip Hallam-Baker. 2009 Domainkeys identified mail (DKIM) service overview, no. RFC 5585: 1–24. https://tools.ietf.org/html/rfc5585.

Humphreys, Lee, Phillipa Gill, and Balachander Krishnamurthy. 2010. How Much Is Too Much? Privacy Issues on Twitter. In *ICA 2010*, 1:1–29, 2010. http://www2.research.att.com/~bala/papers/ica10.pdf.

Ilešič, M. 2014. JUDGMENT OF THE COURT (Grand Chamber) In Case C–131/12, REQUEST for a preliminary ruling under Article 267 TFEU from the Audiencia Nacional (Spain), made by decision of 27 Feb 2012, received at the Court on 9 Mar 2012, in the proceedings Google Spa, (2014).

Jonas, Jeff. 2006. Threat and fraud intelligence, Las Vegas Style. *IEEE Security & Privacy Magazine* 4. doi:10.1109/MSP.2006.169.

Jones, Paul, Joseph Smarr, Gonzalo Salgueiro, and Michael Jones. 2013. WebFinger. *Internet RFC Series*, no. RFC 7033: 1–28. https://tools.ietf.org/html/rfc7033.

Kelbert, Florian, and Alexander Pretschner. 2012. Towards a policy enforcement infrastructure for distributed usage control. In *Proceedings of the 17th ACM Symposium on Access Control Models and Technologies - SACMAT '12*, 119. New York, USA: ACM Press, 2012. doi:10.1145/2295136.2295159.

Kenny, Steve, and Larry Korba. 2002. Applying digital rights management systems to privacy rights management. *Computers & Security* 21(7): 648–664. doi:10.1016/S0167-4048(02)01117-3.

Köpcke, Hanna, and Erhard Rahm. 2010. Frameworks for entity matching: A comparison. *Data & Knowledge Engineering* 69(2): 197–210. doi:10.1016/j.datak.2009.10.003.

Laursen, Lucas. 2015. Google's year of forgetting [News]. *IEEE Spectrum* 52(5): 16–17. doi:10.1109/MSPEC.2015.7095187.

Li, Jiexun, G. Alan Wang, and Hsinchun Chen. 2010. Identity Matching using personal and social identity features. *Information Systems Frontiers* 13(1): 101–13. doi:10.1007/s10796-010-9270-0.

Ligouras, Spyros. 2012. Protecting the social graph: Client–side mitigation of cross–site content forgery attacks. University of Crete. http://elocus.lib.uoc.gr/dlib/9/4/9/metadata-dlib-1354176330-646953-4363.tkl.

Long, C, and L Shi. 2010. Web person name disambiguation by relevance weighting of extended feature sets. *Third Web People Search Evaluation Forum WePS3 CLEF* 2010 (2010): 1–13. http://ceur-ws.org/Vol-1176/CLEF2010wn-WePS-LongEt2010.pdf.

Mantelero, Alessandro. 2013. The EU proposal for a general data protection regulation and the roots of the 'right to be forgotten'. *Computer Law & Security Review* 29(3): 229–235. doi:10.1016/j.clsr.2013.03.010.

Markou, Christiana. 2015. The 'Right to Be Forgotten': Ten reasons why it should be forgotten. In *Reforming European Data Protection Law SE—8*, ed. Serge Gutwirth, Ronald Leenes, and Paul de Hert, 20: 203–26. Law, Governance and Technology Series. Springer, Netherlands, 2015. doi:10.1007/978-94-017-9385-8_8.

Mayer-Schönberger, Viktor. 2011. *Delete: The virtue of forgetting in the digital age*. Princeton: Princeton University Press.

O'Connor, Peter. 2010. Managing a Hotel's image on TripAdvisor. *Journal of Hospitality Marketing & Management* 19(7): 754–72. doi:10.1080/19368623.2010.508007.

Office of the Privacy Commissioner of Canada. 2011. Protect your personal information because the Internet never forgets, Privacy Commissioner of Canada says-27 Jan 2011. Ottawa, 2011. https://www.priv.gc.ca/media/nr-c/2011/nr-c_110127_e.asp.

Ortlieb, Martin. 2014. The anthropologist's view on privacy. *IEEE Security and Privacy* 12(3): 85–87. doi:10.1109/MSP.2014.57.

Pearson, Siani, and Marco Casassa-Mont. 2011. Sticky policies: An approach for managing privacy across multiple parties. *Computer* 44(9): 60–68. doi:10.1109/MC.2011.225.

Pesce, João Paulo, Diego Las Casas, Gustavo Rauber, and Virgílio Almeida. 2012. Privacy attacks in social media using photo tagging networks. In *Proceedings of the 1st Workshop on Privacy and Security in Online Social Media—PSOSM '12*, 1–8. New York, USA: ACM Press. doi:10.1145/2185354.2185358.

Quinn, Sarah (On Device Research). 2013. Facebook Costing 16-34s Jobs in Tough Economic Climate, 2013. http://ondeviceresearch.com/blog/facebook-costing-16-34s-jobs-in-tough-economic-climate.

Sarrouh, Nadim, Florian Eilers, Uwe Nestmann, and Ina Schieferdecker. 2011. Defamation-free networks through user-centered data control. In *Security and Trust Management SE—12*, ed. Jorge Cuellar, Javier Lopez, Gilles Barthe, Alexander Pretschner, 6710: 179–93. Lecture Notes in Computer Science. Berlin, Heidelberg: Springer, 2011. doi:10.1007/978-3-642-22444-7_12.

Schorr, Joanna. 2014. Malicious content on the internet: narrowing immunity under the communications decency act. *St. John's Law Review* 87(2). http://scholarship.law.stjohns.edu/lawreview/vol87/iss2/15.

Shein, Esther. 2013. Ephemeral data. *Communications of the ACM* 56(9): 20. doi:10.1145/2500468.2500474.

Slatkin, Brett. 2013. Bootstrapping WebFinger decentralized discovery with WebFist. *One Big Fluke*, 2013. http://www.onebigfluke.com/2013/06/bootstrapping-webfinger-with-webfist.html.

Squicciarini, Anna C., Mohamed Shehab, and Joshua Wede. 2010. Privacy policies for shared content in social network sites. *VLDB Journal* 19: 777–796. doi:10.1007/s00778-010-0193-7.

Squicciarini, Anna Cinzia, Mohamed Shehab, and Federica Paci. 2009 Collective privacy management in social networks. In *Proceedings of the 18th International Conference on World Wide Web—WWW '09*, 521. New York, USA: ACM Press, 2009. doi:10.1145/1526709.1526780.

Stroud, Scott R. 2014. The dark side of the online self: A pragmatist critique of the growing plague of revenge porn. *Journal of Mass Media Ethics* 29(3): 168–83. doi:10.1080/08900523.2014.917976.

Talukder, Nilothpal, and Sheikh Iqbal Ahamed. 2010. Preventing multi-query attack in location-based services. In *Proceedings of the Third ACM Conference on Wireless Network Security - WiSec '10*, 25. New York, USA: ACM Press, 2010. doi:10.1145/1741866.1741873.

Thompson, David, and Michael Fertik. 2010. *Wild west 2.0: How to protect and restore your online reputation on the untamed social frontier*. amacom, 2010.

Tiffen, Rodney. 2012. Finkelstein report: Volume of media vitriol in inverse proportion to amount of evidence. *Pacific Journalism Review* 18(2): 37–40. http://www.pjreview.info/sites/default/files/articles/pdfs/pjr18(2)_finkelsteinreport_tiffen_pp37-40.pdf.

Tokunaga, Robert S. 2011 Social networking site or social surveillance site? Understanding the use of interpersonal electronic surveillance in romantic relationships. *Computers in Human Behavior* 27(2): 705–13. doi:10.1016/j.chb.2010.08.014.

Weinberg, Tamar. 2009. An introduction to social media marketing. In *The New Community Rules: Marketing on the Social Web*, 1st ed., 1–18. O'Reilly Media, Inc.

Werbin, Kenneth C. 2012. Auto-biography: On the immanent commodification of personal information. *International Review of Information Ethics* 17: 46–53. http://www.i-r-i-e.net/inhalt/017/werbin.pdf.

Werkers, Evi, Eva Lievens, and Peggy Valcke. 2007. A critical analysis of the right of reply in online media. In *Third International Conference on Automated Production of Cross Media Content for Multi-Channel Distribution (AXMEDIS'07)*, 181–88. IEEE, 2007. doi:10.1109/AXMEDIS.2007.11.

Wisniewski, Pamela, Heather Lipford, and David Wilson. 2012. Fighting for My Space. In *Proceedings of the 2012 ACM Annual Conference on Human Factors in Computing Systems—CHI '12*, 609. New York, New York, USA: ACM Press, 2012. doi:10.1145/2207676.2207761.

Do-It-Yourself Data Protection—Empowerment or Burden?

Tobias Matzner, Philipp K. Masur, Carsten Ochs and Thilo von Pape

Abstract Data protection by individual citizens, here labeled do-it-yourself (DIY) data protection, is often considered as an important part of comprehensive data protection. Particularly in the wake of diagnosing the so called "privacy paradox", fostering DIY privacy protection and providing the respective tools is seen both as important policy aim and as a developing market. Individuals are meant to be empowered in a world where an increasing amount of actors is interested in their data. We analyze the preconditions of this view empirically and normatively: Thus, we ask (1) Can individuals protect data efficiently; and (2) Should individuals be responsible for data protection. We argue that both for pragmatic and normative reasons, a wider social perspective on data protection is required. The paper is concluded by providing a short outlook how these results could be taken up in data protection practices.

Keywords Do-it-yourself data protection · Data protection · Responsibilization · Data protection advocates · Data protection practices · Representative study · Privacy paradigm

---

T. Matzner (✉)
Universität Tübingen, Internationales Zentrum für Ethik in den Wissenschaften,
Wilhelmstr. 19, 72074 Tübingen, Germany
e-mail: tobias.matzner@uni-tuebingen.de

P.K. Masur · T. von Pape
Universität Hohenheim Lehrstuhl Für Medienpsychologie (540 F), 70599 Stuttgart (Hohenheim), Germany
e-mail: philipp.masur@uni-hohenheim.de

T. von Pape
e-mail: thilo.vonpape@uni-hohenheim.de

C. Ochs
Universität Kassel Fachbereich 05 Soziologische Theorie,
Nora-Platiel-Str. 5, 34109 Kassel, Germany
e-mail: carsten.ochs@uni-kassel.de

© Springer Science+Business Media Dordrecht 2016
S. Gutwirth et al. (eds.), *Data Protection on the Move*,
Law, Governance and Technology Series 24, DOI 10.1007/978-94-017-7376-8_11

1 Introduction

In current debates, do-it-yourself (DIY) data protection is often conceived as an important element of comprehensive data protection. In particular after the revelations of Edward Snowden and the ensuing distrust in states or legal frameworks, prominent individuals (among them Snowden himself) and NGOs have advocated DIY data protection as the main and most immediate way to protect citizens' data. Here, the term DIY data protection[1] is intended to encompass all measures taken by *individual persons* to protect their data. This includes the use of cryptography and anonymization tools, browser plugins that manage cookies or block tracking and other tools used to minimize data collection. We also include tools which are meant to increase the transparency of data processing, e.g. plugins like Lightbeam which visualize tracking. Apart from tools, data minimization strategies are considered as DIY data protection practices. These include using fake data and profiles, a very conscious and selective provision of data, and not using particular services and technologies at all. There are also some legal actions like requesting the deletion of personal data that can be taken by individuals. These approaches are based on the premise that increasing knowledge about data collection practices and the possible insights that can be derived from data leads to better individual judgments and decisions. Thus fostering knowledge and awareness concerning data is seen as one important contribution to DIY data protection.

In this chapter, we want to take a step back from this premise and question the overall concept of DIY data protection from an empirical and normative perspective: to what extent *can* and *should* our response to data protection problems center on the individual user?

Before responding to these questions, we want to put them into perspective by reconsidering a long lasting debate about another information communication technology (ICT)-related concern: the "digital divide". This discussion still suffers from what Rogers called an "individual-blame-bias"[2]: instead of blaming structural causes for inequalities related to ICT use, the non-adoption of relevant information technology is often attributed to deficits of those "laggards" and "information have-nots"[3] who are on the "wrong" side of the divide because they

---

[1]The term "DIY data protection" was conceived as translation of the German "Selbstdatenschutz", which literally translates as self-data-protection. Thus, the usual connotations of DIY as improvised or alternative to a commercial product are not necessarily intended; the connotations of independence and self-reliance, however, are. The results presented in this article build on a German whitepaper concerning "Selbstdatenschutz" issued by the research project "Privacy Forum" which can be found here: https://www.forum-privatheit.de/forum-privatheit-de/texte/veroeffentlichungen-des-forums/themenpapiere-white-paper/Forum_Privatheit_White_Paper_Selbstdatenschutz_2.Auflage.pdf (accessed 06.03.2015).

[2]Everett M. Rogers, *Diffusion on Innovations* (New York: Free Press, 2005), 118.

[3]National Telecommunications and Information Administration (NTIA), *Falling Through the Net: A Survey of the 'Have-nots' in Rural and Urban America* (Washington, DC: US Department of Commerce), assessed March 10, 2015. http://www.ntia.doc.gov/ntiahome/fallingthru.html.

lack knowledge, social status, or resources.[4] In a similar vein, the deficits in data protection and a lack of implementation today, are often explained through the users' rational and behavioral deficits. They are characterized as a paradox between the users' concerns and attitudes favoring restrictive use of data on the one hand and a very permissive actual use on the other.[5] Failing to see the larger, structural reasons behind individual lacks in privacy protection, this perspective also does not attribute responsibility to the government as the actor who might be able to address structural problems. Finally, even some of those advocates who do blame the government and Internet Service Providers eventually put pressure on the users to take the protection of their privacy into their own hands. For cyber-libertarians such as John Berry Barlow, it would be paradoxical to confide the protection of privacy to the government in principle because this task would go against any government's interest of controlling its citizens.[6]

In consequence, the most discussed explanations focus on deficits of the users or human nature in general: the users are labeled as lacking literacy with respect to privacy,[7] as corruptible by questionable gratifications such as negligible financial rewards or convenience,[8] and as hypocrite to the extent that their apparent concern for privacy may be explained through a social desirability response bias.[9] In short, the problems we perceive with data protection are often presented as simple "user errors". Correcting these errors by fostering DIY data protection is then considered as empowering users. However, as we will argue below, more and more problems with data protection remain, even when users behave through rational and educated decisions. There seems to remain a problem, which should rather be described as a privacy dilemma[10] than a paradox.

This leads us back to the questions we want to answer in the next sections: (1) Can we, the users, actually protect our data? How probable is the emergence of

---

[4]For profound critiques of the term "digital divide" and its applications in public discourse, see Neil Selwyn, "Reconsidering political and popular understandings of the digital divide," *New Media & Society* 6 (2004): 341–362, and David J. Gunkel, "Second Thoughts: Towards a Critique of the Digital Divide," *New Media & Society* 5 (2003): 499–522.

[5]Susan B. Barnes, "A privacy paradox: Social networking in the United States," *First Monday* 11 (2006), accessed March 4, 2015, doi:10.5210/fm.v11i9.1394.

[6]Lincoln Dahlberg, "Cyber-Libertarianism 2.0: a discourse theory/ critical political economy examination. Cultural Politics 6, no. 3 (2010), doi: 10.2752/175174310X12750685679753: 331–356.

[7]Yong J. Park, "Digital Literacy and Privacy Behavior Online," *Communication Research* 40, no. 2 (2013).

[8]Alessandro Acquisti, Leslie K. John and George Loewenstein. "What is privacy worth?," *The Journal of Legal Studies* 42 (2013): 249–274.

[9]e.g., Miriam J. Metzger, "Communication Privacy Management in Electronic Commerce," *Journal of Computer-Mediated Communication* 12 (2007): 351.

[10]Petter Bae Brandtzæg, Marika Lüders, and Jan Håvard Skjetne, "Too many Facebook 'friends'? Content sharing and sociability versus the need for privacy in social network sites," *Intl. Journal of Human–Computer Interaction* 26 (2010): 1006–1030.

DIY DP practices as a mass phenomenon? Can users enable themselves—or be enabled—up to a point where they can take the best decisions in their own interest and can this solve the problem of data protection or only reveal the true dilemmas lying beyond the users' field of action? And—notwithstanding these empirical questions—(2) should we, the users, have to protect data ourselves? Is it normatively desirable to choose the individual user as the main responsible actor to improve the state of data protection?

2 DIY-Data Protection—Can We Do It?

In this section we will deal with the question of how probable the emergence of DIY data protection practices as a mass phenomenon may be in empirical terms. To do so, we will cover three aspects of DIY data protection practices: the question to what extent it is possible for individuals to cultivate such practices (Sect. 2.1); the competing needs and aims which must be taken into account as the context of these practices (Sect. 2.2), and finally the question of DIY data protection practices, as collective activity, being entangled in specific socio-political constellations (Sect. 2.3).

2.1 The Individual Faced with the (Im)possibility of DIY Data Protection

Protecting personal data in online environments is a difficult task for individual users. The exponential growth of "smart" technologies, which quickly move into cultural mainstream, has led to a socio-technological environment in which manifold forms of tracking, data mining, and profiling have emerged.[11] As these data collection practices become more complex and elusive, potential negative consequences of information and communication technology usage are not readily perceivable. Awareness of data collection practices however is a crucial precondition for users to implement DIY data protection practices.[12] Negative outcomes of these practices are mostly not visible or sensible in the daily use of ICT. Grasping the complexities and flows of personal information in the web consequently becomes a rather difficult task, even for interested users or experts.

---

[11]Georgia Skouma and Laura Léonard, "On-Line Behavioral Tracking: What May Change After the Legal Reform on Personal Data Protection," in *Reforming European Data Protection Law*, ed. Serge Gutwirth, Ronald Leenes, and Paul de Hert (Dordrecht: Springer, 2015), 35–62.

[12]George R. Milne and Andrew J. Rohm, "Consumer Privacy and Name Removal across Direct Marketing Channels: Exploring Opt-in and Opt-out Alternatives," *Journal of Public Policy & Marketing* 19, no. 2 (2000): 238–49.

Moreover, in online environments, human communication traverses spheres that are private, public, and social.[13] Previously separated media platforms converge and formerly distinct barriers are blurred. Consequently, data disclosed to one provider might be used by another and resold to third parties. Information communicated to one or several users might be reused, shared or misused by others. Potential threats to informational privacy thus arise from different contexts and dimensions. Furthermore, violations may in particular occur because boundaries of formerly distinct contexts and dimensions become increasingly blurred. Consequently, DIY data protection becomes an even bigger challenge as there is not one globally applicable data protection strategy. In fact, to ensure comprehensive protection against most potential privacy threats, a number of diverse and differently demanding strategies have to be implemented. For most cases, the implementation of a certain practice might require another one, which in turn necessitates another one and so on. For example, if a user wishes to be unrecognizable for online service providers, it is not sufficient to merely opt out from these services. The user furthermore needs to use anonymization tools every time he or she uses the internet and install plugins which hinder online service providers from tracking their surfing activities. Again, the understanding and evaluation of these practices both from a structural and technological perspective demands high competence from individual users.

To categorize DIY data protection practices, it seems fruitful to differentiate measures taken by the individual on a number of different levels. A first distinction refers to the question against whom or what a specific data protection strategy is directed. When sharing data in online environments, several actors with different interests and resources are involved in processing and using the data. On one hand, internet users want to protect their personal data against misuse by other users, but on the other hand, they also want to protect themselves against data collection by companies and institutions. Raynes-Goldie[14] defines the former as *social privacy* and the latter as *institutional privacy*. The protection of social privacy is at least partly realizable by using privacy settings (e.g., restricting visibility, separating audiences, managing disclosures). However, studies have also shown that even social privacy requires different approaches. De Wolf and colleagues for example found that it is not sufficient to imply individual privacy management practices, but also group privacy management practices.[15] To gain an optimal level of social privacy thus involves also the negotiation of common privacy rules. Controlling institutional privacy requires even more sophisticated measures and more general

---

[13]Zizi A. Papcharissi, *A Private Sphere: Democracy in a Digital Age* (Cambridge: Polity Press, 2010).

[14]Katie Raynes-Goldie, "Aliases, creeping, and wall cleaning: Understanding privacy in the age of Facebook," *First Monday* 15, no. 1 (2010).

[15]Ralf De Wolf, Koen Willaert, and Jo Pierson, "Managing privacy boundaries together: Exploring individual and group privacy management strategies on Facebook," *Computers in Human Behavior* 35 (2014).

approaches such as general data parsimony,[16,17] anonymization, pseudonymization and encryption.

Another differentiation refers to passive and active DIY data protection practices. Passive strategies include all strategies relying on withdrawal (opting-out) or data parsimony. As such, they involve the general decision to share or not to share personal information which might be reflected with regard to individual privacy preferences and situational needs (cf. Sect. 2.2). These strategies includes applying general rules of thumb in decisions on sharing, but also the constant monitoring and regulation of disclosures. Active strategies, on the other hand, encompass the use of privacy-enhancing-technologies and taking legal actions. As such, they serve to build a protected sphere, in which users can perform their selves without worrying about potential privacy threats.

DIY data protection practices can further be differentiated into preventive and corrective measures.[18] Whereas most strategies mentioned above can be referred to as preventive measures, there are also a number of actions that users take after a privacy violation has occurred. Among others, these include passive measures such as deleting previously shared content, unlinking or untagging[19] as well as active measures such as taking legal actions (e.g., asking online service providers not to share personal data with other companies or to delete all information about oneself).

Recent studies in the fields of media psychology and communication sciences have examined a number of different DIY data protection practices in the context of social web use and in particular on social network sites. The findings from these studies suggest that users do engage in DIY data protection to prevent attacks on their *social privacy*. These attacks may include inappropriate friend requests,[20] unwanted forwarding or sharing of personal information by other users, discrimination or exposition of sensitive information in public realms. Based on these studies, it can be said that a considerable number of users implement preventive strategies such as faking user names,[21,22,23] using privacy settings to separate

---

[16]Airi Lampinen et al., "We're in It Together: Interpersonal Management of Disclosure in Social Network Sercives," in *Proceedings of the SIGCHI Conference on Human Factors in Computing Systems* (New York, USA: ACM, 2011), 3217–3226.

[17]Philipp K. Masur and Michael Scharkow, "Disclosure Management on Social Network Sites: Individual Privacy Perceptions and User-Directed Privacy Strategies", (in prep).

[18]Lampinen et al., "We're in It Together: Interpersonal Management of Disclosure in Social Network Services." .

[19]Ibid.

[20]Raynes-Goldie, "Aliases, creeping, and wall cleaning: Understanding privacy in the age of Facebook.".

[21]Zeynep Tufekci, "Can You See Me Now? Audience and Disclosure Regulation in Online Social Network Sites," *Bulletin of Science, Technology & Society* 28, no. 1 (2008): 20–36.

[22]Tobias Dienlin and Sabine Trepte, "Is the privacy paradox a relic of the past? An in-depth analysis of privacy attitudes and privacy behaviors," *European Journal of Social Psychology* (2014).

[23]Bernard Debatin et al., "Facebook and Online Privacy: Attitudes, Behaviors, and Unintended Consequences," *Journal of Computer-Mediated Communication* 15, no. 1 (2009): 83–108.

audiences,[24,25] befriending only trusted people[26] and generally restricting the visibility of profile information.[27,28] Furthermore, users also regulate and constantly monitor their disclosing behavior. A study by Masur and Scharkow[29] found that most user actively manage their disclosure by generally sharing less information that they individually perceive as private. Users generally show this type of behavioral pattern, although it is more pronounced in one-to-many communication situations than in one-to-one communications. Results from different studies furthermore revealed that users seem to be more willing to implement specific privacy protection strategies after negative experiences with social privacy violations.[30,31]

Whereas many studies suggest that users seem to safeguard their social privacy at least partially, only a few studies have examined DIY data protection practices in the context of *institutional privacy*. Current societal debates often proclaim that users do not engage in data protection and consequently demand more literacy. In a recent study, Trepte, Masur and Teutsch examined the implementation of DIY data protection practices in the context of institutional privacy.[32] The analysis is based on an online-survey with a representative sample of German internet users ($N = 1932$). The findings revealed that internet users generally do implement some strategies. However, some practices are more widespread than others (see Table 1). In general, a third of the participants engage in passive data protection strategies such as refraining from registering for certain online services (75 %) or stopping to use certain websites (65 %) due to privacy concerns. Also 63 % reported that they have refrained from registering for certain online services after

---

[24]Eden Litt, "Understanding social network site users' privacy tool use," *Computers in Human Behavior* 29, no. 4 (2013): 1649–1656.

[25]Jessica Vitak, "The Impact of Context Collapse and Privacy on Social Network Site Disclosures," *Journal of Broadcasting & Electronic Media* 56, no. 4 (2012): 451–470.

[26]Debatin et al., "Facebook and Online Privacy: Attitudes, Behaviors, and Unintended Consequences.".

[27]Dienlin and Trepte, "Is the privacy paradox a relic of the past? An in-depth analysis of privacy attitudes and privacy behaviors.".

[28]Debatin et al., "Facebook and Online Privacy: Attitudes, Behaviors, and Unintended Consequences.".

[29]Masur and Scharkow, "Disclosure Management on Social Network Sites: Individual Privacy Perceptions and User-Directed Privacy Strategies".

[30]Sabine Trepte, Tobias Dienlin, and Leonard Reinecke, "Risky Behaviors: How Online Experiences Influence Privacy Behaviors," in *Von Der Gutenberg-Galaxis Zur Google-Galaxis. From the Gutenberg Galaxy to the Google Galaxy. Surveying Old and New Frontiers after 50 Years of DGPuK*, ed. Birgit Stark, Oliver Quiring, and Nikolaus Jackob (Wiesbaden: UVK, 2014), 225–246.

[31]Debatin et al., "Facebook and Online Privacy: Attitudes, Behaviors, and Unintended Consequences.".

[32]Philipp K. Masur, Doris Teutsch, and Sabine Trepte, "*Entwicklung der Online-Privatheitskompetenz-Skala*" (in prep).

Table 1 Do-it-yourself data protection practices in the German population (in %)

	Overall	Men	Women	14–19 years	20–29 years	30–39 years	40–49 years	50–59 years	60–69 years
Passive data protection strategies									
Refrained from registering for online services (not wanting to provide personal information)	75	76	74	78	78	77	73	73	70
Stopped using certain websites (because of privacy concerns)	65	68	63	67	67	65	65	66	62
Refrained from registering for an online service (because of its data usage policy)	63	64	61	50	58	65	63	66	67
Refrained from buying certain products online (because of privacy concerns)	57	57	56	55	60	65	57	52	48
Active data protection strategies									
Updates anti-virus-software on a regular basis	95	97	92	89	96	95	94	95	97
Uses anti-malware-software to detect potential threats	85	90	79	85	83	86	86	86	81
Deletes cookies and cache regularly	84	88	79	76	82	89	85	84	80
Deletes browser history regularly	84	88	81	86	83	87	85	84	82
Used pseudonyms when registering for online services	53	58	48	70	72	68	46	40	29

(continued)

Table 1 (continued)

	Overall	Men	Women	14–19 years	20–29 years	30–39 years	40–49 years	50–59 years	60–69 years
Used unidentifiable e-mail address to register for online services	51	57	45	67	64	63	46	39	32
Used anonymization tools to obfuscate identity	35	43	27	40	50	45	33	25	18
Used encryption for e-mail communication	32	37	25	32	36	40	34	24	22
Used anti-tracking-software	32	39	24	35	40	41	31	24	21
Legal data protection strategies									
Asked online service providers not to share personal information with other companies	40	43	36	40	52	48	39	32	24
Asked online service providers to delete personal data	36	40	32	30	47	46	36	31	22

Basis: $N = 1932$

they have read its data usage policy. The implementation of these rather facile strategies does not vary between men and women or young and older people.

With regard to active data protection strategies, the data present a rather mixed picture: whereas simple practices (from a technical point of view) such as updating and using anti-malware-software or deleting browser information are implemented by most users, pseudonymization or anonymization strategies are only used by a few users. Only half of the sample has used a pseudonym when registering for online services (53 %) or has created unidentifiable e-mail-addresses (51 %). In contrast, rather difficult and technically demanding strategies such as using anonymization tools (e.g., TOR, JonDonym) or encryption tools (e.g., PGP) are only implemented by less than a third of the sample. With regard to these practices, differences within the population are visible: male and younger participants were more likely to apply these tools than female and older participants. Corrective measures which require a lot of engagement and expenditure of time are also less prominent within the sample: only less than 40 % have already taken legal steps to safeguard their personal data. For example, only 36 % have asked online service providers to delete their personal information.

These results show that users seem to be willing to adopt simple and easily applicable strategies, but do not use complicated tools which require advanced technical skills or consume a lot of time. This is in particular problematic for the protection of institutional privacy because once users are not able to implement certain active DIY data protection practices such as using anonymization tools or encryption, the only remaining solution for effective data protection on a personal level is opting-out or using passive data protection strategies for that matter. Based on this rationale, it seems logical to assume that promoting online privacy literacy might be a good idea. Online privacy literacy has been said to "support, encourage, and empower users to undertake informed control of their digital identities".[33] Promoting privacy literacy might hence serve as a stopgap between inconsistent privacy attitudes and behaviors.[34] First studies in this field support this assumption: For example, many users feel unable to implement these specific privacy protection tools. For instance, only 35 % of German internet users feel capable of encrypting their e-mail communication.[35] As mentioned above, awareness of data collection and data mining practices presents a precondition for the implementation of DIY data protection practices, yet many users are not aware or at least do not have insights into these practices. A representative study with

---

[33] Park, "Digital Literacy and Privacy Behavior Online," 217.

[34] Sabine Trepte et al., "Do People Know About Privacy and Data Protection Strategies? Towards the 'Online Privacy Literacy Scale' (OPLIS)," in *Reforming European Data Protection Law*, ed. Serge Gutwirth, Ronald Leenes, and Paul de Hert (Dordrecht: Springer, 2015), 333–366.

[35] Deutsches Institut für Vertrauen und Sicherheit im Internet (DIVSI), "DIVSI Studie zur Freiheit versus Regulierung im Internet," (Hamburg, 2013), accessed March 10, 2015. https://www.divsi.de/wp-content/uploads/2013/12/divsi-studie-freiheit-v-regulierung-2013.pdf.

German internet users for example found that 33 % of the participants did not know that website providers combine information from different websites to create user profiles.[36]

Summing up, it can be said that a third of the German population does implement DIY privacy protection strategies, however, the data also show that effective and comprehensive data protection with regard to institutional privacy—which requires to implement also more sophisticated measures—seems to be very difficult to achieve for the most individual users. Being aware of and understanding the technical architecture behind online information flows becomes harder and more complex with the rapid growth of new technologies. Furthermore, data protection itself becomes more and more complex. Although many and singular strategies and tools—which can only help to protect certain aspects of online privacy—exist, a universal remedy in form of a single strategy is not available. Keeping up with new technologies, tools, and strategies, requires time, competence and resources. As such, data protection is at risk of becoming limited to those who can spare the effort to learn handling data protection technologies. Differences in privacy literacy may hence foster a divide between those who are able to ensure data protection and those who are not.

That being said, it is noteworthy to add that absolute data protection (with opting-out as final solution) is mostly not desirable for most users. In many contexts, the sharing of information is appropriate. Depending on contextual factors such as norms, actors, attributes and corresponding transmission principles, user might not feel that their privacy is violated and their contextual integrity is hence preserved.[37] Scholars have found that the use of the social web and other online services satisfies many other needs that have to be taken into account when assessing users' behavior regarding privacy. The following paragraph will hence discuss to what extent the need for privacy in online environments competes with other forms of need satisfaction.

2.2 Competing Needs: Privacy and Data Protection Versus Social Gratifications

To dissolve the seeming paradox between users' privacy concerns and their actual online behavior, many researchers have argued that people refrain from implementing data protection strategies because they benefit from advantages and

---

[36]Trepte et al., "How Skilled Are Internet Users When it Comes to Online Privacy and Data Protection? Development and Validation of the Online Privacy Literacy Scale (OPLIS).".

[37]Helen Nissenbaum, *Privacy in Context: Technology, Policy, and the Integrity of Social Life* (Stanford: Stanford Law Books, 2010).

gratifications that online services have to offer.[38,39] Buying products via online-shops, booking trips via online services, or simply using online-banking is fast, easy, and convenient. Specifically through the use of social web platforms, users are able to obtain a number of gratifications. Self-disclosure can be defined as "the process of making the self known to others"[40] and as such is a basic requirement for social interactions and communications. Disclosing private information to other people fosters social proximity.[41] Sharing personal information in the social web can hence lower barriers of initial interaction, leads to social acceptance and relationship-building, and provides users with feedback regarding their own identity formation.[42] Accordingly, it has been argued that users weigh the risks and benefits of online self-disclosure.[43] It seems plausible that users voluntarily take the risks involved with self-disclosure in order to obtain desired gratifications. However, although these needs might indeed compete with each other at certain times, this balancing is not a zero-sum game: People are in particular open and willing to share personal information if they perceive a situation as private.[44] Creating a safe and secure platform, on which one is able to disclose personal information without fearing privacy violations might hence be more desirable for users than complete withdrawal from social interaction in online realms. The balancing of costs and benefits of use however rests on the assumptions that user always have the choice between using or not using a service. Yet, in many cases, individuals have to engage with certain services or are dependent on them to achieve certain goals (e.g., finding a job, getting or staying on contact with certain people…). In these cases, an individual cost-benefit analysis might be very limited.

Individual aims and concerns that structure the importance and motivation to engage in data protection practices have their equivalent among the advocates of DIY data protection; and yet, the latter's arguments are very much entangled in social and political contexts. Consequently, such contexts are of central relevance

---

[38]Monika Taddicken and Cornelia Jers, "The Uses of Privacy Online: Trading a Loss of Privacy for Social Web Gratifications," in *Privacy Online. Perspectives on Privacy and Self-Disclosure in the Social Web*, ed. Sabine Trepte and Leonard Reinecke (Berlin: Springer, 2011), 143–156.

[39]Trepte et al., "Do People Know About Privacy and Data Protection Strategies? Towards the 'Online Privacy Literacy Scale' (OPLIS)," 338.

[40]Sidney M. Jourard and Paul Lasakow, "Some Factors in Self-Disclosure," *Journal of Abnormal Psychology* 56, no. 1 (1958).

[41]Irwin Altman and Dalmas Taylor, *Social penetration: The development of interpersonal relationships* (New York: Holt, Rinehart and Winston: 1976).

[42]Nicole B. Ellison et al., "Negotiating Privacy Concerns and Social Capital Needs in a Social Media Environment," in *Privacy Online. Perspectives on Privacy and Self-Disclosure in the Social Web*, ed. Sabine Trepte and Leonard Reinecke (Berlin: Springer, 2011), 19–32.

[43]Trepte et al., "Do People Know About Privacy and Data Protection Strategies? Towards the 'Online Privacy Literacy Scale' (OPLIS)," 338.

[44]Sabine Trepte and Leonard Reinecke, "The Social Web as a Shelter for Privacy and Authentic Living," in *Privacy Online. Perspectives on Privacy and Self-Disclosure in the Social Web*, ed. Sabine Trepte and Leonard Reinecke (Berlin: Springer, 2011), 143–156.

when it comes to analyzing the rather moderate success of promoting DIY data protection to date. Next, we will support this claim by presenting a cursory analysis of the German DIY data protection discourse as it is reproduced by some of the most influential participants.

2.3 DIY Data Protection Advocates and Their Socio-Political Entanglements

One way of conceiving DIY data protection—and a rather fruitful one, for that matter—is to view them as a specific form of *sociocultural* practice.[45] Practice, in this context, means that performing DIY data protection is a routinized everyday activity which does not consume much of the social actors' conscious awareness "but goes without saying"; the implicit nature of the knowledge that is involved points to the tacit character of such practical skills.[46] DIY data protection practices, that is, occur as embodied skills collectively developed, performed and maintained by "social worlds."[47]

The collective nature of the practices in question became visible already in the "early days" of DIY data protection. The so-called Cypherpunks, cryptography experts holding libertarian, and thus strong individualistic worldviews, belong to the most profound, and also most enthusiastic proponents of DIY data protection. In "A Cypherpunk's Manifesto", for example, Eric Hughes, one of the most prominent DIY data protection advocates, raised hopes in the early 1990s that "Cryptography will ineluctably spread over the whole globe, and with it the anonymous transaction systems that it makes possible."[48] Whereas such transaction systems, Hughes believed, are a necessary pre-condition for privacy to prevail, privacy itself would be a necessary pre-condition for an "open society".

Whatever one might think of such ideas, there is no doubt that cryptography did *not* spread around the globe; in other words, harnessing cryptography for DIY data protection was not translated into a mass phenomenon,[49] as Hughes stated in

---

[45]Paul Dourish and Ken Anderson, "Collective Information Practice: Exploring Privacy and Security as Social and Cultural Phenomena," *HUMAN-COMPUTER INTERACTION* 21 (2006): 319–342.

[46]In the sense of: Anthony Giddens, *The Constitution of Society: Outline of the Theory of Structuration* (Cambridge: Polity Press, 1984).

[47]Anselm Strauss, "A Social World Perspective," *Symbolic Interaction* 1 (1978): 119–128.

[48]Eric Hughes, "A Cypherpunk's Manifesto", accessed February 23, 2015. http://activism.net/cypherpunk/manifesto.html.

[49]More precisely speaking, cryptography did not spread globally as an everyday practice of average users for the sake of individual privacy protection, though it was, and is, in fact, harnessed by large corporations (business, public authorities) on a global scale to serve IT security ends.

2012—most people do not encrypt, say, emails, as a matter of course[50] (see also Sect. 2.1). Having said this, there is an obvious, yet overlooked reason for the absence of DIY data protection practices: the emergence of such practices presupposes the creation of a social (norms, codes of conduct), cultural (knowledge, skills, frames of meaning), legal (legal rules and regulations), technological (suitable soft- and hardware)—and so on—infrastructure that is to be generated by some collective body. In other words, creating DIY data protection practices is a collective endeavor, no matter how deeply engrained the individualism of DIY data protection proponents may be.

As a result, the creation of DIY data protection practices inevitably is a social process, which is why the dispute concerning cryptography that occurred in the early 1990s did not come as a surprise at all. In this sense, then, the emergence of DIY data protection practices in contemporary societies, whether based on cryptography or else, is likewise contested. For example, even a cursory look at the German discourse on DIY data protection demonstrates that the practices in question, if anything, may emerge in an environment that is a rather hostile one, due to the specific constellation of interests and resources of the actors involved. There are at least four groups participating in the discourse: technology activists, institutionalized data protectionists, political parties building the parliament, and trade associations. Interestingly, the grand majority of these groups, while pursuing rather dissimilar interests, equally call upon *the individual* to build practices of DIY data protection.[51]

For example, *activists* tend to portray public authorities as well as economic enterprises as being motivated to install surveillance techniques—either due to some intrinsic interest in controlling populations, or in maximizing data-driven profits respectively. Thus, it depends on individual citizens to protect themselves against such interests.

Data protectionists call upon individuals as civil right holders. Their perspective is normatively framed by the German right to informational self-determination, which states that, from a legal-normative view, in democratic societies individuals are entitled to know who knows what about them whenever and in any given context. Whereas the jurisdiction transcends the individual in that there are social dimensions taken into account and collective duties being inferred from the centering on the individual, the latter nevertheless builds the normative core of the legal reasoning. Consequently, data protectionists, insofar as they are bound to

---

[50]Ole Reißmann, "Cyptoparty-Bewegung: Verschlüsseln, verschleiern, verstecken," Spiegel-Online, October 9, 2012, accessed February 23, 2015, http://www.spiegel.de/netzwelt/netzpolitik/cryptoparty-bewegung-die-cypherpunks-sind-zurueck-a-859473.html.

[51]Obviously, I am talking about ideal types here that nevertheless coin the discourse on DIY data protection most profoundly.

official jurisprudence, tend to appeal to the individuals to exercise their rights thus trying to activate the individual to act.[52]

Political parties generally have a pretty ambivalent attitude towards data protection, and also towards individuals' performing DIY data protection practices. They necessarily strive to come into power in order to realize their political goals. However, once in power, they represent the state, and it is certainly fair to identify some intrinsic interest of the state in surveillance as regards the populations that public authorities are bound to manage, govern and supervise. Thus, Baumann, for example, when investigating political parties' positions on data protection in the last legislature, found a strong correlation between power and willingness to foster data protection: the more political power a politician is able to execute, the less s/he is interested in data protection.[53] Moreover, there is a perceived conflict of objectives when it comes to administration, insofar as it is the state's duty to safeguard citizens' safety and security, while at the same time being responsible to defend citizens' freedom. Verisimilitude (or lack thereof) of such trade-off argumentations aside, they serve as an instrument for public authorities to have their cake and eat it too: rhetorically they may applaud individuals for developing DIY data protection practices, while at the same time neglecting to take on responsibility for the collective emergence of such practices.[54]

Finally, *the information economy*, insofar as business models are based on harvesting social actors' digital traces, is all but interested in the emergence of DIY

---

[52]I will omit here that to a certain degree data protectionists are caught up in a specific double bind: while they are public authorities and thus subject to the state's agency, they at the same time are bound to protect citizens from illegitimate interventions effected by this very state.

[53]Max-Otto Baumann, "Datenschutz im Web 2.0: Der politische Diskurs über Privatsphäre in sozialen Netzwerken," in *Im Sog des Internets. Öffentlichkeit und Privatheit im digitalen Zeitalter*, ed. Ulrike Ackermann (Frankfurt/M.: Humanities Online, 2013), 47.

[54]In this respect, the infamous statement of the former Minister of the Interior, Hans-Peter Friedrich, speaks volumes: On 16th of July 2013, Friedrich, at the time German Minister of the Interior, was interrogated by the parliamentary board that is supposed to supervise the intelligence service. Friedrich was asked about his state of knowledge concerning the so-called "NSA scandal". After having been interrogated by the board's members he faced the media. In this context Friedrich turned to German citizens, reminding them of their duties, asking them to assume their responsibilities, stating that they were supposed to learn by themselves how to cater for secure internet communication; in particular, Friedrich emphasized that cryptographic techniques and anti-virus software must be brought much more into focus. Also, the by-then Minister stated that people must become aware of the fact that also internet communications need to be protected. Thus we have here a perfect example for the shifting of the focus away from the extremely well-organized collective dimension of the civil rights attack carried out by the intelligence services to the individual's responsibility: DIY data protection serves as a way to individualize the social conflict, and to neglect the collective nature of the practices in question.

data protection practices in the sense of a mass phenomenon.[55] For the time being, it is not very hard for the spokespersons of the information economy—at least as far as the German discourse is concerned—to rhetorically foster the strengthening of the individuals' skills regarding data protection while at the same time giving them a run for their money if it comes to properly navigate privacy settings and so on. Quite obviously, the spread of DIY data protection practices would preclude a manifold of business models being based on harvesting personal information. Thus, businesses following such a model by definition cannot be interested in practices that threaten harvests. At the same time, however, it is very convenient to call upon individual consumers to develop such practices, while knowing that the emergence of these practices is no individual affair at all.

Thus, to summarize, while activists and data protectionists may have an interest in the wide-spread creation of DIY data protection practices, they have no resources to nourish the soci(o-technic)al processes that are required to effectively foster the development of those practices; conversely, the latter two groups do have access to resources,[56] but "by nature" they have no interest in citizens being versed in DIY data protection. For the reasons identified the odds are stacked against the wide-spread emergence of DIY data protection practices. However, as long as the most influential actors do not take on their responsibility in developing the collective, heterogeneous infrastructure that is the *sine qua non* for DIY data protection practices to evolve, the propagation of such practices may have undesired political repercussions, since it allows responsible entities to shift the burden to all those selves who are called upon to do data protection themselves. The normative implications of these shifts will be discussed in the following sections.

---

[55]This is not to say that, say, email service providers did not make use of encryption at all; German webmail service gmx, for example, provides encryption between end user and the company's mail servers, as well as among all the servers belonging to the so-called "E-Mail made in Germany"-network (an association of several Germany based email service providers, such as T-Online and WEB.DE). However, this may be interpreted as a rather superficial strategy to put the minds of worried users at rest, and not at all as the implementation of strong DIY data protection practices. More generally speaking, what I am referring to is the fact that in contemporary socio-technical assemblages it is players belonging to the surveillance *economy* that provide for the infrastructures enabling people to build up sociality. In modern societies, at least as far as European ones are concerned, *the state* used to be the agency that provided populations with the means to construct social structures (telegraph, mail, cable networks, you name it) and it also used to be *the state* that in turn observed the sociality thus built; in recent years, private corporations have become the main providers of key infrastructures of sociality (Online Social Networks serve as a paradigmatic case in point), as well as the main observers of the latter. As return on investment for most of these corporations is fundamentally, totally, absolutely grounded on the observation of the sociality built by "users" (who uses whom here?), the wide-spread emergence of strong DIY data protection practices is not in their interest as a matter of principle.

[56]For example, they could issue laws, install regulating bodies, strengthen relevant education (the state), or develop privacy friendly systems, make their techno-economic structure transparent, and effectively follow suitable business ethics.

3 DIY-Data Protection—Should We Have to Do It?

To answer this question we need to take a step back. In many regulatory frameworks data protection means to prohibit uses of data that would limit the citizens' ability to determine themselves who can access or use personal data and for what purposes. Our paper focuses rather on technological approaches than regulatory frameworks. In a sense, those technologies emphasize personal autonomy even more, since they need not rely on the legal and regulatory instruments; their development and use is often pursued by communities which are quite suspicious of the state (such as, e.g., the "cypherpunk movement", see Sect. 2.3).[57]

Yet, data protection has to be seen in a wider scope. Rather than asking how individuals can protect their data, the question should be: If citizens need data protection in the sense that particular pieces of data should not be accessible to particular actors, who should be responsible for that? This entails that an answer to this question also could change what data protection means or aims at.

This wider scope has several advantages: First of all, we need it to find alternatives for those cases where individual data protection simply is not feasible as Sect. 2.1 shows. But even if—for the sake of argument—these pragmatic concerns could be overcome, the wider scope still would be important.

This is because individual data protection needs a partition of responsibility: who is able to decide about which data? Terms such as "personal data" or "personal identifying information" are used in attempts to give citizens enough control over "their" data without their decisions infringing on others. In times of "Big Data", however, such a partition of responsibility becomes increasingly difficult. Louise Amoore has convincingly shown that not so much personal data but "data derivatives" are at the center of data based surveillance.[58] That is the relations and aggregates of data are more important than individual data sets. These kinds of technologies and data analyses are not only a challenge for individual concepts of privacy and data protection. They even disrupt the partition of privacy norms into wider social contexts as has been famously proposed by Nissenbaum[59]: The problem is that even if a citizen could fully transparently and conscious of the consequences for her or him decide in accordance with the contextual privacy norms, this data can still be used to infringe the privacy of *others*.[60]

---

[57]This, however, does not mean that the use of data protection tools cannot conflict with legal provisions. This can be seen in repeated calls to regulate the use of encryption as well as the legal constraints of the right to privacy, e.g. for the purpose of criminal investigations.

[58]Louise Amoore, "Data Derivatives: On the Emergence of a Security Risk Calculus for Our Times," *Theory, Culture & Society* 28 (2011).

[59]Nissenbaum, "Privacy in Context: Technology, Policy, and the Integrity of Social Life".

[60]Tobias Matzner, "Why Privacy is not Enough Privacy in the Context of 'Ubiquitous Computing' and 'Big Data,'" *Journal of Information, Communication & Ethics in Society* 12 (2014).

As a first step then, a wider, social perspective is necessary in the following sense: Even if the citizens would be responsible for data protection, it must be seen as a *social responsibility* and not as an individual problem. Everyone has to protect data they provide and use—even if it appears to be data "about them" and they think they have "nothing to hide"—because the data can be used to invade the privacy of others. As the results from the study presented in Sect. 2.1 shows, such concerns do not play a role or are even unknown to users. They mostly engage in data protection strategies which serve to protect their social privacy thus concentrating on protecting singular information against misuse.

As this reasoning illustrates, the question: "If citizens need data protection, who should be responsible for that?" opens up many more alternatives than just shifting, as it were, the workload of data protection. It also clarifies that responsibility by the citizens might either mean: "everyone is responsible for their own data protection", or "we are collectively responsible for our data protection." Many of the DIY data protection tools discussed here can be used to support either aim, but are usually advocated just concerning the first perspective.

It is important to note that this turn away from individual self-determination concerning data does not necessarily entail to give up other means of self-determination. To the contrary, it can even support them: Sect. 2.2 shows that data protection often competes with other needs or aims of self-disclosure. Dispensing with data protection or even voluntarily providing data can lead to increased social contacts, better carrier opportunities and many more. Yet, these arguments run the risk of remaining caught within the same logic of subsuming data protection (respectively forgoing it) under the aim of creating individual self-determination. The fact that problems or impediments in protecting data (currently) coincide with the aims of identity management of some or many individuals does not solve the underlying question of responsibility. Accordingly, we need to see the question who should be responsible for data protection in the wider context of distributing responsibilities among individuals, the state, and corporations—and thus also in the context of what individuality or at least individual freedom entail. This problem will be discussed in the next section under the rubric of "responsibilization". Before discussing this concept, however, it is important to remark that this argument concerns widespread data protection for the citizens. DIY-data protection tools are very valuable for particular persons or social actors like whistleblowers, journalists, or NGOs and other activists. Often their activities are important factors for changes on the wider political level that we discuss in the following section. And these activities include or even rely on DIY-data protection technologies—but also on using them in a very experienced and thoughtful manner. Thus, these technologies can be an important tactical tool for political activity as well as an indispensable protection for those who have no other choice.

3.1 Responsibilization

The term "responsibilization" has been coined in governance and criminology discourses and refers to "to the process whereby subjects are rendered individually responsible for a task which previously would have been the duty of another—usually a state agency—or would not have been recognized as a responsibility at all."[61] Usually it is discussed as a neo-liberal mode of governance that has been developed with recourse to Foucault's reflections on governmentality.[62] For example it can be seen in the transformation of the welfare state where citizens increasingly have to make their own provisions for former governmental benefits like health insurance or pension funds.

This perspective of governance is important concerning data protection, when public officials or institutions provide incentives and programs to propagate DIY data protection—as has been described in Sect. 2.3. Yet, we first want to focus on the underlying logic concerning individual actions in a broader sense. Lemke describes this as achieving congruence "between a responsible and moral individual and an economic-rational actor." To be responsible and moral is equated with rational self-determined choices: "As the choice of options for action is, or so the neo-liberal notion of rationality would have it, the expression of free will on the basis of a self-determined decision, the consequences of the action are borne by the subject alone, who is also solely responsible for them."[63]

Bennett and Raab describe the prevailing "privacy paradigm" as based on liberal theory, which "rests on a conception of society as comprising relatively autonomous individuals".[64] The authors show that this yields a particular concept of privacy, which has been criticized from several perspectives and is not without alternatives.[65] Still, though, it is this very concept of privacy that forms the background for most DIY data protection practices. Within the perspective of responsibilization, the named privacy paradigm is, as it were, relegated to a particular space of action for particular individuals that co-depend on social and technical conditions. Importantly, this foucauldian view does not simply say that the liberal privacy paradigm is wrong, but clarifies how it emerges from a particular configuration of states and private actors. Couched in slightly different, albeit cognate terms, such a view makes visible that the privacy paradigm described by Bennett and Raab is the product of a particular socio-technical configuration. A configuration, however, to

---

[61]Pat O' Mailey, "Responsibilization," in *The SAGE Dictionary of Policing*, ed. Alison Wakefield and Jenny Fleming (London: SAGE, 2009), 276.

[62]David Garland, "'Governmentality' and the Problem of Crime: Foucault, Criminology, Sociology," *Theoretical Criminology* 1 (1997).

[63]Thomas Lemke, "'The birth of bio-politics': Michel Foucault's lecture at the Collège de France on neo-liberal governmentality," *Economy and Society* 30 (2001): 201.

[64]Colin J. Bennett and Charles D. Raab, *The Governance of Privacy* (Cambridge: MIT Press, 2006), 4.

[65]Bennett and Raab, *The governance of privacy*, 14.

which it contributes in an essential way: By treating this confined space of action and individuality within the liberal perspective, the conditions producing it are neglected. Thus, the consequences of their actions are conceived as solely the individuals' responsibility.

This logic of responsibilization has several implications for data protection, which will be discussed in the following sections of this paper:

- Not engaging in data protection activities is seen as choice—equal to doing so (Sect. 3.2).
- Data protection becomes a commodity and the protected individuals become consumers (Sect. 3.3).
- Social inequalities concerning data protection cannot be addressed sufficiently, which may lead to victim blaming (Sect. 3.4).

While these points show the problems of locating data protection primarily with the individual, these results must be contextualized within the inherent ties between the logic of responsibilization and surveillance. Thus, paradoxically, individual data protection might seem the only remedy against the implications of responsibilization, if this logic is not addressed on a social-political level. This will be discussed in Sect. 3.5.

3.2 Data Protection as Choice

Positing the possibility for individual choice and data protection as created by socio-technical conditions does not necessarily devaluate self-determination as an aim for policy. It has, however, to be conceived of as the *creation* of possibilities and subject positions. If it is merely seen in the liberal perspective as the shielding from external interferences, it can very well be that even in the absence of any interference the desired action remains impossible. This is the case concerning data protection: Sect. 2.1 shows the purely pragmatic problems of DIY data protection. It is increasingly difficult to grasp the consequences of a person's decisions concerning their data. This is a precondition for responsible actions from the liberal perspective that is as of now almost impossible to attain. The tools which are available are hard to use properly and involve competence and resources. If these structural problems are ignored, the danger arises that data protection regulations shield a space for autonomous decisions that, however, are impossible to carry out. Thus, basic rights to privacy are hollowed out. As a first result, this shows: If the citizens have to protect their privacy on their own, they can only do it based on active provisions by the state and commercial actors—as has already been emphasized at the end of Sect. 2.3—or at least with a considerable extent of self-organization and citizen-led structures.

The logic of responsibilization brings about further ramifications: All kinds of behavior concerning data, and in particular, not engaging in data protection activities, are considered as (rational) choice. This is maybe most salient in refusals to introduce better privacy policies by corporate actors. Often they argue that people who are not content with the level of data protection should just not use their services or products. This presupposes that using or not using a particular service are equal choices. Such an assessment of course depends on the product in question. But generally we can say that this presupposition is often not met on several levels:

The first concerns transparency and coercion: The reasons for not using a service or product are usually buried deeply in license agreements or privacy policies we have to "consent" to before using.[66] The reasons to use them, on the contrary, are promoted by the best advertising agencies in the world. Furthermore, big IT companies are actively advocating the use of their products in education and the workplace,[67] thus spreading a lax data protection regime, which may be compulsory in school or at the workplace. Often such conditions can only be evaded at the high social cost of changing the school or the employer or by organizing resistance and asking for different infrastructure from a dependent position. This leads on to the next problem of framing not to use a service or product as alternative choice. Information and communication technology has pervaded almost every aspect of our lives. In particular smartphones are almost considered as a standard in many contexts in Europe and the USA.[68] Although they are still a commodity that theoretically everyone chooses freely to buy and use, in effect most people who decide to refrain from using them might face more or less severe social costs: less contacts with friends, missing carrier opportunities, more complicated dating, being considered inefficient as a colleague, being considered suspicious at border controls, and many more. Of course, these examples are hardly comparable concerning severity and consequences. But the motley list shows both the variety of aspects of life that are permeated by this technology and the respective breadth of problems that refusing to use a smartphone can cause. In fact, vendors openly advertise the very benefits one will be missing without a smartphone. This reproduces a structure quite common within the logic of responsibilization: an individual/socio-technical asymmetry where the possibilities of the socio-technical changes provided by corporate actors are openly endorsed, whereas the problematic consequences and responsibility lie with the individual alone.

---

[66]On the problematic pragmatics of license agreements, see for example Debatin et al.: "Facebook and online privacy for social networking sites," or Chee et al., "Re-Mediating Research Ethics" concerning games.

[67]See for example Apple's "iPad in Education" website: https://www.apple.com/education/ipad/ (accessed February 19, 2015).

[68]In Europe, more than half of all persons already own a smartphone, with a continually growing market predicted: http://www.statista.com/statistics/203722/smartphone-penetration-per-capita-in-western-europe-since-2000/ (accessed March 4, 2015).

This is by no means a matter of course for widely used commodities. To the contrary, recognizing the importance of ICT for our daily lives can warrant high levels of regulation, like those already in place for many other important goods—their being a commodity on a free market notwithstanding: e.g. food, drugs, or cars.

3.3 Data Protection as Commodity

We have already touched upon many points that could also fall under this rubric in the last section. Here, however, we want to focus less on the implications of certain commodities like smartphones concerning data protection. We rather want to discuss data protection itself becoming a product or at least a price relevant product feature and thus something that is attainable for money. In the context of the infeasibility of completely individual data protection, users of ICT have to entrust some other actor or institution with data protection tasks. The need to build a trustworthy environment has long been considered as an important factor in the IT business[69] but in particular after the revelations by Edward Snowden, data protection has increasingly become a feature for selling products—and the market for data protections as a product by itself is growing. Such products come in many variants: encryption software for many channels of communication (mail, chats, voice), hardware products like encrypted phones or personal servers to run one's own "cloud", subscription services for encrypted and anonymized communications, and many more. Other providers sell privacy as a kind of "add-on" like AT&T's offer not to track their internet subscribers' activities for an additional 29 US dollars.[70]

Of course such products are premised on the condition that the providers are trustworthy in the first place—which is dubitable concerning the revealed powers of secret services to avail themselves of commercially collected and administrated data. Yet, as we note in Sect. 2.1, data protection has many opponents, not only secret services. And concerning many of them, in particular social privacy, commercial data protection products might be a sensible solution. In the end, providers who want to prevail on a market should not be too abusive of the trust of their customers.

This solution, however, replaces the requirements of competences and time, which render DIY data protection impractical, with another requirement: money.[71] Given the omnipresence of IT, this would entail that almost everyone would have to spend some extra money to get data protection. This need arises in a context

---

[69]Bennett and Raab, The Governance of Privacy, 53 et seqq.

[70]http://arstechnica.com/business/2015/02/att-charges-29-more-for-gigabit-fiber-that-doesnt-watch-your-web-browsing/ (accessed February 19, 2015).

[71]Of course, providing competence and time is usually more or less directly related to monetary costs as well.

that is by no means a level playing field for two reasons: money is unequally distributed and data protection needs are unequally distributed. While the first is a matter of course, the second aspect deserves some words: Maybe most problematically, researchers like John Gilliom have shown that many surveillance activities (and thus an increased need for privacy protection) focus on those that do not have much money. Here the responsibilization of data protection becomes entangled with other responsibilization processes concerning welfare. Very often such processes of responsibilization require increased data collection and legitimize surveillance.[72] That does not only mean that those with the least money would have the biggest need to spend—in itself problematic enough—but that such products are ineffective for them since they are under surveillance through other channels.[73]

Many other groups that face the threat of social stigma or discrimination have higher data protection needs as well: e.g. women, homosexuals, migrants, or members of certain religions. Data protection as a commodity thus entails higher financial burdens for those social identities. Thus, we run the risk of privacy becoming a luxury for those who can afford it. And furthermore, this additional cost is especially put on those who already face discrimination or social inequalities.

Considering the argument in Sect. 3 that data protection can only be achieved socially, however, this luxury will not have much worth. If not enough people buy in, there will be sufficient data available to create the data derivatives that are of interest anyway. This shows that customer choice is just a very explicit instance of the logic of individual choice discussed in the last section—and thus reproduces the problems discussed there.

3.4 Data Protection, Social Equality, and Victim Blaming

To discuss this aspect, first of all we have to emphasize that users of social media and other ICTs do care for their privacy—even if they disclose all kinds of information.[74] Some, in particular teenagers and children even perceive online interaction as more private since it is more easily shielded from parents or teachers—their preeminent threat to privacy.[75] Thus, online interaction is structured by complex privacy needs and requirements even where people voluntarily

---

[72] John Gilliom, *Overseers of the Poor* (Chicaco: Chicago University Press, 2001), 130 et seqq.; Nikolas Rose, "Government and Control," *British Journal of Criminology* 40 (2000).

[73] More on this in Sect. 3.5.

[74] See also Sect. 2.

[75] Valerie Steeves, "Data Protection Versus Privacy: Lessons from Facebook's Beacon," in *The contours of privacy*, ed. David Matheson (Newcastle: Cambridge Scholars Publishing, 2009), 187.

provide substantial amounts of data.[76] Steeves argues that a focus on data protection cannot grasp this complexity since it is limited to data and the procedures of its usage, while the actions which yield that data are structured by a wider normative social context.[77] In particular, this focus on the data within the logic of responsibilization means that the data is conceived as provided by choice. As Sect. 2.2 shows, privacy considerations stand in a complex context of other aims and motives but also requirements and coercions. Within the logic of responsibilization, this context only figures insofar as the provision of withholding of data is taken to be the rational choice balancing the various aims and requirements—and if that rational choice did not take place, this is the individual's shortfall.

From an individual point of view, however, interaction is not structured by access and use of data but by the entire complex bundle of norms of action. These norms very well might coerce individuals into disclosing private information or lead to the endorsement of actions that entail providing private data. That does not mean that these persons endorse all the kinds of uses of their data that could be justified by their individual refusal (viz. choice) to keep that data completely private.

Importantly, such privacy norms are not equally distributed. For example, Bailey et al. have researched young women's perception of Facebook profiles. The teenage participants of the study clearly perceived Facebook as a "commoditized environment" where "stereotypical kinds of self-exposure by girls are markers of social success and popularity."[78] For young women, these stereotypes involve providing more private information compared to men: details about their relationships (often including the partner on the profile picture), details about their friends and more intimate pictures, e.g. shots in bikinis. While many of the participants have been critical about such profiles, most have clearly admitted the social success that can be achieved by following these norms. That shows that women face a broader requirement of choices concerning privacy that do not arise for men. If the individuals are held responsible for their use and protection of their data, this means increased burdens for women. Furthermore, when deciding for data protection (which in this case means not providing the data) their socials costs are higher.

Individual responsibility for data protection clearly leads to unequal distribution of effort, material and social costs that materialize along social lines of discrimination—in this example gender. These differences disappear from view when the focus is put on data protection and individual responsibility that mainly asks who did or did not provide which kinds of data. Thus, the responsibility problems or misuse arising from the private or intimate data is attributed to the women, since they did provide the data in the first place, when they could have "simply"

---

[76]This, of course is the rationale of Nissenbaum's approach in "Privacy as Contextual Integrity" that she has developed from reflections on "privacy in public."

[77]Steeves, "Data protection vs. Privacy," 189.

[78]Jane Bailey et al., "Negotiating With Gender Stereotypes on Social Networking Sites: From 'Bicycle Face' to Facebook," *Journal of Communication Inquiry* 37 (2013): 91.

not done so. It is this last supposition that evades the social circumstances and leads to victim blaming.

The logic of individual responsibility concerning data protection thus can contribute to the proliferation of such moral double standards since it is hard to address the "moral climate" in which needs and social costs of data protection arise, when the focus lies only on the individual and the question whether data is accessible or not.

This also precludes emancipatory movements for the freedom to be as explicit and open as one wishes. The act to publish the data despite the moral double standards to appropriate the practice (in this case posting pictures or having private data on facebook) must explicitly posit oneself against the existing norms to not fall prey to the logic of commodification, control and blame.[79]

3.5 Responsibilization, Surveillance, and Politics

The outsourcing of responsibilities and services from the state or corporate actors towards the individual here discussed as "responsbilization" brings about needs for monitoring and surveillance. For example, a common practice in health insurance is to provide incentives for regular medical checks or "healthy" activities like sports or exercises. For this to work, however, the behavior of the clients has to be monitored beyond that which happens in physicians' practices, for example including leisure activities or diets. Of course, when increasingly responsibility is moved to the citizens, also the state's monitoring increases. In fact, the process of responsibilization is closely tied in with surveillance and control.[80] Security then is established within a preemptive logic that tries to sort individuals based on intensive monitoring.[81] Responsible and moral citizens of course will not get into the focus of these practices—only suspect persons will—as security agencies all over the world emphasize. But everybody, again, is responsible for being that particular kind of responsible and moral citizen—also regarding the data they provide and use.

Thus, many of the reasons for an increased need for data protection arise within the logic of responsibilization itself—adding data protection as one further field to look after. Here, however, the logic turns against itself, when the citizens try to inhibit surveillance and thus an intrinsic part of responsibilizationist control. Still, the logic remains intact when states try to support the development and marketing

---

[79]See for example the problems of legislating revenge porn without reproducing the logic of victim blaming or infringing the sexual liberty of women in Henry and Powell, "Beyond the 'sext'".

[80]Rose, "Government and control."

[81]David Lyon, "Surveillance As Social Sorting," in *Surveillance As Social Sorting: Privacy, Risk, and Digital Discrimination*, ed. David Lyon (New York: Routledge, 2003).

of privacy enhancing technologies for the end-consumer market. But states try to establish conditions where DIY data protection can be carried out as a flourishing market but under conditions where these practices do not inhibit state surveillance.[82] The revelations of secret services spying on citizens attest to that. So it is only natural from that point of view that the requirement of government backdoors was immediately voiced when stronger encryption paradigms have recently been rolled out in mobile communications.[83]

In such a climate it may seem rather naïve to entrust anyone but oneself with data protection. Furthermore, since resisting surveillance turns against the logics of responsibilization as just described, it might appear as a valid move of resistance. To an extent, this is true. But it would be mistaking the cause for the symptoms. Much of the states' surveillance is not done by eavesdropping on individuals but by helping themselves to the big databases that accrue in other places like big online enterprises. In a society where most services are commodities and keeping track on customers is part of a business model focused on ever increasing efficiency marketed as individualization, the data which is of interest for commercial actors and security agencies often coincide.[84] In a society where welfare and insurance is detached from communitarian models and broke up into individual provisions again based on circumspect data collection, even more data of interest for secret services is generated. Thus, many of the possibilities to collect data in the first place rise from the commodified and responsibilized societies we life in. Then DIY data protection is an almost vain attempt to fight a functional process of these societies while ignoring the rest—or even keeping it intact and alive by adding data protection as another flourishing branch on the market.

4 Conclusion

A move towards data protection that takes such reflections into account must address the many causes of the accrual of data on a political and social level rather than taking them for granted and trying to evade them where possible. This would entail to call on the state to take its responsibilities in protecting its citizens' data seriously, and not only to enable markets. It also needs to address national security as universal subterfuge from European data protection legislation.

On a more fundamental level, the implications of data protection as a social responsibility have to be assessed. Initiatives to foster data literacy or media literacy can still be a valuable tool, when they include social perspectives and in particular address the unequal distribution of data protection needs in society.

---

[82]See the quote above in note 50 as an example.

[83]http://www.slate.com/blogs/future_tense/2015/01/19/obama_wants_backdoors_in_encrypted_messaging_to_allow_government_spying.html (accessed March 4, 2015).

[84]Jeffrey Rosen, *The naked crowd: Reclaiming security and freedom in an anxious age* (New York: Random House, 2005), Chap. 3.

The imbalance of choosing to use a service versus not using it can be mediated by sensible data protection defaults that emphasize data protection and need an active decision to enable less protective uses. Of course the problems of making that choice in a transparent and reflected manner remain. Albeit, it is better to actively demand accepting that data may be used in ways that are almost impossible to know rather than making it the default. And such defaults mainly address the collection of data but not the processing, sharing, and analysis.

On an institutional level, intermediaries between the citizens on the one side and the state of corporations on the other can organize data protection. Consumer protection models are one possibility. Another way is *social* self-organizing. Many communities in fact have conscious discussions or rules concerning privacy among their members, which include but are not restricted to data protection policies. Often, these are groups that are faced with higher privacy requirements, e.g. online self-help communities. Albeit, practices that are developed by such groups still can be a model for others.

Empirically speaking, the wide-spread emergence of DIY data protection practices is rather improbable, or more precise: As long as (also DIY!) data protection is not considered a collective, profoundly political endeavor, DIY data protection is an ill-fated practice. What's more, without taking on a collective perspective, the advocating of DIY data protection may even create undesired effects, for it allows for neglecting political responsibility, fostering further inequalities between users, and generally asking too much of the individual.

Bibliography

Acquisti, Alessandro, Leslie K. John, and George Loewenstein. 2013. What is privacy worth? *The Journal of Legal Studies* 42: 249–274.
Altman, Irwin, and Dalmas Taylor. 1976. *Social penetration: The development of interpersonal relationships*. New York: Holt, Rinehart and Winston.
Amoore, Louise. 2011. Data derivatives: On the emergence of a security risk calculus for our times. *Theory, Culture & Society* 28: 24–43.
Bailey, Jane, Valerie Steeves, Jacquelyn Burkell, and Priscilla Regan. 2013. Negotiating with gender stereotypes on social networking sites: From "bicycle face" to facebook. *Journal of Communication Inquiry* 37: 91–112.
Barnes, Susan B. 2006. A privacy paradox: Social networking in the Unites States. *First Monday* 11(9). doi:10.5210/fm.v11i9.1394. Accessed 10 Mar 2015.
Baumann, Max-Otto. 2013. Datenschutz im Web 2.0: Der politische Diskurs über Privatsphäre in sozialen Netzwerken. In *Im Sog des Internets. Öffentlichkeit und Privatheit im digitalen Zeitalter*, ed. Ulrike Ackermann, 15–52. Frankfurt/M.: Humanities Online.
Bennett, Colin J., and Charles D. Raab. 2006. *The governance of privacy*. Cambridge: MIT Press.
Brandtzæg, Petter Bae, Marika Lüders, and Jan Håvard Skjetne. 2010. Too many Facebook 'friends'? Content sharing and sociability versus the need for privacy in social network sites. *Intl. Journal of Human-Computer Interaction* 26: 1006–1030.
Chee, Florence M., T.Taylor Nicholas, and Suzanne de Castell. 2012. Re-mediating research ethics: End-user license agreements in online games. *Bulletin of Science Technology & Society* 32: 497–506.

Debatin, Bernhard, Jenette P. Lovejoy, Ann-Kathrin Horn, and Brittany N. Hughes. 2009. Facebook and online privacy: Attitudes, behaviors, and unintended consequences. *Journal of Computer Mediated Communication* 15: 83–108.

Deutsches Institut für Vertrauen und Sicherheit im Internet (DIVSI). 2013. DIVSI Studie zur Freiheit versus Regulierung im Internet. Hamburg. https://www.divsi.de/wp-content/uploads/2013/12/divsi-studie-freiheit-v-regulierung-2013.pdf. Accessed 10 Mar 2015.

De Wolf, Ralf, Koen Willaert, and Jo Pierson. 2014. Managing privacy boundaries together: Exploring individual and group privacy management strategies in Facebook. *Computers in Human Behavior* 35: 444–454.

Dienlin, Tobias, and Sabine Trepte. 2014. Is the privacy paradox a relic of the past? An in-depth analysis of privacy attitudes and privacy behaviors. *European Journal of Social Psychology*.

Dourish, Paul, and Ken Anderson. 2006. Collective information practice: Exploring privacy and security as social and cultural phenomena. *Human-Computer Interaction* 21: 319–342.

Ellison, Nicole B, Jessica Vitak, Charles Steinfield, Rebecca Grey, and Cliff Lampe. 2011. Negotiating privacy concerns and social capital needs in a social media environment. In *Privacy online. Perspectives on privacy and self-disclosure in the social web*, ed. Sabine Trepte, and Leonard Reinecke, 19–32. Berlin: Springer.

Garland, David. 1997. 'Governmentality' and the problem of crime: Foucault, criminology, sociology. *Theoretical Criminology* 1: 173–214.

Giddens, Anthony. 1984. *The constitution of society: Outline of the theory of structuration.* Cambridge: Polity Press.

Gilliom, John. 2001. *Overseers of the poor*. Chicaco: Chicago University Press.

Gunkel, David J. 2003. Second thoughts: Towards a critique of the digital divide. *New Media & Society* 5: 499–522.

Henry, Nicola, and Anastasia Powell. 2014. Beyond the 'sext': Technology-facilitated sexual violence and harassment against adult women. *Australian & New Zealand Journal of Criminology*. doi:10.1177/0004865814524218. Accessed 10 Mar 2015.

Hughes, Eric. 2015. A Cypherpunk's Manifesto. http://activism.net/cypherpunk/manifesto.html. Accessed 23 Feb 2015.

Jourard, Sidney M, and Paul Lasakow. Some factors in self-disclosure. some factors in self-disclosure. *Journal of Abnormal Psychology* 56(1): 91.

Lampinen, Airi, Vilma Lehtinen, Asko Lehmuskallio, and Sakari Tamminen. 2011. We're in it together: Interpersonal management of disclosure in social network services. *Proceedings of the SIGCHI conference on human factors in computing systems*, 3217–3226. New York, USA: ACM.

Lemke, Thomas. 2001. 'The birth of bio-politics': Michel Foucault's lecture at the Collège de France on neo-liberal governmentality. *Economy and Society* 30: 190–207.

Litt, Eden. 2013. Understanding social network site users' privacy tool use. *Computers in Human Behavior* 29(4): 1649–1656.

Lyon, David. 2003. Surveillance as social sorting. In *Surveillance as social sorting: Privacy, risk, and digital discrimination*, ed. David Lyon, 13–30. New York: Routledge.

Masur, Philipp K, and Michael Scharkow. Disclosure management on social network sites: Individual privacy perceptions and user-directed privacy strategies. (in prepartion).

Masur, Philipp K., Doris Teutsch and Sabine Trepte. *Entwicklung der Online-Privatheitskompetenz-Skala*. (in preparation).

Matzner, Tobias. 2014. Why privacy is not enough privacy in the context of 'ubiquitous computing' and 'big data.' *Journal of Information, Communication & Ethics in Society* 12(2):93.

Milne, George R., and Andrew J. Rohm. 2000. Consumer privacy and name removal across direct marketing channels: Exploring opt-in and opt-out alternatives. *Journal of Public Policy & Marketing* 19(2): 238–249.

Metzger, Miriam J. 2007. Communication privacy management in electronic commerce. *Journal of Computer-Mediated Communication* 12: 351–361.

National Telecommunications and Information Administration (NTIA). 2015. Falling through the net: A survey of the 'have-nots' in Rural and Urban America. Washington, DC: US Department of Commerce. http://www.ntia.doc.gov/ntiahome/fallingthru.html. Accessed 10 Mar 2015.

Nissenbaum, Helen. 2010. *Privacy in context: Technology, policy, and the integrity of social life.* Stanford: Stanford Law Books.

O' Mailey, Pat. 2009. Responsibilization. In *The SAGE dictionary of policing*, ed. Alison Wakefield and Jenny Fleming, 276–278. London: SAGE.

Papacharissi, Zizi A. 2010. *A private sphere: Democracy in a digital age.* Cambridge: Polity Press.

Park, Yong J., Scott W. Campbell, and Nojin Kwak. 2012. Affect, cognition and reward: Predictors of privacy protection online. *Computers in Human Behavior* 28(3): 1019–1027.

Raynes-Goldie, Katie. 2010. Aliases, creeping, and wall cleaning: Understanding privacy in the age of Facebook. *First Monday* 15(1). http://firstmonday.org/article/view/2775/2432. Accessed 10 Mar 2015.

Reißmann, Ole. 2012 Cyptoparty-Bewegung: Verschlüsseln, verschleiern, verstecken. *Spiegel-Online*. http://www.spiegel.de/netzwelt/netzpolitik/cryptoparty-bewegung-die-cypherpunks-sind-zurueck-a-859473.html. Accessed 23 Feb 2015.

Rogers, Everett M. 2005. *Diffusion on innovations.* New York: Free Press.

Rose, Nikolas. 2000. Government and control. *British Journal of Criminology* 40: 321–339.

Rosen, Jeffrey. 2005. *The naked crowd: Reclaiming security and freedom in an anxious age.* New York: Random House.

Selwyn, Neil. 2004. Reconsidering political and popular understandings of the digital divide. *New Media & Society* 6: 341–362.

Skouma, Georgia, and Laura Léonard. 2015. On-line behavioral tracking: What may change after the legal reform on personal data protection. In *Reforming European Data Protection Law*, ed. Serge Gutwirth, Ronald Leenes, and Paul de Hert, 35–62. Dordrecht: Springer.

Steeves, Valerie. 2009. Data protection versus privacy: Lessons from Facebook's Beacon. In *The contours of privacy*, ed. David Matheson. Newcastle: Cambridge Scholars Publishing.

Strauss, Anselm. 1978. A social world perspective. *Symbolic Interaction* 1: 119–128.

Taddicken, Monika, and Cornelia Jers. 2011. The uses of privacy online: Trading a loss of privacy for social web gratifications. In *Privacy online. Perspectives on privacy and self-disclosure in the social web*, ed. Sabine Trepte, and Leonard Reinecke, 143–156. Berlin: Springer.

Trepte, Sabine, Tobias Dienlin, and Leonard Reinecke. 2014. Risky behaviors: How online experiences influence privacy behaviors. In *Von Der Gutenberg-Galaxis Zur Google-Galaxis. From the Gutenberg Galaxy to the Google Galaxy. Surveying old and new frontiers after 50 years of DGPuK*, ed. Birgit Stark, Oliver Quiring, and Nikolaus Jackob, 225–246. Wiesbaden: UVK.

Trepte, Sabine, Doris Teutsch, Philipp K. Masur, Carolin Eicher, Mona Fischer, Alisa Hennhöfer, and Fabienne Lind. 2015. do people know about privacy and data protection strategies? Towards the 'Online Privacy Literacy Scale' (OPLIS). In *Reforming European Data Protection Law*, ed. Serge Gutwirth, Ronald Leenes, and Paul de Hert, 333–366. Dordrecht: Springer.

Trepte, Sabine, and Leonard Reinecke. 2011. The social web as a shelter for privacy and authentic living. In *Privacy online. Perspectives on privacy and self-disclosure in the social web*, ed. Sabine Trepte, and Leonard Reinecke, 143–156. Berlin: Springer.

Tufekci, Zeynep. 2008. Can you see me now? Audience and disclosure regulation in online social network sites. *Bulletin of Science, Technology & Society* 28(1): 20–36.

Vitak, Jessica. 2012. The impact of context collapse and privacy on social network site disclosures. *Journal of Broadcasting & Electronic Media* 56(4): 451–470.

Privacy Failures as Systems Failures: A Privacy-Specific Formal System Model

A Systemic and Multi-perspective Approach

Anthony Morton

Abstract There have been numerous cases of adverse publicity concerning the negative effect of technology services—the combination of a technology platform and providing organisation—on people's privacy. Privacy failures represent complex and cross-disciplinary failure situations, encompassing the design and development of technology services, and organisational privacy practice. Investigation of the root causes of privacy failures requires a systemic and multi-perspective approach which views privacy failures as systems failures. Systems thinking, tools and methods have been used for several decades to analyse and model failures, but have not been applied to privacy failures. This chapter introduces the use of a systemic method—the Systems Failures Approach—to study privacy failures. The Systems Failures Approach—founded on Soft Systems Methodology—compares a conceptual model of a failure situation with a Formal System Model (FSM)—a paradigm of a robust system capable of purposeful activity—to identify its causes, and recommend feasible and desirable changes. This chapter describes a Privacy-Specific Formal System Model (PSFSM), as part of the Systems Failures Approach, to identify the actual or potential causes of privacy failures in technology services, and concludes with a brief proof-of-concept application of the PSFSM to the launch of Google Buzz.

Keywords Information privacy · Systems failures approach · Formal system model · Soft systems methodology

---

A. Morton (✉)
Department of Computer Science, University College London,
Gower Street, London WC1E 6BT, UK
e-mail: anthony.morton.09@ucl.ac.uk

1 Introduction

Google Buzz was launched on 9th February 2010, as a competitor to other social networking services, particularly Facebook and Twitter, from which it borrowed features, such as the ability to '*like*' content and '*follow*' other users.[1] Buzz was directly integrated into each user's Gmail account, and without an obvious signup process, automatically populated users' Buzz 'followers' list with their most frequent e-mail and chat contacts, which was publicly visible by default on their Google public profile.[2] One day after the launch of Buzz, privacy concerns were raised in online media and blogs, with particular focus on the auto-following of Gmail contacts.[3] Over the following week, Google responded quickly to these concerns, making various modifications to the Buzz signup process.[4] In March 2011, Google agreed to settle with the Federal Trade Commission (FTC), which charged it with unfair and deceptive acts or practices, contrary to the Federal Trade Commission Act[5]; Buzz was closed in October 2011.[6]

The impact of Google Buzz on people's privacy is not an isolated incident. Recent years have seen numerous cases of adverse publicity, or criticism from privacy advocacy groups, concerning the potential or actual effect of *technology services*[7] on people's privacy. Examples include the use of radio frequency identification (RFID) tags,[8] online video games,[9] peer-to-peer (P2P)

---

[1]BBC, "Google Takes on Facebook and Twitter with Network Site"; Helft and Stone, "With Buzz, Google Plunges Into Social Networking"; Krazit, "Google's Social Side Hopes to Catch Some Buzz."

[2]FTC, "Complaint: In the Matter of Google Inc. a Corporation."

[3]Carlson, "WARNING: Google Buzz Has a Huge Privacy Flaw"; Wood, "Google Buzz."

[4]"Millions of Buzz Users, and Improvements Based on Your Feedback"; Google Inc., "A New Buzz Start-up Experience Based on Your Feedback."

[5]FTC, "FTC Charges Deceptive Privacy Practices in Google's Rollout of Its Buzz Social Network"; FTC, "Decision and Order: In the Matter of Google Inc. a Corporation."

[6]Google Inc., "A Fall Sweep."

[7]A *technology service* is a socio-technical system consisting of a technology platform—referred to as a *technology lens*—and providing organisation (Morton and Sasse, "Privacy Is a Process, Not a PET.").

[8]Associated Press, "Officials: Special Plastic Sleeves May Stop Identity Theft"; D'Innocenzio, "Wal-Mart Plan to Use Smart Tags Raises Privacy Concerns"; Radcliffe, "Tracking Devices Used in School Badges."

[9]Quinn and Arthur, "PlayStation Network Hackers Access Data of 77 Million Users"; Arthur, "Sony Suffers Second Data Breach with Theft of 25 m More User Details."

file sharing,[10] street-level mapping,[11] targeted advertising[12] and smart phones.[13]

In the case of Google Buzz, and the examples cited in the previous paragraph, where technology services have adversely affected people's privacy, the fundamental question is, *"What were the root causes of the privacy failure?"* Answers to this question may include:

- **Failure to understand users' perspectives**—Users of a technology service will have expectations and assumptions about its privacy behaviour. If its privacy behaviour does not match those expectations, users may respond emotionally and reject it.[14] In the case of Buzz, Google appeared to misunderstand users' privacy expectations and assumptions about the use of Gmail contacts to pre-populate Google Buzz, and did not anticipate users' subsequent reaction.[15]
- **Software release management failure**—A privacy failure may be caused by a software release management failure. For example, Google claimed their Street View cars' collection of Wi-Fi data was caused by *"legacy code from an experimental project that had been re-used to programme* (sic) *equipment on the Street View cars"*.[16]
- **Failure to provide users with effective control and feedback of personal information**—Users may inadvertently disclose personal information because of a poorly designed user interface, or one that runs counter to social behaviour,[17] as identified in studies of P2P file-sharing clients.[18]

---

[10]Mennecke, "Pfizer P2P Security Breach"; NBC, "New Warnings on Cyber-Thieves"; Federal Trade Commission, "Widespread Data Breaches Uncovered by FTC Probe."

[11]Macdonald, "Google's Street View Raises Alarms over Privacy."; Barnett, "Google Street View: Survey Raises Privacy Concerns"; Kiss, "Google Admits Collecting Wi-Fi Data through Street View Cars."

[12]Blakeley, "Facebook Shrugs off Privacy Fears with Plan for Targeted Advertising"; Williams, "Google to Build Profiles of Gmail Users for Advertisers."

[13]Allan and Warden, "Got an iPhone or 3G iPad? Apple Is Recording Your Moves"; Panzarino, "It's Not Just the iPhone, Android Stores Your Location Data Too."

[14]Adams and Sasse, "Privacy Issues in Ubiquitous Multimedia Environments: Wake Sleeping Dogs, or Let Them Lie?"; Adams and Sasse, "Privacy in Multimedia Communications: Protecting Users, Not Just Data."

[15]Carlson, "WARNING: Google Buzz Has a Huge Privacy Flaw"; Kolmes, "Google Buzz Alarms a Psychotherapist | Psychologist San Francisco"; Wood, "Google Buzz"; Arthur, "Google Buzz's Open Approach Leads to Stalking Threat."

[16]Kiss, "Google Admits Collecting Wi-Fi Data through Street View Cars."

[17]Bellotti and Sellen, "Design for Privacy in Ubiquitous Computing Environments."

[18]Good and Krekelberg, "Usability and Privacy: A Study of Kazaa P2P File-Sharing."

- **Security failure**—Privacy failures may be the (in)direct result of a security failure, i.e. *"data breaches as a privacy problem"*; three examples of this are ChoicePoint, TJX and Sony.[19]
- **Ethical failure**—Information use and privacy invasions may be caused by unethical use of technology[20]; unauthorised and unexpected collection of personal information may be deliberately engineered, as alleged by some.[21]

The above list is not exhaustive, but its breadth suggests a cross-disciplinary approach is necessary, encompassing the complete organisational 'privacy practice stack', from an organisation's information security to its information privacy ethics, and including the design and development of privacy-aware technology services. To understand the causes of potential and actual privacy failures, it is necessary to consider the totality of a technology service from multiple perspectives, and view technology services as socio-technical systems—in short, *privacy failures must be considered systems failures*.

The ideas presented in this chapter originate from three papers describing the following systemic approaches:

1. **A framework unifying in one model the activities necessary for effective privacy practice**—Morton et al.[22] argue that although technology plays a vital role in safeguarding privacy, a holistic approach is required to deliver effective privacy practice in organisations; privacy practice must be considered and managed as a socio-technical system. They propose a layered framework—the Privacy, Security and Trust (PST) Framework—to represent effective privacy practice within a technology service.
2. **The use of 'tool clinics' to encourage holistic multi-perspective collaboration of a technological solution, research method or artefact**—Morton et al.[23] argue that engineers and researchers must understand the different viewpoints of actors directly or indirectly affected by the solutions they design and implement, and take a holistic view of their proposed solutions and the context in which they operate. They remark that systemic tools and methods, such as Soft Systems Methodology (SSM),[24] have been used for several dec-

---

[19]Culnan and Williams, "How Ethics Can Enhance Organizational Privacy: Lessons from the ChoicePoint and TJX Data Breaches"; Quinn and Arthur, "PlayStation Network Hackers Access Data of 77 Million Users"; Arthur, "Sony Suffers Second Data Breach with Theft of 25 m More User Details."

[20]Bellotti and Sellen, "Design for Privacy in Ubiquitous Computing Environments."

[21]Sydnor, Knight, and Hollaar, "Filesharing Programs and 'Technological Features to Induce Users to Share'"; Kravets, "Lawyers Claim Google Wi-Fi Sniffing 'Is Not an Accident.'".

[22]Morton and Sasse, "Privacy Is a Process, Not a PET."

[23]Morton et al., "4.3 'Tool Clinics'—Embracing Multiple Perspectives in Privacy Research and Privacy-Sensitive Design."

[24]Checkland, *Systems Thinking, Systems Practice*; Checkland and Scholes, *Soft Systems Methodology in Action: Including a 30 Year Retrospective*.

ades to analyse and model failures, but have not been applied in the field of privacy.

3. **The application of systems-related tools and methods to identify actual or potential project failures**—White & Fortune[25] propose the use of the Systems Failures Approach[26] in the study of project failures. The Systems Failures Approach, which is based on SSM, compares a conceptual model of a failure situation with a Formal System Model (FSM)[27]—a paradigm of a robust system capable of purposeful activity—to identify potential and actual weaknesses, and recommend feasible and desirable changes. Similarly, White & Fortune propose a project-specific FSM, which can be compared with a conceptual systems model of the project.

This chapter describes the first application of systems-related concepts to the study of privacy failures. Specifically, it describes a Privacy-Specific Formal System Model (PSFSM) for use with the Systems Failures Approach, to study privacy failures in technology services. This chapter should not be considered exhaustive—it is an introduction of systems thinking, tools and methods to privacy researchers and practitioners. It begins with an overview of the PST Framework, before providing sufficient background on SSM and the Systems Failures Approach for understanding the proposed PSFSM. It describes the PSFSM, and provides a list of the failure modes associated with information privacy systems mapped, where appropriate, to existing privacy research. It concludes with a brief proof-of-concept application of the PSFSM to the launch of Google Buzz.

Identifying potential privacy failures is important, as Spiekermann observes, "[the] *distrust caused by privacy breaches is probably the only real blemish on the image of technology companies such as Google or Facebook*".[28]

2 The PST Framework

The examples cited in the introduction—where the potential or actual effect of technology services on privacy attracted adverse publicity—were principally caused by: (1) ineffective organisational privacy practice; and/or (2) a lack of privacy-sensitive design.

---

[25]White and Fortune, "The Project-Specific Formal System Model."

[26]Bignell and Fortune, *Understanding Systems Failures*; Fortune and Peters, "The Formal System Paradigm for Studying Failures"; Peters and Fortune, "Systemic Methods for the Analysis of Failure"; Fortune and Peters, "Turning Hindsight into Foresight—Our Search for the 'Philosopher's Stone' of Failure"; Fortune and Peters, *Information Systems*.

[27]The formal system model used in the Systems Failures Approach was developed from the one used in SSM.

[28]Spiekermann, "The Challenges of Privacy by Design," 39.

Regarding the first point, privacy research has not helped practitioners, who face the daunting task of reconciling the operational demands of information security, adherence to data protection legislation, and information management, with commercial pressures for the collection and processing of increasing amounts of personal data as a result of business demand for customer-centred interaction.[29] In addition, customers' and regulators' demands for high privacy standards are continually in tension with an organisation's commitment to shareholders to maximise profit, which is often only achievable by using customers' information for commercial gain.

With respect to the second point, principles for privacy-aware systems[30] and Privacy by Design (PbD) guidelines[31] have existed for over a decade, and are championed by data protection regulators in Europe and the United States.[32] However, significant challenges remain in implementing PbD into the design and development process of organisations' information systems.[33]

To address these points, by unifying the insights from research on privacy, information security and trust, with PbD principles, Morton & Sasse[34] propose the Privacy Security and Trust (PST) Framework (Fig. 1). This represents the composition of privacy of the two parties (e.g. a customer and e-commerce vendor) involved in a technology-mediated interaction via a technology platform[35] (e.g. an e-commerce website) as a privacy practice 'stack' consisting of five layers:

1. **Information Privacy Culture**—This layer assists organisations to make decisions about what information assets[36] to collect, store, disseminate and share, and how to use them. It consists of an organisation's information culture, information ethics and information security culture.
2. **Information Use**—This layer defines the use to which an organisation puts its information assets. From an organisation's perspective, this encompasses *actual information use, intended information use* and *advertised information use*. From a user's perspective, this encompasses *expected information use* and *experienced information use*.

---

[29]Boyce, "Beyond Privacy."

[30]Langheinrich, "Privacy by Design—Principles of Privacy-Aware Ubiquitous Systems"; Lederer et al., "Personal Privacy through Understanding and Action."

[31]Cavoukian, *Privacy by Design ... Take the Challenge*; Cavoukian, "Privacy by Design"; Cavoukian, "Privacy by Design—The 7 Foundational Principles."

[32]Rubinstein, "Regulating Privacy by Design"; Spiekermann, "The Challenges of Privacy by Design."

[33]Spiekermann, "The Challenges of Privacy by Design."

[34]"Privacy Is a Process, Not a PET."

[35]Morton & Sasse use the term '*technology lens*' to refer to the technology platform within a technology service to highlight that if an individual views an organisation through a poorly implemented or designed technology platform, they are likely to have a distorted perception of its ability and/or motivation.

[36]Morton & Sasse define '*information asset*' as "*information endowed with value, relevance and purpose for an individual, group or organization*" (p. 93).

Privacy Failures as Systems Failures ... 313

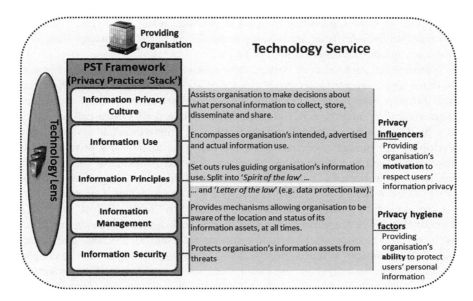

Fig. 1 Privacy, security and trust (PST) framework

3. **Information Principles**—This layer provides the rules which guide an organisation's use of information assets. It includes adherence to data protection legislation (i.e. the 'letter of the law'), where possible exceeding mandatory legal compliance, particularly compliance with fair information principles (i.e. the 'spirit of the law').
4. **Information Management**—This layer provides the tools and processes for the collection, location, archiving, copying, mirroring, sharing, deleting, disseminating, backing up and restoring of information assets.
5. **Information Security**—This layer protects an organisation's information assets from loss, unauthorised access, corruption, interruption or unauthorised disclosure.

Each layer in the PST Framework consumes services provided by the layer below, and in turn, provides services to the layer above. For example, the Information Management layer—necessary for meeting subject access requests, which is a requirement of most data protection legislation—requires the Information Security layer to protect an organisation's information assets. In turn, the Information Management layer enables the Information Principles layer (e.g. facilitating adherence to data protection legislation by allowing an organisation to know the location and status of its information assets).

Four of the five PST Framework layers[37] also exist within the technology lens. For example, a technology lens will have an Information Security layer providing

---

[37]The technology lens does not have an Information Privacy Culture layer, as it cannot possess a culture, albeit the entity designing the technology lens will have a privacy culture influencing its design.

encryption of data transmitted to/from a website using SSL/TLS,[38] and an Information Principles layer, embodying in the design of the technology lens, PbD guidelines and privacy-aware design principles.

Spiekermann's definition of PbD as *"an engineering and strategic management approach that commits to selectively and sustainably minimise information systems' privacy risks through technical and governance controls"*,[39] Cavoukian's SmartPrivacy[40,41] and the UK Information Commissioner's Office call for a privacy by design eco-system within organisations,[42] emphasise that designing and developing information systems which are sensitive to people's privacy needs must be managed as a socio-technical system. Similarly, Morton & Sasse[43] argue that, although technology is important in safeguarding privacy, organisational privacy practice must be considered and managed as a socio-technical system.

Organisational privacy practice is unlikely to tidily follow the PST Framework's layered approach; the PST Framework layers within an organisation's 'privacy practice stack' do not therefore represent identifiable, and isolatable, socio-technical systems responsible for privacy practice. For example, the Information Security layer will contain services, such as security awareness and training for employees, from the information security culture element of the Information Privacy Culture layer.

3 Soft Systems Methodology

3.1 Hard and Soft Systems Thinking—Conceptual Differences

Soft Systems Methodology (SSM) evolved during the late 1960s at the University of Lancaster from systems engineering research by Gwilym Jenkins.[44] Jenkins first proposed the tenets of the *systems approach*—later used in Checkland's SSM[45]—which included: (a) the need for a holistic and interdisciplinary—rather than piecemeal—approach to problem solving—presaging the idea of *'tool*

---

[38]Secure Sockets Layer/Transport Layer Security.

[39]Spiekermann, "The Challenges of Privacy by Design," 38.

[40]Cavoukian, "SmartPrivacy: Lead with Privacy by Design."

[41]Cavoukian's SmartPrivacy augments PbD by including law, regulation, market forces, education and awareness, independent oversight, fair information practices etc.

[42]UK Information Commissioner's Office, "Privacy by Design."

[43]Morton and Sasse, "Privacy Is a Process, Not a PET."

[44]Jenkins, "The Systems Approach."

[45]Checkland, *Systems Thinking, Systems Practice*.

clinics'[46]; (b) the creation of an overall system from subsystems; (c) the unification of disparate specialist techniques to solve complex problems; and (d) the communication of a system's correct objectives to all concerned.[47]

The research group at the University of Lancaster found existing 'hard' systems engineering approaches were not suitable for the ill-defined problems typically faced by managers. What was required was *a system of enquiry*—'soft' systems engineering.[48] SSM—developed as a methodology to analyse complex and messy problems—shares two important characteristics with 'hard' systems engineering: (1) the concept of holism through the application of a methodology that considers all relevant aspects—technological and human—of a system (similar to the PST Framework's 'privacy practice stack'); and (2) the use of concepts linked to the idea of a system.[49]

Checkland & Scholes[50] warn that the use of word *system* in SSM, causes people to assume it refers to a representation of an actual instance of a socio-technical system (e.g. legal system, transport system, etc.). They argue that SSM is a framework for organising the exploration of messy, complex problems as a learning system—not a representational model of reality—it is epistemological, not ontological. SSM views a situation *as if it were* a system—it does not mean *it is* a system. For example, it is impossible to point at something in an organisation and state, "*This is the information privacy practice system*". This is perhaps one of the most significant differences between hard systems thinking, which assumes 'systems' exist in the world and creates a model of them, and soft systems thinking, which assumes "*the process of inquiry into the world can be a consciously organised learning system*".[51] For example, in SSM it is quite correct to state, "*I will treat the provision of information security in an organisation **as if it were** a system*"; this is very different from declaring that it **is** a system. One only has to consider the pervasive influence of an organisation's information security culture in the Information Privacy Culture layer of the PST Framework, to realise the impossibility of identifying and isolating an organisation's 'information security system'.

Checkland & Scholes (see Footnote 50) clarify the distinction between hard and soft systems thinking by suggesting that hard systems thinking assumes the perceived world (*PW*) contains systems—albeit they prefer to call these

---

[46]Morton et al., "4.3 'Tool Clinics'—Embracing Multiple Perspectives in Privacy Research and Privacy-Sensitive Design."

[47]Jenkins, "The Systems Approach."

[48]Checkland and Scholes, *Soft Systems Methodology in Action: Including a 30 Year Retrospective*, 18.

[49]Fortune and Peters, "The Formal System Paradigm for Studying Failures."

[50]Checkland and Scholes, *Soft Systems Methodology in Action: Including a 30 Year Retrospective*.

[51]Checkland and Scholes, *Soft Systems Methodology in Action: Including a 30 Year Retrospective*, A42.

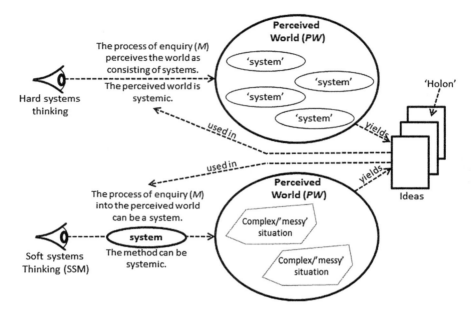

Fig. 2 The process of enquiry (*M*) into the perceived world (*PW*) for hard systems engineering and SSM. Based on Checkland & Scholes (Checkland and Scholes, *soft systems methodology in action: including a 30 year retrospective*)

'holons',[52] whereas soft systems thinking believes the process of enquiry (*M*) can itself be created as a holon, i.e. SSM is a cyclic methodology which is itself a systemic process (Fig. 2). With hard systems thinking the process of enquiry can be systematic, and the perceived world is systemic, whereas with SSM the process of enquiry can be systemic (see Footnote 50).

As part of the development of SSM, Checkland extended the properties of a system proposed by Jenkins[53] for use in his *formal system model*, which is a "*general model of any human activity system*".[54] Comparison between the formal system model and the conceptual model of the problem situation under investigation is an intrinsic part of the SSM process, as it identifies flaws, weaknesses and omissions in the conceptual model, facilitating its improvement.[55] The improved conceptual model may then be compared with the real-world situation to determine which desirable or feasible changes are required

---

[52]'*Holon*' is a term created by Koestler in "*The Ghost in the Machine*" (1967, p. 48). It is a self-reliant and self-organising system made up of other holons. It can operate without recourse to higher authorities, whilst simultaneously being subject to control by those higher authorities. An organisation's information security system is a good example of a holon.

[53]Jenkins, "The Systems Approach."

[54]Checkland, *Systems Thinking, Systems Practice*, 173–174.

[55]Checkland, *Systems Thinking, Systems Practice*.

Checkland (see Footnote 54) stipulated that a system may only be considered a *formal system* if, and only if, it has:

1. an ongoing purpose or mission;
2. a measure of performance;
3. a decision-taking process;
4. components which are themselves systems;
5. components which interact;
6. a boundary separating it from its environment;
7. physical and abstract resources that are at the disposal of the decision-taking process;
8. some guarantee of continuity; and
9. it exists in wider systems and environments with which it interacts.

3.2 The Importance of Weltanschauungen in Systems Thinking

Morton et al.[56] argue that technological determinism and focalism influence scientists and technologists in designing what they believe to be the 'best' solution to a problem. This can result in limited consideration of the viewpoints of stakeholders, potential users, and those affected (positively or negatively) by a proposed solution. To mitigate this, Morton et al. propose the idea of 'tool clinics' to encourage a collaborative (re)consideration of technological solutions, research methods or other artefacts, from multiple perspectives. They describe existing methods that use a multi-perspective and collaborative approach, including 'war games', 'Red Team' reviews, constructive technology assessment (CTA), and SSM—they believe similar holistic approaches should be applied to both theoretical and applied privacy problems. They argue that problems, such as the launch of Google Buzz, and the market failure of the Platform for Privacy Preferences Project (P3P), would have benefited from a critical assessment of their design, development and deployment from multiple perspectives.

One of the multi-perspective and collaborative approaches mentioned by Morton et al. is SSM; Checkland[57] and Ackoff[58] also believe that to fully understand a system it is necessary to consider its purpose from different viewpoints. This pluralism, which represents one of the most important characteristics of SSM, aims to construct a rich picture of a problem, encompassing different viewpoints, rather than the reductionist focus typical of hard systems engineering.

---

[56]Morton et al., "4.3 'Tool Clinics'—Embracing Multiple Perspectives in Privacy Research and Privacy-Sensitive Design."

[57]*Systems Thinking, Systems Practice.*

[58]"The Systems Revolution."

These different viewpoints, or *Weltanschauungen*—the plural of the German word *Weltanschauung*, meaning "*a particular philosophy or view of life; a concept of the world held by an individual or a group*"[59]—make the idea explicit that individuals have different values, assumptions, perceptions and expectations of a system.[60] Similarly, in technology-mediated interactions one party will have assumptions, perceptions and expectations about the privacy behaviour of the other party; if the other party's privacy behaviour deviates from these expectations and assumptions (e.g. collected information is repurposed), the trusting party may respond emotionally, distrusting the other party's motives.[61] In the case of Google Buzz, its developers did not appear to take into account that: (a) users' primary task was to read their e-mail, and hence they would 'swat away' Buzz dialogue boxes without reading them properly[62]; and (b) users' mental models of Gmail is that it is an e-mail service, and not a social networking service.

A recent example of conflicting viewpoints in a privacy-affecting context was the reaction to Google's announcement on 24th January 2012 that it would take more than 70 privacy policies covering its products and services, and consolidate more than 60 of them into its main privacy policy.[63] Two of the reasons for this change given by Google in its announcement, were:

- "*Regulators globally have been calling for shorter, simpler privacy policies—and having one policy covering many different products is now fairly standard across the web*".
- Treating users as a single user across all products (if a user is registered on Google and is signed in), by combining information from one service with information from other services, to "*create a beautifully simple, intuitive user experience across Google*".

During the two days following Google's announcement, concerns were raised about the change to their privacy policy, and its ability to collect data across all its services.[64] An observation by one journalist that Google's change to its privacy policy was either "*simply a matter of adding user (and vendor) convenience, or a gross violation of our privacy*"[65] provides a good example of two different

---

[59]From the Oxford English Dictionary definition of *Weltanschauung* at http://www.oed.com/view/Entry/227763 (accessed 5th March 2015).

[60]Fortune and Peters, *Information Systems*.

[61]Adams and Sasse, "Privacy Issues in Ubiquitous Multimedia Environments: Wake Sleeping Dogs, or Let Them Lie?"; Adams and Sasse, "Privacy in Multimedia Communications: Protecting Users, Not Just Data."

[62]Carlson, "Google Buzz Still Has Major Privacy Flaw."

[63]Google Inc., "Updating Our Privacy Policies and Terms of Service."

[64]DiSalvo, "Google Says Bye Bye to User Privacy"; Gaudin, "Google Stirs up Privacy Hornet's Nest"; Krasnoff, "Google's New Privacy Policy: Checking the Source"; Mills, "Google Wants Ability to 'Combine' Your User Data"; Tsukayama, "Google Faces Backlash over Privacy Changes."

[65]"Google's New Privacy Policy: Checking the Source."

Privacy Failures as Systems Failures ... 319

Weltanschauungen—both equally valid to those who possess them. In October 2012, Google's change to its privacy policy was criticised by European data protection authorities,[66] with Google formally agreeing with the UK's Information Commissioner's Office in January 2015 to improve the information it provides to people about its collection of personal data in the UK.[67]

3.3 Root Definitions and CATWOE

One of the challenges in SSM is the selection of the systems most appropriate to the problem under consideration, and the formulation of names for those systems; in SSM the names of the relevant systems are 'root definitions' that *"express the core purpose of a purposeful activity system"*.[68] The core purpose is expressed as a transformation process where an entity—the 'input'—is transformed into a new form of the same entity—the 'output'. The core purpose of a system can be expressed as a transformation process P (i.e. what to do?), performed by doing Q (i.e. how to do it?), which takes as input some entity, and outputs it in a transformed state, to contribute to achieving an objective R (i.e. why do it?).[69] Using P, Q and R allows root definitions for systems to be formulated as *"a system to do P by Q in order to contribute to achieving R"*.

Checkland & Scholes[70] suggest well-formulated root definitions contain six elements:

1. The customers (C)—beneficiaries or victims—affected by the transformation process T.[71]
2. The actors (A) who perform the transformation process T.
3. The transformation process (T) through which inputs are transformed to outputs.
4. The *Weltanschauung* or worldview (W) that makes the transformation T meaningful in context.

---

[66]Arthur, "Google Privacy Policy Slammed by EU Data Protection Chiefs"; CNIL, "Google's New Privacy Policy : Incomplete Information and Uncontrolled Combination of Data across Services."

[67]Information Commissioner's Office, "Google to Change Privacy Policy after ICO Investigation."

[68]Checkland and Scholes, *Soft Systems Methodology in Action: Including a 30 Year Retrospective*, 33.

[69]Checkland and Scholes, *Soft Systems Methodology in Action: Including a 30 Year Retrospective*, A22.

[70]*Soft Systems Methodology in Action: Including a 30 Year Retrospective*, 35.

[71]Rather confusingly, Checkland & Scholes introduce the letter P (p. A22), the letter T in CATWOE (p. 35), and the letter X (p. 36) to denote the transformation process in '*Soft Systems Methodology in Action: Including a 30 year Retrospective*'.

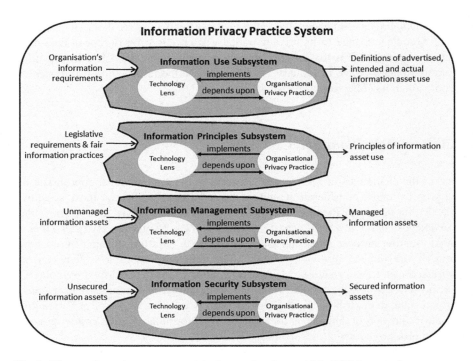

Fig. 3 The transformation processes of the bottom four layers of the PST framework

5. The party who has *ownership* (O) of the system, and power to stop T.
6. The *environmental* (E) constraints outside the system, which it has to take as given.

Checkland & Scholes state, "[it] *is the pairing of the transformation process T and the W, the Weltanschauung or worldview which makes it meaningful*".[72]

Checkland[73] stresses that a system in SSM is not a direct representation of a particular instance of a socio-technical system. Similarly, the PST Framework is not a direct representation of reality, but a simplified and abstracted view. Each of its layers is an abstract representation of a socio-technical system with purposeful activity that contributes to privacy practice within a technology service, e.g. *'an information security system to secure information assets'*. The bottom four layers of the PST Framework can therefore be considered to represent a systemic view of purposeful human activity subsystems operating within a specific technology service's information privacy system, each providing a transformation process T (Fig. 3), with customers (C) and actors (A), owners (O), environmental

---

[72] *Soft Systems Methodology in Action: Including a 30 Year Retrospective*, 35.

[73] Checkland and Scholes, *Soft Systems Methodology in Action: Including a 30 Year Retrospective*.

Table 1 Root definitions and CATWOE of the bottom four layers of the PST framework

Root definition	CATWOE
Information security subsystem	
A system owned by the organisation, provided and operated by its employees and suppliers to protect the organisation's information assets from threats leading to their loss, unauthorised access, corruption, interruption or unauthorised disclosure (P), through the use of security technologies and processes (Q) to avoid harm to the organisation's information assets, protect its competitive advantage and reputation, and facilitate its legal compliance (R)[b] in response to continually changing threats to the organisation's information assets and changes to applicable legislation	C—The organisation and technology service users A—The organisation's employees, suppliers[a] and users T—Transforming unsecured information assets into secured information assets W—Maintaining the organisation's competitive advantage and reputation, and facilitating legal compliance O—The organisation E—Continually changing threats to the organisation's information assets and changes to applicable legislation
Information management subsystem	
A system owned by the organisation, provided and operated by its employees and suppliers to manage the collecting, locating, sharing, archiving, copying, mirroring, deleting, disseminating, backing up and restoring of its information assets (P), through the use of information lifecycle management and storage technologies and processes (Q), to meet the information management requirements of the organisation and users, and facilitate the organisation's adherence to applicable legislation (R), in the face of continually changing principles guiding the organisation's use of information assets	C—The organisation and users A—The organisation employees and suppliers[a], and users T—Transforming unmanaged information assets into managed information assets W—Meeting the organisation's information management requirements O—The organisation E—Continually changing principles guiding the use of the organisation's information assets
Information principles Subsystem	
A system owned by the organisation, provided and operated by its employees to define the rules to guide it in the use of its information assets (P), through the creation and management of privacy policies and company law relating to privacy (Q), to avoid substantive harm[c] to users, adhere to applicable legislation ('letter of the law') and align with fair information practices ('spirit of the law') (R), in the face of changes to applicable legislation, and requirements for different uses for the organisation's existing information assets, or the collection of additional information assets	C—The organisation and users A—The organisation's employees and users T—Transforming the requirements of applicable legislation and fair information practices into an unambiguous set of principles for information asset use W—Adhering to applicable legislation, enacting fair information practices and avoiding substantive harm to users O—The organisation E—Changes to applicable legislation and requirements for different uses for the organisation's existing information assets, or the collection of additional information assets

(continued)

Table 1 (continued)

Root definition	CATWOE
Information use subsystem	
A system owned by the organisation, to define and manage advertised information use, intended information use and actual information use, ensuring advertised information use is a true summary of intended information use, and actual information use is the enacted form of intended information use (P), through clear and unambiguous descriptions of advertised information use, privacy policies setting out intended information use, and accurate and timely feedback of actual information use (Q), aligned with the information principles of the organisation (R), in response to changing information use requirements	C—The organisation and users A—The organisation's employees and users T—Transforming the organisation's requirements into an unambiguous definition of information use, which follows its information principles W—Providing an unambiguous and internally coherent definition of advertised information use, intended information use and actual information use aligned with the organisation's information principles O—The organisation E—Changing information use requirements

[a]Components of an organisation's information security system may be outsourced, with security technologies and processes provided by external parties
[b]For example, responding to Subject Access Requests—required under the 1998 UK Data Protection Act
[c]Bamberger and Mulligan, "Privacy on the Books and on the Ground."

constraints[74] (E), and a worldview (W). Each PST Framework layer views its respective element of overall organisational privacy practice *as if it were* a system—it is impossible to point to a single entity within an organisation and state, *"That is its information security system"*. Root definitions can be constructed for the four privacy subsystems from the *Weltanschauung* of an organisation, and using the CATWOE formulation (Table 1).

Each of the four subsystems within the Information Privacy Practice system reflect the structure of the PST Framework, as they contain a technology lens component that depends on an organisational privacy practice component, which in turn is implemented by the technology lens component (Fig. 3). An example of this relationship is encrypting data transmitted to/from a website using SSL/TLS. To implement this aspect of organisational privacy practice, the technology lens component in the Information Security Subsystem will include SSL/TLS hardware accelerators, SSL/TLS software libraries, certificates, and secure certificate storage. The organisational privacy practice component, on which the technology lens component depends, will include SSL configuration and administration, and certificate lifecycle management.

---

[74]Data protection legislation is an obvious environmental constraint.

4 The Systems Failures Approach

4.1 The Need for a Systems Failures Approach

Most failures arise from highly complex human activity systems consisting of large numbers of interconnected subsystems and components. Attempting to study such failures, and understand the complex interactions between subsystems, requires holistic, or systemic, methods, rather than a reductionist scientific approach, which isolates small groups of individual components to study their interactions.[75] Similarly, the study of privacy failures by focusing on only one layer of the 'privacy practice stack' is not sufficient to understand the root causes of a privacy failure.

A useful way of viewing failure using systems terminology is to consider it as the production of an undesirable output from the system's transformation process.[76] '*Undesirable*' in this context means the system does not meet the designers' objectives, and/or those in system's environment are dissatisfied with its unexpected side-effects.[77] The reduction of an individual's perceived privacy by a technology service is one example of an unexpected side-effect. Whether a particular situation is judged to be a failure is subjective—"*failure is an observation about something, for the failure is not the thing itself, even though we sometimes lapse into this shorthand*".[78] For example, an organisation may achieve its goal of increasing its revenue through effective targeted advertising based on its customers' personal information, but its customers may consider such personalised adverts as intrusive, and therefore a privacy failure.

Table 2 lists the four categories of failure, which are not mutually exclusive, identified by Fortune & Peters,[79] with an example of a privacy failure for each one. They describe Type 3 failures as cases where something is designed to fail, e.g. a fuse blowing. Although there is no direct parallel example of a Type 3 privacy failure, observers have alleged that privacy breaches may be deliberately engineered, e.g. in the design of some P2P file-sharing clients.[80]

---

[75]Fortune and Peters, "The Formal System Paradigm for Studying Failures."

[76]Bignell and Fortune, *Understanding Systems Failures*.

[77]Fortune and Peters, "The Formal System Paradigm for Studying Failures."

[78]Bignell and Fortune, *Understanding Systems Failures*, 7.

[79]Fortune and Peters, *Learning from Failure—The Systems Approach*, 21–23.

[80]Sydnor, Knight, and Hollaar, "Filesharing Programs and 'Technological Features to Induce Users to Share.'".

Table 2 Failure types with examples of privacy failures

	Failure type	Privacy failure	Reason for privacy failure
Type 1	Objectives not met	Sony PlayStation breach	Failure to keep security software up to date, and store users' passwords securely[a]
Type 2	Undesirable side effects	Users' reaction to Google Buzz	Failure to recognise that users would be concerned about the disclosure of their frequent e-mail and chat contacts on their public Google profile
Type 3	Designed failures	Design of P2P file-sharing clients	Failure to respect the privacy of users' personal information on their computers, because of a desire to maximise shared content, as alleged by some observers[b]
Type 4	Inappropriate objectives	Implementation and adoption of P3P	Failure to include in design objectives the need for market incentives, and enforcement through government regulation or industry self-regulation[c]

[a]Information Commissioner's Office, "Sony Fined £250,000 after Millions of UK Gamers' Details Compromised"
[b]Sydnor, Knight, and Hollaar, "Filesharing Programs and 'Technological Features to Induce Users to Share
[c]Morton et al., "4.3 'Tool Clinics'—Embracing Multiple Perspectives in Privacy Research and Privacy-Sensitive Design"

The development of a link between the study of failures, and systems concepts and methods has a long history.[81] Fortune & Peters[82] describe the work of Sheridan,[83] who suggested that control provides a useful framework and discipline for improving industrial safety. Sheridan identified some of the issues of *meta-control* such as, *"what funds is it justified to spend on investigating industrial safety and planning improvements, based on expected payback in loss prevention and ethical standards?"*[84] An analogous question for organisations looking to implement effective privacy practice might be, *"what funds is it justified to spend on investigating technological and process improvements in privacy practice, based on expected payback in data breach prevention, avoidance of prosecution under data protection legislation, ethical standards, and avoidance of reputational damage?"*

---

[81]Peters and Fortune, "Systemic Methods for the Analysis of Failure."

[82]"The Formal System Paradigm for Studying Failures."

[83]T.B. Sheridan, 'Industrial Safety Viewed as a Control Problem', Position paper for World Bank *Workshop on Safety Control and Risk Management*, Washington, DC, 18–20 October 1988.

[84]Sheridan (1988) quoted in Fortune and Peters, "The Formal System Paradigm for Studying Failures," 386.

Privacy Failures as Systems Failures ...

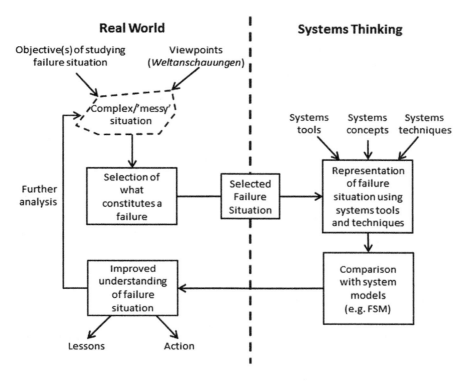

Fig. 4 The systems failures approach. Based on Fortune & Peters (Fortune and Peters, *information systems*, 8)

The study of systems failures for pedagogical purposes by The Open University's Systems Group, led to the development of a systems-thinking approach—the *Systems Failures Approach*[85]—for understanding and predicting failures.[86] The Systems Failures Approach takes a complex real-world situation, decides which aspects may be considered a failure, and represents it as a system, which is compared with system models and other systems-related paradigms. This results in an understanding of the root causes of the failure, leading to the production of a list of lessons learned, action plans and possible directions for further investigation (Fig. 4).

---

[85]Also called 'Systems Failures Method' or 'Failures Method'.

[86]Bignell and Fortune, *Understanding Systems Failures*; Fortune and Peters, "The Formal System Paradigm for Studying Failures"; Peters and Fortune, "Systemic Methods for the Analysis of Failure"; Fortune and Peters, "Turning Hindsight into Foresight—Our Search for the 'Philosopher's Stone' of Failure"; Fortune and Peters, *Information Systems*.

The Systems Failures Approach has seven stages[87]:

1. Pre-analysis of the problem.
2. Identification of significant problems and or failure(s).
3. Selection of the systems to be analysed.
4. Representation of the failure situation as a FSM ("*a model of a robust system that is capable of purposeful activity without failure, and that coordinates a number of key systems concepts within an organized framework*".[88])
5. Testing of a system representation of the failure situation against a FSM, or paradigm.[89]
6. Further analysis through comparison with other systems-related paradigms based on typical failures including control, communication of information, human factors and safety culture.
7. Synthesis to yield a new understanding. In addition to the FSM, stage 4 can include the use of other systems tools such as rich pictures, systems maps and influence diagrams.[90]

4.2 The Formal System Model (FSM)

The FSM[91] used in the Systems Failures Approach—first published in 1984[92]—is adapted from Checkland,[93] and encompasses his nine criteria for a formal system[94]:

- An ongoing purpose that sets the expectations of the system (1).[95]
- A system consisting of:

---

[87]Fortune and Peters, *Learning from Failure—The Systems Approach*.

[88]Stewart and Fortune, "Application of Systems Thinking to the Identification, Avoidance and Prevention of Risk," 283.

[89]In some publications Fortune et al. refer to the FSM used in the Systems Failures Approach as a '*Formal Systems Paradigm*' (e.g. Fortune and Peters, "The Formal System Paradigm for Studying Failures."), whilst other publications refer to it as '*Formal System Model*' (e.g. Fortune and Peters, *Information Systems*.) This chapter will refer to it as a formal system model (FSM).

[90]Fortune and Peters, *Learning from Failure—The Systems Approach*.

[91]Checkland suggests the FSM should be dropped from SSM, because its use of terms such as '*boundary*', '*subsystems*', '*resources*', etc. only served to reinforce people's misconception of SSM as an ontological, rather than epistemological, tool. Nevertheless, it is still used in the Systems Failure Approach.

[92]See Watson L. (1984). *Systems Paradigms*, Open University Press, Milton Keynes, UK, cited in Fortune and Peters, "Turning Hindsight into Foresight—Our Search for the 'Philosopher's Stone' of Failure."

[93]Checkland, *Systems Thinking, Systems Practice*, 173–174.

[94]Stewart and Fortune, "Application of Systems Thinking to the Identification, Avoidance and Prevention of Risk"; Fortune and Peters, "Turning Hindsight into Foresight—Our Search for the 'Philosopher's Stone' of Failure"; Fortune and Peters, *Information Systems*.

[95]The numbers in parentheses refer to Checkland's nine criteria for a formal system.

- A performance-monitoring subsystem to monitor the actual outputs of the system, and compares these with the outputs expected by the decision-making subsystem; any shortfalls are reported to the decision-making subsystem, so it can take corrective action if necessary (2).
- A decision-making subsystem to enable the system's purpose to be achieved. It makes its expectations known to the subsystems and components that carry out the transformation, and to the performance-monitoring subsystem. It also ensures any resources needed by the system are made available (3).
- Subsystems and components that carry out transformations so that the purpose of the system is achieved through the transformation of inputs to outputs (4).
- Subsystems and components that communicate with each other (5).
- A wider system, which is the next hierarchical level upwards from the system. It defines the system's purpose and sets its objectives. It also influences the decision-making subsystem, formulates the system's initial design, provides resources, sets expectations of the system, and monitors its performance (6).
- An environment in which the wider system operates, which the system has no control over (6).
- Boundaries separating the system and the wider system (system boundary), and the system and environment (wider system boundary) (7).
- Resources which the decision-making process allocates (8).
- Some guarantee of continuity (9).

Use of the Systems Failures Approach to identify and conceptualise appropriate systems from within the failure situation, and comparing them with a FSM has highlighted common reasons for systems failures: (a) deficiencies in apparent organisational structure; (b) a lack of a clear statement of purpose from the wider system; (c) poor performance of subsystems; (d) poor communication between subsystems; (e) inadequate design of subsystems; (f) insufficient consideration given to environmental influences; and (g) imbalances in resource allocation.[96] It is therefore likely a similar approach to understanding privacy failures will identify common systemic failings and weaknesses.

5 The Privacy-Specific Formal System Model (PSFSM)

White & Fortune[97] propose a project-specific FSM—a development of Stewart & Fortune's[98] idea of applying systemic methods to project management—as a tool within the Systems Failures Approach for studying project failure, and identifying

---

[96]Fortune and Peters, "The Formal System Paradigm for Studying Failures"; Stewart and Fortune, "Application of Systems Thinking to the Identification, Avoidance and Prevention of Risk."

[97]White and Fortune, "The Project-Specific Formal System Model."

[98]Stewart and Fortune, "Application of Systems Thinking to the Identification, Avoidance and Prevention of Risk."

potential and actual weaknesses in projects. White & Fortune map the causes of project failures reported in project management literature with the components of their project-specific FSM, which they make meaningful to project managers by including failure modes with 'prompts' for considering multiple viewpoints (*Weltanschauungen*). Similarly, the PSFSM System (Fig. 5) can be used within the Systems Failures Approach to investigate privacy failures, and identify potential and actual weaknesses in a technology service's information privacy system.

The PSFSM represents the bottom four layers of the PST Framework as subsystems within a Technology Service Information Privacy System, operating within the wider system of an organisation's Information Privacy Culture System (Fig. 5). Systems can be hierarchical, and therefore the Technology Service Information Privacy System can be part of a wider Technology Service Development/Operation System, and its subsystems can be modelled as a formal system in their own right.

The Information Privacy Culture System (the wider system in Fig. 5), which aligns with the top layer of the PST Framework—the Information Privacy Culture layer influences all activities concerned with an organisation's privacy practice, i.e. the subsystems in the Technology Service Information Privacy System. In turn, the Information Privacy Culture System is disturbed by environmental influences, including users' privacy norms and expectations, users' mental models, data protection legislation, and competitive threats. The Information Privacy Culture System may also attempt to influence areas in the environment, such as users' attitudes to sharing information, and the drafting of data protection law.

Notwithstanding Checkland's advice that his FSM is not prescriptive, and is therefore not a representation of what *ought* to exist in the real world,[99] the Technology Service Information Privacy System in the PSFSM contains four privacy subsystems, each containing a technology lens and an organisational component (Fig. 3), which must exist for effective privacy practice.

The PSFSM assumes there is a separate Technology Service Information Privacy System—containing a Privacy Decision-making Subsystem, Privacy Practice Subsystems, and a Privacy Performance-Monitoring Subsystem—for each technology service designed, developed and operated by an organisation. This recognises that a privacy failure can occur if there are weaknesses in the communication paths between the Information Privacy Culture System and one or more Technology Service Information Privacy Systems. For example, an organisation may have an effective Information Privacy Culture System, but an unexpected competitive threat may cause it to allocate insufficient resources (e.g. time and external privacy expertise) to the development and operation of a new technology service, ultimately leading to a privacy failure in that technology service.

In large organisations with a Chief Privacy Officer and a centralised privacy governance function—part of a wider Information Privacy Culture System—the Privacy Decision-making Subsystem could be a member of the organisation's

---

[99]Checkland, *Systems Thinking, Systems Practice*.

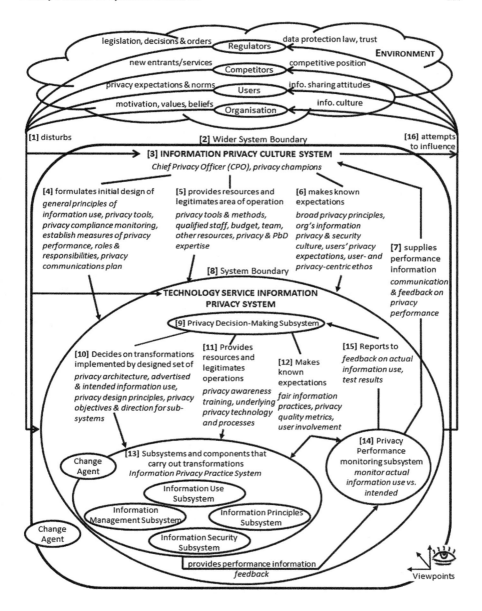

Fig. 5 Privacy-Specific Formal System Model (PSFSM)

privacy governance team seconded to the technology service's development project or service management team. In smaller organisations, the Privacy Decision-making Subsystem could be a project manager or IT service manager who has completed privacy awareness and data protection training, and is therefore able to make decisions about a technology service's privacy practice.

The PSFSM is flexible enough to be used in the design and development, and ongoing operation of technology services. For example, the communication path [15] in Fig. 5 might be for test results during the development of a technology service, or the operational reports to ensure actual information use matches intended information use. In common with White & Fortune's project-specific FSM, the PSFSM includes privacy failure modes—mapped, where appropriate, to privacy literature—as 'prompts' for privacy practitioners (Table 3). Table 3 provides a mapping between the failure modes associated with privacy subsystems and the components of the PSFSM (the bold figures in square brackets in Table 3 refer to elements of the PSFSM in Fig. 5). The list of failure modes and relevant privacy literature in Table 3 is not exhaustive.

This chapter represents an initial step in proposing the use of systems-thinking in general, and the PSFSM in particular, as a tool for gaining a better understanding of privacy failures. Nevertheless, Table 3 identifies areas of privacy research—particularly those concerned with privacy practice—which would benefit from further work. For example, one of the failures identified in the Privacy Performance Monitoring Subsystem in Table 3 is not monitoring that *actual* information use is the same as *intended* information use (i.e. the technology service's privacy policy description of intended information use accurately reflects actual information use). Two of the problems identified with P3P were: (1) its lack of a mechanism to ensure organisations acted in accordance with their privacy policies[100]; and (2) not all organisations carried out corrective maintenance on invalid P3P policies.[101] Valuable lessons can be learned from the failure of P3P, particularly from the *Weltanschauungen* of browser developers, regulators and those who design and host websites.[102]

6 Applying the PSFSM to Google Buzz

The principal objective of this chapter is to describe the application of systemic methods and tools to privacy failures, and introduce the PSFSM. However, to provide an example—albeit limited—proof-of-concept application of the PSFSM, this section describes a brief analysis of Google Buzz using the Systems Failures Approach to identify possible root causes for the failure. It begins with the

---

[100]EPIC, "Pretty Poor Privacy: An Assessment of P3P and Internet Privacy."

[101]Reay, Dick, and Miller, "A Large-Scale Empirical Study of P3P Privacy Policies."

[102]Morton et al., "4.3 'Tool Clinics'—Embracing Multiple Perspectives in Privacy Research and Privacy-Sensitive Design."

Table 3 Mapping of technology service privacy failure modes onto PSFSM[a]

PSFSM component	Failure modes
[1] Environment disturbs	*Failure to:* • Manage uncertainty • Learn from past experience (e.g. adverse publicity and reports by regulators concerning previous technology services) • Consider regulators' position and objectives • Consider regulators' privacy decisions and orders • Deal with changes in the legislative environment • Monitor for competitive threat from new services/entrants; and • Maintain awareness of changes in users' social and privacy norms
[2] Wider system boundary	*Failure to:* • Consider users' privacy expectations and assumptions, and the potential for their emotional reaction to privacy invasion[b] • Understand users' mental models[c] and their social norms and behaviour concerning the sharing of information[d] • Consider users' privacy norms within different contexts[e] • Understand legislation and regulators' expectations[f]; and • Understand organisations' motivation
[3] Wider system (information privacy culture system) (places technology service information privacy system in context)	*Failure to:* • Have broad privacy principles—part of an organisation's technical and business strategy—guiding privacy decisions in different contexts[g] • Consider organisation's values, beliefs and culture which surrounds the development and operation of its technology services • Consider organisation's values and beliefs concerning information collection, processing and use • Appreciate organisation's motives behind requests for information collection, processing and use • Consider effect of organisation's structure, policy and incentives • Consider organisation's information security culture; and • Have a culture of privacy in the organisation[h]

(continued)

Table 3 (continued)

PSFSM component	Failure modes
[4] Formulates initial design of technology service information privacy system	*Failure to*: • Clearly articulate general principles of information use • Identify unambiguous requirements and objectives • Specify design and practice tools and methods to be used • Develop and communicate privacy communications plan • Clearly articulate roles and responsibilities • Produce realistic schedules • Establish privacy compliance monitoring systems; and • Formulate clear measures of performance for information privacy
[5] Provides resources and legitimates area of operation of technology service information privacy system	*Failure to*: • Provide sufficient resources and budget for effective privacy practice and PbD • Provide sufficient and appropriate tools and methods for privacy practice and PbD • Encourage embedding of privacy in the early stages of the design and development process, and the operation of technology services[i] • Employ/use qualified staff • Provide appropriate technology • Adopt new ways of working, new legislation, and recent advances in privacy practice and theory; and • Use external privacy practice and PbD expertise as required
[6] Makes known expectations of technology service information privacy system	*Failure to*: • Clearly communicate organisation's privacy culture, and its principles of information collection, processing and use • Understand and clearly articulate users' privacy expectations concerning the technology service • Maintain balance between being technology-enabled and having value information flows, and privacy-centric and trust-generating concerns[j] • Instil a user- and privacy-centric ethos[k], and • Anticipate and prevent privacy invasive events before they happen[l]
Arrows linking systems/subsystems and feeding back to wider system/environment (communication channels)	*Failure to*: • Provide effective channels of communication; and • Recognise when requirements, design or operation of a technology service conflict with fair information practices

(continued)

Table 3 (continued)

PSFSM component	Failure modes
[7] Supplies performance information from technology service information privacy system to wider information privacy culture system	*Failure to:* • Report progress/service performance accurately; and • Inform organisation's senior management if a technology service needs to be abandoned/redesigned because design, development or operational problems prove to be insurmountable
[8] System boundary	*Failure to:* • Trust opinions provided by wider system • Appreciate different *Weltanschauungen*; and • Appreciate how technology service(s) will be decoded from multiple perspectives[m]
[9] Privacy decision-making subsystem	*Failure to:* • Ensure privacy is embedded into the design, development and operation of the technology service[n] • Manage resources required for the organisation's privacy practice • Decide on privacy awareness and PbD training needs; and • Assign privacy specialists to design, development and operational teams
[10] Decides on transformations implemented by a designed set of subsystems in technology service information privacy system	*Failure to:* • Define the technology service's privacy architecture • Clearly and unambiguously articulate intended information use and advertised information use for technology service[o] • Define the level of information control within the technology service, in line with users' expectations • Provide privacy design principles • Ensure users can take socially interpretable action[p] • Ensure users have anonymity and pseudonymity[q] • Agree attitude to information security risk • Clarify privacy subsystem roles and responsibilities; and • Provide clear privacy objectives and direction for teams
[11] Provides resources and legitimates operations of subsystems in technology service information privacy system	*Failure to:* • Provide privacy awareness training • Define underlying privacy technology and processes • Provide appropriate information management and security processes and technologies; and • Provide usable privacy control mechanisms[r]

(continued)

Table 3 (continued)

PSFSM component	Failure modes
[12] Makes known expectations of subsystems in technology service information privacy system	*Failure to:* • Communicate importance of adherence to principles of fair information practices • Define quality metrics for protection of users' information privacy • Instil importance of showing users actual and potential information flows[s] • Encourage user involvement; and • Clearly articulate intended information use and advertised information use[t]
[13] Information privacy practice system	*Failure to:* • Implement effective and compliant privacy practice
[14] Privacy performance monitoring subsystem	*Failure to:* • Monitor state of subsystems in technology service information privacy system • Perform adequate user acceptance and usability testing • Monitor design and operation of technology service to ensure it does not run counter to accepted social behaviour[u] • Monitor if actual information use breaches intended information use (as specified in the privacy policy) and advertised information use[v]; and • Recognise focalism[w] and 'groupthink', particularly in technology-led organisations
[15] Reports to	*Failure to:* • Quickly and accurately report when design and operation of technology service run counter to accepted social behaviour[x] • Quickly and accurately report if actual information use breaches intended information use and advertised information use; and • Quickly and accurately report results of technology service testing
[16] Attempts to influence environment	*Failure to:* • Influence users' attitude to providing and sharing personal information[y] • Build trust with users through use of trust signals[z] from technology service(s) • Build trust with legislative bodies; and • Persuade legislative bodies that its technology services(s) comply with data protection legislation

(continued)

Table 3 (continued)

a Based on Table 2 in White and Fortune, "The Project-Specific Formal System Model," 44
b Adams and Sasse, "Privacy in Multimedia Communications: Protecting Users, Not Just Data"
c Camp, "Mental Models of Privacy and Security"
d Bellotti and Sellen, "Design for Privacy in Ubiquitous Computing Environments"
e Nissenbaum, "Privacy as Contextual Integrity"
f Bamberger and Mulligan, "Privacy on the Books and on the Ground"
g Bamberger and Mulligan, "Privacy on the Books and on the Ground"
h Culnan and Williams, "How Ethics Can Enhance Organizational Privacy: Lessons from the ChoicePoint and TJX Data Breaches"
i Cavoukian, "Privacy by Design - The 7 Foundational Principles"; Spiekermann, "The Challenges of Privacy by Design"
j Bamberger and Mulligan, "Privacy on the Books and on the Ground"
k Cavoukian, "Privacy by Design - The 7 Foundational Principles"
l Cavoukian, "Privacy by Design - The 7 Foundational Principles"
m Morton et al., "4.3 'Tool Clinics'—Embracing Multiple Perspectives in Privacy Research and Privacy-Sensitive Design"
n Cavoukian, "Privacy by Design - The 7 Foundational Principles"
o Morton and Sasse, "Privacy Is a Process, Not a PET"
p Lederer et al., "Personal Privacy through Understanding and Action"
q Langheinrich, "Privacy by Design—Principles of Privacy-Aware Ubiquitous Systems"
r Langheinrich, "Privacy by Design—Principles of Privacy-Aware Ubiquitous Systems"; Lederer et al., "Personal Privacy through Understanding and Action"
s Lederer et al., "Personal Privacy through Understanding and Action"
t Morton and Sasse, "Privacy Is a Process, Not a PET"
u Bellotti and Sellen, "Design for Privacy in Ubiquitous Computing Environments"
v Morton and Sasse, "Privacy Is a Process, Not a PET"
w Morton et al., "4.3 'Tool Clinics'—Embracing Multiple Perspectives in Privacy Research and Privacy-Sensitive Design," 97
x Bellotti and Sellen, "Design for Privacy in Ubiquitous Computing Environments"
y For example, Mark Zuckerberg, the CEO of Facebook, may have been attempting to influence people's attitude to sharing personal information, with his statement in 2010, "*People have really gotten comfortable not only sharing more information and different kinds, but more openly and with more people. That social norm is just something that has evolved over time.*"(TechCrunchTV, "Crunchies Awards 2010")
z Riegelsberger, Sasse, and McCarthy, "The Mechanics of Trust"

background to Buzz—based on publicly available source material—of its launch, the privacy concerns it raised, Google's response, and the action of regulators.[103] The remaining four subsections follow the stages of the Systems Failures Approach,[104] thus:

1. Creation of a rich picture of the significant aspects of the design and launch of Google Buzz (rich pictures are one of the principal systems-thinking tools used to summarise the output from the pre-analysis of the failure situation in Stage 1).
2. Identification of significant failure(s) and system selection (Stages 2 and 3).
3. System modelling and comparison with the PSFSM (Stages 4 and 5).
4. Identification of lessons learned, remedies and agenda for change (Stage 7).

6.1 Background

Google Buzz, which was directly integrated into Google's Gmail e-mail service, was launched on 9th February 2010,[105] and included features found in Facebook and Twitter, such as the ability to '*like*' content and '*follow*' other Buzz users; Buzz users could also share status updates, comments, photographs and videos through posts, or '*buzzes*'.[106] Following its launch, when Gmail users logged into their e-mail account, they were shown a pop-up dialogue box describing Buzz. Users who chose to find out more were shown a Buzz welcome panel, but this did not inform them that the service would automatically configure a list of people for them to follow on Buzz. Visible as part of their public Google profile, the list was constructed from the e-mail addresses of people they most frequently e-mailed, or chatted with.[107]

One day after the launch of Buzz, privacy concerns were being raised in online media and blogs,[108] with particular focus on the auto-following of Gmail contacts, which caused problems for Gmail users.[109] Use of Twitter's *@reply* convention by Buzz caused further privacy issues, as it made people's private e-mail address in

---

[103] A more detailed description is provided in the appendix to this chapter.

[104] Stage 6 has been omitted from this proof-of-concept application. This stage consists of further analysis of the problem using other systems-related paradigms for analysing failures, including control, communication of information, human factors and safety (e.g. fault trees, common mode and cascade failures).

[105] Google Inc., "Google Buzz in Gmail."

[106] BBC, "Google Takes on Facebook and Twitter with Network Site"; Helft and Stone, "With Buzz, Google Plunges Into Social Networking"; Krazit, "Google's Social Side Hopes to Catch Some Buzz."

[107] FTC, "Complaint: In the Matter of Google Inc. a Corporation."

[108] Carlson, "WARNING: Google Buzz Has a Huge Privacy Flaw"; Wood, "Google Buzz."

[109] Arthur, "Google Buzz's Open Approach Leads to Stalking Threat"; Kolmes, "Google Buzz Alarms a Psychotherapist | Psychologist San Francisco."

Buzz posts available to followers of the person sending the comment.[110] A further privacy flaw—a cross-site scripting vulnerability affecting mobile users—that exposed Buzz users' location was reported on 16th February 2010.[111]

Google responded very quickly to these concerns. Two days after the launch of Buzz it announced a change to the original profile creation screen displayed when a user posted their first item in Buzz, which included a checked-by-default checkbox advising users that the people they were following would be displayed as part of their Google profile, and allowing them to stop the lists of people who were following them, and people they were following, from being accessible.[112] Google also announced two other changes to Buzz to address people's privacy concerns: (1) making it easier to block anyone, by adding '*block*' links to the list of followers; and (2) make it easier for users to see who is on the public list of their followers that everyone else sees.[113] These changes—although welcomed—still attracted criticism, particularly the fact that it was still not clear to users that Buzz would publish people's frequent e-mail and chat contacts.[114] Over the weekend following the launch of Buzz, Google apologised to its users in a blog post, and announced the replacement of the *auto-follow* model with an *auto-suggest* model.[115] However, Buzz still showed people's following/follower lists publicly by default—users still had to manually make these private.[116] A week after the launch of Buzz, Google told BBC News that Buzz—unlike most of its products—had only been tested within the company by its employees, and not with a more representative group of users as part of Google's Trusted Tester program.[117]

On 16th February 2010, the Electronic Privacy Information Center (EPIC) filed a formal complaint with the US Federal Trade Commission (FTC).[118] The day after, the Privacy Commissioner of Canada asked Google to clarify how it had addressed Buzz's privacy issues since its launch, and how it had met the requirements of privacy law in Canada.[119] This was followed on April 19th 2010 by a letter—signed by the heads of data protection authorities in nine other countries—to Eric Schmidt, the chief executive officer of Google, expressing their concern about

---

[110]Schonfeld, "Watch Out Who You Reply To On Google Buzz, You Might Be Exposing Their Email Address | TechCrunch."

[111]Goodin, "Google Buzz Bug Exposes User Geo Location."

[112]"Millions of Buzz Users, and Improvements Based on Your Feedback."

[113]Google Inc., "Millions of Buzz Users, and Improvements Based on Your Feedback."

[114]Carlson, "Google Buzz Still Has Major Privacy Flaw."

[115]Google Inc., "A New Buzz Start-up Experience Based on Your Feedback."

[116]Frommer, "Google Making More Changes To Buzz After Huge Privacy Outcry."

[117]BBC, "Google Admits Buzz Social Network Testing Flaws."

[118]EPIC, "In the Matter of Google, Inc. Complaint, Request for Investigation, Injunction, and Other Relief," 13 [Clause 47]; EPIC, "EPIC Urges Federal Trade Commission to Investigate Google Buzz."

[119]Privacy Commissioner of Canada, "ARCHIVED—News Release."

Google's privacy practices, albeit praising its rapid response to addressing people's privacy concerns about Buzz.[120]

In September 2010, Google settled a class action lawsuit, which consolidated several civil cases claiming the privacy of Gmail users had been violated by the launch of Buzz.[121] In March 2011, Google agreed to settle the FTC complaint,[122] which charged it with unfair and deceptive acts or practices, contrary to the Federal Trade Commission Act.[123] In the settlement of the complaint, which the FTC finally agreed in October 2011,[124] Google agreed to implement a comprehensive privacy program, and to be audited by a qualified, objective and independent third-party professional every two years, for the next twenty years.[125]

In October 2011, Google announced it was closing Buzz and its application programming interface (API).[126] In May 2013, it undertook the final step in closing Buzz, announcing that users' Buzz content would be moved to their Google Drive accounts.[127]

6.2 Rich Picture of Google Buzz Situation (Stage 1)

The first stage of the Systems Failures Approach involves examination of the failure situation, not the failure itself. It is therefore necessary at this stage to keep an open mind, be aware of the objectives of the analysis, and not impose any particular structure on the failure situation. Diagramming techniques are a useful tool to gain a thorough understanding of the failure situation, organise information into a usable format, and identify different viewpoints and perspectives. As Fortune & Peters observe, "[d]*iagrams of various sorts play a big part in the* [Systems Failures] *Approach. During the pre-analysis stage they allow information to be organized and stored and provide working tools for checking that all aspects of the situation are considered and for generating options*".[128]

One of the diagramming techniques that originate from SSM is rich pictures. Rich pictures, which can be considered the whole of the output of the pre-analysis stage of the Systems Failure Approach, attempt to capture in one

---

[120]Privacy Commissioner of Canada, "ARCHIVED—Letter to Google Inc. Chief Executive Officer—April 19, 2010."

[121]Metz, "Google Pays $8.5 m to Settle Buzz Privacy Invasion Suit."

[122]FTC, "FTC Charges Deceptive Privacy Practices in Google's Rollout of Its Buzz Social Network."

[123]FTC, "Complaint: In the Matter of Google Inc. a Corporation."

[124]FTC, "Decision and Order: In the Matter of Google Inc. a Corporation."

[125]FTC, "FTC Gives Final Approval to Settlement with Google over Buzz Rollout"; FTC, "Decision and Order: In the Matter of Google Inc. a Corporation."

[126]Google Inc., "A Fall Sweep."

[127]Lawler, "Google's 'Last Step' in Buzz Shutdown."

[128]Fortune and Peters, *Information Systems*, 98.

diagram—usually using *"cartoon-like encapsulations of key ideas or pieces of information"*[129]—all of the salient features of a situation, both 'hard' information (e.g. facts, computer systems, organisational structures, actors, etc.) and 'soft' information (e.g. relationships, expectations, perceptions, etc.).

The rich picture of the Buzz situation (Fig. 6) captures significant aspects of the design and launch of Buzz, the principal technology services, and the subsequent reaction of users, based on publicly available sources. The principal objective of this proof-of-concept application of the PSFSM is to understand how and why Google appeared to misunderstand users' privacy expectations and assumptions about the use of Gmail contacts to prepopulate Buzz. Figure 6 therefore excludes certain—equally important—aspects of the situation, such as privacy advocates and regulatory authorities.

6.3 Identification of Significant Failure(s) and System Selection (Stages 2 and 3)

Examination of the rich picture (Fig. 6) and source material, suggests the significant failure in Buzz was a Type 2 Failure (undesirable side effects).[130] The principal undesirable side-effect was users' perception of a reduction in their privacy caused by the disclosure of their most frequent e-mail and chat contacts on their Google public profile. Within this failure are a number of sub-failures, and the clauses in the FTC complaint[131] are a useful resource to look for sub-failures. However, these are too focused on outcomes, and benefit from further insight to understand what the systemic sub-failure is. Additional sub-failures can be identified using the four grey clouds with an eye icon (Fig. 6) representing possible *Weltanschauungen*. Some of the sub-failures include:

- The apparent failure of Google to understand users' mental models for e-mail and social networking services.
- Google's assumption that the people a user chats and e-mails most with are people they like, and would therefore like to follow, or have as followers, on Buzz.
- The apparent failure of Google to understand users' reaction to the integration/combination of data across services; there was a similar reaction to the consolidation of many of Google's privacy policies.[132]
- Failure to allow users to easily, and completely, opt-out of Buzz.

---

[129]Fortune and Peters, *Information Systems*, 100.

[130]Fortune and Peters, *Learning from Failure—The Systems Approach*, 23.

[131]FTC, "Complaint: In the Matter of Google Inc. a Corporation."

[132]DiSalvo, "Google Says Bye Bye to User Privacy"; Gaudin, "Google Stirs up Privacy Hornet's Nest"; Krasnoff, "Google's New Privacy Policy: Checking the Source"; Mills, "Google Wants Ability to 'Combine' Your User Data"; Tsukayama, "Google Faces Backlash over Privacy Changes."

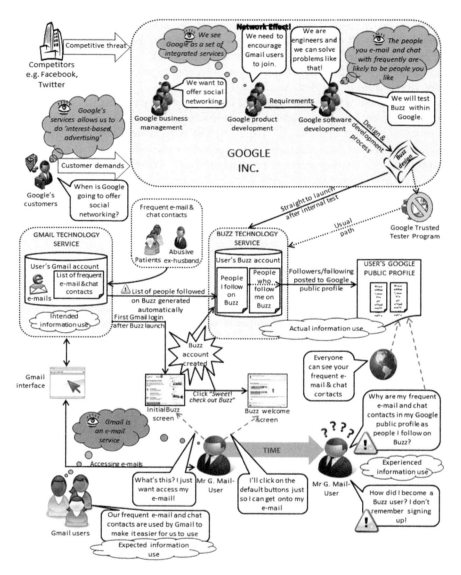

Fig. 6 Rich picture of Google Buzz failure situation

- Failure to make clear to users the actual use of Gmail information assets.
- Obfuscation of potential and actual information flows.

This list of sub-failures is not exhaustive. For example, it excludes Google's apparent failure to understand the implications of not seeking users' consent before using information provided to it for the purpose of providing Gmail, for Buzz, despite its privacy policy at the time stating otherwise.[133]

In addition to the Technology Service Information Privacy System, which has been the focus of this chapter, there are other systems relevant to the Buzz failure situation that can be conceptualized:

- A technology service design and development system.
- A technology service testing system.
- A technology service maintenance and service management system.
- A user system.
- A system to finance technology services.
- A customer system.

Each one of the systems listed above consists of subsystems and components. For example, the technology service design and development system will include consultants, developers, coding standards, user interface standards and guidelines, project management, budgeting system, etc.; users of the system and technology analysts might be considered part of the system's environment.

Any of the systems listed above could be modelled using a FSM, and analysed using the Systems Failures Approach. For the purposes of this example, the system selected for system modelling is the Technology Service Information Privacy System, which will be compared with the PSFSM.

6.4 System Modelling and Comparison with the PSFSM (Stages 4 and 5)

Stage 4 of the Systems Failures Approach involves the construction of diagrammatic models of structure and process of the situation, using tools and techniques, such as input-output diagrams, systems maps and influence diagrams. The ultimate objective of Stage 4 is to facilitate the conceptualisation of the situation as systems in the same format as a FSM. This allows comparison in Stage 5 with the FSM to identify areas of failure, and communications links that are missing from the problem situation. Similarly, a privacy failure situation can be represented as a PSFSM to identify discrepancies and weaknesses in privacy practice. By way of example, Fig. 7 shows a simplified PSFSM for Google Buzz, and Table 4 shows some of the main discrepancies identified from the Buzz situation using the PSFSM.

---

[133]FTC, "Complaint: In the Matter of Google Inc. a Corporation."

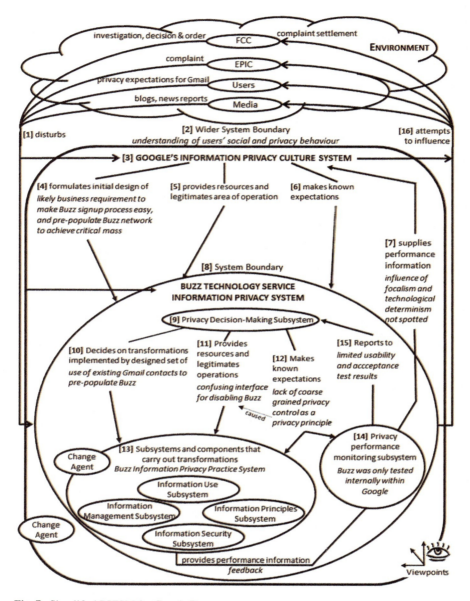

Fig. 7 Simplified PSFSM for Google Buzz

Table 4 The main discrepancies revealed by comparison with PSFSM

Aspect of the PSFSM	Discrepancy
[2] Wider system boundary	Google did not appear to appreciate users' different privacy expectations for e-mail and social networking services. Users do not typically expect be enrolled in a social networking service when checking e-mail. As one journalist observed, "while Google Buzz is in some senses a social network, where follower lists are almost always public, it's still primarily an attachment to people's email accounts—for which people have serious privacy expectations"[a]
	Google's decision to pre-populate Buzz with a user's most frequent e-mail and chat contacts did not take account of the complexity of human relationships. As one journalist observed, *"Google Buzz has demonstrated precisely why—and how—engineers really shouldn't be let loose with human relationships"*[b]
[4] Formulates initial design of	It seems likely there was a business requirement to encourage Gmail users to join Buzz, and therefore the sign-up process was made easy (i.e. click on a few buttons to join by default). To achieve a critical mass of users in Buzz, it was necessary to pre-populate it from people's Gmail chat and -mail contacts
[8] System boundary	There appears to have been insufficient collaborative (re)examination of Buzz from different viewpoints (*Weltanschauungen*) to identify the foreseeable consequences of its design[c]
[14] Privacy performance monitoring subsystem	Buzz was tested within Google with limited types of user, and limited exposure to the complexity of human relationships. This was highlighted by the case of 'Harriet Jacobs'[d]—with one journalist commenting, "[Google] *doesn't have many people who really hate each other internally. Or perhaps none. And of course stalking would be the sort of thing that would lose you your job at Google*"[e]
[7] Supplies performance information	It appears as though Google's engineers were principally focused on solving the problem of pre-populating Buzz, and making the signup process seamless. The wider system should have been more aware of the influence of technological determinism on the technology service information privacy system
[11] Provides resources and legitimates operations	Until Google modified Buzz to include a Buzz tab in Gmail's settings page, *"disabling Buzz and removing a public profile was a multi-step process that confused many users"*[f]. Users were not given a simple top-level mechanism for halting and resuming disclosure of personal information[g]. The information principles for Buzz did not therefore embody usable privacy control mechanisms
[12] Makes known expectations	There did not appear to be a requirement for Buzz to have a coarse grained level of control, so users could quickly opt-out if they wished

[a]Frommer, "Google Making More Changes To Buzz After Huge Privacy Outcry"
[b]Arthur, "Google Buzz's Open Approach Leads to Stalking Threat"
[c]Morton et al., "4.3 'Tool Clinics'—Embracing Multiple Perspectives in Privacy Research and Privacy-Sensitive Design"
[d]This is the pseudonym used in the Guardian article
[e]Arthur, "Google Buzz's Open Approach Leads to Stalking Threat"
[f]Helft, "Anger Leads to Apology From Google About Buzz"
[g]Lederer et al., "Personal Privacy through Understanding and Action"

6.5 Lessons Learned, Remedies and Agenda for Change (Stage 7)

Stage 7 is called 'Synthesis', because it is where the threads of the analysis are pulled together to gain an understanding of the failure as a whole. If a full analysis of the failure situation had been performed, this stage would include the development of further PSFSMs to emphasise salient features, i.e. things that should be changed to prevent the failure from occurring again.

From the brief application of the Systems Failures Approach and the PSFSM to Google Buzz, the four main areas identified for change are:

1. New technology services should be examined from multiple perspectives to minimise any undesirable side-effects (Type 2 Failures) resulting from their use (e.g. a reduction in people's perceived privacy).
2. There should be an increased awareness of the potential dangers of focalism,[134] particularly in technically-biased development teams whose principal objective is solving technical challenges.
3. Privacy-sensitive design principles should be embedded in the development of technology services.
4. There should be an increased awareness of the potential conflict between different *Weltanschauungen*, e.g. the conflict between the business need to pre-populate Buzz, and make the signing up process seamless, and users' desire to control the flow of their personal information.

Some of the observations in the above list may appear *jejune*. This is partly due to the analysis of Google Buzz being a limited proof of concept, and the reliance on publicly available source material. Notwithstanding this criticism, history is littered with instances of apparently simple assumptions, actions, or errors, ultimately leading to a privacy failure. One example of this was the loss in 2007 of 25 million records of child benefit data by Her Majesty's Revenue and Customs (HMRC)—a UK Government department—when a junior official posted two CDs containing the data, using HMRC's postal system.[135] A review of this incident using systems tools and techniques, like the PSFSM, would no doubt ultimately conclude that confidential data should not have been posted using normal courier services. However, the application of systems thinking would look beyond this apparently simple recommendation to discover the root causes of the failure, asking, *"Why did a junior official post the CDs, which inter-system communication paths failed, what privacy subsystems failed, and how did they fail?"*

---

[134]Morton et al., "4.3 'Tool Clinics'—Embracing Multiple Perspectives in Privacy Research and Privacy-Sensitive Design."
[135]BBC, "Timeline."

7 Conclusions and Further Work

Studying and understanding the root causes of privacy failures, such as Google Buzz, requires an holistic approach, encompassing the complete 'privacy practice stack', from 'hard' systems engineering activities, like technical security controls and privacy-sensitive software design, through to 'soft' aspects such as privacy process communication paths, organisational culture, ethics and motivation. To address this challenge, this chapter has described the first application of systems-related concepts, tools and methods to the study of privacy failures. Building on systems-thinking, the systems paradigm of the Formal System Model (FSM),[136] and the Privacy, Security and Trust (PST)[137] Framework, this chapter has described a Privacy-Specific Formal System Model (PSFSM) for use as part of the Systems Failures Approach[138] to study privacy failures. The PSFSM—based on the idea of a project-specific FSM[139]—can be used to analyse privacy practice within technology services, during their design, development and operation. To make the PSFSM meaningful to practitioners, mappings between components of the PSFSM and the failure modes of privacy practice systems have been provided (Table 3); where appropriate, these have been linked with relevant privacy literature.

The proof-of-concept application of the PSFSM in this chapter has two principal limitations: (1) the analysis of the failure of Google Buzz uses only publicly available source material; and (2) the brief application of the Systems Failure Approach, to provide a simple example use of the PSFSM. The first of these limitations means that some aspects of the rich picture in Fig. 6 are inferred, albeit based on publicly available source material. For example, it shows Google's customers as advertisers whose *Weltanschauung* is to view the collection of information by Google as a means of understanding what people are interested in, and hence target advertising more accurately.[140] Similarly, Google's desire to offer a social networking service is informed by the fact that Google was keen to launch a social networking service to compete with Facebook and Twitter; Google had also previously launched social networking services, including Orkut in 2004 and Google Wave in 2009.[141] Regarding the second limitation, a full application of the

---

[136]Checkland, *Systems Thinking, Systems Practice*; Checkland and Scholes, *Soft Systems Methodology in Action: Including a 30 Year Retrospective*.

[137]Morton and Sasse, "Privacy Is a Process, Not a PET."

[138]Bignell and Fortune, *Understanding Systems Failures*; Fortune and Peters, "The Formal System Paradigm for Studying Failures"; Peters and Fortune, "Systemic Methods for the Analysis of Failure"; Fortune and Peters, "Turning Hindsight into Foresight—Our Search for the 'Philosopher's Stone' of Failure"; Fortune and Peters, *Information Systems*.

[139]White and Fortune, "The Project-Specific Formal System Model."

[140]Williams, "Google to Build Profiles of Gmail Users for Advertisers."

[141]BBC, "Google Takes on Facebook and Twitter with Network Site"; Helft and Stone, "With Buzz, Google Plunges Into Social Networking"; Krazit, "Google's Social Side Hopes to Catch Some Buzz."

Systems Failures Approach to Buzz requires access to detailed information, including, but not limited to, Google's design and development processes, how Google makes design decisions, the communication paths between Google's teams, Google's attitude to information use decisions, testing and feedback processes. In addition, a complete study of the failure using the Systems Failures Approach would require considerable work, including additional rich pictures, conceptual systems models, FSMs and PSFSMs, and the use of other systems-related paradigms.

Fortune & White[142] propose the use of the FSM as a framing device to take account of project critical success factors, as it offers the underlying benefits of using critical success factors, whilst overcoming many of the criticisms associated with the unthinking application of a simple checklist approach. Similarly, the PSFSM could be used as a framing device for the critical success factors for effective privacy practice, combined with the methodology used for Privacy Impact Assessments (PIA). This would make PIAs a dynamic process, capturing the relationships between the factors influencing privacy practice, preventing them from being treated as a checklist. Such an approach would also assist in the search for a *'philosopher's stone'* to predict and prevent privacy failure, as has been already suggested for other types of failure.[143]

The ideas presented in this chapter are at an initial stage of development. It is hoped they will encourage future research of the use of systems thinking for the study of potential and actual privacy failures, and viewing privacy failures as systems failures. The mapping between failure modes and privacy literature can be developed further, and used to identify areas for future privacy research—particularly that focused on privacy practice. The PSFSM, with its identification of the wide range of factors which must be considered for effective privacy practice, could serve as meta-framework for organising privacy research, and identifying shortfalls. For example, there has been insufficient development of tools and techniques for ensuring an organisation's *actual* information use matches its *intended* information use—identified as one of the weaknesses of P3P.

In conclusion, privacy failures represent complex and cross-disciplinary failure situations, encompassing the design and development of technology services, and organisational privacy practice. Investigation of the root causes of privacy failures in technology services requires a systemic and multi-perspective approach which must view a privacy failure as a systems failure.

---

[142]Fortune and White, "Framing of Project Critical Success Factors by a Systems Model."

[143]Fortune and Peters, "Turning Hindsight into Foresight—Our Search for the 'Philosopher's Stone' of Failure."

Appendix

Google Buzz—Detailed Background

The Launch of Google Buzz

Google Buzz, which was integrated directly into Google's Gmail e-mail service to capitalise on its existing users, launched on Tuesday 9th February 2010[144] to compete with existing social networking sites—particularly Facebook and Twitter.[145] This was not the first time Google had launched a social network—it launched Orkut in 2004 and Google Wave in 2009.[146] Google Buzz included features found in Facebook and Twitter, such as the ability to '*like*' content and '*follow*' other Buzz users; Buzz users could also share status updates, comments, photographs and videos through posts, or '*buzzes*' (see Footnote 147). For users who opted into Google's Location Services, the location of the sender was added to each status update (see Footnote 147).

To pre-populate people's Buzz network, Google repurposed information provided by users who had previously signed up for Gmail, including their first and last names, and e-mail contacts (see Footnote 147). Following the launch of Google Buzz, when Gmail users logged into their e-mail account, they were shown a pop-up dialogue box describing the new service, which had two options: (1) '*Sweet! Check out Buzz*' (displayed as a large button); and (2) '*Nah, go to my inbox*' (displayed as a small hyperlink).[147] Gmail users who clicked on the button labelled '*Sweet! Check out Buzz*' were shown the Buzz welcome panel, which contained a message at the top stating, '*You're set up to follow the people you email and chat with the most*' (see Footnote 147). If a user clicked on the '*Nah, go to my inbox*' hyperlink, their information could still be shared by the following means (see Footnote 147):

- The user could be '*followed*' by other Gmail users who had enrolled in Buzz.
- If the user already had a public Google profile, they could appear in the public Google profiles of people enrolled in buzz who were following that user.
- A Buzz link would be displayed on the user's Gmail page. If they clicked on that link they were taken to the Buzz welcome panel and automatically enrolled in Buzz, even if they did not click on the '*Okay*' button at the bottom of the panel.

---

[144]Google Inc., "Google Buzz in Gmail."

[145]BBC, "Google Takes on Facebook and Twitter with Network Site"; Helft and Stone, "With Buzz, Google Plunges Into Social Networking"; Krazit, "Google's Social Side Hopes to Catch Some Buzz."

[146]BBC, "Google Takes on Facebook and Twitter with Network Site."

[147]FTC, "Complaint: In the Matter of Google Inc. a Corporation."

The Buzz welcome panel did not inform users that the list of people they would be automatically configured to follow on Buzz—visible as part of their public Google profile—was constructed from the e-mail addresses of people they most frequently e-mailed, or chatted with—making them public unless the user changed their default settings.[148] As Schonfield[149] argued, creating an instant social network around someone's e-mail contacts, blurs the boundaries between what is private and what is public.

One day after the launch of Buzz, privacy concerns were raised in online media and blogs[150,151] with particular focus on the auto-following of Gmail contacts, which caused problems for many Gmail users. For example, in a blog posted three days after the launch of Google Buzz, a woman—'*Harriet Jacobs*'[152]—alleged that because her third most frequent e-mail contact was her abusive ex-husband, Google Buzz had revealed her current relationships to him, and given him access to her comments on Google Reader.[153] Other privacy problems occurred with Gmail users who used it for confidential business-related communication. Dr. Keely Kolmes—a psychotherapist who communicated with her patients via Gmail, and advertised her business on her public Google profile—reported her sudden realisation that she was auto-following her friends and family, and some of her clients on Buzz; in addition, some of her clients were following her.[154] More significant, in terms of patient confidentiality, was that the list of clients who were following Dr. Kolmes were visible on her public Google profile. Even more seriously, public access to people's most frequent e-mail and chat contacts could be used by authoritarian governments to discover unknown links between political activists, or by organisations to discover and supress whistle-blowers.[155]

Google Buzz's use of Twitter's @*reply* convention resulted in another privacy problem, caused by Buzz users directing comments at people—using the @ sign in front of their name—who did not have a Google public profile. If the Buzz user selected the person's private (i.e. non-Gmail) e-mail address from their Gmail contacts, Buzz made the address in the post available to followers of the person sending the comment.[156]

Another privacy flaw with Buzz—affecting Google Buzz for mobile users and reported on 16th February 2010—was a cross-site scripting vulnerability that

---

[148]Carlson, "WARNING: Google Buzz Has a Huge Privacy Flaw"; Wood, "Google Buzz."

[149]"Watch Out Who You Reply To On Google Buzz, You Might Be Exposing Their Email Address | TechCrunch."

[150]Carlson, "WARNING: Google Buzz Has a Huge Privacy Flaw"; Wood, "Google Buzz."

[151]Wood alleged that Google Buzz used a photo on her personal Buzz page, which she had taken with her smartphone, but never uploaded.

[152]This is the pseudonym used in the Guardian article.

[153]Arthur, "Google Buzz's Open Approach Leads to Stalking Threat."

[154]Kolmes, "Google Buzz Alarms a Psychotherapist | Psychologist San Francisco."

[155]"Wrong Kind of Buzz around Google Buzz."

[156]Schonfeld, "Watch Out Who You Reply To On Google Buzz, You Might Be Exposing Their Email Address | TechCrunch."

allowed attackers to access the geographic location of Buzz users who had opted into Google's Location Services. It was claimed that attackers could tamper with victims' accounts by tricking them into visiting a booby-trapped link. There was no evidence that the flaw was exploited, and it was only cookies for Buzz that were at risk; Google fixed the fault later the same day.[157]

The First Week—Google's Immediate Response

Thursday 11th February 2010

In response to privacy concerns about Buzz, Google[158] announced a change to the original profile creation panel displayed when a user posted their first item in Buzz. The modified panel included a checked-by-default checkbox advising users that the people they were following would be displayed as part of their Google public profile, and allowing them to prevent the list of people who were following them, and people they were following, from being accessible. Google announced two other changes to Buzz to address people's privacy concerns: (1) making it easier to block anyone, by adding '*block*' links to the list of followers; and (2) make it easier for users to see who is on the public list of their followers that everyone else sees.[159]

These changes—although welcomed—still attracted criticism, particularly the use of the checked-by-default option in the profile creation screen, and the statement '*Show the list of people I'm following and the list of people following me on my public profile*', which did still did not make it clear that Buzz would publish people's frequent e-mail and chat contacts.[160] As Carlson[161] observed, when '*normal*' people encounter a new service, they will click the '*save and continue*' option—reading as little text as they can—until the service is available to them; similar user behaviour with dialogue boxes that interrupt users' primary task was reported by Krol et al.[162]

Saturday 13th February 2010

Over the weekend following the launch of Buzz, Google apologised to its users in a blog post, and announced the replacement of the *auto-follow* model where Buzz automatically configured users to follow the people they e-mail and chat with the

---

[157]Goodin, "Google Buzz Bug Exposes User Geo Location."
[158]"Millions of Buzz Users, and Improvements Based on Your Feedback."
[159]Google Inc., "Millions of Buzz Users, and Improvements Based on Your Feedback."
[160]Carlson, "Google Buzz Still Has Major Privacy Flaw."
[161]"Google Buzz Still Has Major Privacy Flaw."
[162]"Don't Work. Can't Work?".

most, with an *auto-suggest* model, i.e. providing users with a list of suggested people to follow.[163] Other changes announced by Google included removing Buzz's automatic connection of Reader shared items with public Picasa Web Albums, and adding a Buzz tab to Gmail to allow users to hide Buzz or disable it completely. However, Google Buzz still showed people's following/follower lists publicly by default—users still had to manually make these private.[164] Despite Google's modifications to Buzz, the Electronic Privacy Information Center (EPIC) filed a complaint to the Federal Trade Commission, as *"Gmail users are being driven into a social networking service they didn't sign up for"*.[165]

Tuesday 16th February 2010

A week after the launch of Buzz, Google told BBC News that Buzz—unlike most of its products—had only been tested within Google by its employees, and not with a more representative group of external users as part of its Trusted Tester program.[166]

Response of Legislative and Public Interest Bodies to Buzz

On 16th February 2010, EPIC filed a formal complaint with the US Federal Trade Commission (FTC), citing that Google was *"engaging in unfair and deceptive practices"*—prohibited under the Federal Trade Commission Act.[167] The day after, the Privacy Commissioner of Canada asked Google to clarify how it had addressed Buzz's privacy issues since its launch, and how it had met the requirements of privacy law in Canada.[168] This was followed on April 19th 2010 by a letter—signed by the heads of data protection authorities in nine other countries—to Eric Schmidt, the chief executive officer of Google, expressing their concern about Google's privacy practices, albeit praising its rapid response to addressing people's privacy concerns about Buzz.[169]

In September, 2010 Google paid US$8.5 million to settle a class action lawsuit, which consolidated several civil cases claiming the privacy of Gmail users had been violated by the launch of Buzz.[170] In March 2011, Google agreed to settle

---

[163]Google Inc., "A New Buzz Start-up Experience Based on Your Feedback."

[164]Frommer, "Google Making More Changes To Buzz After Huge Privacy Outcry."

[165]Marc Rotenberg quoted in Helft, "Anger Leads to Apology From Google About Buzz."

[166]BBC, "Google Admits Buzz Social Network Testing Flaws."

[167]Clause 47 in EPIC, "In the Matter of Google, Inc. Complaint, Request for Investigation, Injunction, and Other Relief," 13; EPIC, "EPIC Urges Federal Trade Commission to Investigate Google Buzz."

[168]Privacy Commissioner of Canada, "ARCHIVED—News Release."

[169]Privacy Commissioner of Canada, "ARCHIVED—Letter to Google Inc. Chief Executive Officer—April 19, 2010."

[170]Metz, "Google Pays $8.5 m to Settle Buzz Privacy Invasion Suit."

the FTC complaint[171,172] which charged it with unfair and deceptive acts or practices, contrary to the Federal Trade Commission Act[173]; more specifically the FTC charged Google with:

- Using users' Gmail contacts to populate Buzz, despite its privacy policy at the time representing—expressly or by implication—that the personal information of users who signed up for Gmail would only be used to provide the Gmail service.[174]
- Not seeking users' consent before using information provided to it for the purpose of providing Gmail, for Buzz, despite its privacy policy at the time representing—expressly or by implication—that users' permission would be sought before information was repurposed.[175]
- Enrolling users in certain features of Buzz, even if they declined to join it.[176]
- Failing to disclose adequately that the contacts users e-mailed, or chatted with, most frequently, would become public by default, and *"that user information submitted through other Google products would be automatically broadcast through Buzz"*.[177]
- Not adhering to the US Safe Harbor Privacy Principles of Notice and Choice.[178]

In the settlement of the complaint, which the FTC gave final approval to in October 2011, Google agreed to implement a comprehensive privacy programme, and to be audited by a qualified, objective and independent third-party professional every two years, for the next twenty years.[179]

Closure of Google Buzz

In October 2011, Google announced on its official blog that it was closing Buzz and its API in the *"next few weeks"*, and would focus on Google+.[180] In May 2013, it undertook the final step in closing Buzz, announcing that users' Buzz content would be moved to their Google Drive account.[181]

---

[171]FTC File No. 1023136.

[172]FTC, "FTC Charges Deceptive Privacy Practices in Google's Rollout of Its Buzz Social Network."

[173]FTC, "Complaint: In the Matter of Google Inc. a Corporation."

[174]Clauses 13 and 14 of Complaint (FTC File No. 1023136) FTC, "Complaint: In the Matter of Google Inc. a Corporation."

[175]Clauses 15 and 16 of Complaint (FTC File No. 1023136) FTC, "Complaint: In the Matter of Google Inc. a Corporation."

[176]Clauses 17 and 18 of Complaint (FTC File No. 1023136) FTC, "Complaint: In the Matter of Google Inc. a Corporation."

[177]Clause 19 of Complaint (FTC File No. 1023136) FTC, "Complaint: In the Matter of Google Inc. a Corporation."

[178]Clause 25 of Complaint (FTC File No. 1023136) FTC, "Complaint: In the Matter of Google Inc. a Corporation."

[179]FTC, "FTC Gives Final Approval to Settlement with Google over Buzz Rollout"; FTC, "Decision and Order: In the Matter of Google Inc. a Corporation."

[180]Google Inc., "A Fall Sweep."

[181]Lawler, "Google's 'Last Step' in Buzz Shutdown."

Bibliography

Ackoff, Russell L. 1974. The systems revolution. *Long Range Planning* 7(6): 2–20. doi:10.1016/0024-6301(74)90127-7.

Adams, Anne, and M. Angela Sasse. 2001. Privacy in multimedia communications: Protecting users, not just data. In *Human-Computer Interaction*.

Adams, Anne, and M. Angela Sasse. 1999. Privacy Issues in Ubiquitous multimedia environments: Wake sleeping dogs, or let them lie? In *IFIP Conference on Human-Computer Interaction*.

Allan, Alasdair, and Pete Warden. 2011. Got an iPhone or 3G iPad? Apple is recording your moves. *O'Reilly Radar*, April 20. http://radar.oreilly.com/2011/04/apple-location-tracking.html.

Arthur, Charles. 2010. Google Buzz's open approach leads to stalking threat. *The Guardian Technology Blog*, February 12. http://www.guardian.co.uk/technology/blog/2010/feb/12/google-buzz-stalker-privacy-problems.

Arthur, Charles. 2012. Google privacy policy slammed by EU data protection chiefs. *The Guardian*, October 16, sec. Technology. http://www.theguardian.com/technology/2012/oct/16/google-privacy-policies-eu-data-protection.

Arthur, Charles. 2011. Sony suffers second data breach with theft of 25 m more user details. *Guardian*, May 3. http://www.guardian.co.uk/technology/blog/2011/may/03/sony-data-breach-online-entertainment?intcmp=239.

Associated Press. 2009. Officials: Special plastic sleeves may stop identity theft. *FoxNews.com*, July 13. http://www.foxnews.com/story/0,2933,531787,00.html.

Bamberger, K., and D. Mulligan. 2010. Privacy on the books and on the ground. *Stanford Law Review* 63: 247–316.

Barnett, Emma. 2010. Google street view: Survey raises privacy concerns. *Telegraph.co.uk*, March 12, sec. Technology. http://www.telegraph.co.uk/technology/google/7430245/Google-Street-View-survey-raises-privacy-concerns.html.

BBC. 2010. Google admits buzz social network testing flaws. *BBC News (Technology)*, February 16. http://news.bbc.co.uk/1/hi/technology/8517613.stm.

BBC. 2010. Google takes on Facebook and Twitter with network site. *BBC News (Technology)*, February 9. http://news.bbc.co.uk/1/hi/technology/8506148.stm.

BBC. 2008. Timeline: Child benefits records loss. *BBC*, June 25. http://news.bbc.co.uk/1/hi/7104368.stm.

Bellotti, Victoria, and Abigail Sellen. 1993. Design for privacy in Ubiquitous computing environments. In *Proceedings of the Third European Conference on Computer-Supported Cooperative Work 13–17 September 1993, Milan, Italy ECSCW '93*, edited by Giorgio de Michelis, Carla Simone, and Kjeld Schmidt, 77–92. Netherlands: Springer. http://link.springer.com/chapter/10.1007/978-94-011-2094-4_6.

Bignell, Victor, and Joyce Fortune. 1984. *Understanding systems failures*. England: Manchester University Press.

Blakeley, Rhys. 2007. Facebook shrugs off privacy fears with plan for targeted advertising. *The Times*, September 11. http://technology.timesonline.co.uk/tol/news/tech_and_web/the_web/article2426470.ece.

Boyce, G. 2002. Beyond privacy: The ethics of customer information systems. *Informing Science* 107–25.

Camp, L. Jean. 2006. Mental models of privacy and security. SSRN scholarly paper. Rochester, NY: Social Science Research Network. http://papers.ssrn.com/abstract=922735.

Carlson, Nicholas. 2010. Google Buzz still has major privacy flaw. *Business Insider*, February 12. http://www.businessinsider.com/googles-nice-improvements-to-buzz-dont-correct-major-privacy-flaw-2010-2.

Carlson, Nicholas. 2010. WARNING: Google Buzz has a huge privacy flaw. February 10. http://www.businessinsider.com/warning-google-buzz-has-a-huge-privacy-flaw-2010-2#ixzz1hAXR1Ar1.

Cavoukian, Ann. 2009. *Privacy by design*. Ontario: Office of the Information and Privacy Commissioner. http://www.ipc.on.ca/images/Resources/privacybydesign.pdf.

Cavoukian, Ann. 2009. *Privacy by design ... take the challenge*. Ontario: Office of the Information and Privacy Commissioner.

Cavoukian, Ann. 2011. *Privacy by design—The 7 foundational principles*. Ontario: Office of the Information and Privacy Commissioner, January. http://www.ipc.on.ca/images/resources/7foundationalprinciples.pdf.

Cavoukian, Ann. 2009. SmartPrivacy: Lead with privacy by design. In *Presented at the Conference Board Council of Chief Privacy Officers, Ontario*, September 22. http://www.ipc.on.ca/images/Resources/Conference-Board-Council-of-Chief-Privacy-Officers.pdf.

Checkland, Peter. 1981. *Systems thinking. Systems practice*. Chichester: Wiley.

Checkland, Peter, and Jim Scholes. 1999. *Soft systems methodology in action: Including a 30 year retrospective*. Chichester: Wiley.

CNIL. 2012. Google's new privacy policy : Incomplete information and uncontrolled combination of data across services, October 16. http://www.cnil.fr/linstitution/actualite/article/article/googles-new-privacy-policy-incomplete-information-and-uncontrolled-combination-of-data-across-ser/.

Culnan, M.J., and C.C. Williams. 2009. How Ethics can enhance organizational privacy: Lessons from the choice point and TJX data breaches. *MIS Quarterly: Management Information Systems* 33(4): 673–687.

D'Innocenzio, Anne. 2010. Wal-Mart plan to use smart tags raises privacy concerns. *USA Today*, July 25, sec. Money. http://www.usatoday.com/money/industries/retail/2010-07-25-wal-mart-smart-tags_N.htm.

DiSalvo, David. 2012. Google says bye bye to user privacy. *Forbes*, January 24. http://www.forbes.com/sites/daviddisalvo/2012/01/24/google-says-bye-bye-to-user-privacy/.

EPIC. 2010. EPIC urges federal trade commission to investigate Google Buzz, February 16. http://epic.org/2010/02/epic-urges-federal-trade-commi.html.

EPIC. 2010. In the matter of Google, Inc. complaint, request for investigation, injunction, and other relief, February 16. http://epic.org/privacy/ftc/googlebuzz/GoogleBuzz_Complaint.pdf.

EPIC. 2000. Pretty poor privacy: An assessment of P3P and internet privacy. Electronic Privacy Information Center, June. http://epic.org/reports/prettypoorprivacy.html.

Federal Trade Commission. 2010. Widespread data breaches uncovered by FTC Probe. *Federal Trade Commission*, February 22. http://www.ftc.gov/opa/2010/02/p2palert.shtm.

Fortune, Joyce, and Geoff Peters. 2007. *Information systems: achieving success by avoiding failure*. Chichester: Wiley.

Fortune, Joyce, and Geoff Peters. 1990. The formal system paradigm for studying failures. *Technology Analysis & Strategic Management* 2(4): 383–390.

Fortune, Joyce, and Geoff Peters. 2001. Turning hindsight into foresight—Our search for the 'philosopher's stone' of failure. *Systemic Practice and Action Research* 14(6):791–803.

Fortune, Joyce, and Geoffrey Peters. 1995. *Learning from failure—The systems approach*. Chichester, UK: Wiley. http://www.amazon.com/Learning-Failure-Systems-Approach-Fortune/dp/0471944203.

Fortune, Joyce, and Diana White. 2006. Framing of project critical success factors by a systems model. *International Journal of Project Management* 24(1): 53–65.

Frommer, Dan. 2010. Google making more changes to Buzz after huge privacy outcry. *Business Insider*, February 13. http://www.businessinsider.com/google-making-more-changes-to-buzz-after-privacy-outcry-2010-2.

FTC. 2011. Complaint: In the matter of Google Inc. a corporation, October 13. http://www.ftc.gov/sites/default/files/documents/cases/2011/10/111024googlebuzzcmpt.pdf.

FTC. 2011. Decision and order: In the matter of Google Inc. a corporation, October 13. http://www.ftc.gov/sites/default/files/documents/cases/2011/10/111024googlebuzzdo.pdf.

FTC. 2011. FTC charges deceptive privacy practices in Google's rollout of its Buzz social network. *Federal Trade Commission (News)*, March 30. http://www.ftc.gov/opa/2011/03/google.shtm.

FTC. 2011. FTC gives final approval to settlement with Google over Buzz rollout. *Federal Trade Commission (News)*, October 24. http://www.ftc.gov/opa/2011/10/buzz.shtm.
Gaudin, Sharon. 2012. Google stirs up privacy Hornet's nest. *InfoWorld*, January 26. http://www.infoworld.com/article/2619002/search-engines/google-stirs-up-privacy-hornet-s-nest.html.
Goodin, Dan. 2010. Google Buzz bug exposes user geo location. *The Register*, February 16. http://www.theregister.co.uk/2010/02/16/google_buzz_security_bug/.
Good, Nathaniel S, and Aaron Krekelberg. 2003. Usability and privacy: A study of Kazaa P2P file-sharing. In *Proceedings of the SIGCHI Conference on Human Factors in Computing Systems*, 137–44. CHI '03. New York, NY, USA: ACM.
Google Inc. 2011. A fall sweep. *Google Official Blog*, October 14. http://googleblog.blogspot.com/2011/10/fall-sweep.html.
Google Inc. 2010. A new Buzz start-up experience based on your feedback. *Official Gmail Blog*, February 13. http://gmailblog.blogspot.co.uk/2010/02/new-buzz-start-up-experience-based-on.html.
Google Inc. 2010. Google Buzz in Gmail. *Official Gmail Blog*, February 9. http://gmailblog.blogspot.co.uk/2010/02/google-buzz-in-gmail.html.
Google Inc. 2010. Millions of Buzz users, and improvements based on your feedback. *Official Gmail Blog*, February 11. http://gmailblog.blogspot.co.uk/2010/02/millions-of-buzz-users-and-improvements.html.
Google Inc. 2012. Updating our privacy policies and terms of service. *Official Google Blog*, January 24. http://googleblog.blogspot.com/2012/01/updating-our-privacy-policies-and-terms.html.
Helft, Miguel. 2010. Anger leads to apology from Google about Buzz. *The New York Times*, February 15, sec. Technology/Internet. http://www.nytimes.com/2010/02/15/technology/internet/15google.html.
Helft, Miguel, and Brad Stone. 2010. With Buzz, Google plunges into social networking. *The New York Times*, February 10, sec. Technology/Internet. http://www.nytimes.com/2010/02/10/technology/internet/10social.html.
Information Commissioner's Office. 2015. Google to change privacy policy after ICO investigation, February 24. https://ico.org.uk/about-the-ico/news-and-events/news-and-blogs/2015/01/google-to-change-privacy-policy-after-ico-investigation/.
Information Commissioner's Office. 2013. Sony fined £250,000 after millions of UK Gamers' details compromised, January 24. https://ico.org.uk/about-the-ico/news-and-events/news-and-blogs/2013/01/sony-fined-250-000-after-millions-of-uk-gamers-details-compromised/.
Jenkins, Gwilym M. 1976. The systems approach. In *Systems Behaviour*, 2nd ed., 78–104. London: The Open University, Harper & Row.
Kiss, Jemima. 2010. Google admits collecting Wi-Fi data through street view cars. *The Guardian*, May 15, sec. Technology. http://www.theguardian.com/technology/2010/may/15/google-admits-storing-private-data.
Kolmes, Keeley, Dr. 2010. Google Buzz alarms a psychotherapist | psychologist San Francisco, February 18. http://drkkolmes.com/2010/02/18/google-buzz-alarms-therapists/.
Krasnoff, Barbara. 2012. Google's new privacy policy: Checking the source. *Computerworld*, January 25. http://www.computerworld.com/article/2472135/e-commerce/google-s-new-privacy-policy--checking-the-source.html.
Kravets, David. 2010. Lawyers claim Google Wi-Fi sniffing 'is not an accident.' *Wired News*, June 3. http://www.wired.com/threatlevel/2010/06/google-wifi-sniffing/.
Krazit, Tom. 2010. Google's social side hopes to catch some Buzz. *CNET*, February 9. http://www.cnet.com/uk/news/googles-social-side-hopes-to-catch-some-buzz/.
Krol, K., M. Moroz, and M.-A Sasse. 2012. Don't work. Can't work? Why it's time to rethink security warnings. In *2012 7th International Conference on Risk and Security of Internet and Systems (CRiSIS)*, 1–8. doi:10.1109/CRISIS.2012.6378951.
Langheinrich, Marc. 2001. Privacy by design—Principles of privacy-aware Ubiquitous systems. In *Ubicomp 2001: Ubiquitous Computing*, edited by Gregory D. Abowd, Barry Brumitt, and Steven Shafer, 273–91. Lecture Notes in Computer Science 2201. Berlin: Springer. http://link.springer.com/chapter/10.1007/3-540-45427-6_23.

Lawler, Richard. 2013. Google's 'last step' in Buzz shutdown: Moving all data to Google drive. *Engadget*, May 25. http://www.engadget.com/2013/05/25/google-buzz-shifting-to-google-drive-archives/.

Lederer, S., J. I Hong, A. K Dey, and J. A Landay. 2004. Personal privacy through understanding and action: Five pitfalls for designers. *Personal and Ubiquitous Computing* 8(6): 440–454.

Macdonald, Calum. 2007. Google's street view raises alarms over privacy. *The Herald*, June 4. http://www.heraldscotland.com/google-s-street-view-site-raises-alarm-over-privacy-1.859078.

Mennecke, Thomas. 2007. Pfizer P2P security breach. *Slyck News*, June 20. http://www.slyck.com/story1496_Pfizer_P2P_Security_Breach.

Metz, Cade. 2010. Google pays $8.5 m to settle Buzz privacy invasion suit. *The Register*, September 5. http://www.theregister.co.uk/2010/09/05/google_buzz_suit_settlement/.

Mills, Elinor. 2012. Google wants ability to 'combine' your user data. *CNET*, January 24. http://www.cnet.com/news/google-wants-ability-to-combine-your-user-data/.

Morozov, Evgeny. 2010. Wrong kind of Buzz around Google Buzz. *Foreign Policy Blogs*, February 11. http://neteffect.foreignpolicy.com/posts/2010/02/11/wrong_kind_of_buzz_around_google_buzz.

Morton, Anthony, Bettina Berendt, Seda Gürses, and Jo Pierson. 2013. 4.3 'Tool clinics'—Embracing multiple perspectives in privacy research and privacy-sensitive design. *"My life, shared"—Trust and privacy in the age of Ubiquitous experience sharing*, 96–104. doi:10.4230/DagRep.3.7.74.

Morton, Anthony, and M. Angela Sasse. 2012. Privacy is a process, not a PET: A theory for effective privacy practice. In *Proceedings of the 2012 Workshop on New Security Paradigms*, 87–104. NSPW '12. New York, NY, USA: ACM. doi:10.1145/2413296.2413305.

NBC. 2009. New warnings on cyber-thieves. *TODAY Investigates*. United States: NBC, February 26. http://www.today.com/id/26184891/vp/29405819#29405819.

Nissenbaum, Helen. 2004. Privacy as contextual integrity. *Washington Law Review* 79(1): 119.

Panzarino, Matthew. 2011. It's not just the iPhone, android stores your location data too. *TNW—The Next Web*, April 21. http://thenextweb.com/google/2011/04/21/its-not-just-the-iphone-android-stores-your-location-data-too/.

Peters, Geoff, and Joyce Fortune. 1992. Systemic methods for the analysis of failure. *Systems Practice* 5(5): 529–542.

Privacy Commissioner of Canada. 2010. ARCHIVED—Letter to Google Inc. Chief Executive Officer—April 19, 2010, April 19. http://www.priv.gc.ca/media/nr-c/2010/let_100420_e.asp.

Privacy Commissioner of Canada. 2010. ARCHIVED—News release: Commissioner challenges Google Buzz over privacy concerns—February 17, 2010, February 17. http://www.priv.gc.ca/media/nr-c/2010/nr-c_100217_e.asp.

Quinn, Ben, and Charles Arthur. 2011. PlayStation network hackers access data of 77 million users. *The Guardian*, April 26. http://www.guardian.co.uk/technology/2011/apr/26/playstation-network-hackers-data?intcmp=239.

Radcliffe, Jennifer. 2011. Tracking devices used in school badges. *Houston Chronicle*, October 11. http://www.chron.com/disp/story.mpl/metropolitan/7241100.html.

Reay, Ian, Scott Dick, and James Miller. 2009. A large-scale empirical study of P3P privacy policies. *ACM Transactions on the Web* 3(2): 1–34. doi:10.1145/1513876.1513878.

Riegelsberger, Jens, M. Angela Sasse, and John D. McCarthy. 2005. The mechanics of trust: A framework for research and design. *International Journal of Human-Computer Studies* 62(3): 381–422.

Rubinstein, Ira. 2011. Regulating privacy by design. *Berkeley Technology Law Journal*.

Schonfeld, Erick. 2010. Watch out who you reply to on Google Buzz, you might be exposing their email address | TechCrunch. *TechCrunch (News)*, February 11. http://techcrunch.com/2010/02/11/reply-google-buzz-exposing-email/.

Spiekermann, Sarah. 2012. The challenges of privacy by design. *Communications of the ACM* 55(7): 38–40. doi:10.1145/2209249.2209263.

Stewart, Roger W, and Joyce Fortune. 1995. Application of systems thinking to the identification, avoidance and prevention of risk. *International Journal of Project Management* 13(5): 279–286. doi:10.1016/0263-7863(95)00024-K.

Sydnor, Thomas D., John Knight, and Lee A. Hollaar. 2006. Filesharing programs and 'technological features to induce users to share.' United States Patent and Trademark Office, November 2006. http://www.uspto.gov/ip/global/copyrights/cpright_filesharing_v1012.pdf.

TechCrunchTV. 2010. Crunchies awards 2010. *Mike Arrington Interrogates Mark Zuckerberg*. Las Vegas, January 9. http://static-cdn1.ustream.tv/swf/live/viewer:55.swf?vid=3848950&vrsl=c:170.

Tsukayama, Hayley. 2012. Google faces backlash over privacy changes. *The Washington Post*, January 25. http://www.washingtonpost.com/business/technology/google-faces-backlash-over-privacy-changes/2012/01/25/gIQAVQnMQQ_story.html.

UK Information Commissioner's Office. 2008. Privacy by design. U.K. Information Commissioner's Office, November 26. http://www.docstoc.com/docs/74969464/Privacy-by-design.

White, Diana, and Joyce Fortune. 2009. The project-specific formal system model. *International Journal of Managing Projects in Business* 2(1): 36–52. doi:10.1108/17538370910930509.

Williams, Christopher. 2011. Google to build profiles of Gmail users for advertisers. *The Telegraph*, March 30. http://www.telegraph.co.uk/technology/google/8415769/Google-to-build-profiles-of-Gmail-users-for-advertisers.html.

Wood, Molly. 2010. Google Buzz: Privacy nightmare. *CNET (Molly Rants)*, February 10. http://news.cnet.com/8301-31322_3-10451428-256.html.

A Precautionary Approach to Big Data Privacy

Arvind Narayanan, Joanna Huey and Edward W. Felten

Abstract Once released to the public, data cannot be taken back. As time passes, data analytic techniques improve and additional datasets become public that can reveal information about the original data. It follows that released data will get increasingly vulnerable to re-identification—unless methods with provable privacy properties are used for the data release. We review and draw lessons from the history of re-identification demonstrations; explain why the privacy risk of data that is protected by ad hoc de-identification is not just unknown, but unknowable; and contrast this situation with provable privacy techniques like differential privacy. We then offer recommendations for practitioners and policymakers. Because ad hoc de-identification methods make the probability of a privacy violation in the future essentially unknowable, we argue for a weak version of the precautionary approach, in which the idea that the burden of proof falls on data releasers guides policies that incentivize them not to default to full, public releases of datasets using ad hoc de-identification methods. We discuss the levers that policymakers can use to influence data access and the options for narrower releases of data. Finally, we present advice for six of the most common use cases for sharing data. Our thesis is that the problem of "what to do about re-identification" unravels once we stop looking for a one-size-fits-all solution, and each of the six cases we consider a solution that is tailored, yet principled.

Keywords Re-identification · De-identification · Data · Privacy · Precautionary principle

---

A. Narayanan (✉) · J. Huey · E.W. Felten
Center for Information Technology Policy, Princeton University,
303 Sherrerd Hall, Princeton, NJ 08544, USA
e-mail: arvindn@CS.Princeton.EDU

J. Huey
e-mail: joanna.huey@gmail.com

E.W. Felten
e-mail: felten@cs.princeton.edu

1 Introduction

Once released to the public, data cannot be taken back. As time passes, data analytic techniques improve and additional datasets become public that can reveal information about the original data. It follows that released data will get increasingly vulnerable to re-identification—unless methods with provable privacy properties are used for the data release.

Due to the ad hoc de-identification methods applied to currently released datasets, the chances of re-identification depend highly on the progress of re-identification tools and the auxiliary datasets available to an adversary. The probability of a privacy violation in the future is essentially unknowable. In general, a precautionary approach deals with uncertain risk by placing the burden of proof that an action is not harmful on the person taking the action. Here, we argue for a weak version of the precautionary approach, in which the idea that the burden of proof falls on data releasers guides policies that incentivize them not to default to full, public releases of datasets using ad hoc de-identification methods.

In Sect. 1, we argue that privacy risks due to inference go beyond the stereotypical re-identification attack that links a de-identified record to PII. We review and draw lessons from the history of re-identification demonstrations, including both "broad" and "targeted" attacks. In Sect. 2, we explain why the privacy risk of data that is protected by ad hoc de-identification is not just unknown, but unknowable, and contrast this situation with provable privacy techniques like differential privacy.

Sections 3 and 4 contain our recommendations for practitioners and policy makers.[1] In Sect. 3, we discuss the levers that policymakers can use to influence data releases: research funding choices that incentivize collaboration between privacy theorists and practitioners, mandated transparency of re-identification risks, and innovation procurement. Meanwhile, practitioners and policymakers have numerous pragmatic options for narrower releases of data. In Sect. 4, we present advice for six of the most common use cases for sharing data. Our thesis is that the problem of "what to do about re-identification" unravels once we stop looking for a one-size-fits-all solution, and each of the six cases we consider a solution that is tailored, yet principled.

2 Ill-Founded Promises of Privacy: The Failures of Ad Hoc De-identification

Significant privacy risks stem from current de-identification practices. Analysis methods that allow sensitive attributes to be deduced from supposedly de-identified datasets pose a particularly strong risk, and calling data "anonymous" once

---

[1]Though many of the examples are U.S.-centric, the policy recommendations have widespread applicability.

certain types of personally identifiable information ("PII") have been removed from it is a recipe for confusion. The term suggests that such data cannot later be re-identified, but such assumptions are increasingly becoming obsolete.

The U.S. President's Council of Advisors on Science and Technology ("PCAST") was emphatic in recognizing these risks:

> Anonymization of a data record might seem easy to implement. Unfortunately, it is increasingly easy to defeat anonymization by the very techniques that are being developed for many legitimate applications of big data. In general, as the size and diversity of available data grows, the likelihood of being able to re-identify individuals (that is, re-associate their records with their names) grows substantially.
>
> [...]
>
> Anonymization remains somewhat useful as an added safeguard, but it is not robust against near-term future re-identification methods. PCAST does not see it as being a useful basis for policy.[2]

The PCAST report reflects the consensus of computer scientists who have studied de- and re-identification: there is little if any technical basis for believing that common de-identification methods will be effective against likely future adversaries.

2.1 Privacy-Violating Inferences Go Beyond Stereotypical Re-identification

It is important to consider the full scope of privacy violations that can stem from data releases. The stereotypical example of re-identification is when a name is reattached to a record that was previously de-identified. However, privacy violations often occur through other, less obvious forms of re-identification. In particular, (1) any identifier can affect privacy, not just typical identifiers such as name and social security number, and (2) sensitive attributes of a user can be inferred even when that user cannot be matched directly with a database record.

First, when discussing identifiers, the relevant question is not so much "can this data be linked to PII?" as "can this data be linked to a user?" Account numbers, persistent tags such as device serial numbers, or long-lived tracking identifiers—such as enduring pseudonyms[3]—can all be associated with a collection of information about a user, whether or not they are included in existing definitions of

---

[2]Executive Office of the President, President's Council of Advisors on Science and Technology, *Report to the President: Big Data and Privacy: A Technological Perspective* (Washington, DC: 2014): 38–39.

[3]Ed Felten, "Are pseudonyms 'anonymous'?," *Tech@FTC*, April 30, 2012, https://techatftc.word press.com/2012/04/30/are-pseudonyms-anonymous/.

PII.[4] Nissenbaum and Barocas point out that oxymoronic "anonymous identifiers" such as Google's AdID assigned by an organization to a user do nothing to alleviate the user's privacy worries when interacting with that organization or the universe of applications with which the identifier is shared.[5] A recent example of such problems is Whisper, a social media app that promises anonymity but tracks users extensively and stores their data indefinitely.[6] The false distinction between defined PII and other potential identifiers allows Whisper to monitor the movements of "a sex obsessed lobbyist," noting "[h]e's a guy that we'll track for the rest of his life and he'll have no idea we'll be watching him," while still maintaining that "Whisper does not request or store any personally identifiable information from users, therefore there is never a breach of anonymity."[7]

Second, re-identification affects a user's privacy whenever an inference of a sensitive attribute can be made. Suppose an analyst can narrow down the possibilities for Alice's record in a de-identified medical database to one of ten records.[8] If all ten records show a diagnosis of liver cancer, the analyst learns that Alice has liver cancer. If nine of the ten show liver cancer, then the analyst can infer that there is a high likelihood of Alice having liver cancer.[9] Either way, Alice's privacy has been impacted, even though no individual database record could be associated with her.

---

[4]Paul Ohm, "Broken Promises of Privacy: Responding to the Surprising Failure of Anonymization," *UCLA Law Review* 57 (2010): 1742–43, http://uclalawreview.org/pdf/57-6-3.pdf.

[5]Solon Barocas and Helen Nissenbaum, "Big Data's End Run Around Anonymity and Consent," in *Privacy, Big Data, and the Public Good: Frameworks for Engagement*, ed. Julia Lane, Victoria Stodden, Stefan Bender, and Helen Nissenbaum (New York: Cambridge University Press, 2014), 52–54.

[6]Paul Lewis and Dominic Rushe, "Revealed: how Whisper app tracks 'anonymous' users," *The Guardian*, October 16, 2014, http://www.theguardian.com/world/2014/oct/16/-sp-revealed-whisper-app-tracking-users.

[7]Ibid. A poster self-identified as the CTO of Whisper reiterated this point: "We just don't have any personally identifiable information. Not name, email, phone number, etc. I can't tell you who a user is without them posting their actual personal information, and in that case, it would be a violation of our terms of service." rubyrescue, October 17, 2014, comment on black-Rust, "How Whisper app tracks 'anonymous' users," *Hacker News*, October 17, 2014, https://news.ycombinator.com/item?id=8465482.

[8]This is consistent with the database having a technical property called k-anonymity, with $k = 10$. Latanya Sweeney, "k-anonymity: A Model for Protecting Privacy," *International Journal on Uncertainty, Fuzziness and Knowledge-based Systems* 10, no. 5 (2001): 557–70. Examples like this show why k-anonymity does not guarantee privacy.

[9]Heuristics such as l-diversity and t-closeness account for such privacy-violating inferences, but they nevertheless fall short of the provable privacy concept we discuss in the next section. Ashwin Machanavajjhala et al., "l-diversity: Privacy beyond k-anonymity," *ACM Transactions on Knowledge Discovery from Data (TKDD)* 1, no. 1 (2007): 3; Ninghui Li, Tiancheng Li, and Suresh Venkatasubramanian, "t-closeness: Privacy beyond k-anonymity and l-diversity," in *IEEE 23rd International Conference on Data Engineering, 2007* (Piscataway, NJ: IEEE, 2007): 106–15.

2.2 Re-identification Attacks May Be Broad or Targeted

Two main types of scenarios concern us as threats to privacy: (1) broad attacks on large databases and (2) attacks that target a particular individual within a dataset. Broad attacks seek to get information about as many people as possible (an adversary in this case could be someone who wants to sell comprehensive records to a third party), while targeted attacks have a specific person of interest (an adversary could be someone who wants to learn medical information about a potential employee).

2.2.1 Broad Attacks: Examples and Lessons

Many released datasets can be re-identified with no more than basic programming and statistics skills. But even if current techniques do not suffice, that is no guarantee of privacy—the history of re-identification has been a succession of surprising new techniques rendering earlier datasets vulnerable.

In 2000, Sweeney showed that 87 % of the U.S. population can be uniquely re-identified based on five-digit ZIP code, gender, and date of birth.[10] Datasets released prior to that publication and containing such data became subject to reidentification through simple cross-referencing with voter list information. For example, through comparison with the Social Security Death Index, an undergraduate class project re-identified 35 % of Chicago homicide victims in a de-identified dataset of murders between 1965 and 1995.[11] Furthermore, because research findings do not get put into practice immediately, datasets still are being released with this type of information: Sweeney showed that demographic information could be used to re-identify 43 % of the 2011 medical records included in data sold by the state of Washington,[12] and Sweeney, Abu, and Winn demonstrated in 2013 that such demographic cross-referencing also could re-identify over 20 % of the participants in the Personal Genome Project, attaching their names to their medical and genomic information.[13]

---

[10]Latanya Sweeney, "Simple Demographics Often Identify People Uniquely" (Data Privacy Working Paper 3, Carnegie Mellon University, Pittsburgh, Pennsylvania, 2000), http://dataprivacylab.org/projects/identifiability/paper1.pdf.

[11]Salvador Ochoa et al., "Reidentification of Individuals in Chicago's Homicide Database: A Technical and Legal Study" (final project, 6.805 Ethics and Law on the Electronic Frontier, Massachusetts Institute of Technology, Cambridge, Massachusetts, May 5, 2001), http://mike.salib.com/writings/classes/6.805/reid.pdf.

[12]Latanya Sweeney, "Matching Known Patients to Health Records in Washington State Data" (White Paper 1089-1, Data Privacy Lab, Harvard University, Cambridge, Massachusetts, June 2013), http://dataprivacylab.org/projects/wa/1089-1.pdf.

[13]Latanya Sweeney, Akua Abu, and Julia Winn, "Identifying Participants in the Personal Genome Project by Name" (White Paper 1021-1, Data Privacy Lab, Harvard University, Cambridge, Massachusetts, April 24, 2013), http://dataprivacylab.org/projects/pgp/1021-1.pdf. Sweeney and her team matched 22 % of participants based on voter data and 27 % based on a public records website.

For years, security experts have warned about the failure of simple hash functions to anonymize data, especially when that data has an easily guessable format, such as the nine digits of a social security number.[14] Yet, simple hashing was commonly thought of as an anonymization method, and once again, continues to be used in released datasets. The 2013 dataset released by New York City's Taxi and Limousine Commission after a FOIL request[15] exposed sensitive information in part by using a simple hash function to try to anonymize drivers and cabs, allowing for easy re-identification of taxi drivers:

> Security researchers have been warning for a while that simply using hash functions is an ineffective way to anonymize data. In this case, it's substantially worse because of the structured format of the input data. This anonymization is so poor that anyone could, with less than 2 h work, figure which driver drove every single trip in this entire dataset. It would even be easy to calculate drivers' gross income, or infer where they live.[16]

Additional information in the data leaves the door open to re-identification of riders, which is discussed in the following section.

New attributes continue to be linked with identities: search queries,[17] social network data,[18] genetic information (without DNA samples from the targeted people),[19] and geolocation data[20] all can permit re-identification, and Acquisti, Gross, and Stutzman have shown that it is possible to determine some people's interests

---

[14]Ben Adida, "Don't Hash Secrets," *Benlog*, June 19, 2008, http://benlog.com/2008/06/19/dont-hash-secrets/; Ed Felten, "Does Hashing Make Data 'Anonymous'?," *Tech@FTC*, April 22, 2012, https://techatftc.wordpress.com/2012/04/22/does-hashing-make-data-anonymous/; Michael N. Gagnon, "Hashing IMEI numbers does not protect privacy," *Dasient Blog*, July 26, 2011, http://blog.dasient.com/2011/07/hashing-imei-numbers-does-not-protect.html.

[15]Chris Whong, "FOILing NYC's Taxi Trip Data," March 18, 2014, http://chriswhong.com/open-data/foil_nyc_taxi/.

[16]Vijay Pandurangan, "On Taxis and Rainbows: Lessons from NYC's improperly anonymized taxi logs," *Medium*, June 21, 2014, https://medium.com/@vijayp/of-taxis-and-rainbows-f6bc289679a1.

[17]Michael Barbaro and Tom Zeller, Jr., "A Face Is Exposed for AOL Searcher No. 4417749," *New York Times*, August 9, 2006, http://www.nytimes.com/2006/08/09/technology/09aol.html.

[18]Ratan Dey, Yuan Ding, and Keith W. Ross, "The High-School Profiling Attack: How Online Privacy Laws Can Actually Increase Minors' Risk" (paper presented at the 13th Privacy Enhancing Technologies Symposium, Bloomington, IN, July 12, 2013), https://www.petsymposium.org/2013/papers/dey-profiling.pdf; Arvind Narayanan and Vitaly Shmatikov, "De-anonymizing Social Networks," in *Proceedings of the 2009 30th IEEE Symposium on Security and Privacy* (Washington, D.C.: IEEE Computer Society, 2009): 173–87.

[19]Melissa Gymrek et al., "Identifying Personal Genomes by Surname Inference," *Science* 339, no. 6117 (January 2013): 321–24, doi:10.1126/science.1229566.

[20]Philippe Golle and Kurt Partridge, "On the Anonymity of Home/Work Location Pairs," in *Pervasive '09 Proceedings of the 7th International Conference on Pervasive Computing* (Berlin, Heidelberg: Springer-Verlag, 2009): 390–97, https://crypto.stanford.edu/~pgolle/papers/commute.pdf.

and Social Security numbers from only a photo of their faces.[21] The realm of potential identifiers will continue to expand, increasing the privacy risks of already released datasets.

Furthermore, even staunch proponents of current de-identification methods admit that they are inadequate for high-dimensional data.[22] These high-dimensional datasets, which contain many data points for each individual's record, have become the norm: social network data has at least a hundred dimensions[23] and genetic data can have millions.[24] We expect that datasets will continue this trend towards higher dimensionality as the costs of data storage decrease and the ability to track a large number of observations about a single individual increase. High dimensionality is one of the hallmarks of "big data."

Finally, we should note that re-identification of particular datasets is likely underreported. First, the re-identification of particular datasets is likely to be included in the academic literature only if it involves a novel advancement of techniques, so while the first use of a re-identification method may be published, reuses rarely are. Similarly, people who blog or otherwise report re-identification vulnerabilities are unlikely to do so unless interesting methods or notable datasets are involved. Second, those with malicious motivations for re-identification are probably unwilling to announce their successes. Thus, even if a specific dataset has not been re-identified publicly, it should not be presumed secure.

2.2.2 Targeted Attacks: Examples and Lessons

Another important—but often under-acknowledged—type of re-identification risk stems from adversaries who target specific individuals. If someone has knowledge about a particular person, identifying him or her within a dataset becomes much easier. The canonical example of this type of attack comes from Sweeney's 1997

---

[21] Alessandro Acquisti, Ralph Gross, and Fred Stutzman, "Faces of Facebook: Privacy in the Age of Augmented Reality" (presentation at BlackHat Las Vegas, Nevada, August 4, 2011). More information can be found in the FAQ on Acquisti's website: http://www.heinz.cmu.edu/~acquisti/face-recognition-study-FAQ/.

[22] "In the case of high-dimensional data, additional arrangements [beyond de-identification] may need to be pursued, such as making the data available to researchers only under tightly restricted legal agreements." Ann Cavoukian and Daniel Castro, *Big Data and Innovation, Setting the Record Straight: De-identification Does Work* (Toronto, Ontario: Information and Privacy Commissioner, June 16, 2014): 3.

[23] The median Facebook user has about a hundred friends. Johan Ugander, Brian Karrer, Lars Backstrom, and Cameron Marlow, "The anatomy of the Facebook social graph," (arXiv Preprint, 2011): 3, http://arxiv.org/pdf/1111.4503v1.pdf.

[24] There are roughly ten million single nucleotide polymorphisms (SNPs) in the human genome; SNPs are the most common type of human genetic variation. "What are single nucleotide polymorphisms (SNPs)?," *Genetics Home Reference: Your Guide to Understanding Genetic Conditions*, published October 20, 2014, http://ghr.nlm.nih.gov/handbook/genomicresearch/snp.

demonstration that she could re-identify the medical record of then-governor William Weld using only his date of birth, gender, and ZIP code.[25]

More recently, as mentioned in the previous section, the data from the New York City Taxi and Limousine Commission not only had especially poor de-identification practices that made broad re-identification of all drivers trivial, but also allowed for the re-identification of targeted passengers even though the dataset did not nominally contain any information about passengers. First, it is possible to identify trip records (with pickup and dropoff locations, date and time, taxi medallion or license number, and fare and tip amounts) if some of that information is already known: for example, stalkers who see their victims take a taxi to or from a particular place can determine the other endpoint of those trips.[26] Second, it is possible to identify people who regularly visit sensitive locations, such as a strip club or a religious center.[27] The data includes specific GPS coordinates. If multiple trips have the same endpoints, it is likely that the other endpoint is the person's residence or workplace, and searching the internet for information on that address may reveal the person's identity. Similar analysis can be done on the recently released Transport for London dataset, which includes not only the information in the New York taxi dataset, but also unique customer identifiers for users of the public bicycle system.[28] These violations of the privacy of passengers demonstrate problems that better ad hoc de-identification still would not fix.

Research by Narayanan and Shmatikov revealed that with minimal knowledge about a user's movie preferences, there is an over 80 % chance of identifying that user's record in the Netflix Prize dataset—a targeted attack.[29] In addition, they showed as a proof-of-concept demonstration that it is possible to identify Netflix users by cross-referencing the public ratings on IMDb. Thus broad attacks may also be possible depending on the quantity and accuracy of information available to the adversary for cross-referencing.

---

[25]*DHS Data Privacy and Integrity Advisory Committee FY* (2005)*Meeting Materials* (June 15, 2005) (statement of Latanya Sweeney, Associate Professor of Computer Science, Technology and Policy and Director of the Data Privacy Laboratory, Carnegie Mellon University), http://www.dhs.gov/xlibrary/assets/privacy/privacy_advcom_06-2005_testimony_sweeney.pdf.

[26]Anthony Tockar, "Riding with the Stars: Passenger Privacy in the NYC Taxicab Dataset," *Neustar: Research*, September 15, 2014, http://research.neustar.biz/2014/09/15/riding-with-the-stars-passenger-privacy-in-the-nyc-taxicab-dataset/.

[27]Ibid. Tockar goes on to explain how to apply differential privacy to this dataset.

[28]James Siddle, "I Know Where You Were Last Summer: London's public bike data is telling everyone where you've been," *The Variable Tree*, April 10, 2014, http://vartree.blogspot.com/2014/04/i-know-where-you-were-last-summer.html.

[29]Arvind Narayanan and Vitaly Shmatikov, "Robust de-anonymization of large sparse datasets," in *Proceedings 2008 IEEE Symposium on Security and Privacy, Oakland, California, USA, May 18–21, 2008* (Los Alamitos, California: IEEE Computer Society, 2008): 111–25. The Netflix Prize dataset included movies and movie ratings for Netflix users.

A 2013 study by de Montjoye et al. revealed weaknesses in anonymized location data.[30] Analyzing a mobile phone dataset that recorded the location of the connecting antenna each time the user called or texted, they evaluated the uniqueness of individual mobility traces (i.e., the recorded data for a particular user, where each data point has a timestamp and an antenna location). Over 50 % of users are uniquely identifiable from just two randomly chosen data points. As most people spend the majority of their time at either their home or workplace, an adversary who knows those two locations for a user is likely to be able to identify the trace for that user—and to confirm it based on the patterns of movement.[31] If an adversary knows four random data points, which a user easily could reveal through social media, 95 % of mobility traces are uniquely identifiable.

Many de-identified datasets are vulnerable to re-identification by adversaries who have specific knowledge about their targets. A political rival, an ex-spouse, a neighbor, or an investigator could have or gather sufficient information to make re-identification possible.

As more datasets become publicly available or accessible by (or through) data brokers, the problems with targeted attacks can spread to become broad attacks. One could chain together multiple datasets to a non-anonymous dataset and re-identify individuals present in those combinations of datasets.[32] Sweeney's re-identification of then-Governor Weld's medical record used a basic form of this chaining: she found his gender, date of birth, and ZIP code through a public dataset of registered voters and then used that information to identify him within the de-identified medical database. More recent work by Hooley and Sweeney suggests that this type of chaining remains effective on public hospital discharge data from thirty U.S. states in 2013.[33]

---

[30]Yves-Alexandre de Montjoye, et al., "Unique in the Crowd: The privacy bounds of human mobility," *Scientific Reports* 3 (March 2013), doi:10.1038/srep01376.

[31]Other studies have confirmed that pairs of home and work locations can be used as unique identifiers. Golle and Partridge, "On the anonymity of home/work location pairs;" Hui Zang and Jean Bolot, "Anonymization of location data does not work: A large-scale measurement study," in *Proceedings of the 17th International Conference on Mobile Computing and Networking* (New York, New York: ACM, 2011): 145–156.

[32]A similar type of chaining in a different context can trace a user's web browsing history. A network eavesdropper can link the majority a user's web page visits to the same pseudonymous ID, which can often be linked to a real-world identity. Steven Englehardt et al., "Cookies that give you away: Evaluating the surveillance implications of web tracking," (paper accepted at 24th International World Wide Web Conference, Florence, May 2015).

[33]Sean Hooley and Latanya Sweeney, "Survey of Publicly Available State Health Databases" (White Paper 1075-1, Data Privacy Lab, Harvard University, Cambridge, Massachusetts, June 2013), http://dataprivacylab.org/projects/50states/1075-1.pdf.

3 Quantifiable Risks and Provable Privacy

Current de-identification methods are ad hoc, following a penetrate-and-patch mindset. Proponents ask whether a de-identification method can resist certain past attacks,[34] rather than insisting on affirmative evidence that the method cannot leak information regardless of what the attacker does.

The penetrate-and-patch approach is denounced in the field of computer security[35] because systems following that approach tend to fail repeatedly.[36] Ineffective as the penetrate-and-patch approach is for securing software, it is even worse for de-identification. End users will install patches to fix security bugs in order to protect their own systems, but data users have no incentive to replace a dataset found to have privacy vulnerabilities with a patched version that is no more useful to them. When no one applies patches, penetrate-and-patch becomes simply penetrate.

In addition, ad hoc de-identification makes it infeasible to quantify the risks of privacy violations stemming from a data release. Any such risk calculation must be based on assumptions about the knowledge and capabilities of all potential adversaries. As more data releases occur and more re-identification techniques are honed, such assumptions break down. Yet, accurate risk calculations are a prerequisite for well-informed policy choices, which must weigh the risks to privacy against the benefits of data releases.

These vulnerabilities of de-identification call for a shift in the focus of data privacy research, which currently suffers from ill-defined problems and unproven solutions. The field of privacy can learn from the successes and struggles in cryptography research. The concept of provable security can be translated to this area: "privacy" can be defined rigorously and data practices can be designed to have

---

[34]"Thus, while [Sweeney's re-identification of Governor Weld] speaks to the inadequacy of certain de-identification methods employed in 1996, to cite it as evidence against current de-identification standards is highly misleading. If anything, it should be cited as evidence for the *improvement* of de-identification techniques and methods insofar as such attacks are no longer feasible under today's standards precisely because of this case." Cavoukian and Castro, *De-identification* Does *Work*: 5.
"Established, published, and peer-reviewed evidence shows that following contemporary good practices for de-identification ensures that the risk of re-identification is very small. In that systematic review (which is the gold standard methodology for summarizing evidence on a given topic) we found that there were 14 known re-identification attacks. Two of those were conducted on data sets that were de-identified with methods that would be defensible (i.e., they followed existing standards). The success rate of the re-identification for these two was very small." Khaled El Emam and Luk Arbuckle, "Why de-identification is a key solution for sharing data responsibly," *Future of Privacy Forum*, July 24, 2014, http://www.futureofprivacy.org/2014/07/24/de-identification-a-critical-debate/.

[35]Gary McGraw and John Viega, "Introduction to Software Security," *InformIT*, November 2, 2001, http://www.informit.com/articles/article.aspx?p=23950&seqNum=7.

[36]Anup K. Ghosh, Chuck Howell, and James A. Whittaker, "Building Software Securely from the Ground Up," *IEEE Software* (January/February 2002): 14–16.

provable levels of privacy. In addition, privacy researchers should be careful to avoid the disconnect between theorists and practitioners that has sometimes troubled cryptography[37]—theorists need to develop usable constructs and practitioners need to adopt methods with provable privacy.

3.1 Ad Hoc De-identification Leads to Unknowable Risks

The prominence of ad hoc de-identification has led some authors to endorse ad hoc calculation of re-identification probabilities.[38] However, these calculations are specious and offer false hope about privacy protections because they depend on arbitrary and fragile assumptions about what auxiliary datasets and general knowledge are available to the adversary.

Consider an example recently cited by Cavoukian and Castro: Golle's re-examination of unique identification from U.S. census data.[39] Golle found that, using the census data from 2000, 63.3 % of individuals were uniquely identifiable by year, five-digit ZIP code, and birthdate, 4.2 % when birthdate was replaced by month and year of birth, and 0.2 % when replaced by only birth year. Cavoukian and Castro conclude: "The more effectively the data is de-identified, the lower the percentage of individuals who are at risk of re-identification. The risk of re-identification for weakly de-identified data, such as datasets released with gender, ZIP code, and date of birth, is not the same as for strongly de-identified data."[40] It is true that making data more abstract affects re-identification risk, but the percentages can be misleading standing alone:

- The data will doubtless contain other attributes that the adversary could use for re-identification. A common technique of categorizing columns as useful or not useful for re-identification produces an overly optimistic view of re-identification risk because any column containing nontrivial data poses some risk.
- The focus on whether individuals are uniquely identifiable misses privacy violations through probabilistic inferences.[41]

---

[37]For example, the description for a 2012 conference notes that communication between researchers and practitioners is "currently perceived to be quite weak." "Is Cryptographic Theory Practically Relevant?," Isaac Newton Institute for Mathematical Sciences, http://www.newton.ac.uk/event/sasw07. In addition, "[m]odern crypto protocols are too complex to implement securely in software, at least without major leaps in developer know-how and engineering practices." Arvind Narayanan, "What Happened to the Crypto Dream?, Part 2," *IEEE Security & Privacy* 11, no. 3 (2013): 68–71.

[38]El Emam and Arbuckle, "Why de-identification is a key solution."

[39]Philippe Golle, "Revisiting the Uniqueness of Simple Demographics in the US Population," in *Proceedings of the 5th ACM Workshop on Privacy in Electronic Society* (New York, New York: ACM, 2006): 77–80.

[40]Cavoukian and Castro, *De-identification* Does *Work*: 4.

[41]See Sect. 2.1.

In short, a released dataset without birth day and month will be less vulnerable to re-identification through purely demographic information, but the actual effect removal of that information has on re-identification depends highly on the goals and ever-expanding auxiliary data held by the adversary. Furthermore, with high-dimensional datasets, there are strong limits to how much the data can be generalized without destroying utility, whereas auxiliary information has the tendency to get more specific, accurate, and complete with each passing year.

A more specific example offered by Cavoukian and Castro comes from the Heritage Health Prize, released for a data-mining competition to predict future health outcomes based on past hospitalization (insurance claims) data. The dataset was de-identified by El Emam and his team,[42] and Cavoukian and Castro note that "it was estimated that the probability of re-identifying an individual was 0.0084."[43]

However, El Emam's estimates were derived based on a specific, somewhat arbitrary set of assumptions, such as that "the adversary would not know the exact order of the claims,"[44] in other words, that the adversary would not know that the heart attack occurred before the broken arm. Yet, adversaries could gain detailed timeline information by cross-referencing auxiliary information from online reviews of medical providers or by using personal knowledge of targeted subjects, or by using medical knowledge that certain pairs of conditions or treatments, when they occur together, tend to happen in a particular order.

In his report to the Heritage Health Prize organizers, Narayanan shows the arbitrariness of the re-identification probability calculation by using a different, but equally plausible, set of assumptions. In particular, he assumes that the adversary knows the year but not the month or day of each visit and derives dramatically different re-identification probabilities: up to 12.5 % of members are vulnerable.[45]

Happily for the patients in this dataset, large-scale auxiliary databases of hospital visits and other medical information that could be used for re-identification did not appear to be available publicly at the time of the contest. However, some auxiliary information is available in the form of physician and hospital reviews on Yelp, Vitals, and other sites. Furthermore, in 2014 the Centers for Medicare & Medicaid Services publicly released detailed Medicare physician payment data, including physicians' names and addresses, summaries of services provided, and payments for services.[46] Although the Medicare data is for 2012, it is easy to imagine that

---

[42]Khaled El Emam et al., "De-identification methods for open health data: the case of the Heritage Health Prize claims dataset," *Journal of Medical Internet Research* 14, no. 1 (2012): e33, doi:10.2196/jmir.2001.

[43]Cavoukian and Castro, *De-identification* Does *Work*: 11.

[44]El Emam et al., "Heritage Health".

[45]Arvind Narayanan, "An Adversarial Analysis of the Reidentifiability of the Heritage Health Prize Dataset" (unpublished manuscript, 2011).

[46]The dataset "contains information on utilization, payment (allowed amount and Medicare payment), and submitted charges organized by National Provider Identifier (NPI), Healthcare Common Procedure Coding System (HCPCS) code, and place of service." "Medicare Provider Utilization and Payment Data: Physician and Other Supplier," Centers for Medicare & Medicaid Services, last modified April 23, 2014, http://www.cms.gov/Research-Statistics-Data-and-Systems/Statistics-Trends-and-Reports/Medicare-Provider-Charge-Data/Physician-and-Other-Supplier.html.

A Precautionary Approach to Big Data Privacy 369

such data could have been released for the time period spanned by the contest dataset instead and used to match particular providers with contest records. Physician and hospital reviews could then more easily be matched to those records, and more patients identified. In addition, though this Medicare dataset does not include dates, the safe harbor HIPAA de-identification standards permit inclusion of the year for admission and discharge dates[47]; it is plausible that future releases could include such information and make Narayanan's assumptions clearly more valid than El Emam's.

The later release of publicly available auxiliary information like the Medicare data could enable a broad attack unaccounted for in the initial re-identification probability estimates. The possibility of such future releases can never be ruled out. Even without such a data release, the contest data is vulnerable to targeted attacks by adversaries with specific knowledge about people in the dataset.

It is very tempting to look for assurances about the probability of privacy violations from an ad hoc de-identified dataset, but there is simply no scientific basis for interpreting ad hoc re-identification probability estimates of ad hoc de-identified high-dimensional datasets as anything more than (weak) lower bounds. Ad hoc estimates tend to be based on many assumptions, so that the probability claims must be accompanied by multiple caveats. In practice, the caveats likely will be lost, as they were when Cavoukian and Castro cited El Emam's 0.0084 probability without noting any of the assumptions that El Emam details in his paper. Rigorously quantified privacy risks are only possible when using methods designed to allow for such calculations.

3.2 The Promise of Provable Privacy

As noted earlier, data releases are permanent and re-identification capabilities are improving, making protocols and systems with proven privacy properties an urgent need. The foundation for such protocols and systems are methods of handling data that preserve a rigorously defined privacy, even in the face of unpredicted advances in data analysis, while also permitting useful analysis. At present, algorithms that yield differential privacy are the only well-developed methodology that satisfies these requirements.

One lesson from cryptography research is the importance of getting central definitions correct. Finding a definition of security or privacy that is sound, provable, and consistent with intuitive notions of those terms can be a research contribution in itself. Such a definition enables evaluation of existing and proposed algorithms against a consistent standard.

---

[47]"Guidance Regarding Methods for De-identification of Protected Health Information in Accordance with the Health Insurance Portability and Accountability Act (HIPAA) Privacy Rule," U.S. Department of Health & Human Services, http://www.hhs.gov/ocr/privacy/hipaa/understanding/coveredentities/De-identification/guidance.html.

Differential privacy is based on this type of formal definition: including a particular user's data in a dataset (as opposed to omitting it) must have a strictly limited effect on the output of any differentially private analysis of the data. Differential privacy algorithms[48] typically add "noise"—small, quantified error—to the outputs of analysis and release those blurred outputs, rather than releasing the original input data or unaltered outputs. The effect of including a particular user's data in the dataset can be made arbitrarily small through variations in the type and amount of noise.

Differential privacy is a criterion for privacy. Different algorithms can satisfy this criterion in different ways, and the approach to achieving differential privacy might differ from case to case, although the privacy criterion stays the same.

Like all protective measures, differential privacy algorithms involve a tradeoff between privacy and utility, as the stronger the privacy guarantees are made, the less accurate the estimated statistics from the data must be.[49] Increased noise both improves privacy and reduces the usefulness of the blurred outputs. However, unlike ad hoc de-identification, algorithms implementing differential privacy can quantify the tradeoff between privacy and utility, and do not depend on artificial assumptions about the adversary's capabilities or access to auxiliary information. Their guarantees do not become weaker as adversaries become more capable. No matter how much is known about the targeted person, the information learnable by the adversary due to that person's inclusion in the dataset remains strictly limited.

Given these advantages, differential privacy is a valuable tool for data privacy. Further research is needed on the development and application of differential privacy methods, as well as in the development of other computer science and mathematical techniques aimed at provable privacy.

4 Practical Steps Towards Improved Data Privacy

Given the weaknesses of ad hoc de-identification and the nascent state of provable privacy research, we turn to the difficult policy question of how to handle current datasets: how to balance privacy threats with the benefits fostered by wider access to data. Each dataset has its own risk-benefit tradeoff, in which the expected damage done by leaked information must be weighed against the expected benefit from improved analysis. Both assessments are complicated by the unpredictable effects of combining the dataset with others, which may escalate both the losses and the gains.

---

[48]The following sources contain introductions to differential privacy. Cynthia Dwork et al., "Differential Privacy—A Primer for the Perplexed" (paper presented at the Joint UNECE/Eurostat work session on statistical data confidentiality, Tarragona, Spain, October 2011); Erica Klarreich, "Privacy by the Numbers: A New Approach to Safeguarding Data," *Quanta Magazine* (December 10, 2012); Christine Task, "An Illustrated Primer in Differential Privacy," *XRDS* 20, no. 1 (2013): 53–57.

[49]Ohm, "Broken Promises of Privacy": 1752–55.

In this Section, we explain why releasing datasets to the public using ad hoc de-identification methods should not be the default policy. Then, we consider methods by which policymakers can push the default to be access using provable privacy methods or restricted access to a narrow audience. The individualized nature of each dataset access means that one-size-fits-all solutions must be either incomplete or incorrect—certain broad policies may be useful, but no single rule for dealing with all data access will give good results in every case. We offer policy recommendations below that promote a more cautious and more tailored approach to releasing data: (1) incentivize the development and use of provable privacy methods and (2) encourage narrower data accesses that still permit analysis and innovation. Finally, we argue for increased transparency around re-identification risks to raise public awareness and to bolster the other recommendations.

4.1 Defining a Precautionary Approach

The precautionary principle deals with decision-making and risk regulation in the face of scientific uncertainty. It has many, much-debated formulations, ranging from very weak (for example, that regulation should be permitted when risks are uncertain) to very strong (for example, that any action with an uncertain risk should be barred completely until the actor can prove that the risks are acceptable). We do not wish to engage in the debate over the general formulation of the principle and the breadth of its applicability. Instead, we focus on the specific problem of how to react to the unknowable risks of ad hoc de-identification. Precautionary approaches often shift where the burden of proof for the decision about an action falls when risks are uncertain, and we argue that placing the burden more heavily on data providers will yield better results than the status quo.

The difficulty at the heart of this issue is weighing uncertain privacy risks against uncertain data access benefits. The loss of these benefits—such as potential medical advances or research progress from wider data sharing—is also legitimately characterized as an uncertain risk. The impossibility of completely avoiding both uncertain risks has led to Sunstein's criticism of strong versions of the precautionary principle for creating paralysis by "forbid[ding] all courses of action, including inaction."[50] However, like most proponents of precautionary approaches, we do "not impose a burden on any party to prove zero risk, nor… state that all activities that pose a possible risk must be prohibited."[51] Instead, we see a way forward by altering default behaviors and incentives.

Currently, there is a presumption that data releases to the public are acceptable as long as they use ad hoc de-identification and strip out classes of information

---

[50]Cass R. Sunstein, "The Paralyzing Principle," *Regulation* 25, no. 4 (2002): 33–35.

[51]Noah M. Sachs, "Rescuing the Strong Precautionary Principle from Its Critics," *Illinois Law Review* 2011 no.4 (2011): 1313.

deemed to be PII. This presumption draws a line and the burden of proof shifts when it is crossed: if data providers have used ad hoc de-identification and removed PII, then the burden of proof falls on privacy advocates to show that the particular datasets are re-identifiable or could cause other harms; if data providers have not done so, then they are obliged to demonstrate why data releases that do not conform to standard practices are permissible.

We argue that this line—and the attendant standard practices—should shift. A spectrum of choices for the line exist, with the endpoints completely prioritizing data access or privacy, and current standards lean too far towards data access. Ad hoc de-identification has unknowable risks, and the continued release of ad hoc de-identified data presents the threat of unacceptable widespread re-identification of past datasets. In addition, data providers have the power to limit their data releases and reduce those risks. As such, release of ad hoc de-identified data to the entire public should require justification; it should not be the default behavior. Parties releasing data using ad hoc de-identification methods should have the responsibility, at a minimum, to limit that release to the narrowest possible scope likely to yield the intended benefit.[52]

Ad hoc de-identification is useful to practitioners as an additional layer of defense. However, we join PCAST in urging policymakers to stop relying on it and to stop treating it as a sufficient privacy protection on its own.

Alternatively, data providers could avoid the uncertainty of ad hoc de-identification and the need to take precautionary measures by using provable privacy methods instead. Because provable privacy methods have precisely calculable risks, they allow for traditional risk-benefit analyses and remove the possibility of snowballing re-identification risk that comes with continued unfettered release of data using ad hoc de-identification.

4.2 Researching and Implementing Provable Privacy

Additional funding for provable privacy research is the clearest way to encourage development of provable privacy methods. However, such methods are necessary, but not sufficient, for responsible data practices because once they exist, they still need to be deployed widely. Achieving broad adoption of those methods is as much a social and policy problem as a technical one.

---

[52]Alternatively, a data provider could show that the expected benefit outweighs the privacy cost of complete re-identification of the entire dataset. In other words, the data provider would need to show that there still would be a net benefit from releasing the data even if the names of all individuals involved were attached to their records in the dataset. This standard would be, in most cases, significantly more restrictive.

We emphasize two main goals to help propagate these methods and create more real-world applications of provable privacy like the U.S. Census Bureau's OnTheMap[53] and Google's RAPPOR.[54] First, privacy researchers must communicate with data scientists so that the theoretical privacy work is developed with practical uses in mind. Second, data scientists must accept and use these new methodologies.

Although many levers may be used to influence researchers, funding choices are an essential and practical tool. Much of the work done both by privacy researchers and by data scientists and providers is dependent upon governmental funding streams, so altering allocations to advance provable privacy would be a highly effective motivation to improve practices. It is also a quicker and more flexible path to behavioral change than legislative or regulatory privacy requirements.

Privacy research funding can encourage collaborations with or feedback from practitioners. Data Science funding can favor projects that implement provable privacy methods instead of ad hoc de-identification or no privacy measures. Making the development and application of provable privacy a factor in funding decisions will push practitioners to overcome the inertia that keeps them using existing ad hoc methods involving unproven and risky data privacy practices.

Governments can also encourage development of provable privacy by entering the market for such technologies as a consumer or by making data available under a provably private interface. Innovation procurement—using government demand to drive the development and diffusion of new products or processes—has gained support,[55] particularly in Europe.[56] Provable privacy technologies appear to be a good candidate for this kind of stimulus, as purchasing systems based on these technologies can fulfill both innovation goals and the core goals of obtaining high-quality, useful products for the public sector.[57] Similarly, providing government data through a differential privacy-based interface would serve both innovation and privacy goals by incentivizing data users to learn how to use such interfaces and protecting the people included in the datasets.

---

[53]"OnTheMap," U.S. Census Bureau, http://onthemap.ces.census.gov/; Klarreich, "Privacy by the Numbers.".

[54]Úlfar Erlingsson, Vasyl Pihur, and Aleksandra Korolova, "RAPPOR: Randomized Aggregatable Privacy-Preserving Ordinal Response," in *Proceedings of the 2014 ACM SIGSAC Conference on Computer and Communications Security* (Scottsdale, Arizona: ACM, 2014): 1054–67.

[55]Jakob Edler and Luke Georghiou, "Public procurement and innovation—Resurrecting the demand side," *Research Policy* 36, no. 7 (September 2007): 949–63.

[56]Charles Edquist and Jon Mikel Zabala-Iturriagagoitia, "Public Procurement for Innovation as mission-oriented innovation policy," *Research Policy* 41, no. 10 (December 2012): 1757–69.

[57]Elvira Uyarra and Kieron Flanagan, "Understanding the Innovation Impacts of Public Procurement," *European Planning Studies* 18, no. 1 (2010): 123–43.

4.3 Flexible Options for Narrower Releases of Data

Although we argue that data providers should justify public releases of datasets that use ad hoc de-identification methods, we do not recommend hardening that burden of proof into a single legal or regulatory requirement. Because dataset releases are highly individualized, a universal one-size-fits-all requirement would lead to sub-optimal results in many cases. Instead, the burden-of-proof concept can be considered a guiding principle for an array of more flexible policy choices that can be tailored to particular circumstances, as the case studies in the next part demonstrate. Here we list, for both data custodians and policymakers, some of the considerations—not mutually exclusive—that may help in determining the appropriate scope for the release of datasets:

- Is it possible to use a provable privacy method and thus get an accurate calculation of the privacy risks to weigh against the expected benefit?
- Is it possible to host data on the custodian's system and allow researchers to query it, instead of releasing the dataset?
- Can all or most of the intended benefit of data release be achieved by computing and releasing aggregate statistics instead of raw micro-data?
- Is a limited release similarly useful? Are the people most likely to use the data beneficially a subset of the general public: researchers, affiliates of educational institutions, data analysts with past successes?
- Can multiple forms of the dataset be released so that only those who have demonstrated effectiveness or a need for more vulnerable datasets receive them?
- Can data recipients be required to sign legal contracts restricting their use and transfer of the dataset?
- Can data recipients be required to undergo ethics training?
- Can data recipients be required to provide certain information: identification, a statement of purpose for obtaining the data?

These questions can help determine whether a narrower release of a dataset is wise, and we think that it almost never will be the case that an unlimited release of a dataset to the entire public will be the optimal choice.

4.4 Enabling Transparency of Re-identification Risks

Privacy is, at least in part,[58] an individual right, and as such, transparency about data usage and data flows is a natural response to big data privacy concerns. Such transparency has appeared as a central tenet in governmental pronouncements on

---

[58] Solove, among others, has discussed how privacy is traditionally viewed as an individual right but also has social value. Daniel J. Solove, "'I've Got Nothing to Hide' and Other Misunderstandings of Privacy," *San Diego Law Review* 44 (2007): 760–64.

big data: for example, the U.K.'s Information Commissioner's Office includes transparency among the "practical aspects to consider when using personal data in big data analytics,"[59] and the U.S. White House makes transparency one of the seven rights in its Consumer Privacy Bill of Rights.[60]

This transparency should include informing people about re-identification risks stemming from data collected about them. Knowledge about the possibility of re-identification is necessary "to enabl[e] consumers to gain a meaningful understanding of privacy risks and the ability to exercise Individual Control."[61] We propose that, wherever notice about data collection can be given, a short statement should be included that briefly describes what steps will be taken to protect privacy and notes whether records may be re-identified despite those steps. Users also should be able to access further details about the privacy protection measures easily, perhaps through a link in the notice. Among the available details should be a justification for the protective steps taken, describing why the provider has confidence that re-identification will not occur.

Giving users information about privacy protection measures and re-identification risks helps to even the information asymmetry between them and data collectors.[62] It would allow users to make more informed decisions and could motivate more conscientious privacy practices, including the implementation of provable privacy methods. It is also possible that data collectors could give users options about the privacy protection measures to be applied to their information. Such segmentation would permit personal assessments of the risks and benefits of the data collection: people who have strong desires for privacy could choose heavier protections or non-participation; people who do not care about being identified or who strongly support the potential research could choose lighter, or no, protections.[63] This segmentation is a helpful complement to narrowed releases of data:

---

[59]Information Commissioner's Office, *Big data and data protection* (July 28, 2014): 5–6, 33–37.

[60]The White House, Consumer Data Privacy in a Networked World: A Framework for Protecting Privacy and Promoting Innovation in the Global Digital Economy (Washington, D.C.: February 2012): 47.

[61]Ibid.

[62]Of course, simply providing information can be insufficient to protect users. It may not "be information that consumers can use, presented in a way they can use it," and so it may be ignored or misunderstood. Lawrence Lessig, "Against Transparency," *New Republic*, October 9, 2009. Alternatively, a user may be informed effectively but the barriers to opting out may be so high as to render the choice illusory. Janet Vertesi, "My Experiment Opting Out of Big Data Made Me Look Like a Criminal," *Time*, May 1, 2014. Still, we believe that concise, clear descriptions of privacy protecting measures and re-identification risks can aid users in many circumstances and should be included in the options considered by policymakers.

[63]For example, patients in clinical trials or with rare diseases might wish to have their data included for analysis, even if the risk of re-identification is high or if no privacy protecting measures are taken at all. Kerstin Forsberg, "De-identification and Informed Consent in Clinical Trials," Linked Data for Enterprises, November 17, 2013, http://kerfors.blogspot.com/2013/11/de-identification-and-informed-consent.html.

instead of restricting access to the people who can create the most benefit, segmentation restricts participation to the people who feel the least risk.

5 Specific Advice for Six Common Cases

Now we turn to six of the most common cases in which we believe it is particularly important for data custodians to look beyond ad hoc de-identification for privacy protection. In each case, we present recommendations for data custodians and policymakers, providing real-world applications of the risk-benefit assessments and policy tools described in Sect. 3.

Case 0: "No PII" as a putative justification for data collection.

Companies that track user activities—often without notice or choice—frequently proffer the argument that they do not collect PII in response to privacy concerns. Third-party online tracking is a prime example—U.S. online advertising self-regulation treats PII as the primary dividing line between acceptable and unacceptable tracking.[64] Mobile apps and mall tracking based on WiFi signals are others.

Of course, we should expect that such datasets can be re-identified, and even *accidental* leaks of identity to tracking companies are rampant online.[65] As such, we recommend that policymakers and regulators not consider the absence of deliberate PII collection to be an adequate privacy safeguard. Additional privacy measures include aggregation[66] and data minimization. Requiring affirmative consent for tracking, encouraging the development of easy-to-use opt-out mechanisms, and funding the development of technical defense mechanisms are fruitful policy directions as well.

Online privacy is often a proxy for other worries such as targeting of protected groups and data-driven discrimination.[67] These worries are just as serious whether or not PII is involved or re-identification takes place. In recent years a combination of

---

[64]For example, the Network Advertising Initiative's self-regulatory Code "provides disincentives to the use of PII for Interest-Based Advertising. As a result, NAI member companies generally use only information that is not PII for Interest Based Advertising and do not merge the non-PII they collect for Interest-Based Advertising with users' PII." "Understanding Online Advertising: Frequently Asked Questions," Network Advertising Initiative, http://www.networkadvertising.org/faq.

[65]Balachander Krishnamurthy and Craig E. Wills, "On the Leakage of Personally Identifiable Information Via Online Social Networks," in *Proceedings of the 2nd ACM Workshop on Online Social Networks* (New York, New York: ACM, 2009): 7-12, http://www2.research.att.com/~bala/papers/wosn09.pdf.

[66]Data aggregation replaces individual data elements by statistical summaries.

[67]Cynthia Dwork and Deirdre K. Mulligan, "It's not privacy, and it's not fair," *Stanford Law Review Online* 66 (2013): 35.

press reporting,[68] empirical research,[69] and theory[70] has helped clarify the nature of these dangers. As a result, policy makers' attention has gradually shifted to data use in addition to data collection. While restrictions on collection continue to be important, we encourage the trend toward monitoring data use and developing norms and rules.

Case 1: Companies selling data to one another.

When privacy laws place use limits on customer information, there is typically a carve-out for "anonymized" records. For example, both the EU Data Protection Directive and the proposed General Data Protection Regulation place more stringent restrictions on "personal data": the former defines "personal data" as "information relating to an identified or identifiable natural person"[71]; the latter defines it as "any information relating to a data subject," who is someone who "can be identified, directly or indirectly, by means reasonably likely to be used."[72] These definitions were constructed to provide safe harbors for anonymized data.[73] However, they are only as strong as the anonymization method used. In the case of ad hoc anonymization, re-identification science has shown that such exceptions are not well-founded. It is unclear whether the EU rules will be interpreted to create loopholes or to apply stringent requirements to all data collection and release; other statutes and regulations have more explicit carve-outs for data that omits specific PII, and these rules will create more loopholes.

We call for a move away from such exceptions in future lawmaking and rulemaking, except in cases where strong provable privacy methods are used. Meanwhile, we make two recommendations to minimize privacy risks in domains in which such loopholes do or may exist. First, data custodians must use legal agreements to restrict the flow and use of data—in particular, to prohibit resale of such datasets and specify acceptable uses including limits on retention periods. Second, policymakers should increase the transparency of the data economy by requiring disclosures of "anonymized" data sharing in privacy policies. This change will fix the current information asymmetry between firms and consumers and allow the market to price privacy more efficiently.

---

[68]Julia Angwin, "The web's new gold mine: Your secrets," *Wall Street Journal*, July 30, 2010.

[69]Aniko Hannak et al., "Measuring Price Discrimination and Steering on E-commerce Web Sites," in *Proceedings of the 2014 Conference on Internet Measurement Conference* (Vancouver: ACM, 2014): 305–318.

[70]Solon Barocas and Andrew D. Selbst, "Big Data's Disparate Impact," *California Law Review* 104 (forthcoming); Ryan Calo, "Digital Market Manipulation," *George Washington Law Review* 82 (2014): 995.

[71]Directive 95/46/EC, of the European Parliament and of the Council of 24 October 1995 on the Protection of Individuals with Regard to the Processing of Personal Data and on the Free Movement of Such Data, Art. 2(a), 1995 O.J. (C 93).

[72]Proposal for a Regulation of the European Parliament and of the Council on the Protection of Individuals with Regard to the Processing of Personal Data and on the Free Movement of Such Data, Art. 4(1)-(2) (January 25, 2012).

[73]Ohm, "Broken Promises of Privacy": 1704, 1738–41.

Case 2: Scientific research on data collected by companies.

From telephone call graphs to medical records, customer data collected by private companies has always been tremendously valuable for scientific research. The burgeoning field of computational social science has made great strides in adapting online self-reported data, such as information on social networks, for drawing statistically sound conclusions.[74] Such data were previously considered less useful for research but this thinking is being overturned.

Privacy and re-identification risks are again a vexing concern if these companies are to open their datasets to external researchers. The silver lining is that the largest companies with the most interesting research datasets usually have in-house research teams—AT&T, Microsoft, and more recently, Facebook are good examples. However, there are two problems with relying on in-house research; we now discuss these problems and potential solutions.

First, benefits from published research have large positive externalities, often far exceeding the benefits to the firm, which include improved reputation or increased knowledge about users. So, economic theory would predict that these research teams will be smaller than the public would want them to be. Rather than dealing with this externality by encouraging public release of company data, governments should seek ways to incentivize research publications of this type with fewer privacy implications, such as by sponsoring programs for academic researchers in visiting positions at companies.

Second, in-house research may not be reproducible. However, much of the interesting user research at companies seems to involve interventional experiments on users. For such experiments, publishing data will not enable reproducibility, and the best option for verifying results is for the company to permit outside researchers to visit and re-run experiments on new batches of users. When access to the data would help with reproducibility, we would follow the recommendations laid out below in Case 4 for scientific research in general.

Case 3: Data mining contests.

The ease of data collection means that even small companies that cannot afford in-house research teams often have interesting datasets for scientific research or knowledge discovery—colloquially termed data mining. Data mining contests,

---

[74]Pablo Barberá, "How Social Medial Reduces Mass Political Polarization: Evidence from Germany, Spain, and the U.S." (unpublished manuscript, October 18, 2014), https://files.nyu.edu/pba220/public/barbera-polarization-social-media.pdf; Amaney Jamal et al., "Anti-Americanism and Anti-Interventionism in Arabic Twitter Discourses" (unpublished manuscript, October 20, 2014), http://scholar.harvard.edu/files/dtingley/files/aatext.pdf; Margaret E. Roberts, "Fear or Friction? How Censorship Slows the Spread of Information in the Digital Age" (unpublished manuscript, September 26, 2014), http://scholar.harvard.edu/files/mroberts/files/fearfrictio n_1.pdf.
Computational social scientists can also generate their own self-reported data online. Matthew J. Salganik and Karen E.C. Levy, "Wiki surveys: Open and quantifiable social data collection" (unpublished manuscript, October 2, 2014), http://arxiv.org/abs/1202.0500.

such as the Netflix prize discussed above, have recently gained popularity as a way for such companies to incentivize research that utilizes their data.

Such contests are spurs to innovation, and the most effective scope for data release depends on balancing two factors: having more contestants reduces their motivation because they become less likely to win, but it also increases the chance of having a contestant put forth a rare solution.[75] As such, Boudreau, Lacetera, and Lakhani have concluded that expansive competitions are most useful for problems where the solutions are highly uncertain, including multi-domain problems where it is less clear who would solve them best and how.[76] Jeppesen and Lakhani also suggest that broadening the scope of contestants can bring in people on the margins of the technical fields and social groups primarily associated with the contest problem and that those marginal people are more likely to succeed in these contests.[77]

We make three recommendations for data custodians running contests:

- Consider whether the group of contestants can be narrowed. If the solution desired is less uncertain, perhaps because it lies in a single domain or known methodologies are expected to work, research suggests that a contest between few participants can be more effective. It may also be possible to invite participants with diverse backgrounds and views to provide the advantage from marginal contestants, though we recognize that identifying such people may be difficult because they are on the margins.
- Whenever possible, switch to a model in which data is made available under provable privacy guarantees. We expect that the expense and development effort involved in applying the appropriate data transformations and carrying out privacy analyses will be similar to the current process of data pre-processing and evaluating de-identification methods. Contest organizers are in a good position to effect a behavior change among data scientists because of the financial incentives.
- If de-identified data is released, use a multi-stage process. Early stages can limit the amount or type of data released by releasing data on only a subset of users, minimizing the quantitative risk, or by releasing a synthetic dataset created to mimic the characteristics of the real data.[78] Later stages can permit access to a broader dataset but add some combination of the following restrictions: requiring contestants to sign a data-use agreement; restricting the contest to a shortlist of best performers from the first stage; and switching to an "online computation

---

[75]Kevin J. Boudreau, Nicola Lacetera, and Karim R. Lakhani, "Incentives and Problem Uncertainty in Innovation Contests: An Empirical Analysis," *Management Science* 57, no. 5 (2014): 843–63, doi: 10.1287/mnsc.1110.1322.

[76]Ibid., 860–61.

[77]Lars Bo Jeppesen and Karim R. Lakhani, "Marginality and Problem Solving Effectiveness in Broadcast Search," *Organization Science* 21, no. 5 (2010): 1016–33.

[78]Researchers already have developed methods for creating such synthetic data. Avrim Blum, Katrina Ligett, and Aaron Roth, "A Learning Theory Approach to Non-Interactive Database Privacy," in *Proceedings of the 40th ACM SIGACT Symposium on Theory of Computing* (Victoria, British Columbia: ACM, 2008).

model" where participants upload code to the data custodian's server (or make database queries over its network) and obtain results, rather than download data.

Case 4: Scientific research, in general.

Nearly all scientific research on human subjects would be improved if data could be shared more freely among researchers, enhancing efficiency and reproducibility. These advantages have led to calls for open data, which can be interpreted as advocating the public release of datasets used in research. However, the gains come predominantly from scientists having the data, and so restricted access to a data-sharing system is a good solution in this area.[79] Such a system should implement various gatekeeping functions, such as demanding proof of academic or peer-reviewed standing, requiring ethical training, and designing and overseeing the security of the system.[80] In addition, government research funding can incentivize scientists to use provable privacy methods.

A good example of gatekeeping is the U.S. State Inpatient Databases (SIDs) developed for the Healthcare Cost and Utilization Project sponsored by the Agency for Healthcare Research and Quality (AHRQ). AHRQ wishes this data to be used more broadly than just among scientific researchers, but it is cognizant of the very serious re-identification risk presented by the datasets. Obtaining them involves a number of steps[81]: completing an online Data Use Agreement Training Course; paying a fee; providing information including name, address, and type of organization; describing the intended project, areas of investigation, potential uses of any products created, and reasons for requesting the data; and physically signing a data-use agreement that prohibits the use of the data "to identify any person"—this last requirement could be further strengthened by defining identification to include any use of the data "to infer information about, or otherwise link the data to, a particular person, computer, or other device."[82]

Case 5: Open government data.

In one sense, open government data may be the most difficult case because most of our earlier prescriptions do not apply. First, in most cases there is no ability to

---

[79]"If there are privacy concerns I can imagine ensuring we can share the data in a 'walled garden' within which other researchers, but not the public, will be able to access the data and verify results." Victoria Stodden, "Data access going the way of journal article access? Insist on open data," *Victoria's Blog*, December 24, 2012, http://blog.stodden.net/2012/12/24/data-access-going-the-way-of-journal-article-access/.

[80]Genomics researchers have proposed one such system. Bartha Maria Knoppers, et al., "Towards a data sharing Code of Conduct for international genomic research," *Genome Medicine* 3 (2011): 46.

[81]HCUP, SID/SASD/SEDD Application Kit (October 15, 2014), http://www.hcup-us.ahrq.gov/db/state/SIDSASDSEDD_Final.pdf.

[82]Federal Trade Commission, *Protecting Consumer Privacy in an Era of Rapid Change: Recommendations for Businesses and Policymakers* (Washington, DC: March 2012) 21, http://www.ftc.gov/sites/default/files/documents/reports/federal-trade-commission-report-protecting-consumer-privacy-era-rapid-change-recommendations/120326privacyreport.pdf.

opt out of data collection. Second, while some research could be done in-house by government agencies, it is not possible to anticipate all beneficial uses of the data by external researchers, and the data is not collected for a specific research purpose. Finally, restricting access runs contrary to the transparency goals of improving government by shedding light on its practices.

However, in another sense, re-identification worries are minimal because the vast majority of open government datasets do not consist of longitudinal observations of individuals. Interestingly, for a variety of datasets ranging from consumer complaints to broadband performance measurement, the data is not *intended* to track users longitudinally, but it might *accidentally* enable such tracking if there is enough information about the user in each measurement data point. To prevent such accidental linkability, de-identification is indeed a valuable approach.

Certain aggregate or low-dimensional government data, such as many of the datasets published by the U.S. Census Bureau, seem to avoid privacy violations fairly well by using statistical disclosure control methodologies. However, high-dimensional data is problematic, and there is no reason to expect it cannot be de-anonymized. For these datasets, it seems that the best solution is to implement provable privacy techniques, as the Census Bureau did with its OnTheMap data, or to wait to release such data until provable privacy techniques can be implemented satisfactorily.

These cases illustrate how our various policy recommendations can be applied to practical situations, and the variation among the recommendations demonstrates the importance of a flexible policy response. Data custodians and policymakers will need to make granular decisions about the risks and benefits of releasing specific datasets, and we hope that the factors and examples in this paper will serve as a guide."

Bibliography

Acquisti, Alessandro, Ralph Gross, and Fred Stutzman. 2011. *Faces of facebook: Privacy in the age of augmented reality*. Nevada: Presentation at BlackHat Las Vegas. 4 Aug 2011.

Adida, Ben. 2008. Don't hash secrets. *Benlog*. http://benlog.com/2008/06/19/dont-hash-secrets/. 19 June 2008.

Angwin, Julia. 2010. The web's new gold mine: Your secrets. *Wall Street Journal*. 30 July 2010.

Barbaro, Michael and Tom Zeller, Jr. 2006. A face is exposed for AOL Searcher No. 4417749. *New York Times*. http://www.nytimes.com/2006/08/09/technology/09aol.html. 9 Aug 2006.

Barberá, Pablo. 2014. How social medial reduces mass political polarization: Evidence from Germany, Spain, and the U.S. Unpublished manuscript, 18 Oct 2014. https://files.nyu.edu/pba220/public/barbera-polarization-social-media.pdf.

Barocas, Solon and Andrew D. Selbst. Big Data's Disparate Impact. *California Law Review* 104 (forthcoming).

Barocas, Solon, and Helen Nissenbaum. 2014. Big data's end run around anonymity and consent. In *Privacy, big data, and the public good: Frameworks for Engagement*, ed. Julia Lane, Victoria Stodden, Stefan Bender, and Helen Nissenbaum, 44–75. New York: Cambridge University Press.

Blum, Avrim, Katrina Ligett, and Aaron Roth. 2008. A learning theory approach to non-interactive database privacy. In *Proceedings of the 40th ACM SIGACT symposium on theory of computing*. Victoria, British Columbia: ACM.

Boudreau, Kevin J., Nicola Lacetera, and Karim R. Lakhani. 2014. Incentives and problem uncertainty in innovation contests: An empirical analysis. *Management Science* 57(5): 843–863. doi:10.1287/mnsc.1110.1322.

Calo, Ryan. 2014. Digital market manipulation. *George Washington Law Review* 82: 995–1051.

Cavoukian, Ann and Daniel Castro. 2014. Big Data and innovation, setting the record straight: De-identification does Work. Toronto, Ontario: Information and Privacy Commissioner.

de Montjoye, Yves-Alexandre, César A. Hidalgo, Michel Verleysen, and Vincent D. Blondel. 2013. Unique in the crowd: The privacy bounds of human mobility. *Scientific Reports* 3. doi:10.1038/srep01376.

Dey, Ratan, Yuan Ding, and Keith W. Ross. 2013. The high-school profiling attack: How online privacy laws can actually increase minors' risk. Paper presented at the 13th Privacy Enhancing Technologies Symposium, Bloomington, IN. https://www.petsymposium.org/2013/papers/dey-profiling.pdf. 12 July 2013.

DHS Data Privacy and Integrity Advisory Committee FY 2005 Meeting Materials. 2005. *Statement of Latanya Sweeney, associate professor of computer science, technology and policy and director of the data privacy laboratory*. Carnegie Mellon University. http://www.dhs.gov/xlibrary/assets/privacy/privacy_advcom_06-2005_testimony_sweeney.pdf. 15 June 2015.

Directive 95/46/EC, of the European Parliament and of the Council of 24 October 1995 on the Protection of Individuals with Regard to the Processing of Personal Data and on the Free Movement of Such Data, Art. 2(a), 1995 O.J. (C 93).

Dwork, Cynthia, Frank McSherry, Kobbi Nissim, and Adam Smith. 2011. Differential privacy—a primer for the perplexed. Paper presented at the Joint UNECE/Eurostat work session on statistical data confidentiality, Tarragona, Spain.

Dwork, Cynthia, and Deirdre K. Mulligan. 2013. It's not privacy, and it's not fair. *Stanford Law Review Online* 66: 35–40.

Edler, Jakob, and Luke Georghiou. 2007. Public procurement and innovation—Resurrecting the demand side. *Research Policy* 36(7): 949–963.

Edquist, Charles and Jon Mikel Zabala-Iturriagagoitia. 2012. Public procurement for innovation as mission-oriented innovation policy. *Research Policy* 41(10): 1757–69.

El Emam, Khaled, Luk Arbuckle, Gunes Koru, Benjamin Eze, Lisa Gaudette, Emilio Neri, Sean Rose, Jeremy Howard, and Jonathan Gluck. 2012. De-identification methods for open health data: The case of the Heritage Health Prize claims dataset. *Journal of Medical Internet Research* 14(1): e33. doi:10.2196/jmir.2001.

El Emam, Khaled and Luk Arbuckle. 2014. Why de-identification is a key solution for sharing data responsibly. *Future of Privacy Forum*. http://www.futureofprivacy.org/2014/07/24/de-identification-a-critical-debate/. 24 July 2014.

Englehardt, Steven, Dillon Reisman, Christian Eubank, Peter Zimmerman, Jonathan Mayer, Arvind Narayanan, and Edward W. Felten. 2015. Cookies that give you away: Evaluating the surveillance implications of web tracking. Paper accepted at 24th International World Wide Web Conference, Florence.

Erlingsson, Úlfar, Vasyl Pihur, and Aleksandra Korolova. 2014. RAPPOR: Randomized aggregatable privacy-preserving ordinal response. In *Proceedings of the 2014 ACM SIGSAC conference on computer and communications security*, 1054–67. Scottsdale, Arizona: ACM.

Executive Office of the President, President's Council of Advisors on Science and Technology. 2014. *Report to the President: Big Data and Privacy: A Technological Perspective*. Washington, DC.

Federal Trade Commission. 2012. *Protecting consumer privacy in an era of rapid change: Recommendations for businesses and policymakers*. Washington, DC. http://www.ftc.gov/sites/default/files/documents/reports/federal-trade-commission-report-protecting-consumer-privacy-era-rapid-change-recommendations/120326privacyreport.pdf.

Felten, Ed. 2012a. Are pseudonyms 'anonymous'? *Tech@FTC*. https://techatftc.wordpress.com/2012/04/30/are-pseudonyms-anonymous/. 30 Apr 2012.

Felten, Ed. 2012b. Does hashing make data 'anonymous'? *Tech@FTC*. https://techatftc.wordpress.com/2012/04/22/does-hashing-make-data-anonymous/. 22 Apr 2012.

Forsberg, Kerstin. 2013. De-identification and informed consent in clinical trials. *Linked Data for Enterprises*. http://kerfors.blogspot.com/2013/11/de-identification-and-informed-consent.html. 17 Nov 2013.

Gagnon, Michael N. 2011. Hashing IMEI numbers does not protect privacy. *Dasient Blog*. http://blog.dasient.com/2011/07/hashing-imei-numbers-does-not-protect.html. 26 July 2011.

Ghosh, Anup K., Chuck Howell, and James A. Whittaker. 2002. Building software securely from the ground up. *IEEE Software*.

Golle, Philippe and Kurt Partridge. 2009. On the anonymity of home/work location pairs. In *Pervasive '09 proceedings of the 7th international conference on pervasive computing*, 390–97. Berlin, Heidelberg: Springer. https://crypto.stanford.edu/~pgolle/papers/commute.pdf.

Golle, Philippe. 2006. Revisiting the uniqueness of simple demographics in the US population. In *Proceedings of the 5th ACM workshop on privacy in electronic society*, 77–80. New York, New York: ACM.

Guidance Regarding Methods for De-identification of Protected Health Information in Accordance with the Health Insurance Portability and Accountability Act (HIPAA) Privacy Rule. U.S. Department of Health & Human Services. http://www.hhs.gov/ocr/privacy/hipaa/understanding/coveredentities/De-identification/guidance.html.

Gymrek, Melissa, Amy L. McGuire, David Golan, Eran Halperin, and Yaniv Erlich. 2013. identifying personal genomes by surname inference. *Science* 339(6117): 321–324. doi:10.1126/science.1229566.

Hannak, Aniko, Gary Soeller, David Lazer, Alan Mislove, and Christo Wilson. 2014. Measuring price discrimination and steering on e-commerce web sites. In *Proceedings of the 2014 conference on internet measurement conference*, 305–318. Vancouver: ACM.

HCUP, SID/SASD/SEDD Application Kit. 2014. http://www.hcup-us.ahrq.gov/db/state/SIDSASDSEDD_Final.pdf.

Hooley, Sean and Latanya Sweeney. 2013. Survey of publicly available state health databases. White Paper 1075-1, Data Privacy Lab, Harvard University, Cambridge, Massachusetts. http://dataprivacylab.org/projects/50states/1075-1.pdf.

Information Commissioner's Office. 2014. *Big data and data protection* 28 July 2014.

Is Cryptographic Theory Practically Relevant? Isaac Newton Institute for Mathematical Sciences. http://www.newton.ac.uk/event/sasw07.

Jamal, Amaney, Robert O. Keohane, David Romney, and Dustin Tingley. 2014. Anti-Americanism and Anti-interventionism in Arabic Twitter Discourses. Unpublished manuscript, 20 Oct 2014. http://scholar.harvard.edu/files/dtingley/files/aatext.pdf.

Jeppesen, Lars Bo, and Karim R. Lakhani. 2010. Marginality and problem solving effectiveness in broadcast search. *Organization Science* 21(5): 1016–1033.

Klarreich, Erica. 2012. Privacy by the numbers: A new approach to safeguarding data. *Quanta Magazine*. 10 Dec 2012.

Knoppers, Bartha Maria, Jennifer R. Harris, Anne Marie Tassé, Isabelle Budin-Ljøsne, Jane Kaye, Mylène Deschênes, and Ma'n H Zawati. 2011. Towards a data sharing code of conduct for international genomic research. *Genome Medicine* 3: 46.

Krishnamurthy, Balachander and Craig E. Wills. 2009. On the leakage of personally identifiable information via online social networks. In *Proceedings of the 2nd ACM workshop on online social networks*. New York, New York: ACM. http://www2.research.att.com/~bala/papers/wosn09.pdf.

Lessig, Lawrence. 2009. Against transparency. *New Republic*. 9 Oct 2009.

Lewis, Paul and Dominic Rushe. 2014. Revealed: How whisper app tracks 'anonymous' users. *The Guardian*. http://www.theguardian.com/world/2014/oct/16/-sp-revealed-whisper-app-tracking-users. 16 Oct 2014.

Li, Ninghui, Tiancheng Li, and Suresh Venkatasubramanian. 2007. *t*-closeness: Privacy beyond *k*-anonymity and *l*-diversity. In *IEEE 23rd international conference on data engineering, 2007*, 106–15. Piscataway, NJ: IEEE.

Machanavajjhala, Ashwin, Johannes Gehrke, Daniel Kifer, and Muthuramakrishnan Venkitasubramanian. 2007. *l*-diversity: Privacy beyond *k*-anonymity. *ACM Transactions on Knowledge Discovery from Data (TKDD)* 1(1): 3.

McGraw, Gary and John Viega. 2001. Introduction to software security. *InformIT*. http://www.informit.com/articles/article.aspx?p=23950&seqNum=7. 2 Nov 2001.

Medicare Provider Utilization and Payment Data: Physician and Other Supplier. 2014. Centers for medicare & medicaid services. http://www.cms.gov/Research-Statistics-Data-and-Systems/Statistics-Trends-and-Reports/Medicare-Provider-Charge-Data/Physician-and-Other-Supplier.html. Last modified 23 Apr 2014.

Narayanan, Arvind. 2011. An adversarial analysis of the reidentifiability of the heritage health prize dataset. Unpublished manuscript.

Narayanan, Arvind and Vitaly Shmatikov. 2009. De-anonymizing Social Networks. In *Proceedings of the 2009 30th IEEE symposium on security and privacy*, 173–87. Washington, D.C.: IEEE Computer Society.

Narayanan, Arvind and Vitaly Shmatikov. 2008. Robust de-anonymization of large sparse datasets. In *Proceedings 2008 IEEE symposium on security and privacy*, 111–25, Oakland, California, USA Los Alamitos, California: IEEE Computer Society. 18–21 May 2008.

Narayanan, Arvind. 2013. What happened to the crypto dream? Part 2. *IEEE Security and Privacy* 11(3): 68–71.

Ochoa, Salvador, Jamie Rasmussen, Christine Robson, and Michael Salib. 2001. Reidentification of individuals in Chicago's homicide database: A technical and legal study. Final project, 6.805 Ethics and Law on the Electronic Frontier, Massachusetts Institute of Technology, Cambridge, Massachusetts. http://mike.salib.com/writings/classes/6.805/reid.pdf. 5 May 2001.

Ohm, Paul. 2010. Broken promises of privacy: Responding to the surprising failure of anonymization. *UCLA Law Review* 57: 1742–43. http://uclalawreview.org/pdf/57-6-3.pdf.

OnTheMap. U.S. Census Bureau. http://onthemap.ces.census.gov/.

Pandurangan, Vijay. 2014. On taxis and rainbows: Lessons from NYC's improperly anonymized taxi logs. *Medium*. https://medium.com/@vijayp/of-taxis-and-rainbows-f6bc289679a1. 21 June 2014.

Proposal for a Regulation of the European Parliament and of the Council on the Protection of Individuals with Regard to the Processing of Personal Data and on the Free Movement of Such Data. Art. 4(1)-(2) 25 Jan 2012.

Roberts, Margaret E. 2014. Fear or friction? How censorship slows the spread of information in the digital age. Unpublished manuscript, 26 Sept 2014. http://scholar.harvard.edu/files/mroberts/files/fearfriction_1.pdf.

Rubyrescue. 2014. Comment on blackRust, "How Whisper app tracks 'anonymous' users." *Hacker News*. https://news.ycombinator.com/item?id=8465482. 17 Oct 2014.

Sachs, Noah M. 2011. Rescuing the strong precautionary principle from its critics. *Illinois Law Review* 2011(4): 1313.

Salganik, Matthew J. and Karen E.C. Levy. 2014. Wiki surveys: Open and quantifiable social data collection. Unpublished manuscript, 2 Oct 2014. http://arxiv.org/abs/1202.0500.

Siddle, James. 2014. I know where you were last summer: London's public bike data is telling everyone where you've been. *The Variable Tree*. http://vartree.blogspot.com/2014/04/i-know-where-you-were-last-summer.html. 10 Apr 2014.

Solove, Daniel J. 2007.'I've got nothing to hide' and other misunderstandings of privacy. *San Diego Law Review* 44: 760–64.

Stodden, Victoria. 2012. Data access going the way of journal article access? Insist on open data. *Victoria's Blog*. http://blog.stodden.net/2012/12/24/data-access-going-the-way-of-journal-article-access/. 24 Dec 2012.

Sunstein, Cass R. 2002. The paralyzing principle. *Regulation* 25(4): 33–35.
Sweeney, Latanya, Akua Abu, and Julia Winn. 2013. Identifying participants in the personal genome project by name. White Paper 1021-1, Data Privacy Lab. Harvard University, Cambridge, Massachusetts. http://dataprivacylab.org/projects/pgp/1021-1.pdf. 24 Apr 2013.
Sweeney, Latanya. 2001. k-anonymity: A model for protecting privacy. *International Journal on Uncertainty, Fuzziness and Knowledge-based Systems* 10(5): 557–570.
Sweeney, Latanya. 2013. Matching known patients to health records in Washington State Data. White Paper 1089-1, Data Privacy Lab. Harvard University, Cambridge, Massachusetts. http://dataprivacylab.org/projects/wa/1089-1.pdf.
Sweeney, Latanya. 2000. Simple demographics often identify people uniquely. Data Privacy Working Paper 3. Carnegie Mellon University, Pittsburgh, Pennsylvania. http://dataprivacylab.org/projects/identifiability/paper1.pdf.
Task, Christine. 2013. An illustrated primer in differential privacy. *XRDS* 20(1): 53–57.
The White House. 2012. Consumer data privacy in a networked world: A framework for protecting privacy and promoting innovation in the global digital economy. Washington, D.C.
Tockar, Anthony. 2014. Riding with the stars: Passenger privacy in the NYC taxicab dataset. *Neustar: Research*. http://research.neustar.biz/2014/09/15/riding-with-the-stars-passenger-privacy-in-the-nyc-taxicab-dataset/. 15 Sept 2014.
Ugander, Johan, Brian Karrer, Lars Backstrom, and Cameron Marlow. 2011. The anatomy of the Facebook social graph. arXiv Preprint. http://arxiv.org/pdf/1111.4503v1.pdf.
Understanding Online Advertising: Frequently Asked Questions. Network advertising initiative. http://www.networkadvertising.org/faq.
Uyarra, Elvira, and Kieron Flanagan. 2010. Understanding the innovation impacts of public procurement. *European Planning Studies* 18(1): 123–143.
Vertesi, Janet. 2014. My experiment opting out of big data made me look like a criminal. *Time*, 1 May 2014.
What are single nucleotide polymorphisms (SNPs)? 2014. *Genetics home reference: Your guide to understanding genetic conditions*. Published 20 Oct 2014. http://ghr.nlm.nih.gov/handbook/genomicresearch/snp.
Whong, Chris. 2014. FOILing NYC's Taxi Trip Data. http://chriswhong.com/open-data/foil_nyc_taxi/. 18 Mar 2014.
Zang, Hui and Jean Bolot. 2011. Anonymization of location data does not work: A large-scale measurement study. In *Proceedings of the 17th international conference on mobile computing and networking*, 145–56. New York: ACM.

The Impact of Domestic Robots on Privacy and Data Protection, and the Troubles with Legal Regulation by Design

Ugo Pagallo

Abstract The paper examines a particular class of robotic applications, i.e. "domestic robots," in order to stress that such robots will likely affect current legal frameworks of privacy and data protection. Since most of these machines act, new responsibilities of humans for the behaviour of others should be expected in the legal field. More particularly, focus is on the protection of people's "opaqueness" and the transparency with which domestic robots should collect, process, and make use of personal data. Whilst the aim of the law to govern the process of technological innovation concerns here the regulation of producers and designers of robots through specific sets of norms, or the regulation of users behaviour through the design of their robots, three issues are fated to remain open. They concern: (i) a new expectation of privacy; (ii) the realignment of the traditional distinction between data processors and data controllers; and, (iii) a novel set of challenges to the principle of privacy by design. Although the claim and goal of lawmakers will probably revolve around the protection of individuals against every harm, e.g. psychological problems related to the interaction with domestic robots and the processing of third parties' information, the intent to embed normative constraints into the internal control architecture of such artificial agents entails a major risk. If there is no need to humanize our robotic applications, we should not robotize human life either.

Keywords Agency · Data protection · Design · Human-Robot interaction · Robotics · Privacy

---

U. Pagallo (✉)
Corso Regio Parco N. 8, Turin 10135, Italy
e-mail: ugo.pagallo@unito.it

U. Pagallo
Law School, University of Torino, Lungo Dora Siena 100 A, 10153 Torino, Italy

1 Introduction

There is a panoply of robotics applications out there: robot soldiers and robot scientists, industrial robots and robo-traders, robo-toys and AI personal assistants. This multiplicity makes difficult to determine what a robot is. The UN World 2005 Robotics Report, for example, proposes a general definition of robot as a reprogrammable machine operating in a semi- or fully autonomous way, so as to perform manufacturing operations (e.g. industrial robots), or provide "services useful to the well-being of humans" (e.g. service robots). Sebastian Thrun, director of the AI Laboratory at Stanford, California, argues that robots are machines with the ability to "perceive something complex and make appropriate decisions" (in Singer 2009: 77). While some similarly reckon that robots are machines built basically upon the mainstream "sense-think-act" paradigm of AI research, others stress that robots are those machines able to learn and adapt to changes in environments (Bekey 2005). Such references to the autonomy or intelligence of robots, however, often are a source of misunderstanding. Consider some military robotics applications, as the Global Hawk and the US Navy's anti-ship missile defence system, i.e. the Phalanx CIWS, that operate completely alone. According to the UK Ministry of Defence's Joint Doctrine Note on "unmanned aircraft systems" from 30 March 2011, we are dealing with systems "capable of understanding higher level intent and direction" and moreover, in the words of the Note, "estimates of when artificial intelligence will be achieved (as opposed to complex and clever automated systems) vary, but the consensus seems to lie between more than 5 years and less than 15 years, with some outliers far later than this." Although some find this statement "ludicrous" (Sharkey 2011), we need no Sci-Fi scenarios to admit that certain types of robots are already challenging tenets of social interaction, basic rules among nations, and even cornerstones of the law. This impact naturally varies in connection with the different robotics applications under scrutiny. For instance, going back to the military employment of robotic applications, the legal impact of such machines concerns the 1907 Hague Convention, the four Geneva Conventions from 1949, and the two 1977 additional Protocols, which define the current laws of war and the international framework of humanitarian law. In the case of, say, the civilian use of unmanned aerial vehicles, attention should be drawn to the 1948 Chicago Convention on International Civil Aviation and, in Europe, the EU Regulation 216/2008. Whereas the condition of immunity for the use of robot soldiers contrasts with the strict liability regime for the employment of both industrial and service robots in the civil sector, we should further distinguish in this latter case, between strict liability and vicarious liability rules, between malfunction liability and product liability norms, along with further legal issues on copyright, intellectual property and compulsory insurance, consumer law and environmental regulation, security and data protection, in the fields of criminal law, civil law, administrative law, etc.

Leaving aside the military field, let us focus here on the distinction of the UN Report of Robotics between industrial and service robots in the civilian sector. This latter set of robotics has to be further differentiated between professional and domestic, or personal, uses of service robots. In the first subset, there are robots for professional cleaning, inspection systems, construction and demolition, logistics, medical robots, rescue and security applications, underwater systems, mobile platforms in general use, laboratory robots, public relation robots, etc. In the second subset, we find the personal use of robots for domestic tasks such as iRobot's Roomba vacuum cleaning machines, entertainment robots such as toy robots and hobby systems, personal transportation, home security and surveillance, handicap assistance, and so on. Whilst the "Robotics 2020 Strategic Research Agenda" of the EU Commission refers to this subset of robotic applications as "consumer robots," this paper restricts the focus of the analysis on the sub-set of consumer robots for personal and domestic use in the field of entertainment, hobby, and family or professional assistance. Let us call them "domestic robots" (EU Robotics 2013).

As an illustration of this particular subset of robots, think about a robot toy which spends most of the time at home playing with your children and that, now and then, goes with them to the public garden accompanied by a robot nanny. Also, contemplate a sort of i-Jeeves 2.0 that manages and makes use of the property for your family business, so as to pay bills, entering into binding contracts, hiring robot nannies, buying robot toys, and so forth. All this may seem like a fanciful bit of Sci-Fi and yet, such scenario represents the bread and butter of several scholars that aim to strike a balance between individuals claiming that they should not be ruined by the decisions or behaviour of their robots and the counterparties of such machines, demanding the ability to safely interact with them. In the field of business law, this is the balance that has been aimed at by Tom Allen and Robin Widdison in *Can Computers Make Contracts?* (1996), Ian Kerr in *Ensuring the Success of Contract Formation in Agent-Mediated Electronic Commerce* (2001), Woodrow Barfield in *Issues of Law for Software Agents* (2005), Francisco Andrade et al. in *Contracting Agents: Legal Personality and Representation* (2007), Giovanni Sartor in *Cognitive Automata and the Law* (2009), Samir Chopra and Laurence White in *A Theory for Autonomous Artificial Agents* (2011), down to my own work on the "contract problem" in robotics.

A preliminary step in this kind of research concerns the ways in which such robots may collect, process, and make use of personal data. Contrary to traditional robots with on-board computers, recent applications are increasingly connected to a networked repository on the internet that allows robots to share the information required for object recognition, navigation and task completion in the real world (Pagallo 2013a). Accordingly, we have to take into account the new generation of issues related to the protection of privacy and personal data that domestic robots, much as other applications for health assistance, home security and surveillance, or personal transportation, will provoke in the foreseeable future. By considering that most domestic robots are not a simple sort of "out of the box" machine, what is at stake here does not only concern how to embed data protection safeguards into the informational "processing system" of the robotic application. Moreover, attention should be drawn to the ways in which the behaviour of these robots may crucially depend on how individuals train, treat or manage their artificial agents. In order to offer a hopefully comprehensive view on these issues, the paper is divided into two parts.

The first part of the paper dwells on some technicalities of robotics that are necessary to grasp their impact on tenets of current legal frameworks. Section 2.1 sheds thus light on the autonomy, or even intelligence, of domestic robots, so as to stress that we will be progressively dealing with agents, rather than simple tools of human interaction. Section 2.2 deepens this human-robot interaction ("HRI"), by further differentiating between a human-centred HRI approach and a robot-centred HRI methodology. This demarcation is crucial, because it pinpoints different types of legal responsibility. Section 2.3 examines how the agency of some domestic applications will trigger new kinds of responsibility and accountability of humans for the behaviour of other agents in the legal system. Section 2.4 introduces the second part of the paper, by contextualizing this new type of responsibility in the fields of informational privacy and data protection. Focus of Sect. 3.1 is on what US common lawyers dub as a "reasonable expectation of privacy." Regardless of the differences between the EU and the US legal frameworks, a common expectation of privacy should be expected (not only, but also) in Europe and US, in the basic sense that users and "human masters" of domestic robots will likely assume that some sort of legal protection, restraining the flow of personal information, shall be respected. Section 3.2 explores the guidelines presented by the EU-sponsored RoboLaw project from September 2014, so as to define a new set of issues on data processors and data controllers that will regard how domestic robots should collect and make use of personal data. Whereas, in the words of the RoboLaw Guidelines, "the emerging field of privacy by design can prove useful in making and keeping robots data protection-compliant" (RoboLaw 2014: 19), Sect. 3.3 examines the open issues of legal regulation by design vis-à-vis the different, or even opposite, aims this approach can have. The overall goal of the paper is to provide a guide for how domestic robots may affect current legal frameworks of privacy and data protection. Since robots are here to stay, the aim of the law should be to wisely govern our mutual relationships.

2 A New Kid in Town

On Wednesday 21 January 2015, the University of Turin (Italy) organized the first workshop on robotics, privacy and data protection at CPDP: "A New Kid in Town." As the booklet of the conference informs, the intent of the workshop was "to show that the future is here as several robotic applications—e.g. the imperceptible flying of tiny drones—are already impacting the rules and principles of privacy and personal data protection." Some technicalities of the field concerning the notions of artificial intelligence, adaptability, autonomy, and perception of robots, are needed, in order to appreciate this very impact. Let us start here with the controversial notion of robotic agency. Before examining the two different ways in which the human-robot interaction ("HRI") can be grasped, the first step of the analysis aims to offer a concise illustration of how the field has evolved over the past 55 years, so as to ascertain whether, and to what extent, robots properly "act."

2.1 When Robots Act

At the beginning, they were "cars." The first industry robot was tested within the automobile sector in 1961. Drawing on the ideas of George Devol and Joseph Engelberger, the project culminated in the UNIMATE robot performing spot welding and extracting die-castings in a General Motors factory in New Jersey. It was only 20 years later, in the early 1980s, however, that the use of robotics within the car industry became critical. Japanese industry first began to implement this technology on a large scale in their factories, acquiring strategic competitiveness by decreasing costs and increasing the quality of their products. Western car producers learned a hard lesson and followed Japanese thinking, installing robots in their factories a few years later. This massive trend went on for two decades: remarkably, in the *Editorial* to the World 2005 Robotics Report quoted above in the introduction, Åke Madesäter raised the risk that the robot industry "has become too dominated by car manufacturers and its sub-suppliers. In the period 1997–2003, the automotive industry in Spain received 70 % of all new robot installations. In France, the United Kingdom and Germany the corresponding figure amounted to 68, 64 and 57 %, respectively" (UN 2005: ix).

Still, in the same years as covered by the UN World report, things began to rapidly change. The two decade dependence of robotics on the automobile industry dramatically opened up to diversification, a revolution as phrased by scholars. This occurred with water-surface and underwater unmanned vehicles, or "UUVs," used for remote exploration work and the repairs of pipelines, oil rigs and so on, developing at an amazing pace since the mid-1990s. Ten years later, unmanned aerial vehicles ("UAVs"), or systems ("UAS"), upset the military field (Pagallo 2011). Over the past decade, robots have spread in both the industrial and service fields. Along with robots used in the manufacture of textiles and beverages, refining petroleum products and nuclear fuel, producing electrical machinery and domestic appliances, we also have a panoply of robot surgeons and robot servants, robot nannies and robot scientists, and even divabots, e.g. the Japanese pop star robot singer HRP-4C. The old idea of making machines (e.g. cars) through further machines (e.g. robots), has thus been joined—and increasingly replaced—by the aim to build fully autonomous robots. In the business sector and more particularly, in the trading agent competition context, humans already delegate relevant cognitive tasks to robots that can send bids, accept offers, request quotes, negotiate deals and even execute contracts. As a result, we are progressively dealing with agents, rather than simple tools of human interaction. But, how can we determine the point at which robots really act?

Some, as Wooldridge and Jennings (1995), reckon that robots, as well as any other artificial agent, enjoy such properties as autonomy, reactivity, pro-activeness and social ability to interact with other agents. Likewise, in the analysis of Franklin and Graesser (1997), all kinds of robots are presented as reactive, autonomous, goal-oriented, mobile and temporally continuous, although certain applications can be communicative, flexible and capable of learning and possessing a specific character, as occurs with HRP-4C. In this context, suffice it to emphasize the criteria pointed out by Allen et al. (2000), and further developed by Floridi and Sanders (2004). Drawing on this research, three features of robotic behaviour help us to define the meaning of agency and illustrate why scholars more frequently liken robots to animals (e.g. Latour 2005; McFarland 2008; Davis 2011; etc.), rather than products and things.

First, robots are interactive as they do perceive their environment and respond to stimuli by changing the values of their own properties or inner states.

Second, robots are autonomous, because they modify their inner states or properties without external stimuli, thereby exerting control over their actions without any direct intervention of humans.

Third, robots are adaptable, for they can improve the rules through which their own properties or inner states change.

On this basis, most of today's debate in robotics does not concern whether some applications can act and decide beyond the direct control of humans. Rather, the issue revolves around the type of interaction humans may have with their own robots. All in all, there are two different ways in which we should grasp such human-robot interaction ("HRI") from a legal viewpoint, namely as a "human-centred" HRI vis-à-vis a "robot-centred" HRI stance. Next section explores how far this bifurcation goes.

2.2 A Twofold HRI

There are several ways in which HRI can be understood. Think about forms of cooperative performance, master-apprentice interaction, caregiver and patient relationships, and so on (Coleman 2015; Martinez-Martin and del Pobil 2015: 58; etc.). However, from a legal perspective, suffice it to dwell on a twofold HRI, namely a "human-centred" and a "robot-centred" approaches. In the first case, the idea is to keep robots within limits that people can rationally accept: in the words of *Socially Intelligent Robots*, "human-centred HRI is primarily concerned with how a robot can fulfil its task specification in a manner that is acceptable and comfortable to humans" (Dautenhahn 2007: 684). Vice versa, in the case of a robot-centred HRI approach, the emphasis is on the "robot as a creature, i.e., an autonomous entity that is pursuing its own goals based on its motivations, drives and emotions" (op. cit., 683). In this latter case, although the "social needs" of the robot are defined by the designer and modelled by the internal control architecture of the machine, it is the user that enables the robot to "survive in the environment" by fulfilling its needs. As Cynthia Breazeal illustrated in her seminal work on Kismet, i.e. a robotic head with facial features, the machine is dealt with as an autonomous entity that pursues its own goals based on its motivations, so that humans have to satisfy its social drives by singling out and responding to the robot's internal needs. In the words of *Designing Sociable Robots*, "the robot is treated as a 'baby infant' or 'puppy robot' with characteristic specific and exaggerated child-like features satisfying the 'Kindchenschema' (baby pattern, baby scheme, schema 'bebe'). The Kindchenschema is a combination of features that are characteristic of infants, babies or baby animals, which appeals to the nurturing instinct in people (and many other mammals) and trigger respective behaviours" (Breazeal 2002).

Of course, it is not so difficult to imagine more complex cases, where social interaction with robots may involve emotional, physical and physiological activities that have a cost even for adult human beings. Whether humans will get the same payoff and gratification from their interaction with robots, as they do with other human fellows, is an open question that mostly depends on the cultural context and the type of application with which we are dealing: affective robots, sex tobots, carebots, medibots, AI chauffeurs and so forth. Some wonder if it is "ethically justifiable to aim to create robots that people bond with, e.g., in the case of elderly people or people with special needs" (Dautenhahn 2007: 699). Others, like Peter Sullins in the introduction to *Open Questions in Roboethics* (2011: 236), provocatively affirm that, at least in the field of affective robots, "we might begin to prefer the company of machines." Furthermore, in *Love and Sex with Robots* (2007), David Levy argues that it is somehow inescapable that such machines will soon be widespread in our society, since this technology can fulfil many individuals' dreams and desires. Aside from the moral aspects of the debate, however, how should legal systems govern the use of (some of such) robots? In particular, what about damages caused by a new generation of domestic robots that depend on the fault, or negligence, of the human master?

In accordance with the current state of the art in HRI legal research (Pagallo 2015), let us set the proper level of abstraction (Floridi 2008), namely, that which makes possible an analysis of how the interaction with domestic robots may affect current legal systems. The general idea is to define a set of features representing both the observables and variables of the analysis, the result of which provides a model for the field under examination (Pagallo 2013b: 28–29). By taking into account the parameters of today's research in human-robot interaction, two different types of legal responsibility for the behaviour of robots appear particularly relevant. On the one hand, a human-centred HRI methodology casts light on the responsibility of designers and manufacturers of robots that are considered as "strictly liable" for how the artificial agent fulfils its task specifications through sensors installed in smart houses, RFID, NFC, or QC code-based environment interaction, advanced local and global navigation systems, up to multimodal sensory devices—including brain computer interfaces for robotics—that aim to perceive the physiological and mental state of humans, also exploring new signal processing techniques to develop Electroencephalography (EEG) filters. Admittedly, the devil is in the legal detail, and we should further distinguish the variables of the analysis, i.e. between strict liability and vicarious liability rules, between malfunction liability and product liability norms, etc. More on this in the next section.

On the other hand, the robot-centred HRI approach sheds light on a new type of responsibility for users of such robots. Once "out of the package," the same model of robot will behave quite differently only after a few days or weeks, depending on how humans play their role of caretakers. Hence, human responsibility will often hinge on whether individuals met the social drives of their own robots, detecting and responding to the internal needs of the artificial agent. In addition to the traditional responsibility for robots as means of human interaction, focus should thus be on the duty of care that a reasonable person has to guard others against foreseeable harm. Such a new scenario is deepened in the next section, vis-à-vis cases of strict responsibility for designers and manufacturers of robots.

2.3 On Legal Responsibility

Theoretically speaking, there are three different kinds of legal agency that we have to examine in the field of robotics. Since most of these machines do act, they can be conceived of (i) as proper persons with rights and duties of their own; (ii) as strict agents in the business law-field, e.g. in contract law and negotiations; and, (iii) as a source of responsibility for other agents in the system.

In this context, dealing with the impact of robotics in the fields of privacy and data protection, we can leave the first kind of debate aside (Solum 1992; Pagallo 2013b; etc.); much as scholarly work on robotics and the contract problem, which was already mentioned in the introduction. Focus is then on the third type of legal agency, i.e. robots as a source of responsibility for other agents in the system. This level of abstraction corresponds to a popular point in jurisprudence, according to which robots are neither legal persons nor proper agents, but rather a source of liability for the behaviour of others. Some draw an analogy between strict liability policies for damages caused by animals and human liability for the behaviour of their robots, because the alleged novelty of all these latter cases resembles the responsibility of an owner or keeper of an animal "that is either known or presumed to be dangerous to mankind" (Davis 2011). Others propose the use of the traditional relations between principal and agent, master and servant, parent and child, warden and prisoner, down to keeper and animal, so as to understand how we can figure out the individual's negligent-based liability for the behaviour of (some types of) robots (Chopra and White 2011). Yet, traditional patterns of liability for the behaviour of other agents in the system, e.g. no-fault liability of humans for harm provoked by their animals, children or employees, may fall short in tackling a new kind of legal responsibility. After all, this is the first time ever legal systems will hold individuals accountable for what an artificial state-transition system decides to do (Pagallo 2010).

This accountability will vary in accordance with the type of application with which we are dealing, different types of strict liability norms, and how the burden of proof may work in these cases. Consider for example an ISO 8373 industrial robot, or the da Vinci surgery system in the medical sector. Here, the "human-centred" HRI approach fits like hand to glove with the liability of manufacturers and designers of robots for how such machines are built to fulfil their task specifications. Whilst strict liability rules apply to most producers of today's robots, such a responsibility can be imposed for injuries that either are caused by the defective manufacture or malfunction of the machine, or by defects in its design. Depending on the circumstances, the burden of proof varies as a result. In cases of defective manufacture of the robot, or deficiencies of its design, the burden of proof falls on the plaintiff who has to prove that the product was defective; that such defect existed while the product was under the manufacturer's control; and finally, that the defect was the proximate cause of the injuries suffered by the plaintiff. In cases of strict malfunction liability, responsibility can be imposed although the plaintiff is not able to produce direct evidence on the defective condition of the product or the precise nature of the product's defect. Rather, the plaintiff is to demonstrate that defect through circumstantial evidence of the occurrence of a malfunction, or through evidence eliminating both abnormal use of the product and reasonably secondary causes for the accident. In addition, responsibility may hinge on civil (as opposed to criminal) negligence that concerns the duty to conform to a certain standard of conduct. Accordingly, the plaintiff has to prove that defendants breached that duty, thereby provoking an injury and an actual loss or damage to the plaintiff.

However, in addition to a HRI human-centred approach and strict liability policies for robots conceived as a source of legal responsibility for other agents in the system, we should examine a further class of applications, such as some of the domestic robots mentioned above in the introduction. Pace traditional patterns of responsibility for the behaviour of other agents from a legal viewpoint, matters of liability change vis-à-vis such robots. The more these artificial agents are adaptable, interactive and autonomous, the more users will find it difficult to prove that the manufacturer of the robot did not conform to a certain standard of conduct, or that the supplier did not guard against foreseeable harm. In accordance with a "robot-centred" HRI, especially in the field of tort law, it is likely that responsibility for the behaviour of these robots will increasingly depend on the ways in which end-users train, treat, or manage their artificial companions. Regardless of whether the case will concern negligence-based responsibility or strict liability of humans, the mechanism of attributing to the parties the burden of proof varies with the type of strict liability we should endorse. This variation brings us back to the traditional stances of legal robotics and the aforementioned analogy between robots and, say, animals, children, or employees. For example, we may compare domestic robots with children under the responsibility of their parent, as in American law. Hence, defendants need to prove their machine did not present any dangerous propensity or trait that is not typical of similar applications, even though, for the foreseeable future, little room would be left for defendants to prevent liability. Alternatively, we may compare robots with children under the responsibility of parents as in Italian law. In this case, defendants avoid responsibility when evidence shows that they could not prevent the harmful behaviour of the robot, or that a fortuitous event occurred. But, how about the parallel between domestic robots and AI employees?

In this latter case, the vicarious liability of the user would not let humans evade responsibility, once the plaintiff brings evidence of a legally sufficient condition. This is in agreement with the opinion of those scholars that consider either robots as dangerous animals, or their use as an ultra-hazardous activity (e.g. Davis 2011). Of course, legal systems could also endorse forms of limited liability, so as to prevent the risk that individuals think twice before employing robots that, in the phrasing of the UN Report of Robotics, will provide "services useful to the well-being of humans" (UN World Robotics 2005). Therefore, some propose that we should register such machines just like corporations (Karnow 1996; Lerouge 2000; Weitzenboeck 2001); others suggest that we should bestow robots with capital (Bellia 2001), or that making the financial position of such machines transparent is a priority (Sartor 2009). Whilst further policies are feasible and even indispensable, e.g. insurance models and what I elsewhere called the "digital peculium" of robots, it is nonetheless clear that the aim of the law should be to strike a fair balance between the individual's claim to not be ruined by the decisions of their robots and the claim of a robot's counterparty to be protected when interacting with them. How should we imagine this balance in the fields of privacy and data protection?

2.4 The Invasion of the Sacred Precincts of Private Life?

Domestic robots will know a lot of things about our private lives. Think about smart Roombas equipped with cameras so as to properly clean your flat, or personal artificial assistants connected to the internet so as to help us manage our business, and schedule, say, a set of conferences, lectures and meetings at several European (or US) universities next summer. The amount of personal information, collected and processed by a new generation of domestic robots, will likely depend on the ways in which individuals treat their artificial agents, and what is required for object recognition, navigation, and task completion of robots interacting with humans "out there," in the real world. Although many readers of this chapter may not have met a domestic robot so far, they are familiar with some of the challenges that will be brought on by these applications. Reflect on how a number of mobile devices, such as your smartphone, collect a myriad of different data like images and videos through cameras, motion and activities through gyroscopes and accelerometers, fingerprints through biometric sensors, geo-location data through GPS techniques, and so forth. Whereas some fitness applications, such as Nike+ or Adidas miCoach, track route, pace and time activities of users through GPS and sensors, further risks for user's informational privacy are raised by such real time facial recognition apps as NameTage for Google Glasses. These threats have been stressed time and again over the past years: for instance, as to the risks raised by sensors, contemplate how personal information can be inferred from such data, as occurs with information on mobility patterns, activity and face recognition, health information, and so on. In light of current challenges to user's privacy, what is new with domestic robots is that sensors, cameras, GPS, facial recognition apps, Wi-Fi, microphones and more, will be assembled in a single piece of high-tech. Moreover, as a prolonged epigenetic developmental process, several domestic robots will gain knowledge or skills from their own interaction with the living beings inhabiting the surrounding environment, so that more complex cognitive structures will emerge in the state-transition system of the artificial agent. New expectations of privacy and data protection regulations should be expected as a result and unsurprisingly, scholars have increasingly drawn the attention to how robots may affect these fields of the law (Pagallo 2013a; RoboLaw 2014; etc.). In order to strike a fair balance in distributing responsibility and risk for the ways in which a new generation of domestic robots will collect, process, and make use of personal data, the second part of this paper dwells on what Samuel Warren and Louis Brandeis anticipated more than a century ago. Remarkably, in *The Right to Privacy* (1890), they claimed that "instantaneous photographs and newspaper enterprise have invaded the sacred precincts of private and domestic life; and numerous mechanical devices threaten to make good the prediction that 'what is whispered in the closet shall be proclaimed from the house-tops'" (op. cit., 195). By taking into account current trends of robotics, should we expect a new invasion of the sacred precincts of our intimate life?

3 The Closet and the House-Tops

There are different ways in which we can appreciate the impact of domestic robots on current legal frameworks of privacy and data protection. In this context, let us deepen the twofold kind of responsibility introduced above in the previous Sects. 2.2 and 2.3, by focusing on the aim of the law to govern the process of technological innovation, i.e. the law conceived as a "meta-technology" (Pagallo 2013b). A first level of abstraction has to do with the purposes that law-making can have. Some, as Bert-Jaap Koops (2006), suggest that we should distinguish four main legislative goals, such as: (a) the achievement of particular effects; (b) functional equivalence between online and offline activities; (c) non-discrimination between technologies with equivalent effects; and, (d) future-proofing of the law that should neither hinder the advance of technology, nor require overfrequent revision to tackle such a progress. Others, as Chris Reed (2012), propose to differentiate between (a) technological indifference, i.e. legal regulations which apply in identical ways, whatever the technology, as occurs with the right to authorize communication of a work to the public in the field of copyright law; (b) implementation neutrality, according to which regulations are by definition specific to that technology and yet, they do not favour one or more of its possible implementations, e.g. the signature of e-documents; and, (c) potential neutrality of the law that sets up a particular attribute of a technology, although lawmakers can draft the legal requirement in such a way that even non-compliant implementations can be modified to become compliant.

A second level of abstraction is illustrated by recent work of Ronald Leenes and Federica Lucivero in *Laws on Robots, Laws by Robots, Laws in Robots* (2014). Here, the intent of the law to regulate both human and robot behaviours leads to four different categories, that is, (a) the regulation of human producers and designers of robots through law, e.g. either through ISO standards or liability norms for users of robots; (b) the regulation of user behaviour through the design of robots, that is, by designing robots in such a way that unlawful actions of humans are not allowed; (c) the regulation of the legal effects of robot behaviour through the norms set up by lawmakers, e.g. the effects of robotic contracts and negotiations; and, (d) the regulation of robot behaviour through design, that is, by embedding normative constraints into the design of the artificial agent. This differentiation can be complemented with further work on the regulation of HRI environments and the legal challenges of "ambient law" (Hildebrandt and Koops 2010; Hildebrandt 2011; etc.). Accordingly, attention should be drawn to the set of values, principles, and norms that constitute the normative context in which the consequences of such regulations have to be evaluated.

Against this backdrop, let us assume a third stance, which examines the legal impact of domestic robots in connection with the different fields with which we are dealing, i.e. privacy and data protection, vis-à-vis the intent of the law to govern the process of robotic innovation. This perspective partially overlaps with the previous levels of analysis and yet, it allows us to pinpoint the new observables and variants of the analysis, i.e. what issues ought to be questioned, prioritized and made relevant, so as to stress the legal impact of domestic robots. This field-dependent approach seems moreover necessary, because "privacy" and "data protection" often are used as interchangeable terms of the analysis, although this is not necessarily the case. On the one hand, the many ways in which the notion of privacy has been conceived as a condition of "solitude," or "exclusion," or "secrecy" (Westin 1967; Gavison 1980; Allen 1988; etc.), can be summed up with Hannah Arendt's idea of "opaqueness." In the words of *Vita Activa*, "a life spent entirely in public, in the presence of others, becomes, as we would say, shallow. While it retains its visibility, it loses the quality of rising into sight from some darker ground which must remain hidden if it is not to lose its depth in a very real, non-subjective sense" (Arendt 1958: 71). Nowadays, we can update this idea of opaqueness in informational terms, according to the principles and rules that aim to constraint the flow of information in the environment, so as to keep firm distinctions between individuals and society, agents and system. Principles and rules of the legal system, in other words, determine the degrees of "ontological friction" in the informational sphere, as "the amount of work and efforts required for a certain kind of agent to obtain, filter and/or block information (also, but not only) about other agents in a given environment" (Floridi 2006). The overall idea is that the higher the ontological friction, the lower the accessibility of personal information and thus, the stronger the protection of one's privacy and her opaqueness. Whilst this idea of privacy can entail no data processing at all, e.g. cases of "unwanted fame" or "false light," the intent of the law remains nonetheless the same, i.e. to protect the flow of information that individuals deem appropriate to reveal, share, or transfer, in a given context.

On the other hand, contrary to privacy's "opaqueness," issues of data protection mostly revolve around the transparency with which personal data are collected, processed, and used. In the EU legal system, for example, individuals have the right to know the purposes for which their data are processed, much as the right to access that data and to have it rectified. In the wording of Article 8(2) of the EU Charter of fundamental rights, "such data must be processed fairly... and on the basis of the consent of the person concerned or some other legitimate basis laid down by law." This type of protection through the principles of minimization and quality of the data, its controllability and confidentiality, may of course overlap with the protection of the individual "opaqueness." In such cases, the aim is to constraint the flow of information, and keep firm distinctions between individuals and society, in order to protect what the German Constitutional Court has framed in terms of "informational self-determination" since its *Volkszählungs-Urteil* ("census decision") from 15 December 1983. Yet, there are several cases in which the norms of data protection do not entail the safeguard of any privacy. Together with the mechanism of "notice and consent," laid down by Article 7 of the EU directive 46 from 1995, consider how the processing of personal data can—and at times should—go hand in hand with the strengthening of further rights and interests of individuals, such as freedom of information and the right to knowledge, freedom of expression and access to public documents, up to participatory democracy and the functioning of the internal market with the free circulation of services and information pursuant to the EU directive on the reuse of public sector information, i.e. D-37/2013/EC (Pagallo and Bassi 2013).

In light of the differences between current privacy regulations and data protection frameworks, the goal of the following sections is to emphasize how the spread of domestic robots will entail new legal issues with regard to (i) people's expectation of privacy; (ii) the realignment of the traditional distinction between data processors and data controllers; down to, (iii) novel challenges to the principle of privacy by design. The next three sections focus on each of these issues. Then, the time will be ripe for the conclusions of the analysis.

3.1 A New Expectation of Privacy

Domestic robots will likely affect what US legal scholars dub as a "reasonable expectation of privacy." In a nutshell, the formula means that individuals have the right to be protected against unreasonable searches and seizures under the Fourth Amendment. Pursuant to the jurisprudence of the US Supreme Court from the 1967 *Katz v. United States* case (389 U.S. 347), onwards, this means that the opinion of a person that a certain situation or location is private, must go hand in hand with the fact that society at large would recognize this expectation, so as to protect the latter as a fundamental right. Scholars and also justices of the Supreme Court, however, have stressed time and again that such twofold dimension of this reasonable expectation, both social and individual, can entail a vicious circle, much as "the chicken or the egg" causality dilemma. Moreover, the right to a reasonable expectation of privacy rests on the assumption that both individuals and society have developed a stable set of privacy expectations, whereas technology can dramatically change these very expectations. As Justice Alito emphasizes in his concurring opinion in *United States v. Jones* from 23 January 2012 (565 U.S. __), "dramatic technological change may lead to periods in which popular expectations are in flux and may ultimately produce significant changes in popular attitudes."

The legal framework is different in Europe. According to the EU legal rules and principles of privacy and data protection, the opinion of individuals does not play any normative role, in order to determine the legitimacy of the acts and statutes laid down by the public institutions. On the contrary, what individuals and society can reasonably expect, is that public organizations, multinational corporations, and other private parties, abide by the set of rules and principles established by the EU, or national, legislators. Notwithstanding this approach, it does not follow that social and individual expectations of privacy are totally irrelevant in Europe. Consider the proposal for a new data protection regulation in the EU legal system, presented by the Commission in January 2012. The same day in which the Parliament approved the new set of rules, the Commission was keen to inform us with a press release on 12 March 2014, that the intent to update and modernize the principles enshrined in the 1995 data protection directive is strictly connected with "a clear need to close the growing rift between individuals and the companies that process their data."[1] The source of this "clear need" was provided by the Flash Eurobarometer 359 from June 2011, on the attitudes concerning data protection and electronic identity in the EU. According to this source, 9 out of 10 Europeans (92 %) said they are worried about mobile apps collecting their data without their consent, 7 out of 10 are concerned about the potential use that companies may make of the information disclosed, etc. Whether the new EU regulation will close the rift between individual and companies is, of course, an open issue and yet, it is highly likely that domestic robots will add new worries about radars, sensors or laser scanners of artificial agents collecting data of their human masters, much as companies that may infer personal information from such data on mobility patterns, user's preferences, lifestyles, and the like.

A common expectation of privacy should thus be expected (not only, but also) in Europe and US, in the basic sense that users of domestic robots will likely assume that some "degree of friction," restraining the flow of personal information, should be respected. Clearly, this is not to say that personal choices will have no role in determining different levels of access to, and control over, information. Rather, there are three ways in which we can appreciate the role of these personal choices in keeping firm distinctions between individuals and society, agents and the system. First, in accordance with the robot-centred HRI methodology, the different types of information which robots may properly reveal, share, or transfer, will often hinge on personal preferences of the human master on whether

---

[1] See the press release at http://europa.eu/rapid/press-release_MEMO-14-186_it.htm.

it is appropriate to trace back information to an individual, and how information should be distributed according to different standards in different contexts. Depending on how humans have taken care of their artificial agents, specimens of the same model of domestic robot will accordingly behave in different ways. As mentioned above in Sect. 2.3, those are the cases in which the responsibility of users and "human masters" of robots, rather than issues of strict liability for manufacturers and designers of such artificial agents, will be at stake.

Second, personal choices on both norms of appropriateness and flow will further hinge on the type of domestic robot under scrutiny. The type of information that makes sense to communicate and share with a personal assistant like i-Jeeves 2.0, would be irrelevant or unnecessary to impart to a robot toy. It is thus likely that individuals will modulate different levels of access to, and control over, information, depending on the kind of the artificial interlocutor, the context of their interaction, and the circumstances of the case. As a variant of the "robot-centred" HRI paradigm, users of such robots will remain responsible for the behaviour of their artificial companions.

Third, the overall expectation of informational privacy will probably change as a result of both a human-centred and robot-centred HRI dynamic. What is new with domestic robots, not only regards problems of reliability, traceability, identifiability, trustfulness and generally speaking, how the interaction with such robots and their presence in "the sacred precincts of private and domestic life" may realign both norms of appropriateness and of informational flow (Nissenbaum 2004). In addition, we should expect psychological problems related to the interaction with robots as matters of attachment and feelings of subordination, deviations in human emotions, etc. (Veruggio 2006: 29). Therefore, in accordance with the tenets of the "human-centred" HRI perspective, attention should be drawn to the overall idea that designers and manufacturers of domestic robots should build them in such a way, that the latter can fulfil their tasks within limits that humans can rationally accept and find comfortable or adequate. Although it is the user that enables the robot to "survive in the environment" by fulfilling its needs, such "social needs" of the artificial agent are defined by the designer and modelled by the internal control architecture of the machine. These requirements bring us back to the categories proposed by Leenes and Lucivero on both the regulation of producers and designers of robots through law, and the regulation of user behaviour through the design of their robots. Since most of these artificial agents will increasingly collect, process, and make use of personal data, the next step of the analysis is to assess the level of their impact on current data protection frameworks.

3.2 Robot Data Processors and Human Data Controllers

The reference point for today's state-of-art in roboprivacy is given by the guidelines that a EU-sponsored project, namely "RoboLaw," presented in September 2014. According to this document, the principle of privacy by design can play a key role in making and keeping robots data protection-compliant (RoboLaw 2014: 19). For example, some legal safeguards, such as data security through encryption and data access control, can be embedded into the software and interface of the robot. Likewise, "requirements such as informed consent can be implemented in system design, for example through interaction with users displays and input devices" (ibid). After all, this is what already occurs with some operating systems, such as Android, that require user's consent whenever an application intends to access personal data. Furthermore, robots could be designed in a privacy-friendly way, so that the amount of data to be collected and processed is reduced to a minimum and in compliance with the finality principle. This means that, pursuant to Article 6(1)(b) of the EU data protection directive 46 from 1995, robots should collect data only insofar as it is necessary to achieve a specified and legitimate purpose.

In addition, this set of legal safeguards on data minimization, finality principle, informed consent, etc., shall be pre-emptively checked through control mechanisms and data protection impact assessments, so as to ensure that privacy safeguards are at work even before a single bit of information has been collected. More particularly, in the words of the RoboLaw Guidelines, "as a corollary of a privacy impact assessment, a control mechanism should be established that checks whether technologies are constructed in the most privacy-friendly way compatible with other requirements (such as information needs and security)" (op. cit., 190). Leaving aside specific security measures for particular classes of service robots, such as health robots, personal care robots, or automated cars examined by the EU project, the latter suggests that "the adoption of updated security measures should not be considered only as a user's choice, but also as a specific legal duty. It is clear that the illicit treatment of the data is unlikely to be considered a responsibility of the manufacturer of the robot, but rather a liability of its user, who is the 'holder' of the personal data" (op. cit., 190).

Whether the end-user, or "human master," of the domestic robot should be deemed as the data controller and hence, liable for any illicit treatment of personal data, is however debatable. As stressed above in the previous section, we may admit cases in which the role of personal choices and the "caretaker paradigm" of the robot-centred HRI approach suggest that end-users should be conceived as data controllers and thus, liable for how their artificial agents collect, process, and make use of personal data. But, as occurs today with issues of internet connectivity, or sensors and mobile computing applications, several other cases indicate that the illicit treatment of personal data may depend on designers and manufacturers of robots, internet providers, applications developers, and so forth. After all, the illicit treatment of personal data may be traced back to the malfunctioning of the robot, or to HTTP headers in packets of network traffic data that can be used to determine interests and other personal information about the master of the robot, along with applications that leak identifiable data, such as device ID, GPS, and more. What all these cases make clear is not only hypotheses of illicit treatment of data that do not depend on end-users or masters of domestic robots as data controllers. Additionally, the liability of designers and manufacturers of robots, internet providers, etc., can be problematic in connection with different interpretations of current rules and principles of the data protection legal framework, e.g. the EU 1995 norms on the protection of individuals with regard to the processing of personal data and on the free movement of such data. As stressed by Art. 29 Working Party in the opinion 1/2010 (WP 169), "the concept of controller is a functional concept, intended to allocate responsibility where the factual influence is, and thus based on a factual, rather than a formal analysis," which "may sometimes require an in-depth and lengthy investigation" (op. cit., 9).

Yet, even admitting the conclusions of the Working Party, so that liability of data controllers "can be easily and clearly identified in most situations" (ibid.), we still have to face a major problem. Although normative safeguards can be embedded into the software and interface of domestic robots, significant differences between multiple data protection jurisdictions, e.g. between US and EU, remain. Whereas, in the US, privacy policies of the industry and the agreement between parties mostly regulate matters of data protection in the private sector, we already stressed that the EU has adopted a comprehensive legislation since its 1995 data protection directive. Principles and rules of this directive on data minimization, finality principle, informed consent, etc., set limits to the contractual power of individuals and companies. This divergence will likely increase with the aforementioned regulation proposed by the EU Commission in January 2012 and partially amended by the Parliament in March 2014. Even the RoboLaw Guidelines concede that these "significant differences… could make it difficult for manufacturers catering for the international market to design in specific data protection rules" (op. cit., 19). As a matter of legal fact, which norms and rules should designers and manufactures of domestic robots embed into their products? Should such norms and rules vary according to the specific market (and jurisdiction)? Would this latter option be technically and economically sustainable?

A feasible way-out is pragmatic. Following Anu Bradford's thesis on "the Brussels effect" and how Europe's regulatory model wields unilateral influence across such legal fields, as data protection, antitrust, or health and environmental legislation (Bradford 2012), we may envisage a similar effect in the case of domestic robots. The non-divisibility of data and the compliance costs of multinational corporations dealing with multiple regulatory regimes, may prompt most robot manufacturers to adopt and adapt themselves to the strictest international standards across the board, that is, the EU data protection framework, much as occurred in the case of internet companies vis-à-vis data protection issues. However, compared with traditional privacy regulation, we should not overlook some peculiarities of domestic robots. By affecting what US lawyers dub as a reasonable expectation of privacy, as explored above in the previous section, it is highly likely that such expectation, both individual and social, will be "in flux" for a while. Some insist on this flux to stress that lawmakers, rather than judges or data protection authorities, are in the best position to determine the rules of the game and guide social and individual behaviour (Kerr 2004). But, even in light of the strictest international standards of the EU legislation, it is still vague how we should interpret some of its key assumptions, e.g. the principle of privacy by design. Neither the Commission's proposal for a new data protection regulation in January 2012, nor the amendments of the EU Parliament in March 2014, clarify how to design robots that abide by the law. All in all, we lack a regulatory model that may represent a reference point for international standards on the design, production and commercialization of domestic robots. The aim of next section is thus to deepen current uncertainties on the principle of privacy by design, by fleshing out how such uncertainties are intertwined with the "caretaker paradigm" of the robot-centred HRI view.

3.3 The Troubles with Legal Regulation by Design

Legal design has different and even opposite aims. Think about the latter according to a spectrum: at one end, the purpose is to determine and control both social and individual behaviour through the use of self-enforcing technologies and such automatic techniques, as filtering systems and digital rights management (DRM)-tools, that intend to restrict any form of access, use, copy, replacement, reproduction, etc., of informational resources in the environment. At the other end of the spectrum, design may aim to encourage the change of people's behaviour by widening the range of the choices through incentives based on trust (e.g. reputation mechanisms), or trade (e.g. services in return). In between the ends of the spectrum, design may aim to decrease the impact of harm-generating behaviour through security measures, default settings, user friendly interfaces, and the like. Notwithstanding these different ends, it is noteworthy that legislators and scholars alike often refer to the aim to embed legal constraints into technology, e.g. privacy by design, in a neutral manner, that is, as if the intent of this legal embedding could be impartial and value-free. Consider articles 23 and 30 of the EU Commission's proposal for a new data protection regulation, much as § 3.4.4.1 of the document with which the Commission illustrated the proposal. Here, the formula of "privacy by design" is so broad, or vague, that it can include whatever end design may have. Although, in the amendment 118 of the EU Parliament, the latter refers to "comprehensive procedural safeguards regarding the accuracy, confidentiality, integrity, physical security and deletion of personal data," it is still unclear whether the aim should be to decrease the impact of harm-generating conducts or rather, to widen the range of individual options, or both. In light of these uncertainties, how about the design of domestic robots and the HRI environment through sensors, GPS, facial recognition apps, Wi-Fi, RFID, NFC, or QC code-based environment interaction?

First of all, the principle of privacy by design and the EU Parliament's "comprehensive procedural safeguards" can be grasped in terms of security measures, e.g. data access control and encryption, much as user-friendly default configurations of robotic interfaces. Robots can indeed be designed in such a way that values of design are appropriate even for novice users and still, the robot improves efficiency. Furthermore, the intent can be to seamlessly integrate robots into domestic workflows and IT systems of smart houses via compliant motion control systems and situation awareness technologies, much as flexible and modular systems for the measurement of physical, physiological and electro-physiological variables, that should make the user experience an integral and even natural part of the process. In addition, we should take into account the set of legal safeguards on data minimization, finality principle, or informed consent, that were mentioned in the previous section, so as to tackle the convergence of robotic data processing and the internet (of things, of everything, etc.).

However, a number of further cases suggest that domestic robots could alternatively be designed with the aim to prevent any harm-generating behaviour from occurring. This is not only a popular stance among Western lawmakers in such fields as intellectual property ("IP") protection, data retention, or online security (Pagallo 2013c). Moreover, in the field of robotics, two further reasons may reinforce this design policy. On the one hand, in the phrasing of the EU Parliament, "the accuracy, confidentiality, integrity, physical security and deletion of personal data," processed by domestic robots, will more often concern data of third parties. On the other hand, we must reflect on both the psychological problems related to the very interactions with robots, and the case of human masters that do not properly fulfill their role of caretakers. Lawmakers may thus adopt a stricter version of the principle of privacy by design, in order to preclude any data protection infringement through the use of self-enforcing technologies, e.g. filtering systems, in the name of security reasons. This scenario is not only compatible with the new EU regulation, but has been endorsed by some popular versions of the principle. In Ann Cavoukian's account of privacy by design, for example, personal data should be automatically protected in every IT system as its default position, so that a cradle-to-grave, start-to-finish, or end-to-end lifecycle protection ensures that privacy safeguards are automatically at work even before a single bit of information has been collected (Cavoukian 2010). But, is this automatic version of privacy by design technically feasible and even desirable?

There are several ethical, legal, and technical reasons why we should resist the aim of some lawmakers to protect citizens even against themselves. First, the use of self-enforcing technologies risks to curtail freedom and individual autonomy severely, because people's behaviour and their interaction with robots would be determined on the basis of design rather than by individual choices (Lessig 2004; Zittrain 2007; etc.). Once the normative side of the law is transferred from the traditional "ought to" of rules and norms to what actually is in automatic terms, a modelling of individual conduct follows as a result, namely, that which Kant used to stigmatize as "paternalism" (Pagallo 2012a).

Second, specific design choices (not only, but also) in robotics may result in conflicts between values and furthermore, conflicts between values may impact on the features of design. Since both privacy and data protection may be conceived in terms of human dignity or property rights, of contextual integrity or total control, it follows that privacy by design acquires many different features. In the case of self-enforcing technologies, their use would make conflicts between values even worse, due to specific design choices, e.g. the opt-in vs. opt-out diatribe over the setting of information systems (Pagallo 2011).

Third, attention should be drawn to the technical difficulty of applying to a robot concepts traditionally employed by lawyers, through the formalization of norms, rights, or duties. As stressed by Bert-Jaap Koops and Ronald Leenes, "the idea of encoding legal norms at the start of information processing systems is at odds with the dynamic and fluid nature of many legal norms, which need a breathing space that is typically not something that can be embedded in software" (Koops and Leenes 2014: 167). All in all, informational protection safeguards present highly context-dependent notions that raise several relevant problems when reducing the complexity of a legal system where concepts and relations are subject to evolution (Pagallo 2012b).

At the end of the day, it should be clear that the use of self-enforcing technologies would not only prevent robotic behaviour from occurring. By unilaterally determining how the artificial agent should act when collecting, for example, the information they need for human-robot interaction and task completion from networked repositories, such design policies do impinge on individual rights and freedom. If there is no need to humanize our robotic applications, we should not robotize human life either. The time is ripe for the conclusions of this chapter.

4 Conclusions

The chapter has focused on a particular set of robotic applications, i.e. "domestic robots," and how the latter will likely impact on current legal frameworks of privacy and data protection. Since most of these robots act (Sect. 2.1), in accordance with a "human-centred" methodology and a "robot-centred" approach (Sect. 2.2), new responsibilities of humans for the behaviour of others will emerge in the legal arena (Sect. 2.3). More particularly, we have to be ready for a new set of legal challenges in the fields of privacy and data protection that concern the opinion of that which end-users, or human masters, of domestic robots expect should be preserved as private, much as the goal of manufactures and designers of robots to keep such AI machines within limits that users deem reasonable (Sect. 2.4). The second part of the paper has thus examined the ways in which the law may aim to govern the process of technological innovation through the regulation of the activities of human producers and designers of robots, and the behaviour of users through the design of their artificial companions. Whereas the intent should be to protect people's "opaqueness" (i.e. privacy), and the transparency with which domestic robots will collect, process, and make use of personal data (i.e. data protection), three key problems are fated to remain open.

First, as seen above in Sect. 3.1, a new expectation of privacy should be taken into account, since end-users of domestic robots will likely assume that some "degree of friction," restraining the flow of personal information in HRI, should be respected, notwithstanding the unpredictability of robotic behaviour. In addition, the interaction with robots and the presence of such artificial agents in the "sacred precincts of private life" will not only affect both norms of appropriateness and of informational flow. We should expect psychological problems related to the very interaction with robots that may suggest lawmakers to intervene and impose restrictions on the range of individual choices and possible uses of robotic applications. Lest national and international lawmakers aim to robotize human behaviour, however, it is apparent that the same model of, say, our i-Jeeves 2.0 mentioned above in the introduction, will manage the information that should be kept private in quite divergent ways, following the different opinions and instructions of the "human masters." Moreover, how individuals will modulate different levels of access to, and control over, information, depending on the kind of domestic robot, the context of their interaction, and the circumstances of the case, is highly likely that will be "in flux" for a while.

Second, in Sect. 3.2, attention has been drawn to the set of new legal challenges in the field of data protection. Although normative safeguards can be embedded into the internal control architecture of the artificial agent and the HRI environment, significant differences between multiple data protection jurisdictions, e.g. between US and EU, will affect the set of norms and rules that designers and manufactures of domestic robots should embed into their products. Furthermore, even adopting the strictest international standards across the board, such as the provisions of the EU data protection framework, it can be really tricky to determine who should be held responsible for the illicit treatment of the data collected and processed by an artificial agent. Contrary to the opinion of the EU-sponsored RoboLaw project, which reckons that users, rather than designers and manufacturers of robots, will most of the time be considered responsible as holders of personal data, several other cases suggest that illicit treatment of personal data may depend on designers and manufacturers of robots, together with internet providers, applications developers, and so forth. Who has to be conceived here as the data controller will often entail a factual analysis that, in the words of Art. 29 WP, can be "lengthy."

Third, in Sect. 3.3, focus was on the troubles with legal regulation by design. Admittedly, "the emerging field of privacy by design can prove useful in making and keeping robots data protection-compliant" (RoboLaw 2014: 19); and still, the principle can be interpreted in many different, or even opposite, ways. When the aim is to decrease the impact of harm-generating conducts through, e.g., security measures, such a design policy appears legally and politically sound, because this approach to design prevents threats of paternalism that hinge on the regulatory tools of technology, by respecting collective and individual autonomy. However, to complement the traditional regulation of the law through design entails its own risks, when the intent is to prevent any harm-generating behaviour from occurring. Even though the claim and aim of lawmakers would revolve around the protection of individuals against every harm, such as psychological problems related to the interaction with robots and the processing of third parties' information, this design policy severely threats to impinge on individual rights and freedom, for the "caretaker paradigm" of the robot-centred HRI approach would be transposed into the political arena. The legal challenges brought on by artificial agents in the fields of privacy and data protection may in fact suggest legislators either to regulate user behaviour through the design of robots, or to embed normative constraints into the design of the artificial agent and the HRI environment. In both cases, we should be aware of a crucial menace: the more personal choices are wiped out by legal automation, the bigger the danger of modelling social conduct via design. Waiting for a common international standard for the design and production of domestic robots—and in light of current debate on the EU data protection regulation—let us be cautious against the avuncular legislator.

Bibliography

Allen, Anita. 1988. *Uneasy access: privacy for women in a free society*. Totowa, N.J.: Rowman and Littlefield.
Allen, Colin, Gary Varner, and Jason Zinser. 2000. Prolegomena to any future artificial moral agent. *Journal of Experimental & Theoretical Artificial Intelligence* 12: 251–261.
Allen, Tom, and Robin Widdison. 1996. Can computers make contracts? *Harvard Journal of Law & Technology* 9(1): 26–52.
Andrade, Francisco, Paulo Novais, José Machado, and José Neves. 2007. Contracting agents: Legal personality and representation. *Artificial Intelligence and Law* 15: 357–373.
Arendt, Hannah. 1958. *The human condition*. Chicago: University of Chicago Press.
Barfield, Woodrow. 2005. Issues of law for software agents within virtual environments. *Presence* 14(6): 741–748.
Bekey, George A. 2005. *Autonomous Robots: From Biological Inspiration to Implementation and Control*. Cambridge, London: MIT Press.
Bellia, Anthony J. 2001. Contracting with electronic agents. *Emory Law Journal* 50: 1047–1092.
Bradford, Anu. 2012. The brussels effect. *Northwestern University Law Review* 107(1): 1–68.
Breazeal, Cynthia. 2002. *Designing sociable robots*. Cambridge: MIT Press.
Cavoukian, Ann. 2010. Privacy by design: The definitive workshop. *Identity in the Information Society* 3(2): 247–251.
Chopra, Samir, and Laurence F. White. 2011. *A legal theory for autonomous artificial agents*. Ann Arbor: The University of Michigan Press.
Coleman, Diana (ed.). 2015. *Human-robot interactions: Principles, technologies and challenges*. New York: Nova.
Davis, Jim. (2011). The (common) laws of man over (civilian) vehicles unmanned. *Journal of Law, Information and Science* 21(2). 10.5778/JLIS.2011.21.Davis.1.
Dautenhahn, Kerstin. 2007. Socially intelligent robots: Dimensions of human-robot interaction. *Philosophical Transactions of the Royal Society B: Biological Sciences* 362(1480): 679–704.
EU Robotics. 2013. *Robotics 2020 Strategic Research Agenda for Robotics in Europe*, draft 0v42, 11 Oct 2013.
Floridi, Luciano. 2008. The method of levels of abstraction. *Minds and Machines* 18(3): 303–329.
Floridi, Luciano. 2006. Four challenges for a theory of informational privacy. *Ethics and Information Technology* 8(3): 109–119.
Floridi, Luciano, and Jeff Sanders. 2004. On the morality of artificial agents. *Minds and Machines* 14(3): 349–379.
Franklin, Stan, and Art Graesser. 1997. Is it an agent, or just a program? A Taxonomy for autonomous agents. In *Intelligent Agents III, Proceedings of the Third International Workshop on Agent Theories, Architectures, and Languages*, ed. J.P. Müller, M.J. Wooldridge, and R. Nicholas, 21–35. Berlin: Springer.

Gavison, Ruth. 1980. Privacy and the limits of the law. *Yale Law Journal* 89: 421–471.
Hildebrandt, Mireille. 2011. Legal protection by design: Objections and refutations. *Legisprudence* 5(2): 223–248.
Hildebrandt, Mireille, and Bert-Jaap Koops. 2010. The challenges of ambient law and legal protection in the profiling era. *Modern Law Review* 73(3): 428–460.
Karnow, Curtis E.A. 1996. Liability for distributed artificial intelligence. *Berkeley Technology and Law Journal* 11: 147–183.
Kerr, Ian. 2001. Ensuring the success of contract formation in agent-mediated electronic commerce. *Electronic Commerce Research Journal* 1: 183–202.
Kerr, Orin. 2004. The fourth amendment and new technologies: Constitutional myths and the case for caution. *Michigan Law Review* 102: 801–888.
Koops, Bert-Jaap. 2006 Should ICT Regulation be technology-neutral? In *Starting Points for ICT Regulation: Deconstructing Prevalent Policy One-liners*, ed. B-J. Koops et al., 77–108, The Hague: TMC Asser.
Koops, Bert-Jaap, and Ronald Leenes. 2014. Privacy regulation cannot be hardcoded: A critical comment on the "Privacy by Design" Provision in data protection law. *International Review of Law, Computers & Technology* 28: 159–171.
Latour, Bruno. 2005. *Reassembling the social: an introduction to actor-network-theory*. Oxford: Oxford University Press.
Leenes, Ronald, and Federica Lucivero. 2014. Laws on robots, laws by robots, laws in robots: Regulating robot behaviour by design. *Law, Innovation and Technology* 6(2): 193–220.
Lerouge, Jean-François. 2000. The Use of electronic agents questioned under contractual law: Suggested solutions on a European and American level. *The John Marshall Journal of Computer and Information Law* 18: 403.
Lessig, Lawrence. 2004. *Free culture: The nature and future of creativity*. New York: Penguin Press.
Levy, David. 2007. *Love and sex with robots: The evolution of human-robot relationships*. New York: Harper.
Martinez-Martin, Ester and Angel P. del Pobil. 2015. UJI HRI-BD: A new human-robot interaction benchmark dataset. In *Human-Robot Interactions: Principles, Technologies and Challenges*, ed. D. Coleman, 57–73, New York: Nova.
McFarland, David. 2008. *Guilty robots, happy dogs: The question of alien minds*. New York: Oxford University Press.
Nissenbaum, Helen. 2004. Privacy as contextual integrity. *Washington Law Review* 79(1): 119–158.
Pagallo, Ugo. 2010. Robotrust and legal responsibility. *Knowledge, Technology & Policy* 23(3–4): 367–379.
Pagallo, Ugo. 2011. Designing data protection safeguards ethically. *Information* 2(2): 247–265.
Pagallo, Ugo (2012a) On the principle of privacy by design and its limits: Technology, ethics, and the rule of law. In *European Data Protection: In Good Health?* ed. Serge Gutwirth, Ronald Leenes, Paul De Hert and Yves Poullet, 331–346. Dordrecht: Springer.
Pagallo, Ugo. 2012b. Cracking down on autonomy: Three challenges to design in IT law. *Ethics and Information Technology* 14(4): 319–328.
Pagallo, Ugo. 2013a. Robots in the cloud with privacy: A New threat to data protection? *Computer Law & Security Review* 29(5): 501–508.
Pagallo, Ugo. 2013b. *The laws of robots: Crimes, contracts, and torts*. Dordrecht: Springer.
Pagallo, Ugo. 2013c. Online security and the protection of civil rights: A legal overview. *Philosophy & Technology* 26(4): 381–395.
Pagallo, Ugo. 2015. Teaching "Consumer Robots" respect for informational privacy: A legal stance on HRI. In *Human-Robot Interactions. Principles, Technologies and Challenges*, ed. D. Coleman, 35–55. New York: Nova.
Pagallo, Ugo, and Eleonora Bassi. 2013. Open data protection: Challenges, perspectives, and tools for the reuse of PSI. In *Digital Enlightenment Yearbook 2013*, ed. M. Hildebrand, K. O'Hara, and M. Waidner, 179–189. Amsterdam: IOS Press.

Reed, Chis. 2012. *Making laws for cyberspace*. Oxford: Oxford University Press.

RoboLaw. 2014. Guidelines on regulating robotics. EU project on regulating emerging robotic technologies in Europe: Robotics facing law and ethics, 22 Sept.

Sartor, Giovanni. 2009. Cognitive automata and the law: Electronic contracting and the intentionality of software agents. *Artificial Intelligence and Law* 17(4): 253–290.

Sharkey, Noel. 2011. Automated warfare: Lessons learned from the drones. *Journal of Law, Information and Science* 21(2). 10.5778/JLIS.2011.21.Sharkey.1.

Singer, Peter. 2009. *Wired for war: The Robotics revolution and conflict in the 21st century*. London: Penguin.

Solum, Lawrence B. 1992. Legal personhood for artificial intelligence. *North Carolina Law Review* 70: 1231–1287.

Sullins, John P. 2011. Introduction: Open questions in roboethics. *Philosophy and Technology* 24(3): 233–238.

UN World Robotics. 2005. *Statistics, Market Analysis, Forecasts, Case Studies and Profitability of Robot Investment, edited by the UN Economic Commission for Europe and co-authored by the International Federation of Robotics*. Geneva,Switzerland: UN Publication.

Veruggio, Gianmarco. 2006. Euron roboethics roadmap. In *Proceedings Euron Roboethics Atelier*, 27th Feb–3rd Mar, Genoa, Italy.

Warren, Samuel, and Louis Brandeis. 1890. The right to privacy. *Harvard Law Review* 14: 193–220.

Weitzenboeck, Emily Mary. 2001. Electronic agents and the formation of contracts. *International Journal of Law and Information Technology* 9(3): 204–234.

Westin, Alan F. 1967. *Privacy and freedom*. New York: Atheneum.

Wooldridge, Michael J., and Nicholas R. Jennings. 1995. Agent theories, architectures, and languages: a survey. In *Intelligent Agents*, ed. M. Wooldridge, and N.R. Jennings, 1–22. Berlin: Springer.

Zittrain, Jonathan. 2007. Perfect enforcement on tomorrow's internet. In *Regulating Technologies: Legal Futures, Regulatory Frames and Technological Fixes*, ed. Roger Brownsword, and Karen Yeung, 125–156. London: Hart.

Is the Human Rights Framework Still Fit for the Big Data Era? A Discussion of the ECtHR's Case Law on Privacy Violations Arising from Surveillance Activities

Bart van der Sloot

Abstract Human rights protect humans. This seemingly uncontroversial axiom might become quintessential over time, especially with regard to the right to privacy. Article 8 of the European Convention on Human Rights grants natural persons a right to complain, in order to protect their individual interests, such as those related to personal freedom, human dignity and individual autonomy. With Big Data processes, however, individuals are mostly unaware that their personal data are gathered and processed and even if they are, they are often unable to substantiate their specific individual interest in these large data gathering systems. When the European Court of Human Rights assesses these types of cases, mostly revolving around (mass) surveillance activities, it finds itself stuck between the human rights framework on the one hand and the desire to evaluate surveillance practices by states on the other. Interestingly, the Court chooses to deal with these cases under Article 8 ECHR, but in order to do so, it is forced to go beyond the fundamental pillars of the human rights framework.

Keywords Human rights · Big data · Mass surveillance · Individual harm · Societal interest · Conventionality

---

Bart van der Sloot is a researcher at the Institute for Information Law (IViR), University of Amsterdam, the Netherlands. This research is part of the project "Privacy as virtue", which is financed by the Dutch Scientific Organization (NWO).

B. van der Sloot (✉)
Instituut Voor Informatierecht (IViR), Room B1.16, Korte Spinhuissteeg 3, 1012 CG Amsterdam, Netherlands
e-mail: b.vandersloot@uva.nl

1 Introduction

Human rights are designed to protect humans. Whether one accepts the philosophical idea that they are innate to man even in the state of nature,[1] the theological belief that God has bestowed these rights uniquely onto man,[2] the Habermasian theory of the internal correlation between human rights and democracy,[3] or any other theory, human rights have a unique position in legal discourse. They stand apart from other doctrines and rights in that they are conceived as fundamental, sometimes even non-derogable, and protect the most basic personal needs and interest of every human being, regardless of legal status or background. This focus on the individual is even stronger with regard to the right to privacy, Article 8, than with many other human rights as protected under the European Convention on Human Rights (ECHR). This focus on individual rights of natural persons and their personal interests is quite understandable, as privacy is the most 'private' and 'personal' of all human rights. It should also be recognized that this focus has worked very effectively for decades; it has allowed the European Court of Human Rights (ECtHR) to deal not only with the more traditional privacy violations, such as house searches, wiretapping and body cavity searches, but also with the right to develop one's sexual,[4] relational[5] and minority identity,[6] the right to protect one's reputation and honour,[7] the right to personal development,[8] the right of foreigners

---

[1] Among others: Thomas Hobbes, *Leviathan* (Cambridge: Cambridge University Press, 1996 [1651]). Thomas Paine, *The rights of man: for the benefit of all mankind* (Philadelphia: Webster, 1797 [1791]).

[2] Even in Locke, one might find references to this view: John Locke, *Two treatises of government* (Cambridge: Cambridge University Press, 1988 [1689]).

[3] Jurgen Habermas, 'On the Internal Relation between the Rule of Law and Democracy', *European Journal of Philosophy* 3 (1995).

[4] ECtHR, I.G. v. Slovakia, appl. no. 15966/04, 13 November 2012. ECtHR, V.C. v. Slovakia, appl. no. 18968/07, 08 November 2011. ECtHR, Evans v. the United Kingdom, appl. no. 6339/05, 10 April 2007. ECtHR, Dickson v. the United Kingdom, appl. no. 44362/04, 04 December 2007.

[5] ECtHR, Phinikaridou v. Cyprus, appl. no. 23890/02, 20 December 2007. ECtHR, Mikulic v. Croatia, appl. no. 53176/99, 07 February 2002. ECtHR, Gaskin v. the United Kingdom, appl. no. 10454/83, 07 July 1989.

[6] ECmHR, Lay v. the United Kingdom, appl. no. 13341/87, 14 July 1988. ECmHR, Smith v. the United Kingdom, appl. no. 14455/88, 04 September 1991. ECmHR, Smith v. the United Kingdom, appl. no. 18401/91, 06 May 1993. ECmHR, G. and E. v. Norway, appl. no. 9278/81, 03 October 1983. ECtHR, Chapman v. the United Kingdom, appl. no. 27238/95, 18 January 2001. ECtHR, Aksu v. Turkey, appl. nos. 4149/04 and 41029/04, 27 July 2010.

[7] ECtHR, Pfeifer v. Austria, appl. no. 12556/03, 15 November 2007. ECtHR, Rothe v. Austria, appl. no. 6490/07, 04 December 2012. ECtHR, A. v. Norway, appl. no. 28070/06, 09 April 2009.

[8] ECmHR, X. v. Iceland, appl. no. 6825/74, 18 May 1976. ECtHR, Frette v. France, appl. no. 36515/97, 26 February 2002. ECtHR, Varapnickaite-Mazyliene v. Lithuania, appl. no. 20376/05, 17 January 2012. See further: ECtHR, Biriuk v. Lithuania, appl. no. 23373/03, 25 November 2008. ECtHR, Niene v. Lithuania, appl. no. 36919/02, 25 November 2008. ECtHR, Goodwin v. the United Kingdom, appl. no. 28957/95, 11 July 2002. ECtHR, B. v. France, appl. no. 13343/87, 25 March 1992.

to a legalized stay,⁹ the right to property and even work,¹⁰ the right to environmental protection¹¹ and the right to have a fair and equal chance in custody cases.¹² Although some say that the broadened scope of the ECHR in general and the right to privacy in particular has gone too far,¹³ one thing is clear: the current privacy paradigm under the European Convention on Human Rights works very well when it is applied to cases that revolve around individual rights and individual interests of natural persons.

However, the current developments known as Big Data might challenge this approach.¹⁴ Big Data, for the purpose of this study, is defined as gathering massive amounts of data without a pre-established goal or purpose, about an undefined number of people, which are processed on a group or aggregated level through the use of statistical correlations.¹⁵ The essence of these types of cases is thus that the individual element is lost, although data may originally be linked to individuals and the results of Big Data processes may be applied to individuals or groups of

---

⁹ECtHR, Moustaquim v. Belgium, appl. no.12313/86, 18 February 1991. ECtHR, Cruzvaras and others v. Sweden, appl. no. 15576/89, 20 March 1991. ECtHR, Sen v. the Netherlands, appl. no. 31465/96, 21 December 2001. ECtHR, Slivenko v. Latvia, appl. no. 48321/99, 09 October 2003. ECtHR, Sisojeva and others v. Latvia, appl. no. 60654/00, 15 January 2007. ECtHR, Nasri v. France, appl. no. 19465/92, 13 July 1995.

¹⁰ECtHR, Karner v. Austria, appl. no. 40016/98, 24 July 2003. ECtHR, Sidabras and Dziautas v. Lithuania, appl. nos. 55480/00 and 59330/00, 27 July 2004. ECtHR, Coorplan-Jenni GMBH and Hascic v. Austria, appl. no. 10523/02, 24 February 2005. ECtHR, Ozpinar v. Turkey, appl. no. 20999/04, 19 October 2010.

¹¹ECtHR, Moreno Gomez v. Spain, appl. no. 4143/02, 16 November 2004. ECtHR, Villa v. Italy, appl. no. 36735/97, 14 November 2000. ECtHR, Kyrtatos v. Greece, appl. no. 41666/98, 22 May 2003. ECtHR, Morcuende v. Spain, appl. no. 75287/01, 06 September 2005. ECtHR, López Ostra v. Spain, appl. no. 16798/90, 09 December 1994. ECtHR, Ledyayeva, Dobrokhotova, Zolotareva and Romashina v. Russia, appl. nos. 53157/99, 53247/99, 56850/00 and 53695/00, 26 October 2006.

¹²ECtHR, B. v. the United Kingdom, appl. no. 9840/82, 8 July 1987. See similarly: ECtHR, R. v. the United Kingdom, appl. no. 10496/83, 8 July 1987. ECtHR, W. v. the United Kingdom, appl. no. 9749/82, 8 July 1987. ECtHR, Diamante and Pelliccioni v. San Marino, appl. no. 32250/08, 27 September 2011.

¹³Janneke Gerards, "The prism of fundamental rights", *European Constitutional Law Review*, 8 (2012): 2.

¹⁴See further: Antonella Galetta & Paul De Hert, 'Complementing the Surveillance Law Principles of the ECtHR with its Environmental Law Principles: An Integrated Technology Approach to a Human Rights Framework for Surveillance', *Utrecht Law Review*, 10-1, 2014. Thérèse Murphy & Gearóid Ó Cuinn, 'Work in progress. New technologies and the European Court of Human Rights', *Human Rights Law Review*, 2010.

¹⁵See further: Viktor Mayer-Schönberger and Kenneth Cukier, *Big data: a revolution that will transform how we live, work, and think* (Boston: Houghton Mifflin Harcourt, 2013). Terence Craig and Mary E. Ludloff, *Privacy and Big Data: The Players, Regulators, and Stakeholders* (Sebastopol: O'Reilly Media, 2011). Kate Crawford and Jason Schultz, "Big Data and Due Process: Toward a Framework to Redress Predictive Privacy Harms", *Boston College Law Review* 55 (2014): 93.

individuals. Data are not gathered about a specific person or group (for example those suspected of having committed a particular crime), rather, they are gathered about an undefined number of people during an undefined period of time without a pre-established reason. The potential value of the gathered data becomes clear only after they are subjected to analysis by computer algorithms, not on beforehand.[16] These data, even if they are originally linked to specific persons, are subsequently processed by finding statistical correlations. It may appear, for example, that the data string—Muslim + vacation to Yemen + visit to website X—leads to an increased risk of a person being a terrorist.[17] The data are not based on personal data of specific individuals, but processed on an aggregated level and the profiles are formulated on a group level.[18]

Given this constellation of facts, it becomes more and more difficult for an individual to point out his specific personal interest and personal harm (defined by Feinberg as a setback to interests) in Big Data processes.[19] It should be acknowledged that in the field of privacy, the notion of harm has always been problematic as it is often difficult to substantiate the harm a particular violation has done, e.g. what harm follows from entering a home or eavesdropping on a telephone conversation as such when neither objects are stolen nor private information disclosed to third parties? Even so, the more traditional privacy violations (house searches, telephone taps, etc.) are clearly demarcated in time, place and person and the effects are therefore relatively easy to define. In the current technological environment, however, the individual is often simply unaware that his personal data are gathered by either his fellow citizens (e.g. through the use of their smartphones), by companies (e.g. by tracking cookies) or by governments (e.g. through covert surveillance). Obviously, people unaware of the fact that their data are gathered will not invoke their right to privacy in court.

But even if a person would be aware of these data collections, given the fact that data gathering and processing is currently so widespread and omnipresent,

---

[16]See further: Rob Kitchin, *The Data Revolution: Big Data, Data infrastructures & their consequences* (Los Angeles: Sage, 2014). Andrew McAfee and Eerik Brynjolfsson, "Big Data: The management Revolution: Exploiting vast new flows of information can radically improve your company's performance. But first you'll have to change your decision making culture", *Harvard Business Review* October 2012. Mark Andrejevic, "The Big Data Divide", *International Journal of Communication* 8 (2014).

[17]See for literature on profiling: Toon Calders & Sicco Verwer, "Three Naive Bayes Approaches for Discrimination-Free Classification", *Data Mining and Knowledge Discovery 21(2)*, (2010). Bart H. M. Custers, *The Power of Knowledge; Ethical, Legal, and Technological Aspects of Data Mining and Group Profiling in Epidemiology* (Tilburg: Wolf Legal Publishers, 2004). Mireille Hildebrandt & Serge Gutwirth (eds.), *Profiling the European Citizen Cross-Disciplinary Perspectives* (New York: Springer, 2008). Daniel T. Larose, *Data mining methods and models* (New Yersey: John Wiley & Sons, 2006). Tal Z. Zarsky, "Mine your own business!: making the case for the implications of the data mining of personal information in the forum of public opinion", *Yale Journal of Law & Technology (5)*, 2003.

[18]See further: Chris J. Hoofnagle, "How the Fair Credit Reporting Act Regulates Big Data", <http://papers.ssrn.com/sol3/papers.cfm?abstract_id=2432955>.

[19]Joel Feinberg, *Harm to others* (New York: Oxford University Press, 1984).

and will become even more so in the future, it will quite likely be impossible for him to keep track of every data processing which includes (or might include) his data, to assess whether the data controller abides by the legal standards applicable, and if not, to file a legal complaint. And if an individual does go to court to defend his rights, he has to demonstrate a personal interest, i.e. personal harm, which is a particularly problematic notion in Big Data processes, e.g. what concrete harm has the data gathering by the NSA done to an ordinary American or European citizen? This also shows the fundamental tension between the traditional legal and philosophical discourse and the new technological reality—while the traditional discourse is focused on individual rights and individual interests, data processing often affects a structural and societal interest.

This chapter will discuss how the Court deals with privacy violations by the state through the use of (mass) surveillance under Article 8 ECHR. These are, so far, the only cases under the ECHR that concern mass data gathering, storage and processing (it should be remembered that the Convention can only be invoked against states and not against companies). Section 2 will briefly outline the dominant approach of the Court when it deals with cases under Article 8 ECHR. Sections 3– 5 will point out that the Court is willing to relax its focus on individual rights and interests when cases regard surveillance activities. It does so in three distinct ways. Section 3 will present the cases in which the Court focusses not on actual and concrete harm, but on hypothetical harm through the use of the notion of 'reasonable likelihood'. Section 4 describes under which circumstances the Court is willing to accept a 'chilling effect', or future harm, as basis for a claim. Section 5 discusses the Court's third and final approach to these cases, which is also the most controversial one. Sometimes, it is willing to accept *in abstracto* claims, complaints about the legality and legitimacy of laws or policies as such.

Finally, Sect. 6, containing the analysis, will discuss what this last approach implies for the significance of human rights in the age of Big Data. Given the fact that the notions of individual harm and personal interest are so difficult to uphold in Big Data practices, the abstract assessments of Big Data practices may have a high potential, as the specific characteristic of *in abstracto* claims is that the complainant is not required to show any personal interest. Rather, the complaint regards a general or societal interest and addresses a law or policy as such. However, if it is true that human rights protect humans and their most essential needs and interests, the question is how this type of complaints can be reconciled with the basic pillars of the human rights framework. The more fundamental question is perhaps: can the problems following from mass surveillance activities and Big Data practices by states be qualified as human rights violations or do they rather regard general principles of good governance and due process? And, is it proper to assess the mere legality and legitimacy of governmental policies, without any human right being at stake, under a human rights framework? The main conclusion of this chapter is that it is impossible to address certain problems following from Big Data processes in general and mass surveillance activities in particular under human rights frameworks.

2 The Right to Privacy (Article 8 ECHR)

The right to privacy under the European Convention on Human Rights, Article 8, is focussed on the individual in many ways. To successfully submit an application, a complainant must of course have exhausted all domestic remedies, the application should be submitted within the set time frame and it must fall under the competence of the Court. But more importantly, the applicant needs to demonstrate a personal interest, i.e. individual harm following from the violation complained of. This is linked to the notion of *ratione personae*, the question whether the claimant has individually and substantially suffered from a privacy violation, and in part to that of *ratione materiae*, the question whether the interest said to be interfered falls under the protective scope of the right to privacy. This focus on individual harm and individual interests brings with it that certain types of complaints are declared inadmissible by the European Court of Human Rights, which means that the cases will not be dealt with in substance.[20]

So called *in abstracto* claims are in principle declared inadmissible. These are claims that regard the mere existence of a law or a policy, without them having any concrete or practical effect on the claimant. 'Insofar as the applicant complains in general of the legislative situation, the Commission recalls that it must confine itself to an examination of the concrete case before it and may not review the aforesaid law *in abstracto*. The Commission therefore may only examine the applicant's complaints insofar as the system of which he complains has been applied against him.'[21] *A priori* claims are rejected as well, as the Court will usually only receive complaints about injury which has already materialized. A-contrario, claims about future damage will in principle not be considered. 'It can be observed from the terms "victim" and "violation" and from the philosophy underlying the obligation to exhaust domestic remedies provided for in Article 26 that in the system for the protection of human rights conceived by the authors of the Convention, the exercise of the right of individual petition cannot be used to prevent a potential violation of the Convention: in theory, the organs designated by Article 19 to ensure the observance of the engagements undertaken by the Contracting Parties in the Convention cannot examine—or, if applicable, find—a violation other than a posteriori, once that violation has occurred. Similarly, the award of just satisfaction, i.e. compensation, under Article 50 of the Convention is limited to cases in which the internal law allows only partial reparation to be made, not for the violation itself, but for the consequences of the decision or measure in question which has been held to breach the obligations laid down in the Convention.'[22]

Hypothetical claims regard damage which might have materialized, but about which the claimant is unsure. The Court usually rejects such claims because it is

---

[20]<http://www.echr.coe.int/Documents/Admissibility_guide_ENG.pdf>.
[21]ECmHR, Lawlor v. the United Kingdom, application no. 12763/87, 14 July 1988.
[22]ECmHR, Tauira and others v. France, application no. 28204/95, 04 December 1995.

unwilling to provide a ruling on the basis of presumed facts. The applicant must be able to substantiate his claim with concrete facts, not with beliefs and suppositions. The ECtHR will also not receive an *actio popularis*, a case brought up by a claimant or a group of claimants, not to protect their own interests, but to protect those of others or society as a whole. These types of cases are better known as class actions. 'The Court reiterates in that connection that the Convention does not allow an *actio popularis* but requires as a condition for exercise of the right of individual petition that an applicant must be able to claim on arguable grounds that he himself has been a direct or indirect victim of a violation of the Convention resulting from an act or omission which can be attributed to a Contracting State.'[23]

Furthermore, the Court has held that applications are rejected if the injury claimed following from a specific privacy violation is not sufficiently serious, even although it does fall under the scope of Article 8 ECHR. This can also be linked to the more recent introduction of the so called *de minimis* rule in the Convention, which provides that a claim will be declared inadmissible if 'the applicant has not suffered a significant disadvantage'.[24] With environmental issues, for example, it has been ruled that if the level of noise is not sufficiently high, it will not be considered an infringement on a person's private life or home.[25] Similarly, although data protection partially falls under the scope of Article 8 ECHR, if only the name, address and other ordinary data are recorded about an applicant, the case will be declared inadmissible, because such 'data retention is an acceptable and normal practice in modern society. In these circumstances the Commission finds that this aspect of the case does not disclose any appearance of an interference with the applicants' right to respect for private life ensured by Article 8 of the Convention.'[26] Moreover, an interference might have existed which can be substantiated by the applicant and which was sufficiently serious to fall under the scope of Article 8 ECHR. Still, if the national authorities have acknowledged their wrongdoing and provided the victim with sufficient relief and/or retracted the law or policy on which the violation was based, the person can no longer claim to be a victim under the scope of the Convention.[27]

Then there is the material scope of the right to privacy, Article 8 ECHR. In principle, it only provides protection to a person's private life, family life, correspondence and home. However, the Court has been willing to give a broader interpretation. As discussed in the introduction, it has held, inter alia, that the right to

---

[23]ECtHR, Asselbourg and 78 others and Greenpeace Association-Luxembourg v. Luxembourg, application no. 29121/95, 29 June 1999.

[24]Article 35 paragraph 3 (b) ECHR.

[25]ECmHR, Trouche v. France, application no. 19867/92, 01 September 1993. ECmHR, Glass v. the United Kingdom, application no. 28485/95, 16 October 1996.

[26]ECmHR, Murray v. the United Kingdom, application no. 14310/88, 10 December 1991.

[27]Dean Spielmann, *Bringing a case to the European Court of Human Rights: a practical guide on admissibility criteria* (Oisterwijk: Wolf Legal Publishers, 2014). Theodora A. Christou & Juan Pablo Raymon, *European Court of Human Rights: remedies and execution of judgments* (London: BIICL, British Institute of International and Comparative Law cop. 2005).

privacy also protects the personal development of an individual, it includes protection from environmental pollution and may extend to data protection issues.[28] Still, what distinguishes the right to privacy from other rights under the Convention, such as the freedom of expression, is that it only provides protection to individual interests. While the freedom of expression is linked to personal expression and development, it is also connected to societal interests, such as the search for truth through the market place of ideas and the well-functioning of the press, a precondition for a liberal democracy. By contrast, Article 8 ECHR, in the dominant interpretation of the ECtHR, only protects individual interests, such as autonomy, dignity and personal development (in literature, scholars increasingly emphasize a public dimension of privacy). Cases that do not regard such matters are rejected by the Court.[29]

This focus on individual interests has also had an important effect on the types of applicants that are able to submit a complaint about the right to privacy. The Convention, in principle, allows natural persons, groups of persons and legal persons to complain about an interference with their rights under the Convention. Indeed, the Court has accepted that, under certain circumstances, churches may invoke the freedom of religion (Article 9 ECHR), that press organisations may rely on the freedom of expression (Article 10 ECHR) and that trade unions are admissible if they claim the freedom of assembly and association (Article 11 ECHR). However, because Article 8 ECHR only protects individual interests, the Court has said that in principle, only natural persons can invoke a right to privacy. For example, when a church complained about a violation of its privacy by the police in relation to criminal proceedings, the Commission found that '[t]he extent to which a non-governmental organization can invoke such a right must be determined in the light of the specific nature of this right. It is true that under Article 9 of the Convention a church is capable of possessing and exercising the right to freedom of religion in its own capacity as a representative of its members and the entire functioning of churches depends on respect for this right. However, unlike Article 9, Article 8 of the Convention has more an individual than a collective character [].'[30] This led the Commission to declare the complaint inadmissible, a line which has been confirmed in the subsequent case law of the Court and which it is willing to leave only in exceptional cases.[31] Groups of natural persons claiming a Convention

---

[28]See among others: ECtHR, Leander v. Sweden, application no. 9248/81, 26 March 1987. ECtHR, Amann v. Switserland, application no. 27798/95, 16 February 2000. EctHR, Rotaru v. Roemenia, application no. 28341/95, 04 May 2000. See also: <http://www.echr.coe.int/Documents/FS_Data_ENG.pdf>.

[29]See for one of the earliest examples of the broadening scope of Article 8 ECHR: ECmHR, X. v. Iceland, application no. 6825/74, 18 May 1976.

[30]ECmHR, Church of Scientology of Paris v. France, application no. 19509/92, 09 January 1995.

[31]See among others: ECtHR, Stes Colas Est and others v. France, application no. 37971/97, 16 April 2002. See in more detail: Bart van der Sloot, "Do privacy and data protection rules apply to legal persons and should they? A proposal for a two-tiered system", *Computer Law & Security Review* 31 (2015): 1.

right are also principally rejected by the Court and the possibility of inter-state complaints (Article 33 ECHR) is seldom practiced.[32] This leaves only the individual to submit a complaint about a breach of the right to privacy.

The problem is that this focus on natural persons and individual harm is difficult to uphold in cases that concern practices that do not revolve around specific individuals, but affect large groups in society or potentially everyone. Mass (covert) surveillance is the example par excellence, but Big Data practices in general pose a problem for the victim-requirement of the Court. Given the trend of increasingly big data collection and aggregation systems, the relevance of these types of cases is likely to increase. In these types of cases, the Court is often faced with the choice between sticking to its strict interpretation of the victim-requirement and declaring the cases inadmissible or accepting that the cases fall under its jurisdiction and leaving or stretching its focus on individual harm. The Court typically chooses the latter option in three instances: (1) when there is a reasonable chance that the applicant has been harmed, (2) when it is likely that the applicant will be affected by the practice in the future and (3) when the mere existence of a law or policy as such leads to a violation of Article 8 ECHR. These three approaches will be briefly discussed in the following three sections.

3 Reasonable Likelihood (Hypothetical Harm)

Obviously, a discussion about the victim-requirement and surveillance activities by the state has to start with *Klass and others v. Germany*,[33] which revolved around the claim by the applicants that the contested German legislation permitted surveillance measures without obliging the authorities in every case to notify the persons concerned after the event. They also complained about the lack of remedy before the courts against the ordering and execution of such measures. This led, according to them, to a situation of potentially unchecked and uncontrolled surveillance, as those affected by the measures were kept unaware and would, consequently, not challenge them in a legal procedure. In essence, the case revolved around hypothetical harm, as the applicants claimed that they could have been the victims of surveillance activities employed by the German government, but they were unsure as the governmental services remained silent on this point. The claimants were judges and lawyers, professions which cannot function without respect for secrecy of deliberations or of contacts with clients. Moreover, by virtue of their profession, they are more likely to be affected by the measures than ordinary citizens, at least so the applicants claimed. The government, to the contrary, pointed

---

[32]See further: Bart van der Sloot, "Privacy in the Post-NSA Era: Time for a Fundamental Revision?", *Journal of intellectual property, information technology and electronic commerce law*, 5 (2014a): 1.

[33]ECtHR, Klass and others v. Germany, application no. 5029/71, 06 September 1978.

out that the applicants could not substantiate their claim that they were victims of the contested surveillance activities and consequently, that they were bringing forth an *in abstracto* claim.

The Commission, deciding on the admissibility of the case, referred to Article 25 ECHR, the current Article 34 ECHR, which specifies: 'The Court may receive applications from any person, nongovernmental organisation or group of individuals claiming to be the victim of a violation by one of the High Contracting Parties of the rights set forth in the Convention or the Protocols thereto. The High Contracting Parties undertake not to hinder in any way the effective exercise of this right.' It argued that under this provision 'only the victim of an alleged violation may bring an application. The applicants, however, state that they may be or may have been subject to secret surveillance, for example, in course of legal representation of clients who were themselves subject to surveillance, and that persons having been the subject of secret surveillance are not always subsequently informed of the measures taken against them. In view of this particularity of the case the applicants have to be considered as victims for purposes of Art. 25.'[34]

Before the Court, which dealt with the case in substance, the Delegates of the Commission considered that the government was requiring a too rigid standard for the notion of 'victim'. They submitted that, in order to be able to claim to be the victim of an interference with the exercise of the right to privacy, 'it should suffice that a person is in a situation where there is a reasonable risk of his being subjected to secret surveillance.'[35] The Court took it even one step further and held that 'an individual may, under certain conditions, claim to be the victim of a violation occasioned by the mere existence of secret measures or of legislation permitting secret measures, without having to allege that such measures were in fact applied to him.'[36] In this case, the Court thus accepted an *in abstracto* claim, instead of a hypothetical claim, as the 'mere existence' of a law may lead to an interference with Article 8 ECHR.[37] This contrasts with the test proposed by the Delegates, namely whether there is a 'reasonable likelihood' that the applicants were affected by the measures complained of. In the latter test, the requirement of personal harm remains, though it is not made dependent on actual and concrete proof, but on a reasonable suspicion; in the abstract test, the requirement of personal harm is abandoned, as the laws and policies are assessed as such.

---

[34]ECmHR, Klass and others v. Germany, application no. 5029/71, 18 December 1974.

[35]ECtHR, Klass and others v. Germany, application no. 5029/71, 06 September 1978, § 31.

[36]ECtHR, Klass and others v. Germany, application no. 5029/71, 06 September 1978, § 34.

[37]There is also a discussion about the question whether surveillance in itself entails enough injury to bring a case under the scope of Article 8 ECHR. See among others: ECmHR, Herbecq and the Association Ligue Des Droits de L'Homme v. Belgium, application nos. 32200/96 and 32201/96, 14 January 1998. ECtHR, Perry v. the United Kingdom, application no. 63737/00, 17 July 2003. There is also discussion about in how far redress should go to render claims inapplicable. ECtHR, Rotaru v. Romania, application no. 28341/95, 04 May 2000.

Both approaches have played an important role in the Court's subsequent case law.[38] The abstract test was adopted in *Malone v. the UK*[39] and in *P.G. and J.H. v. the UK*,[40] among other cases. In *Mersch and others v. Luxembourg*, the Commission carefully distinguished between the two tests, applying them to two different types of complaints. The case was declared incompatible with the provisions of the Convention in so far as it regarded a violation of the Convention's provisions on account of measures taken under a legal instrument, as the claimants had not been subjected to surveillance measures. Likewise, the Commission stressed that legal persons, one of the applicants being a legal person, could not complain about such matters as they could not be subjected to monitoring or surveillance ordered in the course of criminal proceedings because legal persons had no criminal responsibility. However, it continued to point out that another part of the claim regarded laws as such, allowing for surveillance not confined to persons who may be suspected of committing the criminal offences referred to therein. With regard to this abstract claim, the Commission accepted all applicants in their claim and declared the case admissible.[41] Vice versa, in *Hilton v. the UK*, the Commission stated that 'the Klass case falls to be distinguished from the present case in that there existed a legislative framework in that case which governed the use of secret measures and that this legislation potentially affected all users of postal and telecommunications services. In the present case the category of persons likely to be affected by the measures in question is significantly narrower. On the other hand, the Commission considers that it should be possible in certain cases to raise a complaint such as is made by the applicant without the necessity of proving the existence of a file of personal information. To fall into the latter category the Commission is of the opinion that applicants must be able to show that there is, at least, a reasonable likelihood that the Security Service has compiled and continues to retain personal information about them.'[42]

Section 5 will explore the use of the abstract test by the Court in more detail. What is important to note with regard to the reasonable likelihood test[43] is that two

---

[38]ECtHR, Case of Association "21 December 1989" and others v. Romania, application nos. 33810/07 and 18817/08, 24 May 2011. ECmHR, Spillmann v. Switzerland, application no. 11811/85, 08 March 1988.

[39]ECmHR, Malone v. the United Kingdom, application no. 8691/79, 13 July 1981. See further: ECtHR, Leander v. Sweden, application no. 9248/81, 26 March 1987. ECtHR, Huvig v. France, application no. 11105/84, 24 April 1990. ECtHR, Kruslin v. France, application no. 11801/85, 24 April 1990.

[40]ECtHR, P.G. and J.H. v. the United Kingdom, application no. 44787/98, 25 September 2001.

[41]ECmHR, Mersch and others v. Luxembourg, application nos. 10439/83, 10440/83, 10441/83, 10452/83, 10512/83 and 10513/83, 10 May1985.

[42]ECmHR, Hilton v. the United Kingdom, application no. 12015/86, 06 July 1988.

[43]ECtHR, Stefanov v. Bulgaria, applicaiton no. 65755/01, 22 May 2008. ECmHR, Nimmo v. the United Kingdom, application no. 12327/86, 11 October 1988.

aspects can lead to the establishment of a reasonable likelihood.[44] First, if the applicant falls under a group or category that is specifically mentioned in the law on which the surveillance activities are based. In these types of cases, the Court is willing to accept that applicants who fall under these categories can demonstrate a reasonable likelihood that they had been affected by the matters complained of. Second, the Court takes into account specific actions by the applicants which make them more likely to be affected by surveillance measures. In *Matthews v. the UK*, for example, the Commission decided that the assumption of the applicants that they were wiretapped was not substantiated by their argument that they heard mysterious clicking noises when telephoning. 'However, in view of the fact that the applicant was active in the campaign against Cruise (nuclear) missiles in the United Kingdom, the Commission will assume for the purposes of this decision that the applicant has established a reasonable possibility that her telephone conversations were intercepted pursuant to a warrant for the purposes of national security.'[45]

4 Chilling Effect (Future Harm)

The chilling effect principle is mostly connected to the freedom of speech and the Court uses it to explain that certain actions by the government, although not directly limiting the freedom of speech of its citizens, may lead to self-restraint: a chilling effect in the lawful use of a right. The chilling effect is the effect which exists when people know that they are watched of know that they might be watched. Afraid of the potential consequences, people will restrain their behavior and abstain from certain acts which they perceive as possibly inciting negative consequences.[46] However, the Court is also willing to accept this doctrine in

---

[44]ECtHR, Senator Lines GmbH v. Austria, Belgium, Denmark, Finland, France, Germany, Greece, Ireland, Italy, Luxembourg, the Netherlands, Portugal, Spain, Sweden and the United Kingdom, application no. 56672/00, 10 March 2004. ECtHR, Segi and others and Gestoras Pro-Amnistia and others v. 15 states of the European Union, application nos. 6422/02 and 9916/02, 23 May 2002. ECmHR, Tauira and 18 others v. France, application no. 28204/95, 04 December 1995. ECtHR, C. and D. and S. and others v. the United Kingdom, application nos. 34407/02 and 34593/02, 31 August 2004. ECtHR, C. v. the United Kingdom, application no. 14858/03, 14 December 2004. ECtHR, Berger-Krall and others v. Slovenia, application no. 14717/04, 12 June 2014. ECmHR, Esbester v. the United Kingdom, application no. 18601/91, 02 April 1993. ECmHR, Hewitt and Harman v. the United Kingdom, application no. 20317/92, 01 September 1993. ECmHR, Redgrave v. the United Kingdom, application no. 20271/92, 01 September 1993. ECmHR, T.D., D.E. and M.F. v. the United Kingdom, application nos. 18600/91, 18601/91 and 18602/91, 12 October 1992.

[45]ECmHR, Matthews v. the United Kingdom, application no. 28576/95, 16 October 1996. ECtHR, Halford v. the United Kingdom, application no. 20605/92, 25 June 1997, § 48.

[46]Jeremy Bentham, *Panopticon; or The inspection-house* (Dublin, 1791). Michel Foucault, *Surveiller et punir: naissance de la prison* (Paris, Gallimard, 1975).

certain cases relating to Article 8 ECHR, primarily when they regard surveillance measures, but also in relation to laws that discriminate or stigmatize certain groups in society. Here, the Court is willing to accept that although no harm has been done yet to an applicant, he may still be received in his (a priori) claim if it is likely that he will suffer from harm in the future, either because he is curtailed in his right to privacy by the government or because he will resort to self-restraint in the use of his right.

An example may be the case of *Michaud v. France*, in which the applicant complained that because lawyers were under an obligation to report suspicious operations, as a lawyer he was required, subject to disciplinary action, to report people who came to him for advice. He considered this system to be incompatible with the principles of lawyer-client privilege and professional confidentiality. The government maintained, however, that the applicant could not claim to be a 'victim' as his rights had not actually been affected in practice, highlighting that he did not claim that the legislation in question had been applied to his detriment, but simply that he had been obliged to organize his practice accordingly and introduce special internal procedures. This would qualify as an *in abstracto* claim, according to the government. It continued to stress that if the Court accepted his status as a 'potential victim', this would open the door for class actions.

The Court pointed out that, indeed, in order to be able to lodge an application in pursuance of Article 34 of the Convention, a person must be able to claim to be a 'victim' of a violation of the rights enshrined in the Convention: to claim to be a victim of a violation, a person must be directly affected by the impugned measure. The ECHR does not envisage the bringing of an *actio popularis* for the interpretation of the rights set out therein, the Court continued, or permit individuals to complain about a provision of national law simply because they consider, without having been directly affected by it, that it may contravene the Convention. Referring to *Marckx v. Belgium*, *Johnston and others v. Ireland*, *Norris v. Ireland* and *Burden v. the UK*, it stressed, however, that it is 'open to a person to contend that a law violates his rights, in the absence of an individual measure of implementation, and therefore to claim to be a "victim" within the meaning of Article 34 of the Convention, if he is required to either modify his conduct or risk being prosecuted, or if he is a member of a class of people who risk being directly affected by the legislation.'[47]

The Court pointed out that if the applicant failed to report suspicious activities as required he would expose himself by virtue of the law to disciplinary sanctions up to and including being struck off. The Court also considered credible the applicant's suggestion that, as a lawyer specialising in financial and tax law, he was even more concerned by these obligations than many of his colleagues and exposed to the consequences of failure to comply. In fact he was faced with a dilemma comparable, mutatis mutandis, to that which the Court already identified in *Dudgeon v. the UK* and Norris: either he applies the rules and relinquishes his

---

[47]ECtHR, Michaud v. France, application no. 12323/33, 06 December 2012, § 51.

idea of the principle of lawyer-client privilege, or he decides not to apply them and exposes himself to disciplinary sanctions and even being struck off. Therefore, the Court accepted that the applicant was directly affected by the impugned provisions and could therefore claim to be a 'victim' of the alleged violation of Article 8. In conclusion, the Court accepted a victim status, not because the applicant had actually suffered from any concrete harm, but because he was likely to be affected by it in the future, either because he would restrict or limit his behaviour or because he would not and face a legal sanction.

The references to the cases of, inter alia, Marckx, Dudgeon and Norris, are particularly telling. The Court is also willing to relax its strict focus on individual harm when cases regard potential discrimination and stigmatization of weaker groups in society. For example, it has accepted that where the national legislator had adopted a prohibition on abortion and the applicant neither was pregnant, nor had been refused an interruption of pregnancy, nor had been prosecuted for unlawful abortion, the claimant could still be received.[48] Likewise, in Marckx, the inheritance laws complained of had not yet been applied to the applicants and presumably would not be applied for a certain period of time, but the Court argued nonetheless that they had a legitimate interest in challenging a legal position, that of an unmarried mother and of children born out of wedlock, which affected them—according to the Court—personally.[49] In Dudgeon and Norris, the case regarded a claim by an applicant about the regulation of homosexual conduct. The Court held that the applicant could be received even without the law being applied to him and without there being any reason to believe that it might be, as 'the very existence of this legislation continuously and directly affects his private life: either he respects the law and refrains from engaging—even in private with consenting male partners—in prohibited sexual acts to which he is disposed by reason of his homosexual tendencies, or he commits such acts and thereby becomes liable to criminal prosecution.'[50]

This approach is becoming increasingly important in cases revolving around surveillance activities by the state, in which the Court is also willing to accept potential future harm and chilling effects. A good example may be the case of *Colon v. the Netherlands*, in which the applicant complained that the designation of a security risk area by the Burgomaster of Amsterdam violated his right to respect for privacy as it enabled a public prosecutor to conduct random searches of people over an extensive period in a large area without this mandate being subject to any judicial review. The government, to the contrary, argued that the designation of a security risk area or the issuing of a stop-and-search order had not in itself

---

[48]ECmHR, Brüggemann and Scheuten v. Germany, application no. 5959/75, 19 May 1976.

[49]ECtHR, Marckx v. Belgium, application no. 6833/74, 13 June 1979, § 27.

[50]ECtHR, Dudgeon v. the United Kingdom, application no. 7525/76, 22 October 1981, § 41. See further: ECtHR, S.A.S. v. France, application no. 43835/11, 01 July 2014. ECtHR, Mateescu v. Romania, application no. 1944/10, 14 January 2014. ECtHR, Ballianatos and others v. Greece, application nos. 29381/09 and 32684/09, 07 November 2013.

constituted an interference with the applicant's private life or liberty of movement. Since the event complained of, several preventive search operations had been conducted; in none of them had the applicant been subjected to further attempts to search him. This was, according to the government, enough to show that the likelihood of an interference with the applicant's rights was so minimal that this deprived him of the status of victim.

The Court stressed again, that in principle, it did not accept *in abstracto* claims or an *actio popularis*. 'In principle, it is not sufficient for individual applicants to claim that the mere existence of the legislation violates their rights under the Convention; it is necessary that the law should have been applied to their detriment. Nevertheless, Article 34 entitles individuals to contend that legislation violates their rights by itself, in the absence of an individual measure of implementation, if they run the risk of being directly affected by it; that is, if they are required either to modify their conduct or risk being prosecuted, or if they are members of a class of people who risk being directly affected by the legislation.'[51] It went on to stress that it was 'not disposed to doubt that the applicant was engaged in lawful pursuits for which he might reasonably wish to visit the part of Amsterdam city centre designated as a security risk area. This made him liable to be subjected to search orders should these happen to coincide with his visits there. The events of 19 February 2004, followed by the criminal prosecution occasioned by the applicant's refusal to submit to a search, leave no room for doubt on this point. It follows that the applicant can claim to be a "victim" within the meaning of Article 34 of the Convention and the Government's alternative preliminary objection must be rejected also.'[52]

Like with the laws prohibiting homosexual conduct, the applicant was left only the choice between two evils: either he avoided traveling to the capital city of the Netherlands or he risked being subjected to surveillance activities. This is enough for the Court to accept a victim-status, which it has reaffirmed in later jurisprudence.[53] Right now pending before the Court is a case regarding mass surveillance activities by the British government and its intelligence services.[54] It will be interesting to see whether in the future, the Court is willing to content that, if governments engage in data retention practices[55] or wiretap all telecommunication coming in or going out of their country, echoing Colon, citizens are left only with the choice either to abstain from legitimately using the internet or other common (electronic) communication channels or face the risk of being subjected to surveillance activities.

---

[51]ECtHR, Colon v. the Netherlands, application no. 49458/06, 15 May 2012, § 60.

[52]Colon, § 61.

[53]ECtHR, Ucar and others v. Turkey, application no. 4692/09, 24 June 2014.

[54]ECtHR, Big Brother Watch and others v. the United Kingdom, application no. 58170/13, 07 January 2014.

[55]ECJ, Digital Rights Ireland, C–293/12 and C–594/12, 8 April 2014.

5 In Abstracto Claims (No Individual Harm)

Although in the cases discussed in the foregoing a relaxation takes place, the Court still holds on to the victim requirement. There are, however, cases, which have been briefly touched upon in Sect. 3, in which the Court allows *in abstracto* claims, regarding laws or policies as such, without them having been applied to the claimant or otherwise having a direct effect on him.[56] Sometimes, the Court, rather artificially, holds on to the victim requirement by holding that everyone living in a certain country is affected by a certain law. For example, in *Weber and Saravia v. Germany*, the applicants claimed that certain provisions of the Fight against Crime Act violated Article 8 ECHR. The Court reiterated that the mere existence of legislation which allows a system for the secret monitoring of communications entails a threat of surveillance for all those to whom the legislation may be applied. 'This threat necessarily strikes at freedom of communication between users of the telecommunications services and thereby amounts in itself to an interference with the exercise of the applicants' rights under Article 8, irrespective of any measures actually taken against them.'[57] In similar fashion, the Court recalled in *Liberty and others v. the UK* its findings 'in previous cases to the effect that the mere existence of legislation which allows a system for the secret monitoring of communications entails a threat of surveillance for all those to whom the legislation may be applied. This threat necessarily strikes at freedom of communication between users of the telecommunications services and thereby amounts in itself to an interference with the exercise of the applicants' rights under Article 8, irrespective of any measures actually taken against them.'[58] The fact that everyone may claim to be a victim means that everyone may submit a claim before the Court, a situation which it hoped to prevent by introducing the prohibition on class actions.

Although in these cases, the Court still holds onto the victim requirement, in most cases revolving around *in abstracto* claims, such as Klass, Malone, P.G. and J.H. and Mersch, the victim requirement is simply abandoned. This fact has had a large influence on the admissibility of cases and complainants more in general. While typical cases under Article 8 ECHR revolve around individual interests such as human dignity, individual autonomy and personal freedom, cases in which the Court accepts *in abstracto* claims revolve around societal interests, such as the abuse of power by the government. Abandoning the victim-requirement means that other hurdles for invoking Article 8 ECHR are also minimized. A number of

---

[56]See further: ECmHR, M.S. and P.S. v. Switserland, application no. 10628/83, 14 October 1985. ECtHR, Tanase v. Moldova, application no. 7/08, 27 April 2010. ECtHR, Hadzhiev v. Bulgaria, application no. 22373/04, 23 October 2012. See further: ECtHR, Goranova-Karaeneva v. Bulgaria, application no. 12739/05, 08 March 2011.

[57]ECtHR, Weber and Saravia v. Germany, application no. 54934/00, 29 June 2006, § 78.

[58]ECtHR, Liberty and others v. the United Kingdom, application no. 58243/00, 01 July 2008, § 56–57.

examples may be provided, three of them will be touched upon here briefly. First, the rejection of the Court of legal persons invoking the right to privacy, second the obligation to exhaust all domestic remedies before submitting a claim under the system of supra-national supervision and third, the requirement that a case must be brought before the European Court of Human Rights within six months after the final decision has been made on the national level.

As has been discussed, in *Mersch and others v. Luxembourg*, the Court was willing to accept a legal person in its claim for the part of the case that regarded the mere existence of laws or policies as such. Besides Mersch, the Court accepted the complaint of a legal person in Liberty and in the case of the Association for European Integration and Human Rights and *Ekimdzhiev v. Bulgaria*. The latter case regarded the authorities' wide discretion to gather and use information obtained through secret surveillance. The applicants suggested that, by failing to provide sufficient safeguards against abuse, by its very existence, the laws were in violation of Article 8 ECHR. The government disputed that the applicants could be considered victims (as they did not claim to be specifically harmed by the matter) and that legal persons should not be allowed to claim a right to privacy in general and in particular in this case because the legal person could not have been harmed itself. The Court, however, pointed to the statutory objectives of the association and found that the 'rights in issue in the present case are those of the applicant association, not of its members. There is therefore a sufficiently direct link between the association as such and the alleged breaches of the Convention. It follows that it can claim to be a victim within the meaning of Article 34 of the Convention.'[59] Essentially the same was held in *Iordachi and others v. Moldova*.[60] This means that legal persons who have statutes that incorporate references to the general protection of privacy and other human rights may have direct access to the court in the future when cases regard mass surveillance activities by the state.

As a second example, reference can be made to the requirement to exhaust all domestic remedies before submitting a claim before the ECtHR, which is also relaxed with *in abstracto* claims. The European Convention on Human Rights, Article 35, regarding the admissibility criteria, specifies that the Court may only deal with a matter after all domestic remedies have been exhausted, according to the general recognized rules of international law. This is connected to the principle that the Court dismisses cases in which the national authorities have acknowledged their mistake and have remedied their misconduct, either by providing compensation and/or by revoking the law or policy on which the abusive practices were based. If the national courts would be passed over by the claimant, national states would be denied this chance. However, the problem with *in abstracto* claims is that, especially when linked to mass surveillance by secret services, the national oversight on surveillance activities is often quite limited. In particular, *in abstracto*

---

[59]ECtHR, Association for European Integration and Human Rights and Ekimdzhiev v. Bulgaria, application no. 62540/00, 08 June 2007, § 59.

[60]ECtHR, Iordachi and other v. Moldova, application no. 25198/02, 10 February 2009, § 33–34.

claims can often not be brought forward by citizens or legal persons on the domestic level. Moreover, the courts and tribunals often simply lack the power to annul laws or policies and can only assess specific individual cases. That is why the ECtHR is often willing to accept claimants which have not exhausted all domestic remedies if the claim regards the mere existence of laws or policies as such.

For example, in *Kennedy v. the UK*, the Court concluded that the applicant had failed to raise his arguments as regarded the overall Convention-compatibility of the Regulation of Investigatory Powers Act 2000 (RIPA) provisions before the Investigatory Powers Tribunal (IPT). However, it also stressed that where the government claimed non-exhaustion it must satisfy the Court that the remedy proposed was an effective one available in theory and in practice at the relevant time, that is to say, that it was accessible, was capable of providing redress in respect of the applicant's complaints and offered reasonable prospects of success. However, if 'the applicant had made a general complaint to the IPT, and if that complaint been upheld, the tribunal did not have the power to annul any of the RIPA provisions or to find any interception arising under RIPA to be unlawful as a result of the incompatibility of the provisions themselves with the Convention. [] Accordingly, the Court considers that the applicant was not required to advance his complaint regarding the general compliance of the RIPA regime for internal communications with Article 8 § 2 before the IPT in order to satisfy the requirement under Article 35 § 1 that he exhaust domestic remedies.'[61] The Court held essentially the same in *M.M. v. the UK*.[62] This means for *in abstracto* claims, that the ECtHR is willing to rule as court of first instance.

To provide a final example, the Convention specifies certain time-restricting principles, which are also put under pressure with *in abstracto* claims, as these do not revolve around specific violations, but the existence of laws or policies as such and are thus not linked to a specific moment in time. The principle of *ratione temporis*, which means that the provisions of the Convention do not bind a national state in relation to any act or fact which took place or any situation which ceased to exist before the date of the entry into force of the Convention or the accession of a state to the ECHR. This means that, for example, if the right to privacy of an individual had been violated by a state before that state entered the Convention, this case will be declared inadmissible by the Court. Obviously, this principle does not apply to *in abstracto* claims, as the infringement continues to exist. The Convention, Article 35, also requires applicants to submit their application within a period of six months from the date on which the final decision on the national level was taken. This principle is also very difficult to maintain with regard to *in abstracto* claims, and the ECtHR has often adopted a flexible approach with this respect.

For example, in *Lenev v. Bulgaria*, the Court made a sharp distinction between the complaint regarding individual harm and the part of the application revolving

---

[61]ECtHR, Kennedy v. the United Kingdom, application no. 26839/05, 18 May 2010.
[62]ECtHR, M.M. v. the United Kingdom, application no. 24029/07, 13 November 2012.

around the mere existence of the law. It stressed that the applicant complained 'more than six months later, on 12 September 2007. The fact that he did not have knowledge of the exact content of the recording is immaterial because the lack of such knowledge could not prevent him from formulating a complaint under Article 8 of the Convention in relation to the secret taping of his interrogation. Nor can the Court accept that the criminal proceedings against the applicant constituted an obstacle to his raising grievances in this respect. It follows that the complaints concerning the secret taping of the applicant's interrogation have been introduced out of time and must be rejected in accordance with Article 35 §§ 1 and 4 of the Convention. By contrast, the concomitant complaints concerning the mere existence in Bulgaria of laws and practices which have established a system for secret surveillance relate to a continuing situation—in as much as the applicant may at any time be placed under such surveillance without his being aware of it. It follows that his complaints in that respect cannot be regarded as having been raised out of time.'[63] Consequently, claims revolving around the mere existence of laws or policies are not bound by the time-limits specified by the Convention. In conclusion, abandoning the victim-requirement has the effect that many threshold for invoking a right under the Convention dissolve.

6 Analysis

To summarize briefly, the following has been shown. The Court focusses on individual harm by natural persons when assessing the admissibility of cases under Article 8 ECHR. According to the Court, this provision guarantees protection only to individual interests such as human dignity, individual autonomy and personal freedom. Cases are declared inadmissible if they do not revolve around individual harm. Examples are: *in abstracto* claims, a priori claims, hypothetical complaints, class actions, claims about minimal harm, claims about harm which has been remedied, claims by legal persons and claims that do not regard strictly personal interests. However, it has also been explained that in certain types of cases, mostly revolving around surveillance activities, the Court is willing to relax its standards. It is sometimes willing to allow for hypothetical complaints if a reasonable likelihood exists that the applicant has been harmed, it is occasionally willing to accept a priori claims, when the applicant is forced to restrict its legitimate use of his right to privacy in order to avoid legal sanctions, and it is even willing to accept claims that revolve around the mere existence of laws and policies as such.

The reason why the Court is willing to relax its stance in these cases specifically is clear. With (mass) surveillance activities, either by secret services or other governmental institutions, the citizen is mostly unaware of the fact that he is being followed or that his data are being gathered, why this is done, by whom, to what

---

[63]ECtHR, Lenev v. Bulgaria, application no. 41452/07, 04 December 2012.

extent, etc. Likewise, especially with regard to laws allowing for mass surveillance and data retention, the fact is that the potential violations do not revolve around a specific person, but affect everyone living under that regime or at least very large numbers of people. Mostly, the issue is simply the presumed abuse of power by national authorities. This is a societal interest, related to the legitimacy and legality of the state.

The reason for discussing these matters in such detail is that these characteristics are shared to a large extent by privacy infringements following from Big Data initiatives. Often, an individual is simply unaware that his personal data are gathered by either his fellow citizens (e.g. through the use of their smartphones), by companies (e.g. by tracking cookies) or by governments (e.g. through covert surveillance). Even if a person would be aware of these data collections, given the fact that data gathering and processing is currently so widespread and omnipresent, and will become even more so in the future, it will quite likely be impossible for him to keep track of every data processing which includes (or might include) his data, to assess whether the data controller abides by the legal standards applicable, and if not, to file a legal complaint. And if an individual does go to court to defend his rights, he has to demonstrate a personal interest, i.e. personal harm, which is a particularly problematic notion in Big Data processes.[64]

Finally, under the current privacy and data protection regimes, the balancing of interests is the most common way in which to resolve cases. In a concrete matter, the societal interests served with the data gathering, for example wiretapping a person's telephone because he is suspected of having committed a murder, is weighed against the harm the wiretapping does to his personal autonomy, freedom or dignity. However, the balancing of interests becomes increasingly difficult in the age of Big Data, not only because the individual interest involved with a particular case is so difficult to substantiate, the societal interest at the other end is also increasingly difficult to specify.[65] For example, it is mostly unclear in how far the large data collections by intelligence services have actually prevented concrete terrorist attacks. This balance is even more difficult if executed on an individual level, i.e. how the collection of personal data of a particular non-suspected person

---

[64]See further: David Bollier, "The Promise and Peril of Big Data", <http://www.emc.com/collateral/analyst-reports/10334-ar-promise-peril-of-big-data.pdf>. Danah Boyd and Kate Crawford, "Six Provocations for Big Data", <http://papers.ssrn.com/sol3/papers.cfm?abstract_id=1926431>. Lawrence Busch, "A Dozen Ways to Get Lost in Translation: Inherent Challenges in Large Scale Data Sets", *International Journal of Communication* 8 (2014). Neil M. Richards & Jonathan H. King, "Three Paradoxes of Big Data", *Stanford Law Review online* 66 (2013): 44.

[65]See further: Kevin Driscoll and Shawn Walker, "Working Within a Black Box: Transparency in the Collection and Production of Big Twitter Data" *International Journal of Communication* 8 (2014). Theresa M. Payton & Theodore Claypoole, *Privacy in the age of Big Data: recognizing threats, defending your rights, and protecting your family* (Rowman & Littlefield: Plymouth, 2014). Cornelius Puschmann and Jean Burgess, "Metaphors of Big Data", *International Journal of Communication* 8 2014. Omer Tene & Jules Polonetsky, "Big Data for All: Privacy and User Control in the Age of Analytics", *Northwestern Journal of Technology and Intellectual Property* 11 (2013): 239.

has ameliorated the national security.⁶⁶ Perhaps more important is the fact that with some of the large scale data collections, there seems not a relative interest at stake, which can be weighed against other interests, but absolute interests. For example, it has been suggested that the data collection by the NSA is so large, is conducted over such a long time span and includes data about so many people that this simply qualifies as abuse of power.⁶⁷ Abuse of power is not something which can be legitimated by its instrumentality towards a specific societal interest—it is an absolute minimum condition of the use of power.

The same problems with applying the current privacy paradigm also count for data protection rules. They too are dependent for their applicability on the material and personal scope, which, like the right to privacy, is linked to the natural person. For example, the Data Protection Directive defines personal data as 'any information relating to an identified or identifiable natural person ('data subject'); an identifiable person is one who can be identified, directly or indirectly, in particular by reference to an identification number or to one or more factors specific to his physical, physiological, mental, economic, cultural or social identity'.⁶⁸ However, if data are processed on an aggregated level and turned into group profiles, it is often impossible to directly identify one particular person on the basis of it. Moreover, like the right to privacy, data protection revolves to a large extent around individual rights, such as the right to access personal data and correct them, the Right to be Forgotten and the right to a legal remedy. The same problems signaled with regard to individual privacy rights consequently apply to the data protection regime.⁶⁹

All notions connected to the victim-requirement, such as the *de minimis* rule, the prohibition on hypothetical, future and abstract harm, the prohibition of class actions and of legal persons instituting a complaint, and the focus on individual interests, seem to be put under pressure by the developments known as Big Data. What seems most suitable for claims regarding privacy infringements following from mass surveillance and Big Data practices is claims about the potential chilling effect (e.g. users being afraid to use certain forms of communication), about hypothetical harm and even abstract assessments of the policies and practices as such. Not the individual seems to be best equipped to file a complaint, but civil

---

⁶⁶See further: Pierre-Luc Dusseault, "Privacy and social media in the Age of Big Data: Report of the Standing Committee on Access to Information, Privacy and Ethics", <http://www.parl.gc.ca/content/hoc/Committee/411/ETHI/Reports/RP6094136/ethirp05/ethirp05-e.pdf>.
Neil M. Richards & Jonathan H. King, "Big Data Ethics", *Wake Forest Law Review* 49 (2014).
Ira Rubinstein, "Big Data: The End of Privacy or a New Beginning?", *NYU School of Law, Public Law Research Paper* No. 12–56. Drury D. Stevenson & Nicholas J. Wagoner, 'Bargaining in the Shadow of Big Data', *Florida Law Review*, 66 (2014): 5.

⁶⁷Bart van der Sloot, "Privacy in the Post-NSA Era: Time for a Fundamental Revision?" *Journal of intellectual property, information technology and electronic commerce law* 5 (2014): 1.

⁶⁸Article 2 sub (a) Directive 95/46/EC of the European Parliament and of the Council of 24 October 1995 on the protection of individuals with regard to the processing of personal data and on the free movement of such data.

⁶⁹See also: <http://ec.europa.eu/justice/data-protection/document/review2012/com_2012_11_en.pdf>.

society groups and legal persons. Not individual interest are at stake in these types of processes, but general and societal interests. Thus, in order to retain the relevance of the rights to privacy and data protection in the modern technological era, the victim-requirement and all its sub-requirements should be relaxed.

And this is exactly what the ECtHR is willing to do in cases that revolve around surveillance activities. It does accept claims about future harm and potential chilling effects, about hypothetical harm, it does receive class actions, abstract claims and legal persons and it does take into account abstract and societal interests. The question is, however, at what price this comes. What is left for the Court, particularly with *in abstracto* claims, to assess in these types of cases is the mere quality of laws and policies as such and the question is whether this narrow assessment is still properly addressed under a human rights framework. The normal assessment of the Court revolves around, roughly, three questions: (1) has there been an infringement of the right to privacy of the claimant, (2) is the infringement prescribed by law and (3) is the infringement necessary in a democratic society in terms of, inter alia, national security, that is, does the societal interest in this particular case outweigh the individual interest. Obviously, the first question does not apply to *in abstracto* claims because there has been no infringement with the right of the claimant. The third question is also left untouched by the Court, because it is impossible, in the absence of an individual interest, to weigh the different interests involved. This means of course that another principle by the Court, namely that it only decides on the particular case before it, is also overturned.

Even the second question is not applicable as such as there is no infringement that is or is not prescribed by law. Although the Court regularly determines in cases, inter alia, whether the laws are accessible, whether sanctions are foreseeable and whether the infringement at stake is based on a legal provision, this does not apply to *in abstracto* claims. There *is* often a law permitting mass surveillance (that is exactly the problem) and these laws *are* accessible and the consequences *are* foreseeable (in the sense that everyone will be affected by it). Rather, it is the mere quality of the policy as such that is assessed—the content of the law, the use of power as such, is deemed inappropriate. The question of abuse of power can of course be addressed by the Court, though not under Article 8 ECHR, but under Article 18 of the Convention, which specifies: 'The restrictions permitted under this Convention to the said rights and freedoms shall not be applied for any purpose other than those for which they have been prescribed.' But as the Court has stressed, this provision can only be invoked if one of the other Convention rights are at stake. Reprehensible as the abuse of power may be, it is only proper to address this question under a human rights framework if one of the human rights contained therein will or have been violated by the abuse. The Court cannot assess the abuse of power as such (a doctrine which it also applies to, inter alia, Article 14 ECHR, the prohibition of discrimination).

However, what is assessed in cases in which *in abstracto* claims regarding surveillance activities have been accepted is precisely the use of power by the government as such, without a specific individual interest being at stake. This is a test of legality and legitimacy, which is well known to countries that have a constitutional court or body, such as France and Germany. These courts can assess the 'constitutionality' of national laws in abstract terms. Not surprisingly, the term 'conventionality' (or 'conventionalité' in French) has been introduced in the cases discussed.[70] For example, in Michaud, the government argued that with a previous *in abstracto* decision, the Court had 'issued the Community human rights protection system with a "certificate of conventionality", in terms of both its substantive and its procedural guarantees.'[71] Referring to the Michaud judgment, among other cases, in his partly concurring, partly dissenting opinion in *Vallianatos and others v. Greece*, judge Pinto De Albuquerque explained: 'The abstract review of "conventionality" is the review of the compatibility of a national law with the Convention independently of a specific case where this law has been applied.'[72]

He argued that the particular interest of the Vallianatos and others case, which revolved around the fact that the civil unions introduced by a specific law were designed only for couples composed of different-sex adults, is that the Grand Chamber performs an abstract review of the "conventionality" of a Greek law, while acting as a court of first instance. 'The Grand Chamber not only reviews the Convention compliance of a law which has not been applied to the applicants, but furthermore does it without the benefit of prior scrutiny of that same legislation by the national courts. In other words, the Grand Chamber invests itself with the power to examine *in abstracto* the Convention compliance of laws without any prior national judicial review.'[73] As explained earlier, when discussing *Lenev v. Bulgaria*, the Court is likewise willing to pass over the domestic legal system and act as court of first instance in cases revolving around mass surveillance. Subsequent to Michaud and Vallianatos, the term 'conventionality' has been used more often,[74] as well as the term 'Convention-compatibility', for example in the case of *Kenedy v. the UK* discussed earlier,[75] and most likely will only gain in dominance as the Court opens up the Convention for abstract reviews of laws and policies.

---

[70]See for the use of the word also: ECtHR, Py v. France, application no. 66289/01, 11 January 2005. ECtHR, Kart v. Turkey, application no. 8917/05, 08 July 2008. ECtHR, Duda v. France, application no. 37387/05, 17 March 2009. ECtHR, Kanagaratnam and others v. Belgium, application no. 15297/09, 13 December 2011. ECtHR, M.N. and F.Z. v. France and Greece, application nos. 59677/09 and 1453/10, 08 January 2013.

[71]Michaud, § 73. See also: ECtHR, Vassis and others v. France, application no. 62736/09, 27 June 2013.

[72]ECtHR, Vallianatos and others v. Greece, application nos. 29381/09 and 32684, 07 November 2013.

[73]Ibid.

[74]See among others: ECtHR, S.A.S. v. France, application no. 43835/11, 01 July 2014. ECtHR, Avotins v. Latvia, application no. 17502/07, 25 February 2014. ECtHR, Matelly v. France, application no. 10609/10, 02 October 2014. ECtHR, Delta Pekarny A.S. v. Czech Republic, application no. 97/11, 02 October 2014.

[75]See among others: ECtHR, Animal Defenders International v. the United Kingdom, application no. 48876/08, 22 April 2013. ECtHR, Emars v. Latvia, application no. 22412/08, 18 November 2014. ECtHR, Kennedy v. the United Kingdom, application no. 26839/05, 18 May 2010. ECtHR, Mikalauskas v. Malta, application no. 4458/10, 23 July 2013. ECtHR, Sorensen and Rusmussen v. Denmark, application nos. 52562/99 and 52620/99, 11 January 2006. ECtHR, Bosphorushava Yollari Turizm ve Ticaret Anonim Sirketi v. Ireland, application no. 45036/98, 30 June 2005. ECtHR, Lunch and Whelan v. Ireland, application nos. 70495/10 and 74565/10, 18 June 2013. ECtHR, Interdnestrcom v. Moldova, application no. 48814/06, 13 March 2012.

What is left in these types of cases is thus the abstract assessment of laws and policies as such, without a Convention right necessarily being at stake. Furthermore, the Court is willing to assess the 'Conventionability' of these laws as court of first instance. Desirable as such an abstract test may be,[76] it is questionable whether it should be conducted under a human rights framework. Of course, in the Big Data era, what is needed is not more individual rights protecting individual interests, but general duties to protect general interests.[77] Accepting *in abstracto* claims and assessing the legality and legitimacy of laws and (Big Data) practices as such fits this purpose. But if it is true that human rights protect humans and their interests, it seems that the Court should only have the competence to address human rights violations. Although it does have the power to assess the abuse of power, under a human rights framework, the abuse of power addressed should at least have an impact on concrete individual rights. When this is not the case, like with cases revolving around the abstract assessment of laws permitting mass surveillance and in the future, potentially, cases revolving around Big Data processes, it seems that the human rights framework is simply not the most appropriate instrument to turn to. When the Court does so nevertheless, although for noble reasons, it seems to overstretch its own competence and change the nature of the ECHR from a human rights instrument to a document resembling a constitution, and its position from a supra-national court overseeing severe human rights violations in last instance, to a first instance court for assessing the legality and legitimacy of laws and policies as such.

Bibliography

Andrejevic, Mark. 2014. The big data divide. *International Journal of Communication* 8.
Bentham, Jeremy. 1791. *Panopticon; or the inspection-house* (Dublin).
Bollier, David. 2010. The promise and peril of big data. http://www.emc.com/collateral/analyst-reports/10334-ar-promise-peril-of-big-data.pdf.
Boyd, Danah and Kate Crawford. (2011) Six provocations for big data. http://papers.ssrn.com/sol3/papers.cfm?abstract_id=1926431.
Busch, Lawrence. 2014. A Dozen Ways to Get Lost in Translation: Inherent Challenges in Large Scale Data Sets. *International Journal of Communication* 8 (2014).
Calders, Toon & Sicco Verwer. 2010. Three naive bayes approaches for discrimination-free classification. *Data Mining and Knowledge Discovery* 21(2).

---

[76]Letting go of the personal and material scope of data protection rules could similarly lead to the application of certain principles *in abstracto*, such as the transparency principle, the requirement of having a clear and defined purpose for the processing, the purpose limitation principle and the obligations to process data safely and confidentially and to keep the data correct and up to date. Again, although this abstract test might be in itself desirable, the question is whether it is appropriate to fit this under the regimes protecting personal data.

[77]See further: Bart van der Sloot, "Do data protection rules protect the individual and should they? An assessment of the proposed General Data Protection Regulation", *International Data Privacy Law* 3 (2014b).

Craig, Terence, and Mary E. Ludloff. 2011. *Privacy and big data: The players, regulators, and stakeholders*. Sebastopol: O'Reilly Media.

Crawford, Kate and Jason Schultz. 2014. Big data and due process: Toward a framework to redress predictive privacy harms. *Boston College Law Review* 55: 93.

Custers, Bart H. M. 2004. *The power of knowledge; ethical, legal, and technological aspects of data mining and group profiling in epidemiology* (Tilburg: Wolf Legal Publishers).

Davis, Kord with David Patterson. 2012. Ethics of big data: Balancing risk and innovation. http://www.commit-nl.nl/sites/default/files/Ethics%20of%20Big%20Data_0.pdf.

der Bart Sloot, Van. 2015. Do privacy and data protection rules apply to legal persons and should they? A proposal for a two-tiered system. *Computer Law & Security Review* 31: 1.

Driscoll, Kevin and Shawn Walker. 2014. Working within a black box: Transparency in the collection and production of big twitter data. *International Journal of Communication* 8.

Dusseault, Pierre-Luc. 2013. Privacy and social media in the age of big data: Report of the standing committee on access to information, privacy and ethics. http://www.parl.gc.ca/content/hoc/Committee/411/ETHI/Reports/RP6094136/ethirp05/ethirp05-e.pdf.

Feinberg, Joel. 1984. *Harm to others*. New York: Oxford University Press.

Foucault, Michel. 1975. *Surveiller et punir: naissance de la prison*. Paris: Gallimard.

Galetta, Antonella and Paul De Hert. 2014. Complementing the surveillance law principles of the ECtHR with its environmental law principles: An integrated technology approach to a human rights framework for surveillance. *Utrecht Law Review* 10-1.

Gerards, Janneke. 2012. The prism of fundamental rights. *European Constitutional Law Review* 8: 2.

Habermas, Jurgen. 1995. On the internal relation between the rule of law and democracy. *European Journal of Philosophy* 3.

Hildebrandt, Mireille, and Serge Gutwirth (eds.). 2008. *Profiling the European citizen cross-disciplinary perspectives*. New York: Springer.

Hobbes, Thomas. 1996. *Leviathan*. Cambridge: Cambridge University Press.

Hoofnagle, Chris J. 2013. How the fair credit reporting act regulates big data. <http://papers.ssrn.com/sol3/papers.cfm?abstract_id=2432955>.

International Working Group on Data Protection in Telecommunications. 2014. Working paper on big data and privacy. Privacy principles under pressure in the age of big data analytics 55th Meeting, 5–6, Skopje.

Kitchin, Rob. 2014. *The data revolution: Big data, data infrastructures & their consequences*. Los Angeles: Sage.

Larose, Daniel T. 2006. *Data mining methods and models* (New Yersey: Wiley).

Locke, John. 1988. *Two treatises of government*. Cambridge: Cambridge University Press.

Mayer-Schönberger, Viktor, and Kenneth Cukier. 2013. *Big data: A revolution that will transform how we live, work, and think*. Boston: Houghton Mifflin Harcourt.

McAfee, Andrew and Eerik Brynjolfsson. 2012. Big data: The management Revolution: Exploiting vast new flows of information can radically improve your company's performance. But first you'll have to change your decision making culture. *Harvard Business Review*.

Murphy, Thérèse and Gearóid Ó Cuinn. 2010. Work in progress. New technologies and the European court of human rights. *Human Rights Law Review*.

Paine, Thomas. 1797. *The rights of man: For the benefit of all mankind*. Philadelphia: Webster.

Payton, Theresa M., and Theodore Claypoole. 2014. *Privacy in the age of big data: Recognizing threats, defending your rights, and protecting your family*. Plymouth: Rowman & Littlefield.

Puschmann, Cornelius and Jean Burgess. 2014. Metaphors of big data. *International Journal of Communication* 8.

Richards, Neil M. & Jonathan H. King. 2013. Three paradoxes of big data. *Stanford Law Review online* 66: 44.

Richards, Neil M. and Jonathan H. King. 2014. Big data ethics. *Wake Forest Law Review* 49.

Rubinstein, Ira. 2012. Big data: The end of privacy or a new beginning?. *NYU School of Law, Public Law Research Paper* No. 12–56.

Stevenson, Drury D. and Nicholas J. Wagoner. 2014. Bargaining in the shadow of big data. *Florida Law Review* 66: 5.
Tene, Omer and Jules Polonetsky. 2013. Big data for all: Privacy and user control in the age of analytics. *Northwestern Journal of Technology and Intellectual Property* 11: 239.
Van der Sloot, Bart. 2014a. Privacy in the Post-NSA Era: Time for a Fundamental Revision?. *Journal of intellectual property, information technology and electronic commerce law* 5: 1.
Van der Sloot, Bart. 2014b. Do data protection rules protect the individual and should they? An assessment of the proposed general data protection regulation. *International Data Privacy Law* 3.
Zarsky, Tal Z. 2003. Mine your own business!: making the case for the implications of the data mining of personal information in the forum of public opinion. *Yale Journal of Law & Technology* 5.

Metadata, Traffic Data, Communications Data, Service Use Information… What Is the Difference? Does the Difference Matter? An Interdisciplinary View from the UK

Sophie Stalla-Bourdillon, Evangelia Papadaki and Tim Chown

Abstract In the wake of the Snowden revelations, it has become standard practice to rely upon the dichotomies metadata/data or metadata/content of communications to delineate the remit of the surveillance and investigation power of law enforcement agencies as well as the range of data retention obligations imposed upon telecommunications operators and in particular Internet service providers (ISPs). There is however no consensual definition of what metadata is and different routes can be taken to describe what metadata really covers. The key question is whether or to what extent metadata should be treated akin to content data for the purposes of identifying the categories of data which shall actually be retained by telecommunications operators and to which law enforcement agencies can have access. In an attempt to answer the question, this paper provides an understanding of what metadata is and what their diversity is by following two steps. First, adopting an interdisciplinary approach, we argue that three types of metadata should be distinguished in relation to the nature of the activity of the service provider processing them and their level in a network communications—network-level, application-level metadata, and service-use metadata—and we identify three types of criteria to classify these metadata and determine whether they should be deemed as akin to content data. Second, we compare these categories with legal concepts and in particular UK legal concepts to assess to what extent law-makers have managed to treat content data and metadata differently.

S. Stalla-Bourdillon (✉) · E. Papadaki
School of Law, University of Southampton, Southampton SO17 1BJ, UK
e-mail: s.stalla-bourdillon@soton.ac.uk

T. Chown
ECS, University of Southampton, Southampton SO17 1BJ, UK
e-mail: tjc@ecs.soton.ac.uk

© Springer Science+Business Media Dordrecht 2016
S. Gutwirth et al. (eds.), *Data Protection on the Move*,
Law, Governance and Technology Series 24, DOI 10.1007/978-94-017-7376-8_16

1 Introduction

In the wake of the Snowden revelations, it has become standard practice to rely upon the dichotomies metadata/data or metadata/content of communication to delineate the remit of the surveillance and investigation power of law enforcement agencies as well as the range of data retention obligations imposed upon telecommunications operators and in particular Internet service providers (ISPs). There is however no consensual definition of what metadata is and different routes can be taken to describe what metadata really covers. Above all metadata is not 'yet' a legal category although law enforcement agencies and commentators including legal commentators more or less implicitly increasingly refer to this notion.

The UK legal framework for example relies upon the notion of 'communications data' distinguished from that of content of communications. Although the UK legislator attempts to breakdown in a systematic manner all the species of communications data[1] without referring to the term of metadata, the latter term is often used as a shortcut to explain what communications data is within the meaning of the legal framework regulating law enforcement access to data retained by telecommunications operators or data retention obligations imposed upon telecommunications operators, i.e. the Regulation of Investigatory Powers Act 2000 (RIPA), the Data Retention and Investigatory Powers Act 2014 (DRIPA) and the Data Retention Regulations 2014 (DRR).[2]

Jemima Stratford QC and Tim Johnston adopt for example a very broad definition of communications data and refer to the term metadata for this purpose:

> The debate triggered by the Snowden revelations has largely used the phrase "metadata" to describe "communications data". The range of information that can be obtained via communications data is extremely broad. Communications data encompasses each individual URL visited, the contents of an individual's Twitter and Facebook address lists and numerous other significant elements of an individual's online private life. It is likely that messages placed on social media sites and Twitter feeds would also fall within the scope of "communications data".[3]

They infer from this definition that the category of content of communications essentially covers the content of emails.[4]

However, depending upon the definition of metadata adopted, equating metadata to communications data can be misleading and have the consequence of unduly broadening the scope of telecommunications operators' data retention obligations or the power of law enforcement agencies wanting to have access to the data retained by these telecommunications operators.

---

[1] See section 2 infra for a definition of this notion.
[2] 2014 No. 2042.
[3] Jemima Stratford QC and Tim Johnston, "The Snowden 'Revelations': Is GCHQ Breaking the Law?", *E.H.R.L.R.* 2 (2014): 132.
[4] Ibid.

The key question that underlines these definitional issues and that will be addressed in this paper is thus whether or to what extent metadata should be treated like content data for the purposes of identifying the scope of both data retention obligations and of the power of law enforcement agencies.

To start with, exceptions to the principle of the confidentiality of communications and the related traffic data to be found in Article 5 of the e-privacy Directive,[5] echoing the right to respect for one's correspondence or communications to be found in Article 8 of the European Convention of Human Rights (ECHR) and in Article 7 of the European Charter of Fundamental Rights, should be based on a clear legal basis with appropriate safeguards.[6] It would seem therefore that the content of communications is not the only type of data protected by the principle of confidentiality.

With this said, the recent case law appears to distinguish between interceptions of content of communications and metering of communications in relation to their respective gravity. In Copland v United Kingdom[7] for example the European Court of Human Rights (ECtHR) appears to opine in this direction, although the Court could maybe have been clearer. The Court draws an analogy between telephone numbers (for communications made by telephone) and email addresses and Internet usage (for communications made by the Internet), the conclusion of which is that information relating to the monitoring of emails and Internet usage is an integral part of electronic communications and thereby deserves the same kind of protection.[8] This said, a few paragraphs further down the Court states that the interference with the right to the respect of private life resulting from the monitoring of communications such as emails and Internet usage is *"of a significantly*

---

[5]Directive 2002/58/EC of the European Parliament and of the Council of 12 July 2002 concerning the processing of personal data and the protection of privacy in the electronic communications sector (Directive on privacy and electronic communications) OJ L 201, 31.7.2002, pp. 37–47 amended two times by Directive 2006/24/EC of the European Parliament and the of the Council of 15 March 2006 and Directive 2009/136/EC of the European Parliament and of the Council of 25 November 2009 [e-privacy Directive].

[6]The Article 29 Data Protection Party stated in a recent working document on surveillance of electronic communications for intelligence and national security purposes that *"Contrary to the general exemptions from the scope of application of the Directive laid down in its Article 3(2), the derogations to specific principles, rights and obligations provided by Article 13(1) or included in other provisions of the Directive assume that the Directive applies in principle to the processing in question. As explicitly required by the Directive such exceptions should then be laid down by Member State's laws, which in many cases also need to provide additional safeguards"*. The Article 29 Data Protection Working Party, "Working Document on surveillance of electronic communications for intelligence and national security purposes", adopted on 5 December 2014, WP 228, accessed December 27, 2014, http://ec.europa.eu/justice/data-protection/article-29/documentation/opinion-recommendation/files/2014/wp228_en.pdf, at [4.4.3]. If we apply this logic to the e-privacy Directive, it should follow that exceptions to the principle of confidentiality of communications (to be found in Article 5) require a clear legal basis with appropriate safeguards.

[7](2007) 45 E.H.R.R. 37 (Copland).

[8]Copland at [43–44].

lower order of seriousness" than that of resulting from the interception of communications.[9]

The Copland case comes after Malone v United Kingdom[10] in which the ECtHR had stated without equivoque that : "[t]*he Court does not accept ... that the use of data obtained from metering, whatever the circumstances and purposes, cannot give rise to an issue under Article 8*".[11] In fact there is a prima facie breach of Article 8(1) when such a technique[12] is used to the benefit of the police. Indeed, "[t]*he records of metering contain information, in particular the numbers dialled, which is an integral element in the communications made by telephone*".[13]

Obviously, the answer to the key question aforementioned depends upon the way metadata is defined. Looking at how law-makers have regulated retention and access to data relating to electronic communications, the UK legal framework comprising RIPA, DRIPA and DRR,[14] is particularly worth examining as it is driven by a clear attempt to distinguish between content data and other types of data, i.e. communications data. This can probably be explained by the specificity of the procedure set forth for the acquisition of communications data by law enforcement agencies, which does not require as a matter of principle the issuance of a warrant.[15] It is difficult to find trace of similar discussion in other European jurisdictions, making an analysis of this sort particularly interesting.[16] In France for example at least three reasons explain why discussions on how to legally distinguish content data and metadata have not taken place yet. First of all, the judicial authority plays a significant role in the process of acquisition of data retained by telecommunications operators.[17] Second, even if the judicial authority is not involved, content data and data relating to electronic communications tend to be treated in the same way.[18] Third, the scope of data retention obligations is quite

---

[9]Copland at [54].

[10](1985) 7 E.H.R.R. 14 (Malone).

[11]Malone at [84].

[12]This is how the Court defines the technique of metering in this case: "*the use of a device (a meter check printer) which registers the numbers dialled on a particular telephone and the time and duration of each call*". Malone at [84].

[13]Malone at [84].

[14]2014 No. 2042.

[15]There is an exception to this principle for local authorities which must receive prior judicial approval. See s. 23A and 23B of RIPA (as amended by the Protection of Freedoms Act 2012). Note that for interceptions (revealing the content of communications), it is the Secretary of State who issues the warrant. See s. 5 of RIPA.

[16]Australia would also be worth examining, as the recent decision of the Privacy Commissioner in Ben Grubb and Telstra Corporation Limited (2015) AICmr 35 shows it. The purpose of this paper was however to shed light upon one specific interpretation and implementation of the EU legal framework.

[17]See for example articles 60-2 and Article 77-1-2 of the French Code of Penal Procedure.

[18]See for example Article L246-1 of the French Code of Internal Security.

broad at least *rationae personae* as over-the-top service providers such as hosting providers have been asked quite early at the same time as ISPs acting as Internet access providers to retain data relating to electronic communications.[19] It is true nonetheless that under French law, for the purposes of determining the scope of data retention obligations, the distinction between content data and more precisely the content of correspondence and browsed information and *"technical data"* is crucial.[20]

Comparing legal and technical definitions of the main components of electronic communications,[21] we argue in this paper that it is helpful to distinguish between three types of metadata, taking into account the nature of the activity of the service provider processing them and their level in a communications network: network-level metadata, application-level metadata, and service-use metadata. Indeed, in comparison to network-level metadata, the collection of application-level metadata by ISPs requires the implementation of intrusive technologies such as Deep Packet Inspection (DPI) technologies.[22] In addition, while network-level metadata is first used to answer the question who speaks with whom, application-level metadata can be used to answer the questions what is said, or what is thought. As a result, application-level metadata can directly reveal sensitive information such as political, religious or philosophical opinions or beliefs, as well as information concerning health or sex life.[23] Furthermore, network-level metadata as such do not directly identify individuals, whereas application-level metadata can directly do so (in many cases email addresses contain real names as well as subject lines). It is the combination of network-level metadata with customer information that makes

---

[19]See Article 6.II of the Loi No. 2004-575 du 21 juin 2004 pour la confiance dans l'économie numérique and Article 1 of Décret No. 2011-219 du 25 février 2011 relatif à la conservation et à la communication des données permettant d'identifier toute personne ayant contribué à la création d'un contenu mis en ligne.

[20]See Article L34-1 (VI) of the Code of the Post and Electronic Communications. This comes from the transposition of Article 1 of the data retention Directive. See also the Spanish Act 25/2007 on the retention of data related to electronic communications and public communications networks, which applies to traffic and location data of both legal entities and natural persons and to the related data necessary to identify the subscriber or registered users (Article 1).

[21]For the sake of clarity it is important to note that we understand data as numbers, characters, symbols that can be processed by a computer. Data can thus be stored and/or transmitted through the means of a communication process which in our case takes the form of an electronic communication issued by a sender to a recipient. Data becomes information when it is possible to ascribe a semantic meaning to it, e.g. when it is possible to derive the identity of the sender or recipient, or when it is possible to derive what is said or thought by the sender or recipient.

[22]For a full analysis of DPI technologies see Sophie Stalla-Bourdillon, Evangelia Papadaki and Tim Chown, "From Porn to Cybersecurity Passing by Copyright: How Mass Surveillance Technologies Are Gaining Legitimacy... The case of Deep Packet inspection Technologies", *Computer Law & Security Review* 30 (2014): 670–686.

[23]Article 8 of the Data Protection Directive. Directive 95/46/EC of the European Parliament and of the Council of 24 October 1995 on the protection of individuals with regard to the processing of personal data and on the free movement of such data, OJ L 281, 23.11.1995, pp. 31–50 lists the different categories of sensitive data.

identification possible. To be sure, this holds true to the extent customer information is not available via open sources. In principle ISPs are the sole holders of such information.[24] Finally, service-use metadata stored by web or application servers can mirror both network-level and application-level metadata.

In the end, we suggest that in order to determine whether metadata should be treated like content data a first set of questions must be asked:

1. Does the collection of this data require the implementation of deep inspection technologies?
2. Can this data directly identify individuals?
3. Can this data directly reveal sensitive information?[25]
4. Can this data single out individuals and can their its systematic collection be deemed as amounting to profiling?

As the answer to these four questions is affirmative for application-level metadata there is an argument that they should be protected in the same way as the content of communications. At the very least the systematic retention of application-level metadata should be prohibited as it would allow the creation of extensive profiles about individuals. A similar case could be built for corresponding service-use metadata as they can directly identify individuals as well as reveal sensitive information. There is also an argument for making sure additional safeguards are in place (e.g. prior judicial approval) when network-level metadata is combined with subscriber information.

We start this paper with a technical analysis of IP packets to provide a definition of network-level metadata, application-level metadata and service-use metadata.

We then show that the UK legal framework is based on a legitimate attempt to treat content data and certain types of metadata differently. However, it appears that the UK category of communications data comprises both network-level metadata and application-level metadata, which is problematic in terms of privacy protection if ISPs acting as Internet access providers are invited to retain application-level metadata and if law enforcement authorities can have access to both categories in the same way. In addition, even assuming ISPs should not retain application-level metadata, by extending *rationae personae* the scope of data retention obligations to target over-the-top service providers without carefully delineating the category of service use information not to be retained, the end result might be very similar. Yet, what would be needed is a stricter regulation[26] of

---

[24]In this sense metadata relating to electronic communications are different from traditional metering information such as phone numbers.

[25]To be sure, drawing a distinction between these two types of question is not without problem as it might be possible to infer the content of communications (i.e. what is said) from "merely" who speaks with whom.

[26]See however Neil Brown, "An Assessment of the Proportionality of Regulation of 'Over the Top' Communications Services under Europe's Common Regulatory Framework for Electronic Communications Networks and Services", Computer Law & Security Review 30 (2014): 368, arguing that "*there appears to be an obvious case for the extension of the requirement of data retention to over the top providers*".

the collection and retention of service-use metadata, which by the way is not governed by the Privacy and Electronic Communications (EC Directive) Regulations 2003[27] transposing the e-privacy Directive (although obviously the Data Protection Act of 1998 transposing the Data Protection Directive[28] remains applicable).

2 Three Categories of Metadata

The purpose of this section is to define what metadata is from a technical perspective and shed light upon its variety in order to justify the fact that all types of metadata are not equal and cannot be treated in the same manner.

Broadly understood, the term metadata means 'data about data'; metadata provides descriptive information about other data. It can be something as simple as the information stored in a picture (i.e. size, colour depth, resolution) or as complex as the information that can be parsed out of the TCP/IP traffic (i.e. source/destination addresses and ports, email address, website name etc.).[29] In this sense metadata can either be part of the content of communications or can be seen as information obtained from the metering of communications.

Adopting a more restrictive approach and distinguishing communications in transit from stored communications, it is first possible to identify two categories of metadata, both relating to communications in transit: 'network-level metadata' and 'application-level metadata'. A third category of metadata is collected and stored at the end of the communications by the servers: we label them service-use metadata.

2.1 Network-Level Metadata

Metadata can be viewed at different levels in a communications network. The network-level metadata comprise information about the flow of TCP/IP packets between the sender and receiver—often referred to as 'network flow data'—and cover a standard form of session data that details the 'who, what, when and where' of network traffic, specifically which devices are communicating (by IP address), a hint of which applications are involved (by port number), and the packet count, data volume and duration of such a flow.

---

[27]2003 No. 2426.

[28]Directive 95/46/EC of the European Parliament and of the Council of 24 October 1995 on the protection of individuals with regard to the processing of personal data and on the free movement of such data OJ L 281, 23/11/1995, pp. 31–50.

[29]Qosmos, "DPI and Metadata for Cybersecurity Applications", White Paper, January 2012, accessed October 10, 2014, http://www.accumuli.com/pages/files/datasheets/DPI-and-Metadata-for-Cybersecurity-Applications_Qosmos.pdf.

When contents, e.g. the contents of a web page, are transferred over the Internet, they are broken down into multiple *packets*—units of binary data capable of being routed through a computer network—and then reassembled to the original data chunk once they reach their destination. Each packet transmitted includes both a *header* and a *payload*; however, the structure of a packet varies depending on its type and on the protocol used.[30] The network layer header contains transmission-related information, which tells routers how to handle and forward the packet from the sending host along to the destination host, and the transport layer header indicates which application processes are sending and receiving data on the hosts, e.g. a web browser and a web server. Header information at the different layers includes such fields as the packet's total length, originating address (where the packet came from), destination address (where the packet is going), sequence number (which packet this is in a sequence of packets), port number (a hint as to what type of application packet is being transmitted, e.g. e-mail, web page, or a chat protocol).[31]

A transport layer port number is only an indication of the applications that are communicating. While there are 'well-known' (default) port numbers that are usually used, as listed in the Internet Assigned Numbers Authority (IANA) registry,[32] users and software developers can choose to use different port numbers. As an example, a standard web server using the Hypertext Transfer Protocol (HTTP) runs on port 80, so a packet going to port 80 is typically being sent to a web server, but a user may run a different application on that port number if he/she chooses.

A brief analysis of the structure of the most widely used communication standard, TCP/IP,[33] is necessary to fully understand the specificities of network-level metadata. The TCP/IP protocol suite consists of a layered architecture, where each layer depicts some functionality necessary for end-to-end transmission. In addition to the physical layer, over which the TCP/IP networking model runs, it is composed of the following four layers. The *data link* (or network interface) layer formats the packet so that it can be sent either directly to its destination, or where necessary—and more commonly—to the next router towards the destination; the *network* (or Internet) layer is responsible for handling the movement (routing) of data on network, and uses IP addresses as both identifiers and locators for the communicating hosts; the *transport* layer organises the data transmission process in several sequential steps by dividing the data from upper levels into appropriate

---

[30]Nadeem Unuth, "What is a Data Packet?", accessed October 5, 2014, http://voip.about.com/od/glossary/g/PacketDef.htm.

[31]Qosmos, "DPI and Metadata for Cybersecurity Applications".

[32]IANA, "Service Name and Transport Protocol Port Number Registry", accessed October 18, 2014, http://www.iana.org/assignments/service-names-port-numbers/service-names-port-numbers.xhtml.

[33]TCP stands for Transmission Control Protocol and IP for Internet Protocol.

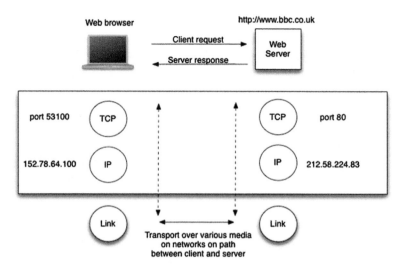

Fig. 1 A conceptual view of the TCP/IP layers

sized chunks and then passing them onto the network; the *application* layer represents the software applications that are exchanging data.[34]

The first three layers—link, network and transport—each adds a header to the packet, whereas the application layer manages the packet payload.[35] The transport layer lies between the network layer and the application layer, and thus a transport header exists between the IP packet header and the packet payload. The transport header indicates the source and destination applications at the communicating endpoints, e.g. a web browser and a web server, and these are identified by 'port numbers', with certain applications generally (but not always) using well-known port numbers, such as aforementioned port 80 for unencrypted web (HTTP protocol) traffic.[36]

A conceptual view of the TCP/IP layers in action is shown in Fig. 1.

The user's view is of a web browser displaying content from a web page. Their browser sends requests for the web server content, which are returned and displayed, in the example above for the BBC web site.

Underneath, the two communicating host network stacks are using the TCP/IP protocols to exchange data. The transport layer protocol, TCP, uses source and destination port numbers, and the network layer uses source and destination IP addresses. The IP packets are carried from the client to the server over—typically—many

---

[34]Andrew S. Tanenbaum and David J. Wetherall, *Computer Networks*, 5th Edition (US: Pearson, 2010), 45.

[35]Alison Cooper, "Doing the DPI Dance: Assessing the Privacy Impact of Deep Packet Inspection", in *Privacy in America: Interdisciplinary Perspectives,* ed. William Aspray and Phillip Doty (Maryland: Scarecrow Press, 2011), 139.

[36]Gary Kessler, "An Overview of TCP/IP Protocols and the Internet", accessed October 12, 2014, http://www.garykessler.net/library/tcpip.html.

router hops. The link layer is used to carry the IP packet between routers on the path; at each router, the IP packet is taken from the link layer, and a new link layer header added to send the IP packet to the next hop. While the IP packet is (usually) carried unaltered, the link layer header is different for each hop.

Metadata for network traffic is therefore the so-called '5-tuple' which is commonly used in various networking contexts to identify specific application flows. A 5-tuple refers to a set of values that comprise a TCP/IP connection and is the combination of source and destination IP addresses and ports, together with the protocol in use (usually TCP or UDP).

It is the IP packet being carried, and the transport header inside it, that define the 5-tuple of source and destination IP and port numbers, with the protocol. Thus in the example above, the 5-tuple is

Source IP: 152.78.64.100
Destination IP: 212.58.224.83
Source port: 53100
Destination port: 80
Protocol: TCP

All packets matching those properties belong to the same network flow between the communicating hosts, and are network layer communication metadata.

2.2 Application-Level Metadata

There is also an application-layer view of metadata, e.g. for an email the sender and receiver email addresses, and a subject line. Or for a web request, the specific file name or image being requested from the server. Such application data is only contained in the payload of the TCP/IP packets. **While network flow metadata can only provide summary data about a communication, application-level metadata comprises much greater specifics about the communication.** A network flow record might show which email servers are talking to each other, but only the application-level metadata would contain the email addresses involved.

The payload, also called the body or data of a packet, is the cargo of a data transmission. The packet payload contains the actual application content, which in some cases may be encrypted, e.g. web traffic might be sent unencrypted over the HTTP protocol, to port 80, or encrypted via the Hypertext Transfer Protocol Secure (HTTPS) protocol, to port 443. The application data might include the text of an e-mail, a URL, website content, a chat message, video content, image content etc.). But application-level metadata is also contained in the payload e.g. the subject line and sender and receiver e-mail addresses for an email, or specific URLs for a web browser request, etc.).[37] In order to send the application data over

---

[37]Christian Fuchs, "Societal and Ideological Impacts of Deep Packet Inspection (DPI) Internet Surveillance for Society", *Information Communication & Society* 16(8) (2013): 1334.

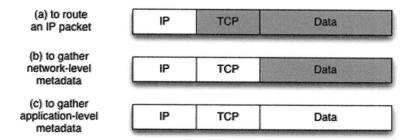

Fig. 2 TCP-based application

a network, network packet headers are prepended to the payload for transport and then discarded when the packet arrives at its destination.

It is important to distinguish between network-level metadata and application-level metadata since the set of tools used by an ISP or operator to gather that information differs in relation to the metadata level. Capturing network-flow data or network-level metadata, e.g. using the Cisco Netflow protocol, is relatively simple. It is a few lines of configuration on a network router that causes flow data to be sent to a collector device, which adds the flow metadata to a database that can later be queried for flow analysis purposes.

In contrast, capture of application-level metadata requires more detailed inspection of the traffic, specifically the payload of the communication. This is what is typically referred to as DPI.[38] This is a more CPU[39] intensive task, and more challenging at the higher data rates found when the capture is performed to the core of the Internet backbone.

In Fig. 2, we illustrate the case in the example of a TCP-based application, for the following functions: IP packet routing, gathering network-level metadata, and gathering application-level metadata.

An IP router (a) only needs to inspect the IP destination address to make a routing decision. The TCP header and payload are opaque to the router. This should be the most common configuration.

A system, such as a router, exporting/gathering network flow information (b) would need the IP and port number data from the IP and TCP headers, but the payload would remain opaque.

However, where application-level metadata is being gathered, the payload is then inspected (which implies full DPI), and thus the full content of the packet is inspected.

---

[38]For a full analysis of DPI technologies see Stalla-Bourdillon, Papadaki and Chown, "From Porn to Cybersecurity Passing by Copyright".

[39]CPU stands for Central Processing Unit.

2.3 Service-Use Metadata

By comparison to network-level metadata and application-level metadata that are data in transit, at the end of the communications servers store service-use metadata, which can mirror both network-level metadata and application-level metadata.

One example of service-use metadata for an application would be the case of an e-mail relay (mail transfer agent, or MTA) that is responsible for handling email for a given organisation. In a typical deployment, such a relay would be used to relay email messages in and out of the organisation. It would thus receive email sent from clients within the organisation and deliver them to the email server of the receiver's organisation, and it would in turn also receive emails sent from external senders to recipients within the organisation, forwarding them on to the local mail server to be viewed by the recipient. The email relay would log all transactions, noting such metadata as the sender's and receiver's email addresses, the date/time, the message size, and the unique Message-ID for the email.

In general, the content of an email being sent/received through such a relay is not stored, and the logs would typically be used to present a summary of service use, e.g. the volume of emails being processed by the relay, and—if the relay is performing a spam detection function—the percentage of email received that has been classified as spam. But the metadata may also assist in identifying a specific email stored elsewhere, or its presence in other service use logs, through searching for the unique Message-ID (which also appears as a mail header in the email, though such headers are not by default presented to users when reading their email).

Another example of service-use metadata would be the case of a web server that logs all accesses made to the server from its users' web browsers. They are often called web logs. In this case the log might include the date/time, the IP address or host name of the client running the browser, the login name and the full name of the user who owns the account that is making the request,[40] the specific web page (URL) being fetched and the visiting path (the path taken by the user while visiting the web site) and/or the path traversed (the path taken by the user within the web site suing the various link). Typically an organisation will use web analytics software to profile its web visitors, in terms of the most popular content being accessed, the time a visitor spends on site, the unique 'hits' per day, or the geographic spread of the visitors. These log files are usually used for technical site auditing and troubleshooting, but can also be of great interest for web mining and traffic analysis.[41]

---

[40]Most remote sites do not give out this information for security reasons. If this field is disabled by the host, there is a dash (–) instead of the login/full name. If the server requires a user ID in order to fulfil an HTTP request, the user ID will be placed in this field.

[41]Daniel Butler, "Log File Analysis: The Ultimate Guide", accessed January 16, 2015, http://builtvisible.com/log-file-analysis/.

Web logs can thus replicate both network-level metadata and application-level metadata, as well as application data (e.g. URLs).

Obviously, the logs can also potentially be used to correlate a specific user's activity, if the IP addresses recorded in the logs are correlated between different web server logs, and the user is repeatedly using the same IP address.

Notably although in many cases application endpoints are talking client to server, like a web browser to a web server, there are also examples of 'middle boxes' that process communications in some way. These middle boxes can record application-level metadata (or in many cases all the data, though that is rare due to the storage requirements).

A web cache for example will see all client web requests, fetch the data from the intended server, and relay that data back to the client. An email relay (like smtp.soton.ac.uk) will take emails from email clients in an organisation, and send the email on to the destination site's email relay (usually where it is spam-processed and then forwarded on to the receiver's mail server for reading/collection). A firewall or an Identification Detection System (IDS) are both devices that sit usually at the edge of a site and inspect traffic to make decisions on whether it is allowed to pass, or whether there may be malicious content—the firewall usually looking only at network/transport layer data, the IDS usually doing full DPI to look for 'suspicious' content. Both can of course log everything they do as metadata for communications they are making decisions about.

From the foregoing it should appear that application-level metadata and corresponding service-use metadata can directly reveal what has been said or thought.[42] Besides, they can also directly reveal the identity of the speakers themselves (e.g. emails). As a result, the systematic retention of application-level metadata and/or service-use metadata can allow the creation of extensive user profiles. By comparison, strictly speaking, although network-level metadata can be used to single out individuals, they directly identify devices and not individuals, even if one device is used by one individual. As such IP addresses are not direct identifiers of individuals (in particular when IP addresses are shared, e.g. Network Address Translation (NAT) is a way to link many computers to one IP address and this can happen with unrelated computers).[43] Strictly speaking once again the identification of individuals takes place when two sets of data are combined together: network-level metadata and subscriber information.

---

[42]There is an argument that emails or user names could be considered as belonging to the same category as network-level metadata as they can be used to determine who speaks more than what is said. However, as emails addresses or user names are to be found in the payload and if used to determine who is speaking they can directly identify individuals, they could be deemed closer to content data.

[43]It is true however that IP addresses can then be combined with port numbers, protocols and eventually MAC addresses and thereby allow the reaching of specific devices, even when IP addresses are shared.

Now that it is clear that the term metadata can cover a variety of data, be it in transit or stored, it is necessary to confront the aforementioned technical definitions with the legal concepts. The distinction between network-level metadata, application-level metadata and service-use metadata is not usually found within legislative texts as such. UK law however is worth examining at this stage as it attempts to draw a clear line between data relating to electronic communications—i.e. 'communications data'—and content data by identifying three types of communications data: traffic data, service use information and subscriber information. Confronting the technical and legal definitions of metadata and its subcategories will enable us to make the case for clearer and above all narrower legal categories.

3 Communications Data Within the Meaning of UK Law

Under UK law, telecommunications operators can be required to retain 'relevant communications data', while law enforcement agencies and public authorities can get access under certain conditions to 'communications data' without a warrant. The general definition of communications data is to be found within the act regulating access to communications data by law enforcement agencies and public authorities, i.e. RIPA.

Under s. 21(4) of RIPA communications data comprises:

(a) any traffic data comprised in or attached to a communication (whether by the sender or otherwise) for the purposes of any postal service or telecommunication system by means of which it is being or may be transmitted;

(b) any information which includes none of the contents of a communication (apart from any information falling within paragraph (a)) and is about the use made by any person—

   (i) of any postal service or telecommunications service; or
   (ii) in connection with the provision to or use by any person of any telecommunications service, of any part of a telecommunication system;

(c) any information not falling within paragraph (a) or (b) that is held or obtained, in relation to persons to whom he provides the service, by a person providing a postal service or telecommunications service.

Communications data is thus divided up into three subcategories: traffic data (s. 21(4)(a)), service use information (s. 21(4)(b)) and subscriber information (s. 21(4)(c)).

3.1 Three Categories of Communications Data

Traffic data is the first subcategory of communications data and it means:

(a) any data identifying, or purporting to identify, any person, apparatus or location to or from which the communication is or may be transmitted,

(b) any data identifying or selecting, or purporting to identify or select, apparatus through which, or by means of which, the communication is or may be transmitted,

(c) any data comprising signals for the actuation of apparatus used for the purposes of a telecommunication system for effecting (in whole or in part) the transmission of any communication, and

(d) any data identifying the data or other data as data comprised in or attached to a particular communication, but that expression includes data identifying a computer file or computer program access to which is obtained, or which is run, by means of the communication to the extent only that the file or program is identified by reference to the apparatus in which it is stored.[44]

To delineate the breadth of the definition of traffic data it is helpful to have a look at the Code of practice for the acquisition and disclosure of communications data,[45] a set of guidelines produced by the Home Office on the procedures that should be followed when communications data is accessed or disclosed under RIPA. In the words of the drafters of the first version of the Code *"data comprised in or attached to a communication"* *"includes data which is found at the beginning of each packet in a packet switched network that indicates which communications data attaches to which communication"*.[46] In addition it is specified that *"data identifying a computer file or a computer program to which access has been obtained, or which has been run, by means of the communication"* means in relation to Internet communications data identifying *"a server or domain name (web site) but not a web page"*.[47] Full URLs are therefore excluded (i.e. only the part before the first slash can be inspected). This is confirmed by the new version of the Acquisition of Data Code of Practice.[48]

From a network flow perspective, traffic data could seem to correspond to network-level metadata as defined in Sect. 2.1. They are however broader in scope as they also cover application-level metadata. At the network level the metadata available is only the source and destination IP addresses and ports (and the use of TCP). By purely inspecting this data gathered from the network and transport layer, we can make no firm assumption of the specific 'domain name (web site)' and a fortiori 'web page' being accessed. It is possible that many different web servers are running on the same IP address, and indeed this practice is encouraged by ISPs due to the exhaustion of the IPv4 address space as of February 2011.[49]

---

[44]S. 21(6).

[45]Home Office, "Code of practice for the acquisition and disclosure of communications data", published on September 8, 2010, accessed October 10, 2014, https://www.gov.uk/government/publications/code-of-practice-for-the-acquisition-and-disclosure-of-communications-data (Acquisition of Data Code of Practice).

[46]Acquisition of data Code of practice, [2.19].

[47]Acquisition of data Code of practice, [2.20].

[48]Home Office, "Code of practice for the acquisition and disclosure of communications", December 9, 2014, accessed March 25, 2015, https://www.gov.uk/government/publications/code-of-practice-for-the-acquisition-and-disclosure-of-communications-data, [2.24–2.25] (New acquisition of data Code of practice).

[49]NRO, "Free Pool of IPv4 Address Space Depleted", February 3, 2011, accessed October 15, 2014, https://www.nro.net/news/ipv4-free-pool-depleted.

Fig. 3 An example of a specific file (image) being retrieved as part of an HTTP-based web browser request to a server

In order to determine the web site or domain name being accessed, or the specific page or file being requested, inspection of the packet payload is required. A fortiori the same holds true for the precise file being requested.

The screenshot in Fig. 3 illustrates an example of a specific file (image) being retrieved as part of an HTTP-based web browser request to a server. The screenshot shows full packet capture. The destination port alone (80) does not indicate the server or file being requested; that information is only available from inspecting the payload.

In addition, the drafters of the old and new version of the Acquisition of data Code of practice seem to confuse two distinct concepts: metadata in transit and service-use metadata. They state that traffic data comprises "*routing information identifying apparatus through which a communication is or has been transmitted (for example, dynamic IP address allocation, file transfer logs and e-mail headers—to the extent that content of a communication, such as the subject line of an e-mail, is not disclosed)*".[50] E-mail headers are elements of the payload and are usually retained not as traffic data but as data relating to the use of a service or 'service use information' (i.e. service-use metadata to use our terminology). It is likely that mail server logs at the sender and receiver's organisations will include such information, but for this information to be captured in transit across a network, the application payload must be inspected, i.e. DPI must be performed. Said otherwise, an Internet access provider which does not offer an email service would need to use DPI technology reaching the payload of packets to extract an e-mail header from traffic flowing through its network.

As aforementioned, communications data also comprise other categories of data and in particular subscriber information and service use information.

Subscriber information is defined as data "*in relation to persons to whom he provides the service, by a person providing a postal service or telecommunications service*"[51] to the exclusion of traffic data and service use information. Subscriber

---

[50]New acquisition of data Code of practice, [2.26].
[51]RIPA s. 21(4)(c).

information is defined by the Acquisition of data Code of practice as *"information held or obtained by a CSP [communications service providers] about persons to whom the CSP provides or has provided a communications service. Those persons will include people who are subscribers to a communications service without necessarily using that service and persons who use a communications service without necessarily subscribing to it"*.[52]

Service use information is more interesting for our purpose and is defined as data about the *"use made by a person ...of a telecommunications service or data relating to the provision of a telecommunication service"*[53] to the exclusion of traffic data and includes according to the Acquisition of data Code of practice among other things: itemised records of connections to internet services, itemised timing and duration of service usage, information about amounts of data downloaded and/or uploaded.[54] One should probably also add identifiers allocated to the subscriber of the service as well as when it comes to Internet-telephony and Internet-emails the identifier of the intended recipient of the communication.

Notably, when it comes to service use information there is no distinction in relation to the level of metadata and nothing expressly excludes from this category browsed information such as full URLs.

In the end, under RIPA traffic data covers both network-level metadata and application-level metadata to the exclusion of full URLs (which are in fact application data, i.e. content data), while service-use metadata is potentially even broader in scope. Assuming the distinction between content data and other data relating to electronic communications makes sense, it would however be more appropriate to distinguish between different subcategories of metadata in relation to the nature of the activity of online services providers and state that for example ISPs acting as Internet access providers can only be required to retain/supply network-level metadata, which do not require the implementation of DPI technologies for their capture.

Another concern is that of the breadth of the category of service use information. One of the questions is here whether the contents of an individual's Twitter and Facebook address lists and/or the messages placed on these social media sites should be considered as service use information and thereby communications data to which law enforcement agencies could have access without a warrant. As the category of service use information is defined limitatively and is therefore a closed list it would be very difficult to include these types of data in this category. The only category that would remain available and which is open-ended is that of subscriber information. But as these types of data can directly identify an individual and can directly reveal what is said or thought there is an argument to say that they should be considered as tantamount to content data.

---

[52]New acquisition of Data Code of practice, [2.21].
[53]RIPA s. 21(4)(b).
[54]New acquisition of data Code of practice, [2.29].

It should be clear at this stage why gaining access to communications data even simply network-level metadata (e.g. IP addresses) should imply requesting the access to telecommunications operators. This in itself constitutes a safeguard for telecommunications operators' subscribers, who should be able to assert their data subjects' rights against communications service providers.[55] Going further the tapping of these data without appropriate legal safeguards (i.e. outside any clearly defined procedure) should amount to a breach of Article 8 as recognised by the ECtHR in its case law starting with the Malone case and culminating with the Copland case.

3.2 Relevant Communications Data

The list of categories of data to be retained by telecommunications operators under UK law, i.e. 'relevant communications data', is to be found in the Schedule to the DRR, which is a servile copy of the Schedule to the Data Retention (EC Directive) Regulations 2009[56] literally transposing the defunct data retention Directive.[57] The list comprises:

- *"Data necessary to trace and identify the source of a communication"*, i.e. user ID and telephone number allocated to the communication, the name and address of the subscriber or registered user to whom an IP address, telephone number or user ID was allocated
- *"Data necessary to identify the destination of a communication"*, i.e. user ID, telephone number, name and address of the subscriber or registered user at the other end
- *"Data necessary to identify the data, time and duration of a communication"*, i.e. IP address, user ID and date and time of the log-in and log-off

---

[55]See e.g. New acquisition of data Code of Practice, [7.4–7.5]: *"There is no provision in RIPA preventing CSPs from informing individuals about whom they have been required by notice to disclose communications data in response to a Subject Access Request made under section 7 of the DPA. However a CSP may exercise certain exemptions to the right of subject access under Part IV of the DPA. Section 28 provides that data are always exempt from section 7 where such an exemption is required for the purposes of safeguarding national security".*

[56]2009 No. 859.

[57]Directive 2006/24/EC of the European Parliament and of the Council of 15 March 2006 on the retention of data generated or processed in connection with the provision of publicly available electronic communications services or of public communications networks and amending Directive 2002/58/EC OJ L 105, 13.04.2006, pp. 54–63. See for the declaration of invalidity by the Court of Justice of the European Union (CJEU) Joined cases C-393/12 and C-594/12 Digital Rights Ireland Ltd v Minister for Communications, Marin and Natural Resources et al. and Kärntner Landesregierung, Micheal Seitlinger, Christof Tschohl and others of 8 April 2014 (Digital Rights Ireland).

- "*Data necessary to identify the type of communication*", i.e. the internet service used
- "*Data necessary to identify users' communication equipment*", i.e. calling telephone number or DSL or other end point of the originator of the communication

Relevant communications data is thus a mix of traffic data, subscriber information and service use information, although it appears to be a closed list of data and in this sense is narrower than the category of communications data. Interestingly, the UK legislator recently enlarged the category of data to be retained by telecommunications operators by adding to that of relevant communications data the category of 'relevant Internet data', of which aim is to allow identifying devices behind IP addresses (and ultimately users) by combining IP addresses with other types of information such as port numbers, protocols and MAC (media access control) addresses.[58] It should be noted that it is not sure whether the legislative amendment could allow the systematic retention of list of browsed websites (to the exclusion of full URLs as explained below).[59]

Once again it would be better to distinguish between different subcategories of metadata in relation to the nature of the activity of online services providers and state, for example, that ISPs acting as Internet access providers cannot retain application-level metadata such as email addresses, email subject lines or even application data such as URLs, to make sure DPI technologies are not systematically used. The Counter-Terrorism and Security Act 2015 attempts to do so in part by excluding from the category of relevant Internet data URLs at least for Internet access providers. This said the formulation adopted by the legislator remains confusing, although it echoes s. 21(6) of RIPA. Section 21(3)(c) provides that data "*used to identify an internet communications service to which a communication is transmitted through an internet access service for the purpose of obtaining access to, or running, a computer file or computer program*" is excluded from the category of relevant Internet data as long as it is generated or processed by an Internet access provider.

---

[58]This was one the purposes of the Counter-Terrorism and Security Bill 2014-2015, (HC Bill 127), and in particular Part 3, available at http://services.parliament.uk/bills/2014-15/counterterrorismandsecurity/documents.html. See also the Explanatory Notes at http://www.publications.parliament.uk/pa/bills/cbill/2014-2015/0127/en/15127en.htm, which states at [50] that "*Part 3 enables the Secretary of State to require communications service providers (CSPs) to retain data that would allow relevant authorities to link a public internet protocol (IP) address to the person or device using it at any given time*". See also Home Office, "Internet protocol address resolution: the Addendum to the retention of communications data code of practice, Draft for public consultation", December 9, 2014, accessed December 17, 2014, https://www.gov.uk/government/uploads/system/uploads/attachment_data/file/383403/Draft_Data_Retention_Code_of_Practice_-_IP_resolution_addendum_-_for_pub....pdf. The Counter-Terrorism and Security Act 2015 was enacted on 12 February 2015, and its section 21 expands the category of relevant communications data by adding the category of relevant Internet data.

[59]One could argue that this should not happen as section 2 (section 1 supplementary) of DRIPA excludes data which "*may be used to identify an internet communications service to which a communication is transmitted through an internet access service for the purpose of obtaining access to, or running, a computer file or computer program*".

One significant change brought by DRIPA is the reformulation of the category of the debtors of data retention obligations and potential addressees of data acquisition requests. They now include providers of services that consist in or include *"facilitating the creation, management or storage of communications transmitted, or that may be transmitted, by means of such a system"*.[60] Reading the explanatory notes of the Bill it seems that the intention was to make sure that services such as webmail were included within the definition. In addition, the new Retention of Communications Data Code of Practice[61] explains what Internet email means under DRIPA: *"any text, voice, sound or image message sent over a public electronic communications network which can be stored in the network or in the recipient's terminal equipment until it is collected by the recipient and includes messages sent using a short message service"*.[62] Social networking websites could thus seem to be concerned, even though strictly speaking communications through the means of these platforms do not take the path of public telecommunication systems.[63]

If this means that data retention obligations are to be extended to many if not all over-the-top service providers such as social networking websites, there is a stronger argument to clearly delineate the category of service use information. This is all the more true that even if over-the-top service providers do not have an obligation to retain communications data, they do so at their own initiative for more or less legitimate reasons. And they retain a lot,[64] although their retention

---

[60]See s. 5. Prior to DRIPA, the majority view was that data retention obligations did not concern over-the-top service providers as the 2009 Regulations referred to the definition to be found in section 151 of the Communications Act 2003(1). See s. 2(e).

[61]Home Office, "Retention of communications data code of Practice", accessed March 25, 2015, https://www.gov.uk/government/publications/code-of-practice-for-the-acquisition-and-disclosure-of-communications-data (Retention of communications data code of Practice).

[62]Retention of communications data code of Practice, [2.21].

[63]See Chambers v DPP (2012) EWHC 2157 (Admin) which interpreting s. 127 of the Communications Act 2003 seems to have a broad understanding of public electronic communications networks. But see the restrictive view of the Experts Group, "The platform for electronic data retention for the investigation, detection and prosecution of serious crime" established by Commission Decision 2008/324/EC, DATRET/EXPGRP (2009) 2 FINAL—03 12 2009, http://ec.europa.eu/home-affairs/doc_centre/police/docs/position_paper_1_annexe_09_12_03_en.pdf. S. 2(1) of RIPA defines telecommunication system as "any system (including the apparatus comprised in it) which exists (whether wholly or partly in the United Kingdom or elsewhere) for the purpose of facilitating the transmission of communications by any means involving the use of electrical or electro-magnetic energy".

[64]See Geek.com, "Facebook stores up to 800 pages of personal data per user account", accessed January 10, 2015, http://www.geek.com/news/facebook-stores-up-to-800-pages-of-personal-data-per-user-account-1424807/; "LinkedIn Privacy Policy", accessed January 10, 2015, https://www.linkedin.com/legal/privacy-policy#info-collected; "Skype Privacy Policy", accessed January 10, 2015, http://www.skype.com/en/legal/privacy/#collectedInformation.

behaviour should be confined by data protection law and in particular by the data protection Directive as transposed in national law.[65]

Importantly s. 2 of DRIPA provides that relevant communications data is data generated or processed by telecommunications operators in the UK, just like the defunct data retention Directive[66] and its previous national transposition. This was meant to act as a restrictive condition. Indeed, in principle, the range of data processed by ISPs acting as Internet access providers is 'relatively' limited.

More precisely, strictly speaking the term 'generated' could be seen as misleading as an ISP generally just ships data (application traffic, largely) between its customers and its own backbone, and between its backbone and other ISPs. To do that an ISP routes and forwards IP packets, generally based on destination IP addresses. When doing this, the packets are in principle unaltered. Therefore while there is 'processing' of packets by routers (looking at destination IP, forwarding to next router based on the value), nothing is generated per se.

As a result the range of data processed by ISPs does not necessarily include all types of network-level metadata, i.e. all the 5-tuple. In principle ISPs do not need to process data beyond the network layer in order to route packets on the network. This would mean that they should not be asked to systematically retain the 5-tuple, as they do not need to process them.

Second, the range of the data processed does not include browsed information (i.e. the list of domain names and webpages visited) or the content of messages posted on social networking websites.

However, in practice for both traffic management and network security purposes, ISPs are now processing the 5-tuple.[67] Some Quality of Service (QoS) mechanism being applied to traffic (to give special treatment to certain application flows) relies on routers being configured to match traffic against the necessary 5-tuple to be given preferential treatment. Truly, it is also possible to apply Quality of Service, albeit in a less fine-grained fashion, by simply using one value in a field of the IP header only. The former method is generally referred to as Integrated Services, or IntServ, the latter as Differentiated Services, or DiffServ. When using the 5-tuple for QoS purposes, there should be no need to retain the 5-tuple information beyond the lifetime of the flows. Where network flow records (5-tuple, plus duration, packet volume, etc.) are kept for network monitoring purposes, ISPs may typically retain the records for some period of time, and may typically seek to search those records for analysis of reported security incidents,[68] for

---

[65]The e-privacy Directive should not be applicable since under Article 3 "[t]*his Directive shall apply to the processing of personal data in connection with the provision of publicly available electronic communications services in public communications networks*".

[66]Article 1.

[67]See Stalla-Bourdillon, Papadaki and Chown, "From Porn to Cybersecurity Passing by Copyright", 672.

[68]It might be argued that at this stage data is generated as the ISPs process the logs to produce derived data, unless one considers that data is generated each time it is logged.

example. ISPs are also using intrusive practices such as DPI[69] for the same purposes. Nevertheless, *"in practice the sheer volume of traffic passing through a DPI system may make it impractical to record all network data"*.[70] There is thus little retention in these cases. Does this mean however that ISPs have to retain all these data on the ground of data retention legislation? The answer should be negative as long as the list of data to be retained is conceived as a closed list.

In any case, irrespective of network security practices, as data retention obligations cover data generated *or processed* in the UK it is in fact unclear whether that condition is really restrictive given the very broad definition of processing.[71] This could maybe justify more or less implicit 'invitations' sent to ISPs to implement DPI technologies to collect what has been coined 'third party data'. Yet this was the very intention of the Communications Data Bill of 2012.[72] It is unclear whether DRIPA really aims at avoiding the implementation of DPI practices by ISPs.[73] The sole exclusion is that ISPs acting as Internet access providers cannot be required to retain URLs.

At the end of the day, the legal safeguards designed to frame the collection, retention and transfer of metadata should vary in relation to the nature of the activity of the online service providers meant to retain these data. Not only do the collection, retention and transfer of metadata in transit deserve scrutiny but also the collection, retention and transfer of stored metadata, i.e. service-use metadata, should be taken into account since even if DPI practices are banned, server logs remain potentially available. To note, the e-privacy Directive only covers the grounds for processing traffic data (and traffic data seems to equate network-level metadata[74]) on the part of ISPs. One would thus have to go back to the data protection Directive for the rest, unless it means that *a contrario* ISPs cannot process application-level metadata without consent.

---

[69]Ibid. 675.

[70]Ibid. 671.

[71]See e.g. Court of Justice of the European Union (CJEU), Case C-101/01, Bodil Lindqvist, 6 November 2003, ECLI:EU:C:2003:596 at [25] (*"According to the definition in Article 2(b) of Directive 95/46, the term processing of such data used in Article 3(1) covers any operation or set of operations which is performed upon personal data, whether or not by automatic means..."*).

[72]Cm 8359. The solution the Government proposed was to agree with the UK telecommunications operators to place data probes on their networks to collect the required communications data as it traversed to the end user. The probes would be programmed to generate information from network links within the communication service provider's network, while deep packet inspection would be used to isolate key pieces of information from data packets in a communication service provider's network traffic.

[73]But see s. 2(1).

[74]As it is defined as *"any data processed for the purpose of the conveyance of a communication on an electronic communications network or for the billing thereof"* under Article 2(b), although the adjunct *"or the billing thereof"* is problematic.

4 Conclusion

It has already been demonstrated that the tapping of both content data and communications data by agencies dedicated to intelligence and information gathering such as the UK Communications Government Headquarters without a warrant or with a s. 8(4)[75] warrant is problematic if not unlawful.[76] What seemed to be less clear after the Snowden revelations is which categories of data shall actually be retained by ISPs and over-the-top services and to which categories of data law enforcement agencies can have access to (by requesting the data to service providers), at least in Europe including in the UK, and in particular without prior judicial approval.[77] Ultimately, no clear answer has been given to the question whether and to what extent metadata should be treated akin to content data. This is acknowledged by the recent report of the UK Interception of Communications Commissioner's Office.[78]

To tackle these questions, the first step is to have a technical understanding of what metadata is and to what extent this is a uniform category. The second step is to compare technical definitions with legal concepts. Principally, there are two classes of metadata in transit: one is network-level metadata, typified by network-flow data, and the other is application-level metadata, which invariably requires DPI to capture. Although the UK legislation (RIPA and DRIPA) does not seem to draw a distinction between the two, we have explained why it would make sense to distinguish between them to prevent the normalisation of DPI practices. In this sense some of the recent findings of the Intelligence and Security Committee of Parliament in relation to the definition of communications data should be welcome and in particular the need to distinguish between 'communications data' and 'communications data plus' comprising in particular browsed websites.[79]

---

[75]Of RIPA.

[76]See e.g. Stratford QC and Johnston, "The Snowden 'Revelations"; Sophie Stalla-Bourdillon, "Privacy vs Security... Are we done Yet?", in *Privacy vs security*, ed. Sophie Stalla-Bourdillon, Joshua Phillips and Mark D. Ryan (London: Springer, 2014), 1–90. Compare with Liberty et al. v CGHQ et al. (2014) UKIPTrib 13_77-H and Liberty et al. v The Secretary of State for Foreign and Commonwealth Affairs et al. (2015) UKIPTrib 13 77-H.

[77]Note that judicial approval is not always considered to be an appropriate safeguard. See Interception of Communications Commissioner's Office, "Evidence for the Investigatory Power Review", December 5, 2014, p. 35 (ICCO's Report), accessed January 21, 2015, http://www.iocco-uk.info/docs/IOCCO%20Evidence%20for%20the%20Investigatory%20Powers%20Review.pdf.

[78]Ibid. 18–19.

[79]See Intelligence and Security Committee of Parliament, "Privacy and Security: A modern and transparent legal framework", Presented to Parliament pursuant to section 3 of the Justice and Security Act 2013, Ordered by the House of Commons to be printed on 12 March 2015, accessed March 25, 2015, http://isc.independent.gov.uk/committee-reports/special-reports at [143] ("*'Communications Data Plus'—this goes further than the basic 'who, when and where' of CD. So, for example, this would encompass details of web domains visited or the locational tracking information in a smartphone*").

Table 1 A breakdown of metadata

Types of data	Does their collection require the implementation of deep packet technologies?	Can they directly identify individuals?	Can they directly reveal sensitive information?	Can they single out individuals and can their systematic collection be deemed as amounting to profiling?
Network-level metadata Ex: Source IP, Destination IP, Total length, Source port, Destination port, Protocol, Sequence number	No, the packet payload is not inspected	No, they need to be combined with subscriber information	No, but they can give some hints	If only network-level metadata is collected, they can single out individuals but it is arguable whether the collection of these data *alone* could be deemed as amounting to profiling as they only give hints, e.g. as to what type of application is being used
Application-level metadata Ex: application programme version, email address sender/receiver, email subject line, domain name of website visited (a URL is akin to the content of an email)	Yes, the packet payload is inspected	Yes, they can	Yes, they can	Yes, they can
Content data or application data Ex: email text, URL, website content, chat message, video content, image content, etc.	Yes, the packet payload is inspected	Yes, they can	Yes, they can	Yes, they can
Service-use metadata Ex: they can mirror network-level metadata, application-level metadata (and even content data)	No	If they mirror network-level metadata, no If they mirror application-level metadata, yes	If they mirror network-level metadata, no If they mirror application-level metadata, yes	Yes, they can

A third category of metadata can be added to the list: stored service-use metadata. Clearly isolating it from metadata in transit would allow acknowledging the importance of regulating the data retention practices of over-the-top service providers. This is all the more crucial if data retention obligations are extended to over-the-top service providers and if extra-territorial legislation is now being adopted to reach foreign online service providers.

Table 1 breaks down the different types of metadata and classifies them in relation to the four key criteria that emerge from our analysis.

From the foregoing it results that there is an argument that application-level metadata should be protected in the same way as the content of communications. At the very least the systematic retention of application-level metadata should be prohibited as it would allow the creation of extensive profiles about individuals including browsed information such as websites and emails' subject lines. A similar case could be built for corresponding service-use metadata as they can directly identify individuals as well as reveal sensitive information.

What the foregoing also shows is that there should be an argument for making sure appropriate safeguards are in place when network-level metadata is combined with subscriber information.

To conclude, using appropriate terminology is always better to understand what is really at stake.[80] Only then will it be possible to appropriately adapt or modernise the current regulatory framework. In this line, future work should go at least in two directions. First, work should be done to clarify that ISPs are not required to use DPI technologies to process and subsequently retain application-level metadata. A domain-name should not be considered as traffic data within the meaning of RIPA. Second, measures should be adopted to effectively minimise the amount of service-use metadata retained by Internet access providers as well as over-the-top service providers.

Acknowledgments We would like to thank the anonymous reviewers.

Bibliography

Brown, Neil. 2014. An assessment of the proportionality of regulation of 'over the top' communications services under Europe's common regulatory framework for electronic communications networks and services. *Computer Law & Security Review* 30: 357–374.

Butler, Daniel. 2015. Log file analysis: the ultimate guide. http://builtvisible.com/log-file-analysis/. Accessed 16 Jan 2015.

Bygrave, Lee A. 2015. Information concepts in law: generic dreams and definitional daylight. *Oxford Journal of Legal Studies* 35: 91–120.

---

[80]Such a finding thus echoes Lee A. Bygrave's claim. This author argues that legal definitions of basic but crucial information concepts on which the entire regulatory edifice is based are often too poorly understood and thereby interpreted. See Lee A. Bygrave, "Information Concepts in Law: Generic Dreams and Definitional Daylight", *Oxford Journal of Legal Studies* 35 (2015): 91–120.

Chirgwin, Richard. Data retention: encryption won't protect you much. *The Register*, 28 April 2014. http://www.theregister.co.uk/2014/04/28/data_retention_encryption_wont_protect_you/. Accessed 12 Oct 2014.

Cooper, Alison. 2011. Doing the DPI dance: assessing the privacy impact of deep packet inspection. In *Privacy in America: interdisciplinary perspectives*, ed. William Aspray, and Phillip Doty, 139–166. Maryland: Scarecrow Press.

Facebook. 2013. Secure browsing by default. 31 July 2013. https://www.facebook.com/notes/facebook-engineering/secure-browsing-by-default/10151590414803920. Accessed 18 Oct 2014.

Firewall.cx. 2014. Understanding VPN IPSec tunnel mode and IPSec transport mode—What's the difference? http://www.firewall.cx/networking-topics/protocols/870-ipsec-modes.html. Accessed 12 Oct 2014.

Fisher, Denis. 2012. Final report on DigiNotar hack shows total compromise of CA servers. 31 Oct 2012. http://threatpost.com/final-report-diginotar-hack-shows-total-compromise-ca-servers-103112/77170#sthash.zgQpsWlP.dpuf. Accessed 18 Oct 2014.

Fuchs, Christian. 2013. Societal and ideological impacts of deep packet inspection (DPI) internet surveillance for society. *Information Communication & Society* 16(8): 1328–1359.

Geek.com. 2015. Facebook stores up to 800 pages of personal data per user account. http://www.geek.com/news/facebook-stores-up-to-800-pages-of-personal-data-per-user-account-1424807/. Accessed 10 Jan 2015.

IANA. 2014. Service name and transport protocol port number registry. http://www.iana.org/assignments/service-names-port-numbers/service-names-port-numbers.xhtml. Accessed 18 Oct 2014.

IEFT. 2014. Security architecture for the internet protocol, Nov 1998. https://www.ietf.org/rfc/rfc2401.txt. Accessed 15 Oct 2014.

Intelligence and Security Committee of Parliament. 2015. Privacy and security: a modern and transparent legal framework. Presented to Parliament pursuant to Section 3 of the Justice and Security Act 2013, Ordered by the House of Commons to be printed on 12 March 2015. http://isc.independent.gov.uk/committee-reports/special-reports. Accessed 25 March 2015.

Interception of Communications Commissioner's Office. 2015. Evidence for the investigatory power review, 5 Dec 5, 2014. http://www.iocco-uk.info/docs/IOCCO%20Evidence%20for%20the%20Investigatory%20Powers%20Review.pdf. Accessed 21 Jan 2015.

Kessler, Gary. 2014. An overview of TCP/IP protocols and the internet. http://www.garykessler.net/library/tcpip.html. Accessed 12 Oct 2014.

LinkedIn. 2015. LinkedIn privacy policy. https://www.linkedin.com/legal/privacy-policy#info-collected. Accessed 10 Jan 2015.

Mason, Andrew. 2001. *CISCO secure virtual private networks*. Indianapolis: Cisco Press.

NRO. 2014. Free pool of IPv4 address space depleted, 3 Feb 2011. https://www.nro.net/news/ipv4-free-pool-depleted. Accessed 15 Oct 2014.

Palmer, Chris and Yan Zhu. 2013. How to deploy HTTPS correctly. EFF, 12 Dec 2013. https://www.eff.org/https-everywhere/deploying-https. Accessed 16 Oct 2014.

Qosmos. 2012. DPI and metadata for cybersecurity applications. White paper, Jan 2012. http://www.accumuli.com/pages/files/datasheets/DPI-and-Metadata-for-Cybersecurity-Applications_Qosmos.pdf. Accessed 10 Oct 2014.

Sander, Chris. 2010. Understanding man-in-the-middle attacks—Part 4: SSL Hijacking, 2 June 2010. http://www.windowsecurity.com/articles-tutorials/authentication_and_encryption/Understanding-Man-in-the-Middle-Attacks-ARP-Part4.html. Accessed October 15, 2014.

Skype. 2015. Skype privacy policy. http://www.skype.com/en/legal/privacy/#collectedInformation. Accessed 10 Jan 2015.

Stalla-Bourdillon, Sophie. 2014. Privacy vs security… Are we done yet? In *Privacy vs security*, ed. Sophie Stalla-Bourdillon, Joshua Phillips, and Mark D. Ryan, 1–90. London: Springer.

Stalla-Bourdillon, Sophie, Evangelia Papadaki, and Tim Chown. 2014. From porn to cybersecurity passing by copyright: how mass surveillance technologies are gaining legitimacy… The case of deep packet inspection technologies. *Computer Law & Security Review* 30: 670–686.

Stratford QC, Jemima and Tim Johnston. 2014. The Snowden 'Revelations': Is GCHQ breaking the law? *E.H.R.L.R.* 2: 129–141.

Tanenbaum, Andrew S., and David J. Wetherall. 2010. *Computer networks*, 5th ed. Upper Saddle River: Pearson.

Tyson, Jeff. 2014. How encryption works. http://computer.howstuffworks.com/encryption4.htm. Accessed 10 Oct 2014.

Unuth, Nadeem. 2014. What is a data packet? http://voip.about.com/od/glossary/g/PacketDef.htm. Accessed 5 Oct 2014.

Global Views on Internet Jurisdiction and Trans-border Access

Cristos Velasco, Julia Hörnle and Anna-Maria Osula

Abstract This paper offers insights and perspectives on the jurisdiction of law enforcement authorities (LEAs) under international law and reviews current approaches to the territoriality principle and trans-border access to data for LEAs to conduct criminal investigations; controversial topics that are currently in the center of discussions, both at the international and national level. The views and perspectives offered in this paper seek to contribute to the international debate on cross-border access to data by LEAs and how the principles on internet jurisdiction should evolve in order to turn the administration of the criminal justice system more efficient, dynamic and compliant with the needs to obtain and secure evidence while respecting data protection safeguards.

Keywords Internet jurisdiction · Cross-border access · Extraterritoriality · Mutual legal assistance · International law · Data protection

---

Anna-Maria Osula is a Researcher at NATO CCD COE Law & Policy Branch. This contribution contains the opinion of the respective author only, and does not necessarily reflect the policy or the opinion of NATO CCD COE, NATO or any agency or any government.

C. Velasco (✉)
Protección Datos México (ProtDataMx), Ciberdelincuencia.Org, Francia no. 38 Casa 12, Col. Florida, C.P., 01030 Mexico, D.F., Mexico
e-mail: cristosv@protecciondatos.mx
URL: http://protecciondatos.mx; http://ciberdelincuencia.org

J. Hörnle
The Queen Mary School of Law, CCLS, Queen Mary University of London, 67-69 Lincoln's Inn Fields, London WC2A 3JB, UK
e-mail: j.hornle@qmul.ac.uk

A.-M. Osula
Tina 21-14, 10126 Tallinn, Estonia
e-mail: anna-maria.osula@ccdcoe.org

1 Introduction

'*Jurisdiction*' has different meanings depending on the context that the term is used, whether in the context of international law, private international law or criminal law and also depending on the legal system and tradition of a country. The scope of jurisdiction may vary widely from one state to another, however the term '*jurisdiction*' usually includes two main aspects. The first aspect is connected to state sovereignty and designates the power of a state and its agents over the territory, country, region, state or province. The second aspect concerns the exercise of authority and powers of a national court or judicial authority to apply and execute national procedural laws that are within their sphere of competence in order to attract and investigate a particular case based on existing principles, legislation and precedents or jurisprudence in a certain area of law.[1]

Internet jurisdiction has been one of the most controversial areas of Internet Governance[2] fundamentally because there is no '*one size fit all approach*' for each state to resolve the cross-border problems of the inherent in the use of ICTs and the internet. Internet jurisdiction intersects with different areas of law and a number of national courts around the world have issued landmark judgments and jurisprudence in order to resolve legal issues regulating the activities of companies with internet presence or individuals located in different territories that have experienced damage or loss of property or assets as result of their interaction and use of internet.[3]

It is well known that the internet is borderless and it has no geographic boundaries. However, laws and policies are still mostly subject to the territory and scope of the national boundaries of each state and the judgments issued by national courts usually—unless under very specific exemptions and circumstances—have no extra-territorial effects in other countries as further discussed in this paper. This is one of the main reasons why there is a wide number of legal approaches regarding the application of national laws to conduct in cyberspace.[4]

---

[1]Velasco, Cristos. *La Jurisdicción y Competencia sobre Delitos Cometidos a través de Sistemas de Cómputo e Internet* (Tirant lo Blanch 2012), 207–209. For a perspective on the classification and types of jurisdiction under public international law, see pp. 209–215.

[2]For a perspective of internet Jurisdiction in the context of internet Governance, see: Kurbalija, Jovan. *Internet Governance* (Diplo Foundation 2014), Section 3 Jurisdiction, pp. 92–96.

[3]For instance in the area of internet content and freedom of speech, see: *Yahoo! Inc. v. La Ligue Contre Le Racisme et l'antisemitisme (LICRA)* 433 F.3d 1199 (9th Cir. 2006) http://law.justia.com/cases/federal/appellate-courts/F3/433/1199/546158/ and in the area of internet defamation *Down Jones & Company Inc v. Gutnick, Joseph* [2002] HCA 56 10 December 2002, full text of the High Court of Australia available at: http://www.austlii.edu.au/au/cases/cth/HCA/2002/56.html.

[4]For national perspectives on cybercrime jurisdiction, see: Bert-Jaap Koops and Susan W. Brenner. *Cybercrime and Jurisdiction. A Global Survey* (Asser Press 2006). For a perspective on cyberspace jurisdiction under public international law, see: Henrik Spang-Hansen. *Cyberspace and International Law on Jurisdiction* (DJOF Publishing 2004).

For instance, one of the main problems in the area of criminal law is to know the exact place and time where the crime was perpetrated and the location of the party or parties involved in the commission of such crime; a situation that is by all means uncertain on the internet, precisely because of the ubiquity of that medium, the difficulty to collect and secure electronic evidence by law enforcement authorities (hereinafter LEAs), as well as the availability of technologies and means used by perpetrators to conceal their identity,[5] situations which make it extremely difficult for LEAs to know the exact geographical location of perpetrators to launch a particular criminal investigation.

This paper provides views and perspectives on some of the jurisdictional challenges discussed during the panel on "*Internet Jurisdiction and Law Enforcement*" at the Computers, Privacy and Data Protection 2015 Conference in Brussels.[6]

The views presented in this paper are mainly academic and do not neither represent consensus on the subject matter nor official views, opinions or policies of the institutions, organizations affiliated with each of the authors.

2 The Principle of Territoriality and Trans-border Access

The territoriality principle is a fundamental principle of international law and effectively limits LEAs' powers to act within the territory of their state.[7] However it is argued here that it is not entirely clear what the territoriality principle means in the modern internet connected world.[8] An example where the legal boundaries of the principle of territoriality become especially blurred is when LEAs need for investigation purposes access data that is located extraterritorially.

All measures used and employed by LEAs to access data extraterritorially must be in accordance with the legal limits as set in both national and international law.[9] Notably, according to the established principle of jurisdiction to enforce, also known as the Lotus principle, established by the International Court of Justice (ICJ), states are prohibited to "*exercise its power in any form in the territory of another state*" unless there are specific grounds to do so deriving from international custom or agreements.[10] This may include, for example, the general prohibition of conducting an investigation on the territory of another state. Failure to do

---

[5]See for instance the Tor network https://www.torproject.org/.
[6]The video of this session is available at: https://www.youtube.com/watch?v=NL4nNlzyqmQ&feature=youtu.be.
[7]Malcolm Shaw *International Law* (5th edition Cambridge University Press 2003) 579–584.
[8]Uta Kohl *Jurisdiction and the Internet* (Cambridge University Press 2007) 96–102.
[9]See Article 15 of the *Convention on Cybercrime*.
[10]S.S. Lotus, Fr. v. Turk., 1927 P.C.I.J. (ser. A) No. 10, at 4 (Decision No. 9), 45 (Permanent Court of International Justice 1927).

so may be considered as a breach of the sovereignty of the other state, and may lead to undesired escalation of retaliation activities.[11]

This concurs with the fundamental presumption against the extraterritorial expansion of enforcement powers based on national, domestic law. The consequence of the territoriality principle has been that a state who required intelligence or evidence stored abroad in the context of criminal investigations or prosecutions would have to use recognized international co-operation procedures, such as *letters rogatory* or Mutual Legal Assistance (MLA), the latter of which is based on bi-lateral or multi-lateral treaties.[12]

As an indication of states attempting to keep up with the territorial limits of jurisdiction to enforce, it has been reported that 70 % of the cases where there is a need to access evidence located extraterritorially, mutual legal assistance mechanisms have been used.[13] At the same time, recent studies have concluded that the format and procedures involved in mutual legal assistance treaties are not suitable for the volatile nature of digital evidence.[14] There are many reasons for this. MLA is usually considered slow and bureaucratic as it depends on the workings of diplomatic channels and is frequently hampered by political considerations and the principle of reciprocity.[15] Oftentimes the mutual legal assistance does not contain the required clauses to be considered valid or the lack of mutual legal agreements entered and ratified among the countries involved.[16]

It has therefore been argued that traditional MLA does not fit for the internet age, where cybercrime crosses borders on a massive scale and cloud computing[17] means that data is stored and controlled remotely. Thus, the internet age causes

---

[11]For a more detailed analysis of the Lotus Case, see: Paul de Hert, "Cybercrime and Jurisdiction in Belgium and the Netherlands. Lotus in Cyberspace-Whose Sovereignty is at Stake?" in *Cybercrime Jurisdiction. A Global Survey*. Edited by Bert-Jaap Koops and Susan W. Brenner, pp. 97–98.

[12]Susan Brenner *Cybercrime and the Law* (North Eastern University Press 2012) 171–188.

[13]UNODC, *Comprehensive Study on Cybercrime*, February 2013. <http://www.unodc.org/documents/organizedcrime/UNODC_CCPCJ_EG.4_2013/CYBERCRIME_STUDY_210213.pdf>.

[14]Council of Europe Cybercrime Convention Committee (T-CY), *The mutual legal assistance provisions of the Budapest Convention on Cybercrime*. Adopted by the T-CY at its 12th Plenary (2–3 December 2014) e.g. p 123. <http://www.coe.int/t/dghl/cooperation/economiccrime/Source/Cybercrime/TCY/2014/T-CY(2013)17_Assess_report_v50adopted.pdf>.

[15]For further views on Mutual Legal Assistance and cooperation provisions in international and regional cybercrime instruments, see UNODC, *Comprehensive Study on Cybercrime*, Op. cit. 13. pp. 197–208.

[16]For a comprehensive overview, see ibid.

[17]For views on cloud computing and cybercrime jurisdiction see: Cristos Velasco. *Jurisdictional Aspects of Cloud Computing* (Paper presented at the Octopus 2009 Conference on Cooperation against Cybercrime of the Council of Europe February 2009) available at http://www.coe.int/t/dghl/cooperation/economiccrime/cybercrime/Documents/Reports-Presentations/2079%20if09%20pres%20cristos%20cloud.pdf and Council of Europe. *Cloud Computing and cybercrime investigations: Territoriality vs. the power of disposal?* (Council of Europe 31 August 2010), available at: http://www.coe.int/t/DGHL/cooperation/economiccrime/cybercrime/Documents/Internationalcooperation/2079_Cloud_Computing_power_disposal_31Aug10a.pdf.

massive challenges for law enforcement. LEAs exercise coercive powers domestically to force the disclosure of communications data and/or the simultaneous interception of data in transit, both in respect of content data and meta-data[18] but the extraterritorial application of the same powers may become problematic.

In addition to MLA treaties (that may sometimes cover regions such as the European Union), informal cooperation with the foreign LEAs, or using the 24/7 points of contact networks may also be a way for obtaining relevant data.[19] However, there are two approaches to accessing extraterritorially located data that have recently been most actively discussed: first, access to data based on specific agreements such as the Council of Europe Convention of Cybercrime facilitating the cooperation between its Parties[20]; and second, obtaining data through contacting the Service Provider. Latest developments regarding these two options will briefly be commented below.

2.1 Council of Europe Convention of Cybercrime

Despite current ongoing proposals seeking to amend mutual legal assistance treaties to better satisfy the needs of modern cyber crime investigations[21] and coordination and cooperation between regional judicial and police enforcement bodies like EuroJust and EuroPol,[22] countries are seeking alternative approaches. For example, an explicit need to explore other options besides traditional MLA occurs in situations where it simply is not possible to identify the location of the data, like when the perpetrator makes use of anonymising or techniques to conceal their identity or data storage service features offered by cloud service providers, which may include storing data simultaneously in several databases, or distributed storage platforms worldwide.[23]

Transborder access has been the subject of analysis of an ad hoc working group of the Cybercrime Convention Committee (TC-Y) of the Council of Europe since 2001. Perhaps the most well-known example of the "exception of the traditional territoriality principle"[24] is the Council of Europe's Convention of Cybercrime

---

[18]This paper will not discuss the details of domestic powers and different categories of communications data.

[19]For a good overview, see Ian Walden, *Accessing Data in the Cloud: The Long Arm of the Law Enforcement Agent* (Rochester, NY: Social Science Research Network, 14 November 2011) <http://papers.ssrn.com/abstract=1781067>.

[20]The Council of Europe Convention on Cybercrime (CETS 185) is available at: http://www.coe.int/en/web/conventions/full-list/-/conventions/rms/0900001680081561

[21]Ibid. pp 128–134.

[22]See Joint Investigative Teams, EUROPOL, <https://www.europol.europa.eu/content/page/joint-investigation-teams-989>.

[23]*Supra* Footnote 29, p. 9.

[24]Cybercrime Convention Committee (T-CY) Ad hoc Subgroup on Transborder Access and Jurisdiction Council of Europe, 'T-CY Guidance Note #3: Transborder Access to Data (Article 32)' (December 2013) <http://www.coe.int/t/dghl/cooperation/economiccrime/Source/Cybercrime/TCY/2014/T-CY(2013)7REV_GN3_transborder_V12adopted.pdf>.

that includes a separate article on "Transborder access to stored computer data with consent or where publicly available" (Article 32). Article 32 of the Council of Europe Convention on Cybercrime allows the access to data located extraterritorially without the authorisation of another Party if it is publicly available (open source) or if the data is located in the territory of another Party and the Party seeking to obtain access to such data obtains "lawful and voluntary consent of the person who has the lawful authority to disclose the data to the Party through that computer system." Old debates over the scope and exact meaning of said provision of the Convention on Cybercrime have led to explore a proposal for an Additional Protocol to further outline options for accessing data extraterritorially.[25] However, the limits of the scope of interpretation of Article 32 have not fully been agreed upon among the states that ratified the Budapest Convention, and the prospect of the adoption of an Additional Protocol has been halted due to the lack of consensus among state governments and other relevant stakeholders.[26]

There is ambiguity and legal uncertainty regarding the extraterritorial powers of law enforcement authorities to access data remotely in other countries in order to collect and secure evidence for purpose of criminal investigations. The scope of interpretation and the application limits of Article 32 of the Convention are very broad and countries have implemented this provision very different on a practical basis. The lack of consensus to create an Additional Protocol to the Budapest Convention will not prevent the regulation of transborder access to data in other countries. However, we strongly believe that the traditional concept of jurisdiction to enforce and its territorial application should evolve and be transformed by the states through the adoption of modern legal frameworks and transborder access to data practices that offer both, certainty and transparency for the states involved in cross-border investigations.

2.2 Direct Communication with Service Providers

Given the relative inflexibility of the MLA mechanisms as well as the ambiguity deriving from the interpretation of the Convention of Cybercrime, states are looking for alternatives. The pressing question today is whether LEAs should have

---

[25]Council or Europe, *(Draft) elements of an Additional Protocol to the Budapest Convention on Cybercrime regarding transborder access to data*. Proposal prepared by the Ad hoc Subgroup on Transborder Access (9 April 2013) <http://www.coe.int/t/dghl/cooperation/economiccrime/Source/Cybercrime/TCY/TCY%202013/T-CY%282013%2914 transb_elements_protocol_V2.pdf>.

[26]Council of Europe Cybercrime Convention Committee (T-CY), *Transborder access to data and jurisdiction: Options for further action by the T-CY. Report prepared by the Ad hoc Subgroup on Transborder Access on Jurisdiction*. Adopted by the 12th Plenary of the TC-Y (2–3 December 2014) <http://www.coe.int/t/dghl/cooperation/economiccrime/Source/Cybercrime/TCY/2014/T-CY(2014)16_TBGroupReport_v17adopted.pdf>.

the power to request communications data directly (i) from foreign service providers (i.e. those established or headquartered in a foreign country) or (ii) from local service providers (i.e. those established on domestic soil), where the data is physically stored remotely on a foreign server.

The former scenario arose when the Belgian Public Prosecutor requested Yahoo Inc. in 2008 to disclose subscriber data in relation to Yahoo email accounts supposedly used to commit and execute computer fraud and forgery affecting local residents located in Belgium.

The latter scenario arose in 2014, when Microsoft was ordered by a US court warrant to disclose content data physically stored on a data center located in Ireland operated by a wholly owned subsidiary of Microsoft.

Such powers to request communications data would be governed by the LEA's domestic law. For some states the exercise of these powers is restricted by statute for others it is not. This immediately raises serious concerns about the data subject's privacy, as privacy protections for communications data vary enormously between countries.

Therefore it is important to consider what the territoriality principle means for networked computing (in particular cloud computing) and law enforcement. The question to be answered is which is the most appropriate link to a territory to determine jurisdiction and it is argued here that there are four basic possibilities (with additional variations), which are illustrated in the following chart:

| Data subject's Country A | Territory of the person whose data is requested | Based on:
– location when sending/transmitting
– location when data is sought
– domicile
– nationality | Arguably accords with data subject's expectations of protection, but would restrict LEAs activities to local data subjects, which is an extremely narrow interpretation of the territoriality principle—may mean that privacy-friendly jurisdictions become a haven for cybercriminals, encouraging cross-border cybercrime providing victims with no protection | May be same as Country C or at least the same region as Country C; in a cloud computing environment data is frequently stored as local to users as possible in order to deal with data latency issues |

(continued)

(continued)

Service provider(s) Country B	Territory of establishment of the person/entity effectively controlling the data (being able to access and disclose)	May be more than one in different countries; in a cloud computing environment there may be a chain of different service providers	The data controller has de facto ability to control the relevant data; data subject may have a contractual relationship with that service provider (not always, e.g. sender of email stored by recipient's service provider); data controller may have to comply with data subject's law (e.g. EU Data Protection Directive 1995/46/EC)	May provide some protection for data subject, but in many scenarios it will not; service provider may be in any jurisdiction as services are provided remote; data subject/user may not be aware of location of service provider
		May be none, if the data is encrypted by user		
Data storage Country C	Territory of the location where the data is physically stored	May be single location; but also possibility that data spread over several server locations; sometimes impossible to determine even by the service provider	This is the traditionally recognized territorial link, since in the offline world control was at the place of storage. Some element of control still exists at the physical location of a server (to put it drastically: that country could move in with bulldozers), but not necessarily direct access to data and disclosure	Likely to be close to the data subject and hence likely to provider better protection to the data subject than Country B
Law enforcement, authority Country D	Territory of the state investigating and prosecuting a crime; gathering intelligence	LEA proceeds under local domestic law; may be politically motivated, discriminatory or in breach of internationally recognized human rights standards	This has never been a recognised principle of territoriality and is the classic example of extraterritorial application of the law	Insufficient protection for data subject (the LEAs domestic protections, if any, may or may not apply to the data subject); most efficient for law enforcement purposes and prevention of cross-border cybercrime

Having presented the different *theoretical* possibilities for law enforcement jurisdiction and potential connection factors to territory, the next section will briefly describe three case examples where either the courts have sanctioned direct law enforcement access to foreign communications data or the national legislation provides for direct law enforcement requests to foreign service providers.

The *Belgian Yahoo case* has been widely discussed and criticized.[27] This case concerned a criminal prosecution of fraud committed through the use of Yahoo email accounts. The Public Prosecutor requested subscriber information from Yahoo under Art 46 bis of the Belgian Code of Criminal Procedure to identify the perpetrators of the fraud. Yahoo refused to comply with the request, arguing that the request must be served by US authorities under the Electronic Communications Privacy Act (ECPA).[28]

At the time, Yahoo did not have an office or establishment in Belgium. At first instance, the Dendermonde Court ordered Yahoo to disclose the information requested in 2009 and resolved to levy a fine of EUR 55,000 and a EUR 10,000 penalty for each day of non-compliance. The Belgian Court found jurisdiction on the basis of commercial presence: Yahoo was commercially present in Belgium through the provision of internet services to persons located in Belgian territory. Yahoo appealed the case and after long and complex appealing proceedings, the Belgian Supreme Court[29] found on 4 September 2012 that the direct order requesting subscriber information sent by the Belgian Public Prosecutor had been validly made to Yahoo (upholding the original decision of 2009).[30]

In the final judgment of the Court of Appeals of Antwerp of November 20, 2013, the justices confirmed the opinion of the Court of First Instance of Dendermonde and found: (i) that Yahoo had a territorial presence in Belgium, (ii) that Yahoo is and should be considered a provider of electronic communications services within the meaning of Article 46 bis of the Code of Criminal Procedure, and therefore, (iii) that Yahoo should collaborate with investigative authorities in the facilitation of the information requested and (iv) levied a penalty of 44,000 euros against the company.[31]

---

[27]See for example http://whoswholegal.com/news/features/article/30840/the-yahoo-case-end-international-legal-assistance-criminal-matters and P. de Hert, M. Kopcheva, "International Mutual Legal Assistance in Criminal Law Made Redundnant" (2011) 27 *Computer Law & Security Review* 291–297.

[28]For a synthesis of the scope of *ECPA*, see the website of the United states Department of Justice, Office of Justice Programs, available at: https://it.ojp.gov/default.aspx?area=privacy&page=1285.

[29]Supreme Court, September 4th, 2012, A.R. P.11.1906.N/2.

[30]There was also an issue as to whether *Yahoo* was an electronic communication service provider, but this is not relevant for purposes of this paper.

[31]As of the time of the publication of this paper, the final judgment of the Court of Appeals of Antwerp is not final and it is still pending to be enforced against Yahoo in Belgium.

In *Re Warrant to Search a Certain Email Account Controlled and Maintained by Microsoft Corp*[32] the Magistrate ordered by way of a warrant under the Stored Communications Act[33] that Microsoft disclose the content of emails in connection with a criminal investigation, even though the emails were stored on a data center in Ireland by Microsoft's wholly owned subsidiary.[34] The court order was affirmed by the Federal District Court for Southern District of New York. Microsoft had already disclosed meta-data stored on its servers in the US but had refused to disclose content data physically stored in a data center located in Ireland, citing presumption against extraterritorial reach of laws. The Federal District Court did not accept this argument and held that this was not an extraterritorial application of the law, as it was sufficient that Microsoft had (remote) control over the data in the US.[35] A distinction was made between a "normal" search and seizure warrant which was limited by its nature to US territory[36] and a warrant under the Stored Communications Act which allows for electronic disclosure and is therefore more akin to a subpoena for the disclosure of documents which is not limited to US territory.[37]

Microsoft appealed the judgment of the Magistrate Judge of the United States District Court for the Southern District of New York on December 18, 2014 and the matter is yet pending to be decided in the United States Court of Appeals for the Second Circuit.[38]

In the UK, new data retention legislation was passed in 2014. Section 4 of *the Data Retention Investigatory Powers Act* (DRIPA)[39] provides for express extraterritorial powers for LEAs to make a direct request to foreign communication service providers without going through MLA procedures. These direct requests can be made in respect of interception of content, interception capabilities and meta-data. This legislation was rushed through the UK Parliament just before the summer break of 2014 and its provisions have been heavily criticized.[40] The government claims that the existing law already contained a power to request

---

[32] 15 FSupp 3d 466 (S.D.N.Y 2014).

[33] 18 U.S.C. §§2701–2712.

[34] 15 FSupp 3d 466, 477 (S.D.N.Y 2014).

[35] 15 FSupp 3d 466, 472 (S.D.N.Y 2014).

[36] Rule 41 Federal Rules of Criminal Procedure.

[37] 15 FSupp 3d 466, 471 (S.D.N.Y 2014); see also Case Review in 128 *Harv. L.Rev 1019*.

[38] See: Brief for Appellant in the Matter of a Warrant to Search a Certain E-mail Account Controlled and Maintained by Microsoft Corporation on Appeal from the United states District Court for the Southern District of New York, (14-2985-cv December 18, 2014), available at: http://digitalconstitution.com/wp-content/uploads/2014/12/Microsoft-Opening-Brief-120820141.pdf.

[39] http://www.legislation.gov.uk/ukpga/2014/27/crossheading/investigatorypowers.

[40] See: The Guardian. "*Academics: UK 'Drip' data law changes are 'serious expansion of surveillance*" (15 July 2014) available at: http://www.theguardian.com/technology/2014/jul/15/academics-uk-data-law-surveillance-bill-rushed-parliament.

communications data from foreign service providers providing services to the UK and that this new Act only clarifies the position (for the protection of the participating communication service providers).[41]

As a matter of the analysis made in this section, we can conclude that there are sufficient legal precedents for LEAs to request communications data directly from foreign service providers or from local service providers where the data is stored on a foreign territory. Some states accept this practice while others differ and avoid conducting such practice, in other words there is not a uniform established accepted practice on cross-border access to data. We strongly believe that there is a need to make a more detailed analysis of the conformity of this practice under the scope of international law and particularly, to consider the creation of an international standard or additional safeguards to protect privacy and data when LEAs deal with or conduct direct access to data in other countries.

The questions raised above suggest reviewing the traditional jurisdictional approaches from a more fundamental perspective. As one possible solution to these issues, the final section of the paper suggests considering alternative perspectives on jurisdiction.

3 Final Remarks

The use of Mutual Legal Assistance Mechanisms continue to generate controversy since most of those instruments are subject to the reciprocity of states, the cooperation of government authorities which is usually slow, and often delay or hampered the investigation for not having immediate access to data that could be used as evidence in a criminal investigation.

Cross-border access to data and jurisdictional approaches are complex and difficult issues. Given the examples contained in this paper, it seems unlikely that states reach a mutually agreeable solution in the near future. The development and consensus on an Additional Protocol on Cross-Border Access to Data to the Convention against Cybercrime will be a lengthy process, and at its current stage of ambiguity, we do not believe that such instrument would help to improve the current practices of law enforcement authorities for accessing and securing data in foreign countries. One option to take these matters further is to provide additional guidance on the scope of interpretation of Article 32 of the Convention against Cybercrime with full attention to both, the operational needs of law enforcement as well as respect of fundamental human rights of privacy and data protection.

Examples of case law on LEAs' requests to foreign service providers shown in this paper suggest that States have a wide spectrum of possibilities to assert

---

[41]Ibid. See also Response by the Interception of Communications Commissioner Office (24 July 2014), available at: http://www.iocco-uk.info/docs/IOCCO%20response%20to%20new%20 reporting%20requirements.pdf.

jurisdiction under international law, the national constitutional framework and local laws. The views hereby presented are real and may encourage states to combine and apply jurisdictional principles according to their own needs and not only focus solely in the application of the territoriality principle. We strongly believe that there is a need for a more detailed analysis of the legal limits of LEAs' requests to foreign service providers under the scope of international law. In particular, states should consider the creation of an international standard or a set of additional safeguards to protect privacy and data when LEAs deal with or conduct direct access to data in other countries.

The cross-border access to data debate will continue in the following years since it involves a number of controversial regulatory aspects for the states namely, national sovereignty issues, the use of MLA channels, conflicts of laws in the field of data protection and the protection of the fundamental right to privacy pursuant to the current international and regional instruments in the subject matter. With this in mind, we hope that the views presented in this paper will be helpful for further debates, and in particular to states, that need to be prepared to confront these issues on a more proactive and expeditious basis.

Printed by Printforce, the Netherlands